Second Edition

ELECTRONICS MATH

BILL DEEM
San Jose City College
San Jose, California

PRENTICE-HALL
Englewood Cliffs, New Jersey 07632

Library of Congress-in-Publication Data

DEEM, BILL R.
   Electronics math.

   Includes index.
   1. Electronics—Mathematics.   I. Title.
TK7835.D36 1986   512′.1′0246213   85-19257
ISBN 0-13-252321-3

Editorial/production supervision: *Kathryn Pavelec*
Interior design: *Virginia Huebner*
Manufacturing buyer: *Gordon Osbourne*

© 1986, 1981 by **Prentice-Hall**
A Division of Simon & Schuster, Inc.
Englewood Cliffs, New Jersey 07632

Printed in the United States of America

10 9 8 7 6 5 4 3 2 1

ISBN 0-13-252321-3   01

PRENTICE-HALL INTERNATIONAL (UK) LIMITED, *London*
PRENTICE-HALL OF AUSTRALIA PTY, LIMMITED, *Sydney*
PRENTICE-HALL CANADA INC., *Toronto*
PRENTICE-HALLL HISPANOAMERICANA, S.A., *Mexico*
PRENTICE-HALL OF INDIA PRIVATE LIMITED, *New Delhi*
PRENTICE-HALL OF JAPAN, INC., *Tokyo*
PRENTICE-HALL OF SOUTHEAST ASIA PTE, LTD., *Singapore*
EDITORA PRENTICE-HALL DO BRASIL, LTDA., *Rio de Janeiro*
WHITEHALL BOOKS LIMITED, *Wellington, New Zealand*

CONTENTS

PREFACE

This text is written for students in high schools, community colleges, technical institutes and for technicians in the field of electronics. The mathematical topics chosen are those that are most useful in solving electronics problems.

This math text places greater stress upon certain areas of the discipline than does abstract math. For example, binary numbers are used in computers. Accordingly, binary notation and binary operations are developed along with decimal notation and operations. Also, because technicians utilize mathematical techniques that start and end with concrete numbers, concrete numbers are explained to the extent required by the beginning student.

Practice problems immediately followed by solutions, are presented throughout the text. At the end of each topic a self test is given so that the student can determine if he has mastered the material. Additional practice problems are found at the end of each chapter. Answers to the odd-numbered problems can be found in the back of the book. A solutions manual that includes answers to even-numbered problems is available to instructors.

The beginning chapters deal with numbers, powers of ten and prefixes. These are followed by chapters on algebra including linear equations and factoring. At the conclusion of these chapters the student utilizes the skills he has developed by solving DC circuit problems.

In chapters 15 through 20 those elements of algebra and trigonometry necessary to the solution of AC circuit problems are covered. Throughout these chapters, practical applications are given.

Chapters 21, 22, and 23 cover both common and natural logarithms and their applications. Chapter 24 discusses the basic logic functions inherent in all logic circuits and presents those theorems, laws, and postulates used in the simplification of logic expressions.

In this edition, most end-of-chapter problems have been expanded to include additional, more challenging problems.

Throughout the text, selected topics have been expanded or altered to improve clarity and understanding of the subject. Additional applications have been included in Chapters 5 and 14. A section on AC networks has been added to Chapter 19 to show the student how to solve complex AC circuit problems.

It is expected that students will use calculators when necessary throughout the text. To assist the student, sample problems and their solutions using a calculator are included in Appendix A. This appendix has been expanded to show a greater variety of problems and their solutions using two of the latest model calculators.

The author wishes to acknowledge the assistance given him by the editorial and production staffs of Prentice-Hall, and the many students and teachers who aided and assisted him in the preparation of this Edition. In particular, a special thanks to Clyde Herrick and his students for working out all the problems; and a big thank you to Barbara Snyder who was always willing to read and critique the manuscript and who made many contributions and suggestions which were incorporated into this Edition.

San Jose, California   BILL DEEM

ELECTRONICS MATH

CHAPTER 1

THE DECIMAL NUMBER SYSTEM

The decimal number system is the number system that we have always used. We all grew up using this system for all computations. Many of the concepts we will discuss in this chapter we have applied for years without really thinking about them. We are going to spend some time discussing the decimal numbering system in order to help us better understand working with numbers and to pave the way for a quicker understanding of other number systems. These other number systems such as the *binary number system* and the *hexadecimal number system* are important number systems used in the computer world and will be discussed in Chapter 4.

1.1 DECIMAL NUMBER SYSTEM

Our number system is called the decimal number system. *Decimal* means ten. In the decimal number system there are ten symbols. These symbols are called *digits*. The ten digits are 0, 1, 2, 3, 4, 5, 6, 7, 8, 9. Zero is the digit having the least value. In counting, when the count reaches 9, a limit has been reached because 9 is the digit of greatest value. An additional count produces a *carry*. This carry is equal to 10 and is said to occupy the tens position; the 0 occupies the units position. This process continues each time 9 is reached in the units position. When a 9 appears in the tens position, the next carry to the tens position produces a carry to the hundreds position.

It is important to note that a 1 in the tens position possesses a value (*weight*) 10 times that of a 1 in the units position. The same is true for each position of greater weight. Because of this relationship, the decimal numbering system is called a *place value system* and its base is 10.

TABLE 1-1

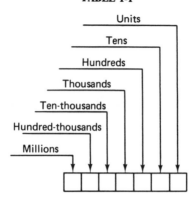

Let's examine a number in the decimal system, for example, the number 687. This number is made up of three digits: 6, 8, and 7. The *least significant digit* (LSD) is 7 and its value is 7. The next significant digit is 8 and its value is 80 (8×10). The 6 is the *most significant digit* (MSD) and its value is 600 (6×100). The 7 occupies the units position, the 8 occupies the tens position, and the 6 occupies the hundreds position. That is, there are six 100's, eight 10's, and seven 1's.

In Table 1-1 are listed the names for places up to millions.

How do we treat zero when it is a digit? Consider the number 2073. We can see that there are 2 thousands, 0 hundreds, 7 tens, and 3 units. The zero is in the hundreds position and simply tells us how many hundreds there are. It does the same job as any other digit and is just as important. We certainly couldn't drop the zero. If we did we would no longer have the number 2073 (two thousand seventy-three) but would have the number 273 (two hundred seventy-three). We cannot drop the zero without changing the value of the number. In the number 70630 the zeros tell us that there are no thousands and no units.

In the number 48,607 the MSD (most significant digit) is 4. The LSD (least significant digit) is 7. The digit 4 occupies the ten-thousands position; the 8 occupies the thousands position; the 6 occupies the hundreds position; the 0 occupies the tens position; and the 7 occupies the units position.

PRACTICE PROBLEMS 1-1

1. In the number 4762
   (a) Which digit is the most significant digit (MSD)?
   (b) Which digit is the least significant digit (LSD)?
   (c) Which digit occupies the hundreds position?

2. In the number 7,623,418
   (a) How many ten-thousands are there?
   (b) How many tens are there?
   (c) How many hundred-thousands are there?
   (d) How many units are there?

3. In the number 893,462
   (a) What weight does the 9 have?
   (b) What weight does the 4 have?
   (c) How many thousands are there?

Solutions:

1. (a) 4   (b) 2   (c) 7   **2.** (a) 2   (b) 1   (c) 6   (d) 8

3. (a) 9   ten-thousands   or   90,000
   (b) 4   hundreds or 400   (c) 3

Additional practice problems are at the end of the chapter.

1.2   DECIMAL FRACTIONS

In order to discuss decimal digits, let's review some names used in dealing with fractions. In the fraction $\frac{7}{8}$ the number above the line (7) is called the *numerator*. The number below the line (8) is called the *denominator*. The line itself is called the *vinculum*.

A decimal fraction is a fraction whose denominator is 10 or a multiple of 10 (100, 1000, 10,000, etc.). If the denominator is 10, the fraction is read one-tenth if the number is $\frac{1}{10}$, two-tenths if the number is $\frac{2}{10}$, etc.

Here are some decimal fractions and how they are read:

$$\frac{3}{100} \qquad \text{3 hundred}ths$$

$$\frac{1}{1000} \qquad \text{1 thousand}th$$

$$\frac{12}{10,000} \qquad \text{12 ten-thousand}ths$$

Notice that we read the denominator just as we read a whole number except that we add *th* or *ths*. Similarly, if we write "4 thousandths," we know this is a fraction (the *ths* at the end of thousand tells us this); so we know the fraction is $\frac{4}{1000}$. For example:

$$7 \text{ tenths} = \frac{7}{10}$$

$$23 \text{ hundredths} = \frac{23}{100}$$

$$243 \text{ ten-tnousandths} = \frac{243}{10,000}$$

Table 1-2 gives the names of places from tenths to millionths.

TABLE 1-2

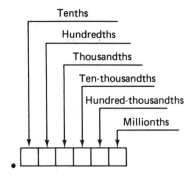

Decimal fractions can be converted to decimal numbers. For example:

$$3 \text{ tenths} = \frac{3}{10} = 0.3$$

$$3 \text{ hundredths} = \frac{3}{100} = 0.03$$

$$3 \text{ thousandths} = \frac{3}{1000} = 0.003$$

$$23 \text{ thousandths} = \frac{23}{1000} = 0.023$$

To convert decimal fractions to decimal numbers, we use this simple rule:

Determine the value of the denominator. Place the LSD of the numerator in this position.

For example; $\frac{17}{100} = 0.17$. The value of the denominator is one hundredth. The LSD in the numerator is 7; therefore, the 7 is placed in the hundredths position. Of course, this puts the 1 in the tenths position and the number is 0.17. Change the fraction $\frac{17}{1000}$ to a decimal number. The 7 must appear in the thousandths position. The decimal number would be written 0.017. Notice that in each case a zero was placed to the left of the decimal point. This is done so that we remember where the decimal point belongs.

Let's reverse the process; suppose we want to convert a decimal number to a fraction. For example, let's convert 0.007 to a fraction. Since the 7 appears in the thousandths position, $0.007 = \frac{7}{1000}$. Let's convert 0.0023 to a fraction. The digit 3 appears in the ten-thousandths position. Therefore, $0.0023 = \frac{23}{10,000}$.

PRACTICE PROBLEMS 1-2

1. Convert the following fractions to decimal numbers:
   (a) $\frac{4}{100}$ (b) $\frac{23}{1000}$ (c) $\frac{203}{100,000}$

2. Write the following numbers as decimal fractions:
   (a) 0.05   (b) 0.00073   (c) 0.00009

3. Write each of the following first as a decimal fraction and then as a decimal number:
   (a) 7 tenths   (b) 37 hundredths   (c) 7 ten-thousandths   (d) 417 hundred-thousandths
   (e) 6 millionths

4. In the number 0.00246, in which place does the 2 appear? The 6?

5. Write the names of the following numbers:
   (a) 0.03   (b) 0.0005   (c) 0.00073

Solutions:

1. (a) 0.04   (b) 0.023   (c) 0.00203

2. (a) $\frac{5}{100}$ (b) $\frac{73}{100,000}$ (c) $\frac{9}{100,000}$

3. (a) $\frac{7}{10} = 0.7$ (b) $\frac{37}{100} = 0.37$ (c) $\frac{7}{10,000} = 0.0007$
   (d) $\frac{417}{100,000} = 0.00417$ (e) $\frac{6}{1,000,000} = 0.000006$

4. The 2 appears in the thousandths position.
   The 6 appears in the hundred thousandths position.

5. (a) 3 hundredths   (b) 5 ten-thousandths
   (c) 73 hundred-thousandths

Additional practice problems are at the end of the chapter.

Now let's look at numbers in which there is a whole-number part and a decimal-number part. To do this let's prepare a table which is a combination of Tables 1-1 and 1-2. The result is shown in Table 1-3.

TABLE 1-3

Consider the number 13.36. This would be read "thirteen *and* thirty-six hundredths." The word *and* is used in a number to define the position of the decimal point. We read the whole-number part, add the work *and*, and then read the decimal-number part. We could also write the number as a *mixed* number: $13\frac{36}{100}$. This indicates the addition of two numbers: a whole number and a fraction. $13.36 = 13\frac{36}{100} = 13 + \frac{36}{100}$. Typically though, we wouldn't use the $+$ sign but would simply write the number as a decimal number (13.36) or as a mixed number ($13\frac{36}{100}$).

Consider the number 432.243. This number is read "four hundred thirty-two *and* two hundred forty-three thousandths." Remember the word *and* is used to connect the whole-number and decimal-number parts. As a mixed number we would write:

$$432\frac{243}{1000}$$

PRACTICE PROBLEMS 1-3

1. Convert each of the following decimal numbers to mixed numbers:
   (a) 24.007 (b) 706.024 (c) 4.00017

2. Convert each of the following mixed numbers to decimal numbers:
   (a) $1076\frac{7}{100}$ (b) $8\frac{23}{10,000}$ (c) $76\frac{14}{1000}$

3. Express the following numbers as decimal fractions and as mixed numbers:
   (a) Fifty-six and seventy-eight one hundredths;
   (b) fifteen and thirty-five thousandths; (c) one hundred six and eight ten-thousandths.

Solutions:

1. (a) $24\frac{7}{1000}$ (b) $706\frac{24}{1000}$ (c) $4\frac{17}{100,000}$ 2. (a) 1076.07 (b) 8.0023 (c) 76.014

3. (a) $56.78 = 56\frac{78}{100}$ (b) $15.035 = 15\frac{35}{1000}$

   (c) $106.0008 = 106\frac{8}{10,000}$

Additional practice problems are at the end of the chapter.

1.3 ROUNDING WHOLE NUMBERS

In problem solving it is common practice to simplify multi-digit numbers. This simplification is done by replacing non-zero digits with zeros. For example, the number 72,348 could be simplified to 72,350 to the nearest ten; 72,300 to the nearest hundred; 72,000 to the nearest thousand; or 70,000 to the nearest ten-thousand. This simplification is called *rounding.*

Rounding numbers is done to make problem solving easier. Of course, some accuracy is lost as a result of the rounding. The more rounding we do, the less accurate the number. In electronics we work with components that typically have 5% or 10% tolerances. We normally make measurements with instruments whose tolerances are within the 5% or 10% range (or worse). Because of these reasons, great accuracies in problem solving are not necessary.

Let's look at the number 72,348 again. When we rounded to the nearest ten, we rounded to 72,350. We had the choice of rounding down to 72,340 or rounding up to 72,350. Since 48 is closer to 50 than to 40, we rounded up to 72,350. When we rounded to the nearest hundred, we rounded to 72,300. We had the choice of rounding down to 72,300 or rounding up to 72,400. Since 348 is closer to 300 than to 400, we rounded down to 72,300. The same logic was used in rounding to the nearest thousand. Since 2348 is closer to 2000 than to 3000, we rounded to 72,000. Finally, 72,000 was rounded to 70,000 because 72,000 is closer to 70,000 than to 80,000. The rule is:

> *Round up if the digit is 5, 6, 7, 8, or 9. To round up, the non-zero digit is replaced with zero and one is added to the next significant digit. Round down if the digit is 1, 2, 3, 4. To round down, the nonzero digit is replaced with zero.*

Although the digit 5 could be either way, it is typically rounded *up.*

Let's round the number 2735 to the nearest ten, hundred, and thousand. To round to the nearest ten, we examine the digit in the units position. Since the digit is 5, we round *up* to 2740 by replacing 5 with 0 and adding one to the next significant digit. To round to the nearest hundred, we note that the digit in the tens position is 3; so we round *down* to 2700 by replacing 3 with zero. (Of course, the digit in the units position is also changed to zero since we are rounding to the nearest hundred.) Since 7 is the digit in the hundreds position, we round *up* to the nearest thousand and the answer is 3000.

PRACTICE PROBLEMS 1-4   Round the following numbers to the nearest ten, hundred, and thousand:

1. 2714

2. 6,526

3. 43,258

4. 76,547

5. 82,803

6. 26,764

7. 78,226

8. 18,999

9. 4050

10. 243,671

Solutions:

To Nearest Tens	To Nearest Hundreds	To Nearest Thousands
1. 2710	2700	3000
2. 6530	6500	7000
3. 43,260	43,300	43,000
4. 76,550	76,500	77,000
5. 82,800	82,800	83,000
6. 26,760	26,800	27,000
7. 78,230	78,200	78,000
8. 19,000	19,000	19,000
9. 4050	4100	4000
10. 243,670	243,700	244,000

In problem 5, since a zero already exists in the tens position, rounding to the nearest hundred is not necessary. In problem 8, when we round to the nearest ten we round up because the units digit is 9. Remember that we round up by adding one in the next significant digit position (the tens position). Adding 1 to 9 in the tens position produces a zero and a carry into the hundreds position. A carry into the hundreds position results in zero and a carry into the thousands position resulting in the answer 19,000.

In problem 9, since a zero already exists in the units position, rounding to the nearest ten is not necessary. The 5 in the tens position causes a carry to the hundreds resulting in the answer 4100 to the nearest hundred.

Additional practice problems are at the end of the chapter.

1.4 ROUNDING NON-WHOLE NUMBERS

Non-whole numbers are rounded by using the same rules as for whole numbers. As with whole numbers, we work from the least significant to the most significant digits. Consider the number 12.736. We could round to the nearest hundredth, tenth, unit, or ten. To round to the nearest hundredth, we would note the digit in the thousandths position. Since the digit is 6, we change the digit to 0 and add one to the next significant digit making the number 12.74. We don't write the number as 12.740. The zero has no meaning (no place value) since we rounded to the nearest hundredth. Therefore, we just drop the zero. To round to the nearest tenth, we note the digit in the hundredths position. Since the number is 4, we simply drop it and the answer is 12.7. To round to the nearest unit, we note the digit in the tenths position. Because it is a 7, a number which is 5 or greater, we drop it and add one to the number in the units position. The answer is 13. To convert to the nearest ten, we change the 2 to a 0. The answer is 10. Remember that in rounding whole numbers the zero is a place holder and cannot be dropped. In a decimal fraction, if a zero exists in the LSD position as a result of rounding, the zero is dropped because it has no meaning and no value.

PRACTICE PROBLEMS 1-5 Round the following numbers to the nearest hundredth, tenth, unit, and ten:

1. 73.647

2. 26.401

3. 17.0419

4. 30.6908

5. 50.4736

6. 48.047

7. 33.781

8. 68.147

9. 78.6671

10. 15.5554

Solutions:

Hundredth	*Tenth*	*Unit*	*Ten*
1. 73.65	73.6	74	70
2. 26.40	26.4	26	30
3. 17.04	17.0	17	20
4. 30.69	30.7	31	30
5. 50.47	50.5	50	50
6. 48.05	48.0	48	50
7. 33.78	33.8	34	30
8. 68.15	68.1	68	70
9. 78.67	78.7	79	80
10. 15.56	15.6	16	20

Additional practice problems are at the end of the chapter.

SELF-TEST 1-1

1. In the number 20,378
(a) Which digit is the MSD?
(b) Which digit is the LSD?
(c) Which digit occupies the thousands position? (d) Which digit occupies the tens position? (e) What weight does the digit 3 have?
(f) How many units are there?

2. Convert the following fractions to decimal numbers:
(a) $\frac{41}{100}$ (b) $\frac{9}{1000}$ (c) $\frac{1783}{100,000}$

3. Write each of the following first as a decimal fraction and than as a decimal number:
(a) Three tenths   (b) Eighty-five hundredths
(c) Eighteen ten-thousandths

4. Write the names of the following numbers:
(a) 0.33   (b) 0.004

5. Convert each of the following decimal numbers to mixed numbers:
(a) 7.46   (b) 18.006

6. Convert each of the following mixed numbers to decimal numbers:
(a) $76\frac{14}{100}$ (b) $6\frac{23}{1000}$

7. Express the following numbers as decimal fractions and as mixed numbers:
   (a) Three and seven hundredths (b) Twenty-eight and sixty-three thousandths

8. Round to the nearest ten, hundred, and thousand:
   (a) 4765 (b) 9705 (c) 15,789 (d) 37,046

9. Round to the nearest tenth, hundredth, and thousandth:
   (a) 0.0746 (b) 0.4605 (c) 0.4056
   (d) 0.3748

10. Round to the nearest ten, unit, tenth, and hundredth:
    (a) 17.486 (b) 23.462 (c) 36.547
    (d) 20.706

Answers to Self-Test 1-1 are found at the end of the chapter.

1.5 SIGNED NUMBERS

In mathematics, numbers may be either positive (+) or negative (−). The plus sign is not usually written. If we saw the numbers 6, −4, 15, −9 written, we would recognize that the numbers 6 and 15 are positive and the numbers 4 and 9 are negative.

In this section we introduce the symbols < and >. The symbol < means *less than* and the symbol > means *greater than*.

In Figure 1-1 we have drawn a straight line which starts at zero and extends both right and left. Notice that numbers starting at zero and extending to the right are positive. Numbers starting at zero and extending to the left are negative. In comparing signed numbers we see that the rightmost number is the greater and the leftmost number is the lesser. For example, compare 4 and −3. We could say that 4 is greater than −3 (4 > −3) or we could say that −3 is less than 4 (−3 < 4).

Consider the numbers −7, −2, and 6. We could say that −7 < −2 < 6. Or we could say that 6 > −2 > −7. Notice that in comparing −2 and −7 we said that −2 > −7. Remember, on the number scale −2 is to the right of −7; therefore it is greater. −2 is more positive than −7. −7 is more negative than −2. If we compare 3 and 11, we would say that 3 < 11. We could also say that 3 is more negative than 11.

The *absolute* value of a number is that number without regard to sign. The absolute value of +5 and −5 is 5 and is symbolized as |5|. This tells us that no matter if 5 is − or +, the distance from the origin (0) is 5.

Figure 1-1.

In addition to using the symbols $+$ and $-$ to denote the sign of a number, we also use these symbols to indicate the operations of addition and subtraction.

Anyone who has trouble visualizing adding signed numbers should refer to Figure 1-1. Adding two or more positive numbers does not present any problem with signs. $2+3+5=10$. Let's add two negative numbers: $(-2)+(-3)$. -2 and -3 are enclosed in parentheses to indicate that the $-$ is the sign of the number and not a sign of subtraction. $(-2)+(-3)=-5$. $(-6)+7+(-9)=-8$. We could simplify the problem before finding the sum by adding the negative numbers together first:

$$-6+7+(-9)=-15+7=-8$$

In each case we move left on the number scale if the number is $-$ and we move right if the number is $+$.

In subtraction we call the first number the *minuend*. The second number is the *subtrahend*. The answer is the *difference*.

$$
\begin{array}{rl}
6 & \text{minuend} \\
-3 & \text{subtrahend} \\
\hline
3 & \text{difference}
\end{array}
$$

To subtract we change the sign of the subtrahend and then *add* as before. $6-(+3)=6+(-3)=3$.

$$
\begin{array}{rlcrl}
6 & & & 6 & \\
-3 & = & + & -3 & \\
\hline
3 & & & 3 &
\end{array}
$$

When more than two numbers are subtracted; the same rule applies. Whenever subtracting is indicated, we change the sign of the following numbers and add. $17-(+6)-(+3)=17+(-6)+(-3)=17+(-9)=8$. When addition and subtraction are both indicated, we follow the rules for both.

$$25-(+6)+4=25+(-6)+4=23$$

$$-7+16-(-4)=-7+16+4=13$$

PRACTICE PROBLEMS 1-6

1. $6+(-4)$ 　　　　　　　　　　　**2.** $-14+(-6)$

3. $16-(+4)$ 　　　　　　　　　　**4.** $9-(-3)$

5. $-10-(-6)$ 　　　　　　　　　　**6.** $-15-(+4)$

7. $25+(-3)-(+4)$ 　　　　　　　**8.** $4-(-6)+(-15)$

9. $-6-(-3)-(-10)$ 　　　　　　**10.** $40+(-30)-(-6)$

Solutions:

1. 2   **2.** −20

3. 12   **4.** 12

5. −4   **6.** −19

7. 18   **8.** −5

9. 7   **10.** 16

1.7   MULTIPLICATION AND DIVISION

Both $\times$ and $\cdot$ are symbols used to indicate multiplication. The symbol $\div$ is used to indicate division. Each part of a multiplication or division has a name as shown in Figure 1-2.

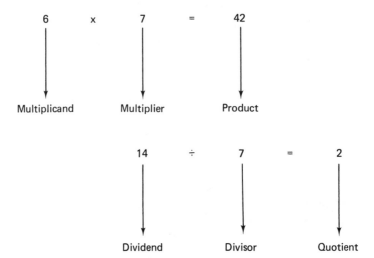

Figure 1-2. Names of the parts of a product or quotient.

Signed numbers are multiplied and divided by using the same methods learned in arithmetic. The rules are:

If all numbers are positive, the answer is positive. If there is an even number of negative signs, the answer is positive. If there is an odd number of negative signs, the answer is negative.

$$(+)\cdot(+) = +$$
$$(+)\cdot(-) = -$$
$$(-)\cdot(-) = +$$

$$9 \times 8 = 72 \qquad\qquad (+)\cdot(+) = +$$
$$7 \times (-6) = -42 \qquad (+)\cdot(-) = -$$
$$-8 \times (-6) = 48 \qquad (-)\cdot(-) = +$$
$$16 \div 4 = 4 \qquad\qquad (+)\div(+) = +$$
$$12 \div (-4) = -3 \qquad (+)\div(-) = -$$
$$-20 \div 5 = -4 \qquad (-)\div(+) = -$$
$$-35 \div -7 = 5 \qquad (-)\div(-) = +$$

PRACTICE PROBLEMS 1-7

1. $17 \times (-5)$ **2.** $-14 \times (-16)$

3. -9×12 **4.** -12×14

5. $-14 \div 2$ **6.** $-42 \div 6$

7. $172 \div (-4)$ **8.** $98 \div (-14)$

9. $-121 \div (-11)$ **10.** $-144 \div (-12)$

Solutions:

1. -85 **2.** 224

3. -108 **4.** -168

5. -7 **6.** -7

7. -43 **8.** -7

9. 11 **10.** 12

Additional practice problems are at the end of the chapter.

1.8 MATHEMATICAL EXPRESSIONS AND TERMS

Many problem-solving situations in electronics involve all four operations: addition, subtraction, multiplication, and division. Therefore, it is important that we understand the order in which these operations are performed.

Before we talk about these problem-solving situations, we must define some new words. Here and in future chapters we are going to be talking about mathematical *expressions* and *terms*. A *term* is a number preceded by a + or − sign. A mathematical *expression* is made up of one or more terms. Thus $4 + 7$ is an expression containing two terms. Of course, this expression can be reduced to *one* term (11). $4 + 7 \times 3$ is a two-term expression. 4 is one term and 7×3 is the other. Remember, *terms* are separated by + and − signs. Signs of multiplication and division are signs within a term. In problem solving, all operations within a term are performed first and then the terms are combined. Said another way, *to simplify a mathematical expression perform the multiplications and divisions first and then the additions and subtractions.* Here are some examples:

$$7 \times 6 + 4 \times 2 = 42 + 8 = 50$$

$$-14 \div 7 - (-6) \times 3 = -2 - (-18) = -2 + (+18) = 16$$

$$4 + (-3) - (6) \times (-4) = 4 + (-3) - (-24) = 4 + (-3) + (+24) = 25$$

My colleague, Barbara Snyder, uses the phrase "**M**y **D**ear **A**unt **S**ally" to help her students remember which function is performed first—**M**ultiply, **D**ivide, **A**dd, and **S**ubtract.

PRACTICE PROBLEMS 1-8   Simplify the following expressions to one term:

1. $8 + 6 \times 3$

2. $-3 + 12 \div (-2)$

3. $6 \times 3 - 2 \times 4$

4. $16 \div (-2) - (-4) + 6 \times (-6)$

5. $(-6) \times (-4) + 7 \times (-3)$

6. $6 \times (-4) \div 2 - 4 + (-6)$

7. $15 \times 3 \div 3 + 4$

8. $8 \times 4 - (-6) \times (-4)$

Solutions:

1. 26

2. -9

3. 10

4. -40

5. 3

6. -22

7. 19

8. 8

Additional practice problems are at the end of the chapter.

Signs of grouping are another way of indicating the order in which problems are solved. These signs of grouping include parentheses ( ), brackets [ ], braces { }, and the vinculum (or bar) ———. Signs of grouping are also a way of indicating multiplication. For example, to multiply two numbers we could symbolize the multiplication by using parens. (7)(6) or 7(6) means the same as 7×6 or $7 \cdot 6$. $7 \times (6 + 2)$ could be written $7(6 + 2)$. In both cases we are finding the product of 7 and 8. $7(6 + 2) = 7(8) = 56$.

When signs of grouping are used, we first perform all the operations within the group and then we combine terms. Here are some examples:

$$7(6 + 3) = 7 \times 9 = 63$$

$$4 - (-6 + 3) \times 2 = 4 - (-3) \times 2 = 4 - (-6) = 4 + 6 = 10$$

When more than one sign of grouping is used, we work from the inside out.

$$3[4(3 + 2) - 4] = 3(4 \times 5 - 4) = 3(20 - 4) = 3(16) = 48$$

$$-3 - [2 + 4 \times (-6)] = -3 - [2 + (-24)] = -3 - (-22) = -3 + 22 = 19$$

PRACTICE PROBLEMS 1-9

1. $-6-(-3)+3(3-4)$

2. $6\times(-3)-[6(-3)]$

3. $14-(-4+18\div3)$

4. $(4+3)(6-3)$

5. $4-[6(7-3)]$

6. $6[18+(-6+3)]+7\times4$

7. $10-(6+3)\times4$

8. $18-3[(3+4)(8-12)]$

9. $4[7-(-3)+(2-6)(3-7)]$

10. $4-3\{-2[6(2+3)]-6\}$

Solutions:

1. -6

2. 0

3. 12

4. 21

5. -20

6. -8

7. -26

8. 102

9. 104

10. 202

SELF-TEST 1-2

1. $14-(+4)+(-5)$

2. $-16-(-4)-(+6)$

3. -8×6

4. $-9\times(-4)$

5. $4\times(-3)$

6. $51\div(-3)$

7. $-64\div(-4)$

8. $-36\div4$

9. $5\times(-3)+(-6)\times7$

10. $-9+3+4\times(-6)$

11. $16-(-6)+27\div(-3)$

12. $4-(-6)+4(5-2)$

13. $7\times(-2)-[4(-6)]$

14. $7[15+(-3+6)]+8\div4$

15. $6-5\{-4[8(4+2)]-4\}$

Answers to Self-Test 1-2 are at the end of the chapter.

END OF CHAPTER PROBLEMS 1-1

1. In the number 67,403
   (a) What is the MSD?
   (b) What is the LSD?
   (c) What digit occupies the thousands position?
   (d) What digit occupies the tens position?
   (e) What weight does the 7 have?

2. In the number 8042
   (a) What is the MSD?
   (b) What is the LSD?
   (c) What digit occupies the hundreds position?
   (d) What digit occupies the units position?
   (e) What weight does the 8 have?
   (f) How many tens are there?

END OF CHAPTER PROBLEMS 1-2

1. Convert the following fractions to decimal numbers:
   (a) $\frac{3}{10}$ (b) $\frac{16}{1000}$ (c) $\frac{278}{100,000}$
   (d) $\frac{1763}{10,000}$

2. Convert the following fractions to decimal numbers:
   (a) $\frac{495}{10,000}$ (b) $\frac{6}{100}$ (c) $\frac{75}{1000}$
   (d) $\frac{6}{100,000}$

3. Write the following numbers as decimal fractions:(a) 0.007   (b) 0.0432   (c) 0.174
   (d) 0.000065

4. Write the following numbers as decimal fractions:
   (a) 0.173   (b) 0.0004   (c) 0.00473
   (d) 0.000806

5. Write each of the following first as a decimal fraction and then as a decimal number:
   (a) 17 thousandths   (b) 4 hundredths
   (c) 460 ten-thousandths   (d) 27 millionths

6. Write each of the following first as a decimal fraction and then as a decimal number:
   (a) 6 tenths   (b) 83 hundred-thousandths
   (c) 780 millionths
   (d) 27 thousandths

7. In the number 0.01642, in which place does the 1 appear? The 2? The 6?

8. In the number 0.008706, in which place does the 7 appear? The 6? The 8?

9. Write the names of the following:
   (a) 0.006   (b) 0.147   (c) 0.00092
   (d) 0.000007

10. Write the names of the following:
    (a) 0.175   (b) 0.000035   (c) 0.00463
    (d) 0.00056

END OF CHAPTER PROBLEMS 1-3

1. Convert each of the following decimal numbers to mixed numbers:
   (a) 7.14   (b) 50.02   (c) 710.143   (d) 9.099
   (e) 73.653   (f) 207.7834   (g) 28.00736
   (h) 8.0706

2. Convert each of the following decimal numbers to mixed numbers:
   (a) 47.3   (b) 8.75   (c) 56.075
   (d) 9.475   (e) 450.00125   (f) 37.4073
   (g) 48.00045   (h) 10.10738

3. Convert each of the mixed numbers to decimal numbers:
   (a) $5\frac{68}{100}$ (b) $25\frac{7}{1000}$ (c) $7\frac{165}{10,000}$
   (d) $70\frac{4}{10}$ (e) $473\frac{25}{1000}$ (f) $80\frac{743}{100,000}$
   (g) $2475\frac{35}{1,000,000}$ (h) $307\frac{8}{100,000}$

4. Convert each of the following mixed numbers to decimal numbers:
   (a) $45\frac{3}{10}$ (b) $365\frac{83}{100}$
   (c) $3\frac{14}{10,000}$ (d) $17\frac{85}{1,000,000}$
   (e) $819\frac{6}{10,000}$ (f) $83\frac{38}{1000}$
   (g) $8\frac{17}{100}$ (h) $25\frac{275}{100,000}$

5. Express the following numbers as decimal fractions and then as mixed numbers:
   (a) Ninety three and seven tenths
   (b) Thirty and four hundredths
   (c) Eleven and one ten-thousandths
   (d) Two hundred seventy three and twenty-five hundred-thousandths
   (e) Seven hundred four and seven hundred four millionths
   (f) Three and thirty-three thousandths

6. Express the following numbers as decimal fractions and then as mixed numbers:
   (a) 78 and three tenths
   (b) 10 and twenty-five thousandths
   (c) 50 and seventeen hundredths
   (d) Eight hundred forty-five and sixty-five ten-thousandths
   (e) Fifty-three and 6 millionths
   (f) Eighty-five and forty-eight hundred-thousandths

Round the following numbers to the nearest ten:

1. 17

2. 72

3. 45

4. 33

Round the following numbers to the nearest ten and hundred:

5. 273

6. 708

7. 356

8. 843

9. 137

10. 674

Round the following numbers to the nearest ten, hundred, and thousand:

1. 4817

2. 6706

3. 85,468

4. 27,046

5. 78,673

6. 84,615

7. 2784

8. 10,778

9. 35,486

10. 20,066

END OF CHAPTER PROBLEMS 1-5

Round the following numbers to the nearest hundredth, tenth, unit, and ten:

1. 163.782

2. 89.076

3. 9.464

4. 273.926

5. 88.888

6. 25.546

7. 749.493

8. 20.506

9. 39.278

10. 76.567

11. 63.7478

12. 74.2636

13. 478.6706

14. 215.2535

END OF CHAPTER PROBLEMS 1-6

Perform the indicated operations:

1. $12-(+4)+(-3)$

2. $-3-(+4)-(-6)$

3. $16+4+(-6)$

4. $-15-(-7)-(+10)$

5. $-20-(-4)-(-10)$

7. $30+(-3)+7$

9. $-10-(+4)+(-14)$

11. $7-(-7)-(+7)$

13. $-5-(-3)-(+12)$

15. $14+(-3)-(-7)$

17. $9-(-4)-(-8)$

19. $-20-(+6)+(-7)$

6. $-30-(-6)-(+10)$

8. $40+(-13)+6$

10. $-14-(+6)+(-18)$

12. $14+(-7)-(+7)$

14. $-6-(-4)-(+16)$

16. $20+(-5)-(-8)$

18. $7-(-10)-(-12)$

20. $-25-(+4)+(-8)$

END OF CHAPTER PROBLEMS 1-7

Perform the indicated operations:

1. $14\times(-3)$

3. $-9\times(-6)$

5. -12×7

7. $28\div(-4)$

9. $-144\div6$

11. $-72\div(-6)$

2. -15×4

4. $25\times(-6)$

6. $-15\times(-15)$

8. $165\div(-5)$

10. $-256\div16$

12. $-625\div(-25)$

END OF CHAPTER PROBLEMS 1-8

Perform the indicated operations:

1. $(-7)\times3+4$

3. $6\times(-3)-6\times(-3)$

5. $6+3\times5-4$

7. $6\times4-6\div2$

9. $(-7)-(-4)\times6+1$

2. $-6-(-3)+3\times3-4$

4. $14-(-4)+18\div3$

6. $21+4\times3-21\div7$

8. $(-5)+6\times(-4)-3$

10. $(-18)\div3+4\times6$

END OF CHAPTER PROBLEMS 1-9

Perform the indicated operations:

1. $-4-(-2)+6(6-3)$

3. $7(-3)-[4(-7)]$

5. $(5+4)(-6+2)$

7. $5-2-3[4(3+4)]-2$

9. $4[24\div(-8+4)]+7\times4$

2. $-9-(-4)+5(7-9)$

4. $5(-4)+[-3(-4)]$

6. $(-4+9)(7-3)$

8. $-6-4-[4(-3+1)]-7$

10. $8[27\div(-9+6)]+6\times3$

11. $15-4[(-3+6)(9-12)]$
13. $5[6-(-4)+(3-7)(2-6)]$
✐ **15.** $4[-6-(-3)(4)]+7\times3$
17. $[9+(3\times-7)][18-(2\times3)]$

12. $12-5[(4-9)(-9-3)]$
14. $3[5-(-5)+(2-7)(4-9)]$
16. $(4-5)(-3\times4)-[-6(2-4)]$
18. $[24-(4\times5)][14-(7\times4)]$

ANSWERS TO SELF-TESTS

Self-Test 1-1

1. (a) 2 (b) 8 (c) 0 (d) 7 (e) 300 (f) 8

2. (a) 0.41 (b) 0.009 (c) 0.01783

3. (a) $\frac{3}{10}=0.3$ (b) $\frac{85}{100}=0.85$ (c) $\frac{18}{10,000}=0.0018$

4. (a) Thirty-three hundredths (b) Four thousandths

5. (a) $7\frac{46}{100}$ (b) $18\frac{6}{1000}$

6. (a) 76.14 (b) 6.023

7. (a) $3.07=3\frac{7}{100}$ (b) $28.063=28\frac{63}{1000}$

8.

	Ten	Hundred	Thousand
(a)	4770	4800	5000
(b)	9710	9700	10,000
(c)	15,790	15,800	16,000
(d)	37,050	37,000	37,000

9.

	Tenth	Hundredth	Thousandth
(a)	0.1	0.07	0.075
(b)	0.5	0.46	0.461
(c)	0.4	0.41	0.406
(d)	0.4	0.37	0.375

10.

	Ten	Unit	Tenth	Hundredth
(a)	20	17	17.5	17.49
(b)	20	23	23.5	23.46
(c)	40	37	36.5	36.55
(d)	20	21	20.7	20.71

Self-Test 1-2

1. 5
3. -48
5. -12
7. 16
9. -57
11. 13
13. 10
15. 986

2. -18
4. 36
6. -17
8. -9
10. -27
12. 22
14. 37

CHAPTER 2

POWERS OF TEN

2.1 POSITIVE EXPONENTS

To simplify working with large numbers, we use a math shorthand called *powers of ten or scientific notation*. Let's see what these terms mean.

In Table 2-1 we have converted regular numbers to powers of ten. We could continue the table indefinitely but this table should be long enough to understand the concept.

TABLE 2-1

$$1 \times 10 = 10 = 10^1$$
$$10 \times 10 = 100 = 10^2$$
$$10 \times 10 \times 10 = 1000 = 10^3$$
$$10 \times 10 \times 10 \times 10 = 10,000 = 10^4$$
$$10 \times 10 \times 10 \times 10 \times 10 = 100,000 = 10^5$$
$$10 \times 10 \times 10 \times 10 \times 10 \times 10 = 1,000,000 = 10^6$$

Let's look at $10 \times 10 \times 10 = 10^3$. In the expression 10^3 the number 10 is referred to as the base and 3 is referred to as the exponent. When a multiple of 10 is expressed in this way, it is said to be in the powers of ten form. 10^3 is read "ten to the third power."

The exponent tells us how many times 10 is used as a factor. We may also think of the exponent as telling us how many zeros are included to the right of the digit 1.

10^4 is read "ten to the fourth power." 4 is the exponent. Therefore, 10 is multiplied by itself four times: $10 \times 10 \times 10 \times 10 = 10,000$. Again, this is the same as saying that the answer is one followed by four zeros. For positive exponents the rule is:

In converting from powers of ten form to a number, write the digit one (1) and follow with as many zeros as indicated by the exponent.

PRACTICE PROBLEMS 2-1   Find the following numbers:

1. 10^2

2. 10^1

3. 10^5

4. 10^0

5. 10^6

6. 10^3

7. 10^4

8. 10^7

9. 10^9

10. 10^8

Solutions:

1. 100

2. 10

3. 100,000

4. 1

5. 1,000,000

6. 1000

7. 10,000

8. 10,000,000

9. 1,000,000,000

10. 100,000,000

Notice in problem 4 that the exponent tells us to add no zeros after the 1. Therefore, the answer is 1.

Now let's consider a problem in which we wish to convert a number to powers of ten form. The power to which 10 is raised (the exponent) is determined by the number of zeros following the digit 1. $100 = 10^2$ because there are two zeros following the digit 1. $100,000 = 10^5$ because there are five zeros following the digit 1. Looking at it another way: $100 = 10 \times 10$. 100 equals 10 multiplied by itself two times; therefore the exponent is 2. $100,000 = 10 \times 10 \times 10 \times 10 \times 10$. 100,000 equals 10 multiplied by itself five times; therefore the exponent is 5. The rule is:

In converting some multiple of 10 to powers of ten, count the number of zeros. This number is the exponent. The answer is 10 raised to that power.

PRACTICE PROBLEMS 2-2   Convert the following numbers to powers of ten form:

1. 100

2. 100,000

3. 1

4. 10,000,000

5. 1000

7. 1,000,000

9. 1,000,000,000

6. 10

8. 10,000

10. 100,000,000

Solutions:

1. 10^2

3. 10^0

5. 10^3

7. 10^6

9. 10^9

2. 10^5

4. 10^7

6. 10^1

8. 10^4

10. 10^8

The answer to problem 6 could be written simply as 10 since $10^1 = 10$. In the future when the number is raised to the first power we will not write the exponent.

Additional practice problems are at the end of the chapter.

2.2 NEGATIVE EXPONENTS

When we were dealing with numbers that were multiples of 10, we saw that the exponents were 1 or greater. The number 1 expressed in powers of ten form was 10^0. Numbers whose values are less than 1 have negative exponents. Table 2-2 shows some of these numbers and their exponents.

TABLE 2-2

$$0.1 = 10^{-1}$$
$$0.01 = 10^{-2}$$
$$0.001 = 10^{-3}$$
$$0.0001 = 10^{-4}$$
$$0.00001 = 10^{-5}$$
$$0.000001 = 10^{-6}$$

The exponent tells us how many places to the right of the decimal point the 1 is located. The numbers in Table 2-2 could also be written as decimal fractions. This may help us see the relationship between negative powers of ten, decimal numbers, and decimal fractions.

$$10^{-2} = 0.01 = \frac{1}{100} = \frac{1}{10^2}$$

$$10^{-3} = 0.001 = \frac{1}{1000} = \frac{1}{10^3}$$

$$10^{-4} = 0.0001 = \frac{1}{10,000} = \frac{1}{10^4}$$

Here are the rules:

In converting from powers of ten to decimal fractions, the negative exponent tells how many zeros follow the 1 in the denominator. In converting from powers of ten to decimal numbers, the negative exponent tells how far the 1 is located to the right of the decimal point.

PRACTICE PROBLEMS 2-3   Convert the following powers of ten to (a) decimal numbers and (b) decimal fractions:

1. 10^{-4} **2.** 10^{-2}

3. 10^{-1} **4.** 10^{-8}

5. 10^{-3} **6.** 10^{-7}

7. 10^{-6} **8.** 10^{-5}

Solutions:

1. (a) 0.0001   **2.** (a) 0.01

 (b) $\dfrac{1}{10,000}$ (b) $\dfrac{1}{100}$

3. (a) 0.1   **4.** (a) 0.00000001

 (b) $\dfrac{1}{10}$ (b) $\dfrac{1}{100,000,000}$

5. (a) 0.001   **6.** (a) 0.0000001

 (b) $\dfrac{1}{1000}$ (b) $\dfrac{1}{10,000,000}$

7. (a) 0.000001   **8.** (a) 0.00001

 (b) $\dfrac{1}{1,000,000}$ (b) $\dfrac{1}{100,000}$

Let's turn it around and consider problems in which we wish to convert decimal numbers and decimal fractions to powers of ten form.

$$0.01 = \frac{1}{100} \quad = \frac{1}{10^2} = 10^{-2}$$

$$0.001 = \frac{1}{1000} \quad = \frac{1}{10^3} = 10^{-3}$$

$$0.0001 = \frac{1}{10,000} \quad = \frac{1}{10^4} = 10^{-4}$$

The rules are:

In converting from a decimal fraction to powers of 10 *form, count the number of zeros in the denominator. This number is the negative power to which* 10 *is raised.*

$$\frac{1}{10,000} = 10^{-4}$$

$$\frac{1}{100} = 10^{-2}$$

In converting from a decimal number to powers of ten form, count how far the digit 1 *is located to the right of the decimal point. This is the value of the negative exponent.* (*Examples of this conversion are shown in Table 2-2.*)

PRACTICE PROBLEMS 2-4   Convert the following decimal numbers to powers of ten form:

1. 0.001   **2.** 0.1

3. 0.000001   **4.** 0.01

5. 0.00001   **6.** 0.0000001

7. 0.0001   **8.** 0.00000001

Convert the following decimal fractions to powers of ten form:

9. $\frac{1}{1000}$ **10.** $\frac{1}{100,000}$

11. $\frac{1}{10}$ **12.** $\frac{1}{1,000,000}$

13. $\frac{1}{100}$ **14.** $\frac{1}{10,000}$

Solutions:

1. 10^{-3} **2.** 10^{-1}

3. 10^{-6} **4.** 10^{-2}

5. 10^{-5} **6.** 10^{-7}

7. 10^{-4} **8.** 10^{-8}

9. 10^{-3} **10.** 10^{-5}

11. 10^{-1} **12.** 10^{-6}

13. 10^{-2} **14.** 10^{-4}

2.3 MULTIPLICATION IN POWERS OF TEN FORM

To multiply numbers in powers of ten form, we merely add the exponents.

$$10^2 \times 10^3 = 10^{2+3} = 10^5$$

That is, $100 \times 1000 = 100,000 = 10^5$

$$10^6 \times 10^{-2} = 10^{6+(-2)} = 10^4$$

or $1,000,000 \times 0.01 = 10,000 = 10^4$

or $1,000,000 \times \dfrac{1}{100} = 10,000 = 10^4$

$$10^{-4} \times 10^3 = 10^{-4+3} = 10^{-1}$$

or $0.0001 \times 1,000 = 0.1 = 10^{-1}$

or $\dfrac{1}{10,000} \times 1,000 = \dfrac{1}{10} = 10^{-1}$

$$10^{-5} \times 10^{-2} = 10^{-5+(-2)} = 10^{-7}$$

or $\dfrac{1}{100,000} \times \dfrac{1}{100} = \dfrac{1}{10,000,000} = 10^{-7}$

or $0.00001 \times 0.01 = 0.0000001 = 10^{-7}$

PRACTICE PROBLEMS 2-5   Multiply the following numbers. Express the products in powers of ten form.

1. $10^2 \times 10^3$

2. $10^4 \times 10^6$

3. $10^{-1} \times 10^{-3}$

4. $10^{-3} \times 10^{-5}$

5. $10^5 \times 10^{-2}$

6. $10^6 \times 10^{-4}$

7. $10^2 \times 10^{-3}$

8. $10^3 \times 10^{-3}$

9. $10^{-4} \times 10^7$

10. $10^{-2} \times 10^6$

Solutions:

1. 10^5

2. 10^{10}

3. 10^{-4}

4. 10^{-8}

5. 10^3

6. 10^2

7. 10^{-1}

8. 10^0

9. 10^3

10. 10^4

Additional practice problems are at the end of the chapter.

2.4 DIVISION IN POWERS OF TEN FORM

To divide numbers in powers of ten form, we *subtract* the exponents.

$$\frac{10^5}{10^2} = 10^{5-2} = 10^3$$

That is,

$$\frac{100,000}{100} = 1000 = 10^3$$

$$\frac{10^{-5}}{10^2} = 10^{-5-2} = 10^{-7}$$

or

$$\frac{0.00001}{100} = 0.0000001 = 10^{-7}$$

$$\frac{10^5}{10^{-2}} = 10^{5-(-2)} = 10^{5+2} = 10^7$$

or

$$\frac{100,000}{0.01} = 10,000,000 = 10^7$$

$$\frac{10^{-5}}{10^{-2}} = 10^{-5-(-2)} = 10^{-5+2} = 10^{-3}$$

or

$$\frac{0.00001}{0.01} = 0.001 = 10^{-3}$$

PRACTICE PROBLEMS 2-6   Perform the indicated operations. Express your answers in powers of ten form.

1. $\dfrac{10^4}{10^2}$

2. $\dfrac{10^6}{10^3}$

3. $\dfrac{10^2}{10^5}$

4. $\dfrac{10}{10^6}$

5. $\dfrac{10^2}{10^{-2}}$

6. $\dfrac{10^3}{10^{-5}}$

7. $\dfrac{10^{-2}}{10^{-4}}$

8. $\dfrac{10^{-3}}{10}$

9. $\dfrac{10^{-2}}{10^{-4}}$

10. $\dfrac{10^{-1}}{10^{-4}}$

11. $\dfrac{10^{-5}}{10^{-3}}$

12. $\dfrac{10^{-9}}{10^{-6}}$

13. $\dfrac{1}{10^2}$

14. $\dfrac{1}{10^{-4}}$

15. $\dfrac{1}{10^6}$

Solutions:

1. 10^2 2. 10^3
3. 10^{-3} 4. 10^{-5}
5. 10^4 6. 10^8
7. 10^2 8. 10^{-4}
9. 10^2 10. 10^3
11. 10^{-2} 12. 10^{-3}
13. 10^{-2} 14. 10^4
15. 10^{-6}

Additional practice problems are at the end of the chapter.

2.5 COMBINED MULTIPLICATION AND DIVISION IN POWERS OF TEN FORM

Multiplication and division may be performed in the same problem. We must take care to observe the signs of the exponents as we add or subtract.

$$\frac{10^4 \times 10^2}{10^3} = \frac{10^{4+2}}{10^3} = \frac{10^6}{10^3} = 10^{6-3} = 10^3$$

Notice that we performed the indicated multiplication by adding the exponents. Next we performed the indicated division by subtracting the exponent in the denominator from the exponent in the numerator.

$$\frac{10^{-5} \times 10^3}{10^{-2}} = \frac{10^{-5+3}}{10^{-2}} = \frac{10^{-2}}{10^{-2}} = 10^{-2+2} = 10^0 = 1$$

Remember $10^0 = 1$. In this case $\frac{0.01}{0.01} = 1$.

Let's try two more.

$$\frac{10^3 \times 10^4}{10^{-2} \times 10^6} = \frac{10^{3+4}}{10^{-2+6}} = \frac{10^7}{10^4} = 10^{7-4} = 10^3$$

$$\frac{1}{10^2 \times 10^4} = \frac{1}{10^{2+4}} = \frac{1}{10^6} = 10^{-6}$$

We arrive at the answer 10^{-6} because the numerator (1) can be written in powers of ten form as 10^0. Then we have:

$$\frac{10^0}{10^6} = 10^{0-6} = 10^{-6}$$

A very useful rule to remember is:

Whenever a power of ten is in the denominator, it can be moved to the numerator by merely changing the sign of the exponent.

Consider the example above in which we had

$$\frac{10^3 \times 10^4}{10^{-2} \times 10^6}$$

We could have solved by this method:

$$\frac{10^3 \times 10^4}{10^{-2} \times 10^6} = 10^3 \times 10^4 \times 10^2 \times 10^{-6} = 10^3$$

Notice that 10^{-2} in the denominator became 10^2 when it was moved to the numerator. 10^6 in the denominator became 10^{-6} in the numerator. When all powers of ten are in the numerator, we simply add the exponents.

PRACTICE PROBLEMS 2-7   Perform the indicated operations. Express your answers in powers of ten form.

1. $\dfrac{10^2 \times 10^{-3}}{10^4}$

2. $\dfrac{10^{-1} \times 10^{-6}}{10^3}$

3. $\dfrac{10 \times 10^0 \times 10^6}{10^{-3}}$

4. $\dfrac{10^{-3} \times 10^4 \times 10^{-6}}{10^{-3}}$

5. $\dfrac{10^{-2} \times 10^4}{10^3 \times 10^2}$

6. $\dfrac{10^2 \times 10^6}{10^3 \times 10^{-1}}$

7. $\dfrac{1}{10^2 \times 10^4 \times 10}$

8. $\dfrac{1}{10^{-2} \times 10^6 \times 10^{-8}}$

9. $\dfrac{10^4 \times 10^{-1} \times 10^6}{10^2 \times 10^{-4} \times 10^5}$

10. $\dfrac{10^{-6} \times 10^{-3} \times 10^3}{10^{-9} \times 10^6 \times 10^{-3}}$

Solutions:

1. 10^{-5}

2. 10^{-10}

3. 10^{10}

4. 10^{-2}

5. 10^{-3}

6. 10^6

7. 10^{-7}

8. 10^4

9. 10^6

10. 10^0

Additional practice problems are at the end of the chapter.

SELF-TEST 2-1   Convert the following numbers to powers of ten form:

1. 1000

2. 1,000,000

3. 0.001

4. 0.0000001

5. 0.0001

6. 0.00001

Determine the numbers represented by the following powers of ten:

7. 10^3

8. 10^5

9. 10^2

10. 10^0

11. 10^{-4}

12. 10^{-1}

13. 10^{-5}

14. 10^{-8}

Convert the following powers of ten to decimal fractions and to decimal numbers:

15. 10^{-2}

16. 10^{-3}

17. 10^{-5}

18. 10^{-1}

Perform the indicated operations. Express your answers in powers of ten form.

19. $10^2 \times 10^5$

20. $10^3 \times 10^{-4}$

21. $10^6 \times 10^{-2}$

22. $10^{-3} \times 10^{-6}$

23. $\dfrac{10^2}{10^5}$

24. $\dfrac{10^4}{10^{-6}}$

25. $\dfrac{10^{-3}}{10^6}$

26. $\dfrac{10^{-2}}{10^{-5}}$

27. $\dfrac{10^{-2} \times 10^4}{10^{-3}}$

28. $\dfrac{10 \times 10^{-5}}{10^4 \times 10^{-9}}$

29. $\dfrac{10^{-4} \times 10^{-1} \times 10^6}{10^6 \times 10^{-3} \times 10}$

30. $\dfrac{10^{-2} \times 10^{-3} \times 10^9}{10^2 \times 10^3 \times 10^4}$

31. $\dfrac{10^4 \times 10^{-4} \times 10^6}{10^3 \times 10^{-3} \times 10^{-2}}$

32. $\dfrac{10 \times 10^{-7} \times 10^{-2}}{10^{-3} \times 10^{-4} \times 10^2}$

33. $\dfrac{1}{10^2 \times 10^{-3} \times 10^4}$

Answers to Self-Test 2-1 are at the end of the chapter.

2.6 SCIENTIFIC NOTATION

In science and engineering it is common practice to round answers to two, three, or four significant figures and then to change the number to a number between 1 and 10 times some power of ten. When a number is written in this manner, it is said to be expressed in scientific notation. For example, 1.76×10^4, 4.067×10^{-2}, and 9.2×10^6 are values expressed in scientific notation.

As we mentioned in Chapter 1, great accuracy in problem solving is not realistic in electronics because we usually work with components whose values are given to two significant figures (color bands on resistors, for example) and then may be as much as $\pm 10\%$ different than that value. In addition, measuring instruments such as voltmeters and ohmmeters, if of the analog type, like a multimeter, do not provide accuracy beyond two figures. Even digital meters at best provide 1% accuracies.

Therefore, throughout the text, with few exceptions, we will not require accuracies beyond three places—even though your calculator can provide eight or more.

Consider the number 764,832. Let's express this number in scientific notation accurate to three significant digits or simply "three places." Let's first round to three places. The number becomes 765,000. We rounded up because the fourth digit was 8. Now let's change the number to a number between 1 and 10. The answer is 7.65. To get 7.65 we had to move the decimal point five places *to the left*. In order to have an answer that equals 765,000, we must multiply 7.65 by 100,000 or 10^5. $765,000 = 7.65 \times 10^5$. For each place we move the decimal point left, we must multiply by 10. Since we move the decimal point five places to the left we must multiply by 10^5. 4703 in scientific notation (accurate to three places) would be 4.70×10^3. The rule is:

> *Count the number of places the decimal point moves left. This number is the exponent and is positive.*

PRACTICE PROBLEMS 2-8   Round the following numbers to three significant figures and express your answers in scientific notation:

1. 1730

2. 26,745

3. 1,654,736

4. 42.69

5. 173,426

6. 47,394

7. 79.063

8. 678,091

9. 189.83

10. 6746

Solutions:

1. 1.73×10^3

2. 2.67×10^4

3. 1.65×10^6

4. 4.27×10

5. 1.73×10^5

6. 4.74×10^4

7. 7.91×10

8. 6.78×10^5

9. 1.90×10^2

10. 6.75×10^3

Additional practice problems are at the end of the chapter.

Consider the number 0.00715. In this case we must move the decimal point three places to the *right*. For the numbers to be equal, we must multiply by 0.001 or 10^{-3}. The answer is 7.15×10^{-3}. For each place we move the decimal point *right*, we must multiply by 0.1. Remember that moving the decimal point *right* results in a negative power of ten and moving the decimal point *left* results in a positive power of ten. The absolute value of the exponent is determined by the number of places the point is moved. The rule is:

Count the number of places the decimal point moves right. This number is the exponent and is negative.

PRACTICE PROBLEMS 2-9   Round the following numbers to three significant figures and express your answers in scientific notation:

1. 0.0167

2. 0.640732

3. 0.0003988

4. 0.00607

5. 0.1086

6. 0.00000706

7. 0.00001763

8. 0.009096

9. 0.07646

10. 0.000567

Solutions:

1. 1.67×10^{-2}

2. 6.41×10^{-1}

3. 3.99×10^{-4}

4. 6.07×10^{-3}

5. 1.09×10^{-1}

6. 7.06×10^{-6}

7. 1.76×10^{-5}

8. 9.10×10^{-3}

9. 7.65×10^{-2}

10. 5.67×10^{-4}

Additional practice problems are at the end of the chapter.

2.6.1 Multiplication and Division. Let's work the following problems together:

$$A. \quad 76.3 \times 446 = 3.40 \times 10^4$$

$$B. \quad \frac{0.00743 \times 10^2 \times 14}{5384 \times 0.04} = 4.83 \times 10^{-2}$$

$$C. \quad \frac{25 \times 3300}{39,645} = 2.08 \times 10^0$$

The problems are solved by entering them into the calculator. Most calculators are capable of displaying the answers in scientific notation. If your calculator does not have this feature, then you must manipulate the decimal point to arrive at the answer in the desired form. For example, problem A equals 34,000 accurate to three significant figures. This becomes 3.40×10^4 when the decimal point is moved four places to the left.

In problem B there is no preferred choice of which operation to perform first. The answer is 0.0483 accurate to three significant figures. This becomes 4.83×10^{-2} when the decimal point is moved two places to the right. In problem C no adjustment of the decimal point is necessary.

Work the following problems by using your calculator. Because all the operations are multiplications and divisions, there is no preferred choice of the order in which they are performed.

PRACTICE PROBLEMS 2-10   Perform the indicated operations. Express your answers in scientific notation accurate to three places.

1. 47×106

2. 789×206

3. 24.6×0.00173

4. $473 \times 10^{-3} \times 26$

5. 2700×0.743

6. $33 \times 10^3 \times 65 \times 10^{-4}$

7. $17 \times 43 \times 0.003$

8. $7800 \times 10^{-3} \times 27 \times 10^{-2}$

9. $680 \times 10^3 \times 473 \times 10^{-7}$

10. $1.8 \times 10^4 \times 2.3 \times 10^{-6}$

11. $9 \div 0.000742$

12. $25 \div 5600$

13. $15 \div 24 \times 10^{-3}$

14. $12 \div 680 \times 10^2$

15. $\dfrac{20 \times 22 \times 10^3}{720}$

16. $\dfrac{40 \times 213 \times 10^{-6}}{27 \times 10^{-4}}$

17. $\dfrac{18 \times 10^3 \times 27 \times 10^3}{45 \times 10^3}$

18. $\dfrac{18 \times 0.00473}{673 \times 10^{-6}}$

19. $\dfrac{6700 \times 876}{914 \times 732}$

20. $\dfrac{743 \times 10^2 \times 0.00365}{65 \times 10^{-1} \times 0.403}$

21. $\dfrac{0.000467 \times 10^4 \times 7680}{2.6 \times 10^{-3} \times 526 \times 10^2}$

22. $\dfrac{14 \times 5600}{61.6 \times 10^4}$

23. $\dfrac{1}{6.28 \times 12 \times 10^3 \times 470 \times 10^{-12}}$

24. $\dfrac{1}{6.28 \times 25 \times 10^3 \times 0.02 \times 10^{-9}}$

25. $\dfrac{1}{6.28 \times 50 \times 10^2 \times 0.5 \times 10^{-6}}$

Solutions:

1. 4.98×10^3

2. 1.63×10^5

3. 4.26×10^{-2}

4. 1.23×10

5. 2.01×10^3

6. 2.15×10^2

7. 2.19×10^0

8. 2.11×10^0

9. 3.22×10

10. 4.14×10^{-2}

11. 1.21×10^4

12. 4.46×10^{-3}

13. 6.25×10^2

14. 1.76×10^{-4}

15. 6.11×10^2 **16.** 3.16×10^0

17. 1.08×10^4 **18.** 1.27×10^2

19. 8.77×10^0 **20.** 1.04×10^2

21. 2.62×10^2 **22.** 1.27×10^{-1}

23. 2.82×10^4 **24.** 3.18×10^5

25. 6.37×10

Often an answer like the answer to number 7 (2.11×10^0) is simply written 2.11, since $10^0 = 1$ $(2.11 \times 10^0 = 2.11 \times 1 = 2.11)$.

Additional practice problems are at the end of the chapter.

2.6.2 Addition and Subtraction. We can add or subtract numbers in powers of ten form or when using scientific notation *provided that the exponents are the same.* For example, let's perform the following addition: $3 \times 10^2 + 4 \times 10^3$. The decimal numbers are 300 and 4000. If we add 300 and 4000, we get 4300 or 4.3×10^3. To perform the addition using powers of ten we must first move the decimal point of one number (it doesn't matter which) so that the exponents are the same. Let's change 3×10^2 so that the exponent is 3. $3 \times 10^2 = 300$. For the exponent to equal 3, we must move the decimal point three places to the left. $300 = 0.3 \times 10^3$. Now we can perform the addition:

$$
\begin{array}{r}
0.3 \times 10^3 \\
+4 \times 10^3 \\
\hline
4.3 \times 10^3
\end{array}
$$

Addition or subtraction is performed by adding the numbers together and multiplying by the indicated power of ten. The *exponents* are not added. The exponents merely tell us that we are adding 0.3 thousands to 4 thousands to get 4.3 *thousands.* We could have changed 4×10^3 to 40×10^2 and added.

$$
\begin{array}{r}
3 \times 10^2 \\
40 \times 10^2 \\
\hline
43 \times 10^2
\end{array}
$$

$$43 \times 10^2 = 4.3 \times 10 \times 10^2 = 4.3 \times 10^3$$

The result is the same but there are more steps involved.

Let's subtract 6×10^{-3} from 10^{-2}.

$$
\begin{array}{r}
1 \times 10^{-2} \\
(-)6 \times 10^{-3} \\
\hline
\end{array}
$$

10^{-2} is written in scientific notation form $(10^{-2} = 1 \times 10^{-2})$ because addition or subtraction of the decimal numbers is to be performed. In this problem we must again change one of the numbers so that both exponents are the same. Let's change 6×10^{-3} so that the exponent is -2. First, $6 \times 10^{-3} = 0.006$. For the exponent to equal -2, we move the decimal point two places to the right. $0.006 = 0.6 \times 10^{-2}$. Now:

$$
\begin{array}{r}
1\ \times 10^{-2} \\
(-)0.6 \times 10^{-2} \\
\hline
0.4 \times 10^{-2}
\end{array}
$$

$$0.4 \times 10^{-2} = 4 \times 10^{-1} \times 10^{-2} = 4 \times 10^{-3}$$

We could have changed 1×10^{-2} to a number whose exponent is -3. $1 \times 10^{-2} = 0.01 = 10 \times 10^{-3}$. Because we wanted the exponent to equal -3, we had to move the decimal point three places to the right.

$$
\begin{array}{r}
10 \times 10^{-3} \\
(-)6 \times 10^{-3} \\
\hline
4 \times 10^{-3}
\end{array}
$$

We can check our work by performing the subtraction in decimal.

$$
\begin{array}{ll}
1 \times 10^{-2} = 0.01 & 0.01 \\
6 \times 10^{-3} = 0.006 & (-)0.006 \\
& \overline{0.004}
\end{array}
$$

$$0.004 = 4 \times 10^{-3}$$

In the following problems we will practice moving the decimal point to change the exponent. Then we will add or subtract numbers. Remember that in adding or subtracting the exponents must be the same.

PRACTICE PROBLEMS 2-11   Change the following numbers so that the exponent is 4:

1. 3.78×10^3

2. 76.4×10^2

3. 178,000

4. 273×10

5. 67.2×10

Change the following numbers so that the exponent is -2:

6. 4.76×10^{-3}

7. 4.76×10^{-1}

8. 0.476×10^{-4}

9. 871×10^{-1}

10. 871×10^{-3}

Perform the indicated operations. Express your answers in scientific notation accurate to three places.

11. $10^2 + 10^3$

12. $10^1 + 10^{-1}$

13. $10^3 - 10^2$

14. $2 \times 10^2 + 3 \times 10^3$

15. $47 \times 10^{-2} + 560 \times 10^{-3}$

16. $40 \times 10^3 - 300 \times 10^2$

17. $3 \times 10^{-2} - 45 \times 10^{-4}$

18. $270 \times 10^{-1} + 46.3 \times 10$

19. $6.8 \times 10^3 + 6.8 \times 10^4 + 56,000$

20. $213 \times 10^{-6} + 0.000043 + 51.3 \times 10^{-5}$

Solutions:

1. 0.378×10^4 2. 0.764×10^4

3. 17.8×10^4 4. 0.273×10^4

5. 0.0672×10^4 6. 0.476×10^{-2}

7. 47.6×10^{-2} 8. 0.00476×10^{-2}

9. 8710×10^{-2} 10. 87.1×10^{-2}

11. 1.10×10^3 12. 1.01×10

13. 9.00×10^2 14. 3.2×10^3

15. 1.03 16. 1×10^4

17. 2.55×10^{-2} 18. 4.90×10^2

19. 1.31×10^5 20. 7.69×10^{-4}

Additional practice problems are at the end of the chapter.

2.7 PROBLEMS WITH COMPLEX DENOMINATORS

A typical Ohm's law problem might look like this:

$$\frac{25 \times 3.3 \times 10^3}{3.3 \times 10^3 + 4.7 \times 10^4}$$

To solve this problem, we must first perform the indicated addition in the denominator.

$$3.3 \times 10^3 = 0.33 \times 10^4$$
$$+ 4.7 \times 10^4 = 4.7 \ \times 10^4$$
$$\overline{5.03 \times 10^4}$$

Now:

$$\frac{25 \times 3.3 \times 10^3}{5.03 \times 10^4} = 1.64$$

Of course, all the math may be done by using the calculator as shown in Appendix A. The important rule to remember is to perform the *addition* first in order to simplify the denominator.

2.8 RECIPROCALS

The *reciprocal* of a number is that number divided into 1. $\frac{1}{3}$ is the reciprocal of 3, $\frac{1}{50}$ is the reciprocal of 50, and so on. In electronics it is often necessary to find the reciprocal of numbers and then perform some other mathematical operation with that reciprocal. For example:

$$\frac{1}{22,000} + \frac{1}{47,000} = 4.55 \times 10^{-5} + 2.13 \times 10^{-5} = 6.67 \times 10^{-5}$$

Notice that we first found the reciprocal of each number and then performed the indicated addition. Going one step further, we get:

$$\frac{1}{\dfrac{1}{680} + \dfrac{1}{560}} = \frac{1}{1.47 \times 10^{-3} + 1.79 \times 10^{-3}}$$

$$= \frac{1}{3.26 \times 10^{-3}} = 3.07 \times 10^{2}$$

Notice that we first found the reciprocal of each part of the denominator, added them together to simplify the denominator, and then found the reciprocal of that number.

PRACTICE PROBLEMS 2-12 Perform the indicated operations. Express your answers in scientific notation accurate to three places.

1. $\dfrac{1}{470}$

2. $\dfrac{1}{33 \times 10^{4}}$

3. $\dfrac{1}{68 \times 10^{-9}}$

4. $\dfrac{1}{3.7 \times 10^{-3}}$

5. $\dfrac{1}{2760}$

6. $\dfrac{1}{5.6 \times 10^{3}}$

7. $\dfrac{1}{50 \times 10^{3}}$

8. $\dfrac{1}{370 \times 10^{-6}}$

9. $\dfrac{1}{4 \times 10^{-3}}$

10. $\dfrac{1}{10 \times 10^{3}}$

11. $\dfrac{1}{2700} + \dfrac{1}{1800}$

12. $\dfrac{1}{68 \times 10^{3}} + \dfrac{1}{47 \times 10^{3}}$

13. $\dfrac{1}{12 \times 10^{3}} + \dfrac{1}{7.5 \times 10^{3}}$

14. $\dfrac{25 \times 4700}{4700 + 47 \times 10^{3}}$

15. $\dfrac{20 \times 3.3 \times 10^{3}}{3.3 \times 10^{3} + 3.3 \times 10^{4}}$

16. $\dfrac{9 \times 10^{4}}{10^{3} + 10^{4}}$

17. $\dfrac{1}{\dfrac{1}{27 \times 10^{3}} + \dfrac{1}{18 \times 10^{3}}}$

18. $\dfrac{1}{\dfrac{1}{6800} + \dfrac{1}{5600}}$

19. $\dfrac{1}{27.6 \times 10^{3}} - \dfrac{1}{47 \times 10^{3}}$

20. $\dfrac{1}{\dfrac{1}{270} + \dfrac{1}{470} + \dfrac{1}{560}}$

Solutions:

1. 2.13×10^{-3}
2. 3.03×10^{-6}
3. 1.47×10^{7}
4. 2.7×10^{2}
5. 3.62×10^{-4}
6. 1.79×10^{-4}
7. 2.00×10^{-5}
8. 2.70×10^{3}
9. 2.50×10^{2}
10. 1×10^{-4}
11. 9.26×10^{-4}
12. 3.6×10^{-5}
13. 2.17×10^{-4}
14. 2.27×10^{0}
15. 1.82×10^{0}
16. 8.18×10^{0}
17. 1.08×10^{4}
18. 3.07×10^{3}
19. 1.5×10^{-5}
20. 1.31×10^{2}

Additional practice problems are at the end of the chapter.

SELF-TEST 2-2 Perform the indicated operations. Express your answers in scientific notation accurate to three places.

1. 27×65
2. 0.00173×3900
3. $47,000 \times 0.0000423$
4. $1765 \times 24,673$
5. $33 \times 10^4 \times 27 \times 10^3$
6. $5.6 \times 10^3 + 4.7 \times 10^4$
7. $147 \times 10^{-7} + 0.833 \times 10^{-5} + 21.3 \times 10^{-6}$
8. $2 \times 3.14 \times 2500 \times 500 \times 10^{-9}$
9. $\dfrac{47.3 \times 10^2}{0.00846}$
10. $\dfrac{0.025 \times 10^{-6}}{483 \times 10^{-9}}$
11. $\dfrac{7800 \times 0.00273}{1.28 \times 10^{-4} \times 8000}$
12. $\dfrac{0.0000064 \times 84 \times 10^2}{6730 \times 0.045 \times 10^{-3}}$
13. $\dfrac{12 \times 3900}{56,000 + 3900}$
14. $\dfrac{1}{270} + \dfrac{1}{390}$
15. $\dfrac{25 \times 4700}{4700 + 47,000}$
16. $\dfrac{1}{3900} + \dfrac{1}{56 \times 10^3} + \dfrac{1}{20 \times 10^3}$
17. $\dfrac{1}{6.28 \times 12 \times 10^3 \times 470 \times 10^{-12}}$
18. $\dfrac{25 \times 2700}{2.7 \times 10^3 + 3.3 \times 10^4}$
19. $\dfrac{1}{\dfrac{1}{560 \times 10^2} + \dfrac{1}{47 \times 10^3}}$
20. $\dfrac{1}{\dfrac{1}{2700} + \dfrac{1}{3300} + \dfrac{1}{4700}}$

Answers to Self-Test 2-2 are at the end of the chapter.

2.9 POWERS AND ROOTS IN BASE TEN

When we wrote 10^3 we learned that the exponent (3) told us to multiply 10 by itself three times. $10^3 = 10 \times 10 \times 10$. If we wrote $(10^3)^2$, we would be raising 10^3 to the second power. That is $(10^3)^2 = 10^3 \times 10^3 = 10^6$. We need the parens to identify the quantity we are acting on. $(10^{-2})^3 = 10^{-2} \times 10^{-2} \times 10^{-2} = 10^{-6}$. Notice in each case the exponent can be thought of as a multiplier. $(10^3)^2 = 10^{3 \times 2} = 10^6$. $(10^{-2})^3 = 10^{-2 \times 3} = 10^{-6}$. $(10^4)^{-3} = 10^{4 \times (-3)} = 10^{-12}$. $(\frac{1}{10^4})^2 = \frac{1}{10^{4 \times 2}} = \frac{1}{10^8}$. In the last example the numerator (1) is also squared but $1^2 = 1$. Therefore, the numerator is unchanged. The rule is:

When raising a power to a power, multiply the exponents.

PRACTICE PROBLEMS 2-13   Perform the indicated operations. Express your answers in powers of ten form.

1. $(10^2)^3$

2. $(10^4)^2$

3. $(10^3)^3$

4. $(10^{-1})^2$

5. $(10^{-4})^3$

6. $(10^{-6})^2$

7. $(10^{-2})^{-3}$

8. $(10^{-3})^{-1}$

9. $(10^{-1})^{-4}$

10. $(10^3)^{-5}$

11. $(10^2)^{-1}$

12. $(10^6)^{-2}$

13. $\left(\dfrac{1}{10^2}\right)^4$

14. $\left(\dfrac{1}{10^{-3}}\right)^2$

15. $\left(\dfrac{1}{10^3}\right)^{-3}$

Solutions:

1. 10^6

2. 10^8

3. 10^9

4. 10^{-2}

5. 10^{-12}

6. 10^{-12}

7. 10^6

8. 10^3

9. 10^4

10. 10^{-15}

11. 10^{-2}

12. 10^{-12}

13. 10^{-8}

14. 10^6

15. 10^9

Additional practice problems are at the end of the chapter.

To find the square root or cube root of powers of ten, we multiply the exponents. $(10^4)^{1/2} = 10^{4 \times 1/2} = 10^2$. $(10^{12})^{1/3} = 10^{12 \times 1/3} = 10^4$.

$$\left(\frac{1}{10^6}\right)^{1/2} = \frac{1}{10^{6 \times 1/2}} = \frac{1}{10^3}$$

PRACTICE PROBLEMS 2-14 Perform the indicated operations. Express your answers in powers of ten form.

1. $(10^6)^{1/2}$

2. $(10^9)^{1/3}$

3. $(10^{-6})^{1/2}$

4. $(10^{-12})^{1/3}$

5. $\left(\dfrac{1}{10^{-3}}\right)^{1/3}$

6. $(10^{-6})^{1/3}$

7. $(10^3)^{1/3}$

8. $\left(\dfrac{1}{10^4}\right)^{1/2}$

9. $(10^{-15})^{1/3}$

10. $(10^{10})^{1/2}$

Solutions:

1. 10^3

2. 10^3

3. 10^{-3}

4. 10^{-4}

5. 10

6. 10^{-2}

7. 10

8. 10^{-2}

9. 10^{-5}

10. 10^5

Additional practice problems are at the end of the chapter.

Let's square some numbers other than base 10 and express the answers in scientific notation accurate to three places. $14.7^2 = ?$ This problem requires the use of another function key on the calculator. This key is usually labeled x^2. If your calculator doesn't have an x^2 key, then you simply perform the indicated operation, which is 14.7×14.7. $14.7^2 = 2.16 \times 10^2$. $(0.0236 \times 10^2)^2 = 5.57$. In this problem we are squaring *both* 0.0236 and 10^2. $\dfrac{(278)^2}{16} = 4.83 \times 10^3$. Here 278 is squared but 16 is *not* squared.

Let's try a few to see how you do.

PRACTICE PROBLEMS 2-15 Perform the indicated operations. Express your answers in scientific notation accurate to three places.

1. 6.3^2

2. 43^2

3. 0.0372^2

4. 0.00047^2

5. $(6.73 \times 10^2)^2$

6. $(1.04 \times 10^{-3})^2$

7. $(796 \times 10^{-2})^2$

8. 8764^2

9. 0.973^2

10. $(0.00567 \times 10^{-4})^2$

11. $(0.0000873 \times 10^4)^2$

12. $(1260 \times 10^{-3})^2$

13. $(0.0073)^2 \times 4700$

14. $(27 \times 10^{-6})^2 \times 68,000$

15. $20^2 + 3300$

16. $9^2 + 270$

Solutions:

1. 3.97×10

2. 1.85×10^3

3. 1.38×10^{-3}

4. 2.21×10^{-7}

5. 4.53×10^5

6. 1.08×10^{-6}

7. 6.34×10

8. 7.68×10^7

9. 9.47×10^{-1}

10. 3.21×10^{-13}

11. 7.62×10^{-1}

12. 1.59

13. 2.50×10^{-1}

14. 4.96×10^{-5}

15. 1.21×10^{-1}

16. 3.00×10^{-1}

Additional practice problems are at the end of the chapter.

The following problems require using the $\sqrt{x}$ key on the calculator. A problem like $367^{1/2} = 1.92 \times 10$ simply requires entering 367 and pressing $\sqrt{x}$. More complex problems require a sequence of operations. For example, $\dfrac{1}{6.28(0.05 \times 5 \times 10^{-9})^{1/2}}$ requires that we find the square root of $0.05 \times 5 \times 10^{-9}$, multiply that quantity by 6.28, and then take the reciprocal of that product to get the answer, which is 1.01×10^4.

PRACTICE PROBLEMS 2-16   Perform the indicated operations. Express your answer in scientific notation accurate to three places.

1. $36.5^{1/2}$

2. $0.925^{1/2}$

3. $(635 \times 10^{-2})^{1/2}$

4. $(1870 \times 10^3)^{1/2}$

5. $(0.000675 \times 10^2)^{1/2}$

6. $(27.3 \times 10^2)^{1/2}$

7. $(5.46 \times 10^{-4})^{1/2}$

8. $\left(\dfrac{734 \times 10^2}{423} \right)^{1/2}$

9. $\left(\dfrac{0.00726 \times 10^{-3}}{77 \times 10^{-6}} \right)^{1/2}$

10. $\dfrac{1}{6.28(0.2 \times 3 \times 10^{-9})^{1/2}}$

Solutions:

1. 6.04

2. 9.62×10^{-1}

3. 2.52

4. 1.37×10^3

5. 2.60×10^{-1}

6. 5.22×10

7. 2.34×10^{-2}

8. 1.32×10

9. 3.07×10^{-1}

10. 6.50×10^3

Additional practice problems are at the end of the chapter.

SELF-TEST 2-3 Perform the indicated operations. Express your answers in scientific notation accurate to three places.

1. $(10^3)^5$

2. $(10^2)^{-4}$

3. $\left(\dfrac{1}{10^{-2}}\right)^3$

4. $(10^{-6})^{1/2}$

5. $(10^6)^{1/3}$

6. $(10^{-3})^{1/3}$

7. 78.3^2

8. 0.000293^2

9. $(8796 \times 10^3)^2$

10. $(0.0173)^2 \times 1800$

11. $(39 \times 10^{-4})^2 \times 120$

12. $\dfrac{15^2}{1500}$

13. $273^{1/2}$

14. $(2.65 \times 10^{-4})^{1/2}$

15. $\left(\dfrac{0.0000467 \times 10^3}{63}\right)^{1/2}$

16. $\left(\dfrac{645 \times 10^2}{14}\right)^{1/2}$

17. $\dfrac{(24.5 \times 10^{-3})^{1/2}}{17}$

18. $(300 \times 10^{-3} \times 220)^{1/2}$

19. $\left(\dfrac{273 \times 10^4}{0.0012}\right)^{1/2}$

20. $\dfrac{1}{6.28(50 \times 10^{-3} \times 2 \times 10^{-6})^{1/2}}$

Answers to Self-Test 2-3 are at the end of the chapter.

END OF CHAPTER PROBLEMS 2-1

Convert the following numbers to powers of ten form:

1. 10

2. 100,000

3. 1000

4. 10,000,000

5. 0.1

6. 0.000001

7. 0.01

8. 0.0001

Determine the numbers represented by the following powers of ten:

9. 10^4

10. 10^6

11. 10^3

12. 10^2

13. 10^{-1}

14. 10^{-3}

15. 10^{-6}

16. 10^{-9}

Convert the following powers of ten to (a) decimal fractions and to (b) decimal numbers:

17. 10^{-4}

18. 10^{-6}

19. 10^{-9}

20. 10^{-7}

END OF CHAPTER PROBLEMS 2-2

Perform the indicated operations. Express your answers in powers of ten form.

1. $10^4 \times 10^6$

2. $10^3 \times 10^2$

3. $10^5 \times 10^{-1}$

4. $10^6 \times 10^{-3}$

5. $10^{-1} \times 10^4$

6. $10^{-2} \times 10^6$

7. $10^{-3} \times 10^{-9}$

8. $10^{-2} \times 10^{-5}$

9. $10^2 \times 10^4$

10. $\dfrac{1}{10^4}$

11. $\dfrac{1}{10^{-3}}$

12. $\dfrac{1}{10^6}$

13. $\dfrac{1}{10^{-7}}$

14. $\dfrac{10^2 \times 10^6}{10^4}$

15. $\dfrac{10^{-3} \times 10^{-6}}{10^{-2}}$

16. $\dfrac{10^{-4} \times 10^9}{10^3}$

17. $\dfrac{1}{10^2 \times 10^{-4} \times 10^{-9}}$

18. $\dfrac{1}{10^{-2} \times 10^4 \times 10^{-7}}$

19. $\dfrac{1}{10^{-3} \times 10^{-6} \times 10^{-1}}$

20. $\dfrac{1}{10^2 \times 10^6 \times 10^{-3}}$

21. $\dfrac{10^{-4}}{10^6 \times 10^{-2} \times 10^4}$

22. $\dfrac{10^3}{10^4 \times 10^6 \times 10^{-1}}$

23. $\dfrac{10^{-3} \times 10^2 \times 10^4}{10^2 \times 10^{-7}}$

24. $\dfrac{10^{-1} \times 10^2 \times 10^4}{10^4 \times 10^6}$

25. $\dfrac{10^4 \times 10^6 \times 10^3}{10^2 \times 10^{-7} \times 10^5}$

26. $\dfrac{10^{-1} \times 10^{-7} \times 10^6}{10 \times 10^0 \times 10^{-6}}$

27. $\dfrac{10^3 \times 10^{-5} \times 10^6}{10^4 \times 10^{-3} \times 10^7}$

28. $\dfrac{10^{-2} \times 10^{-6} \times 10^3}{10^3 \times 10^6 \times 10^{-1}}$

29. $\dfrac{10^{-3} \times 10^{-6} \times 10^{-9}}{10^3 \times 10^6 \times 10^9}$

30. $\dfrac{10^{-9} \times 10^3 \times 10^{-3}}{10^4 \times 10^{-1} \times 10^7}$

END OF CHAPTER PROBLEMS 2-3

Round the following numbers to three significant figures and express your answers in scientific notation:

1. 27,640

2. 17,896

3. 47.8

4. 276.7

5. 1,765,400

6. 3,706,400

7. 273.46

8. 45.986

9. 173,460

10. 407,600

11. 0.004783

12. 0.17600

13. 0.7474

15. 0.01637

17. 0.000002778

19. 0.170983

21. 15,673

23. 0.006678

25. 1,670,900

27. 0.0000004679

29. 0.005836

14. 0.0002087

16. 0.00001037

18. 0.0005278

20. 0.07096

22. 15.706

24. 0.000899

26. 78.88

28. 4,767

30. 0.0000204

END OF CHAPTER PROBLEMS 2-4

Perform the indicated operations. Express your answers in scientific notation accurate to three places.

1. 273×78

3. 0.0000273×4700

5. $\dfrac{50}{0.00475}$

7. $\dfrac{176 \times 473}{27,800}$

9. $\dfrac{0.00790 \times 835 \times 10^2}{9840 \times 10^{-2}}$

11. $\dfrac{651 \times 0.00179}{73.4 \times 0.874}$

13. $\dfrac{1}{680} + \dfrac{1}{560}$

15. $\dfrac{1}{56 \times 10^3} + \dfrac{1}{47 \times 10^3}$

17. $\dfrac{1}{\dfrac{1}{4700} + \dfrac{1}{5600}}$

19. $\dfrac{1}{\dfrac{1}{56 \times 10^3} + \dfrac{1}{68 \times 10^3}}$

21. $\dfrac{25 \times 3300}{3300 + 56,000}$

23. $\dfrac{12 \times 4.7 \times 10^3}{4.7 \times 10^4 + 4.7 \times 10^3}$

25. $\dfrac{25 \times 10^{-3} \times 213 \times 10^{-6}}{213 \times 10^{-6} + 303 \times 10^{-6}}$

27. $6.28 \times 25 \times 10^3 \times 50 \times 10^{-3}$

29. $\dfrac{1}{6.28 \times 50 \times 10^3 \times 30 \times 10^{-9}}$

2. 793×467

4. 0.00675×0.00147

6. $\dfrac{12}{0.0000273}$

8. $\dfrac{4630 \times 127}{4360}$

10. $\dfrac{76 \times 10^{-2} \times 906 \times 10^4}{0.000559}$

12. $\dfrac{127 \times 10^2 \times 0.024 \times 10^{-4}}{79 \times 10^{-6} \times 410 \times 10^4}$

14. $\dfrac{1}{470} + \dfrac{1}{330}$

16. $\dfrac{1}{27 \times 10^4} + \dfrac{1}{10 \times 10^4}$

18. $\dfrac{1}{\dfrac{1}{270} + \dfrac{1}{680}}$

20. $\dfrac{1}{\dfrac{1}{2.2 \times 10^3} + \dfrac{1}{750}}$

22. $\dfrac{20 \times 4700}{33,000 + 4700}$

24. $\dfrac{15 \times 1.8 \times 10^3}{5.6 \times 10^4 + 1.8 \times 10^3}$

26. $\dfrac{130 \times 10^{-6} \times 2 \times 10^{-3}}{2 \times 10^{-3} + 3 \times 10^{-3}}$

28. $6.28 \times 2 \times 10^3 \times 5 \times 10^{-6}$

30. $\dfrac{1}{6.28 \times 500 \times 10 \times 10^{-6}}$

END OF CHAPTER PROBLEMS 2-5

Perform the indicated operations. Express your answers in scientific notation accurate to three places.

1. $(10^2)^3$

2. $(10^2)^4$

3. $(10^5)^2$

4. $(10^{-2})^1$

5. $(10^3)^{-4}$

6. $(10^2)^{-6}$

7. $(10^{-4})^{-3}$

8. $\left(\dfrac{1}{10^3}\right)^2$

9. $\left(\dfrac{1}{10^{-4}}\right)^{-2}$

10. $(10^8)^{1/2}$

11. $(10^6)^{1/3}$

12. $(10^{-2})^{1/2}$

13. $\left(\dfrac{1}{10^5}\right)^3$

14. $\left(\dfrac{1}{10^{-8}}\right)^{1/2}$

15. 79.3^2

16. 246^2

17. 0.00873^2

18. 0.000185^2

19. $(45\times10^3)^2$

20. $(364\times10^2)^2$

21. $(0.296\times10^{-3})^2$

22. $(0.0033\times10^{-2})^2$

23. $(50\times10^{-3})^2\times2700$

24. $(375\times10^{-9})^2\times1200$

25. $93^{1/2}$

26. $873^{1/2}$

27. $0.00705^{1/2}$

28. $0.065^{1/2}$

29. $(270\times10^5)^{1/2}$

30. $(43.2\times10^4)^{1/2}$

31. $(0.00923\times10^{-3})^{1/2}$

32. $(0.176\times10^4)^{1/2}$

33. $\left(\dfrac{146\times10^3}{17}\right)^{1/2}$

34. $\left(\dfrac{560\times10^4}{26}\right)^{1/2}$

35. $\dfrac{1}{6.28(20\times10^{-3}\times20\times10^{-9})^{1/2}}$

36. $\dfrac{1}{6.28(0.0047\times470\times10^{-12})^{1/2}}$

37. $\dfrac{1}{6.28(0.150\times680\times10^{-12})^{1/2}}$

38. $\dfrac{1}{6.28(100\times10^{-3}\times30\times10^{-9})^{1/2}}$

ANSWERS TO SELF-TESTS

Self-Test 2-1

1. 10^3

2. 10^6

3. 10^{-3}

4. 10^{-7}

5. 10^{-4}

6. 10^{-5}

7. 1000

8. 100,000

9. 100

10. 1

11. 0.0001

12. 0.1

13. 0.00001

14. 0.00000001

15. $\frac{1}{100} = 0.01$

16. $\frac{1}{1000} = 0.001$

17. $\frac{1}{100,000} = 0.00001$

18. $\frac{1}{10} = 0.1$

19. 10^7

20. 10^{-1}

21. 10^4

22. 10^{-9}

23. 10^{-3}

24. 10^{10}

25. 10^{-9}

26. 10^3

27. 10^5

28. 10

29. 10^{-3}

30. 10^{-5}

31. 10^8

32. 10^{-3}

33. 10^{-3}

Self-Test 2-2

1. 1.76×10^3

2. 6.75

3. 1.99

4. 4.35×10^7

5. 8.91×10^9

6. 5.26×10^4

7. 4.43×10^{-5}

8. 7.85×10^{-3}

9. 5.59×10^5

10. 5.18×10^{-2}

11. 2.08×10

12. 1.78×10^{-1}

13. 7.81×10^{-1}

14. 6.27×10^{-3}

15. 2.27

16. 3.24×10^{-4}

17. 2.82×10^4

18. 1.89

19. 2.56×10^4

20. 1.13×10^3

Self-Test 2-3

1. 10^{15}

2. 10^{-8}

3. 10^6

4. 10^{-3}

5. 10^2

6. 10^{-1}

7. 6.13×10^3

8. 8.58×10^{-8}

9. 7.74×10^{13}

10. 5.39×10^{-1}

11. 1.83×10^{-3}

12. 1.5×10^{-1}

13. 1.65×10

14. 1.63×10^{-2}

15. 2.72×10^{-2}

16. 6.79×10

17. 9.21×10^{-3}

18. 8.12

19. 4.77×10^4

20. 5.04×10^2

UNITS AND PREFIXES

3.1 UNITS

Table 3-1 is a partial list of the International System of Units (SI). Only those units of interest to electronics students are included. Most of the units we deal with are either quite large or quite small. In order to deal with these quantities more easily, prefixes are used. The prefixes relate to powers of ten which are multiples of 3 or -3. 10^3, 10^6, 10^{-3}, and 10^{-6} are examples. Before we begin to

TABLE 3-1

Quantity	Unit of measure	Unit abbreviation
Capacitance (C)	Farad	F
Admittance (Y)	Siemens	S
Conductance (G)	Siemens	S
Current (I, i)	Ampere	A
Energy (w)	Joule	J
Frequency (f)	Hertz	Hz
Impedance (Z)	Ohm	Ω
Inductance (L)	Henry	H
Power (P)	Watt	W
Reactance (X)	Ohm	Ω
Resistance (R)	Ohm	Ω
Susceptance (B)	Siemens	S
Time (t)	Second	s
Wavelength (λ)	Meter	m
Electromotive force (E)	Volt	V
Difference in potential (V)	Volt	V

use these prefixes, let's review powers of ten by changing numbers to numbers times some power of ten that is a multiple of 3 or -3.

For example, change 6700 to a number times 10^3. The exponent is positive and is 3. Therefore, the decimal point must move three places to the left.

$$6700 = 6.7 \times 10^3$$

Change 670×10^2 to a number times 10^3.

We can solve this one two ways. First, we could change 670×10^2 to a regular number. $670 \times 10^2 = 67,000$. Then applying the same rule as before, $67,000 = 67 \times 10^3$. Another approach would be to note that the exponent is already 2. The difference between 10^2 and 10^3 is 10^1. Therefore, we need to move the decimal point one place to the left.

$$670 \times 10^2 = 67 \times 10^3$$

The rule is:

> *If we make the exponent more positive, the decimal point moves to the left.*

This is simply manipulation using powers of ten. $6700 = 670 \times 10 = 67 \times 10^2 = 6.7 \times 10^3 = 0.67 \times 10^4$ and so on.

Change 330,000 to some number times 10^3 and then to some number times 10^6.

$$330,000 = 330 \times 10^3$$
$$330,000 = 0.33 \times 10^6$$

Again, we moved the decimal point three places to the left because the exponent is 3. Then we moved the decimal point 6 places to the left because the exponent is 6. Once we had 330×10^3 we could have done this:

$$330 \times 10^3 = 0.330 \times 10^6$$

The exponent increases from 3 to 6, a change of 3 in a positive direction. Therefore, the decimal point moves three places to the left.

The rule for changing a number to a number times some *negative* power of ten is:

> *If the exponent is made more negative, the decimal point moves to the right.*

For example, change 1.73 to some number times 10^{-3}. The exponent is negative and equals 3. Therefore, we move the decimal point three places to the right. It is the same as $1,730 \times 0.001 = 1.73$.

$$1.73 = 1,730 \times 10^{-3}$$

Change 27.7×10^{-4} to some number times 10^{-6}. We can solve the problem by changing 27.7×10^{-4} to a regular number and then moving the decimal point six places to the right.

$$27.7 \times 10^{-4} = 0.00277 = 2770 \times 10^{-6}$$

We could also solve the problem by noting that the exponent is already -4. Increasing its value to -6 is a change of 2 in a negative direction. Therefore, the decimal point moves two places to the right. If we make the exponent more negative, the decimal point moves to the *right*.

$$27.7 \times 10^{-4} = 2770 \times 10^{-6}$$

Change 0.05×10^{-7} to some number times 10^{-9} and then to some number times 10^{-6}. When we change the power of ten from 10^{-7} to 10^{-9}, we are making the exponent more negative. The difference is -2. That is, $10^{-7-2} = 10^{-9}$. Therefore, the decimal point must move to the *right*.

$$0.05 \times 10^{-7} = 5 \times 10^{-9}$$

Next we need to change the exponent so that it is -6. We must add 1 to -7. $-7 + 1 = -6$. That is, $10^{-7+1} = 10^{-6}$. According to our first rule, if we *add* 1 to the exponent, we must move the decimal point to the *left*.

$$0.05 \times 10^{-7} = 0.005 \times 10^{-6}$$

Change 4.7 to some number times 10^3 and then to some number times 10^{-3}.

$$4.7 = 0.0047 \times 10^3$$

Since the exponent is 3, the decimal point moves to the *left*.

$$4.7 = 4700 \times 10^{-3}$$

Since the exponent is -3, the decimal point moves to the *right*.

Expressing numbers as numbers between 1 and 10 times the proper power of ten is called *Scientific* notation. Expressing numbers as numbers times powers of ten that are multiples of 3 is called *Engineering* notation.

PRACTICE PROBLEMS 3-1   Change the following numbers to numbers times 10^3 and to numbers times 10^6:

1. 27,000	**2.** 330,000
3. 5600	**4.** 390×10^2
5. 68×10^4	**6.** 1,200,000
7. 180×10^4	**8.** 1500×10^5
9. 1800×10^2	**10.** 51×10^5

Change the following numbers to numbers times 10^{-3} and to numbers times 10^{-6}:

11. 0.000423	**12.** 0.00716
13. 0.000014	**14.** 28.3×10^{-4}
15. 173×10^{-2}	**16.** 8.3×10^{-1}
17. 173×10^{-5}	**18.** 1.73×10^{-1}
19. 17.3×10^{-4}	**20.** 0.0706

Change the following numbers to numbers times 10^{-9} and to numbers times 10^{-12}:

21. 0.0046×10^{-4}

22. 0.413×10^{-7}

23. 6.73×10^{-10}

24. 6.73×10^{-8}

25. 67.3×10^{-11}

26. 0.0173×10^{-6}

27. 0.0563×10^{-8}

28. $76,500 \times 10^{-15}$

29. 6.43×10^{-10}

30. 906×10^{-10}

Change the following numbers to regular numbers, to numbers times 10^{-3}, and to numbers times 10^{3}:

31. 17×10^{-1}

32. 0.637×10^{2}

33. 716×10^{-2}

34. 47×10^{1}

35. 27.3×10^{1}

36. 3.93×10^{2}

37. 3.93×10^{-2}

38. 0.043×10^{4}

39. 0.706×10^{-1}

40. 89.1×10^{2}

Solutions:

1. $27,000 = 27 \times 10^{3} = 0.027 \times 10^{6}$

2. $330,000 = 330 \times 10^{3} = 0.330 \times 10^{6}$

3. $5600 = 5.60 \times 10^{3} = 0.0056 \times 10^{6}$

4. $390 \times 10^{2} = 39 \times 10^{3} = 0.039 \times 10^{6}$

5. $68 \times 10^{4} = 680 \times 10^{3} = 0.680 \times 10^{6}$

6. $1,200,000 = 1,200 \times 10^{3} = 1.20 \times 10^{6}$

7. $180 \times 10^{4} = 1800 \times 10^{3} = 1.8 \times 10^{6}$

8. $1500 \times 10^{5} = 150,000 \times 10^{3} = 150 \times 10^{6}$

9. $1800 \times 10^{2} = 180 \times 10^{3} = 0.180 \times 10^{6}$

10. $51 \times 10^{5} = 5.1 \times 10^{6} = 5100 \times 10^{3}$

11. $0.000423 = 0.423 \times 10^{-3} = 423 \times 10^{-6}$

12. $0.00716 = 7.16 \times 10^{-3} = 7160 \times 10^{-6}$

13. $0.000014 = 0.014 \times 10^{-3} = 14 \times 10^{-6}$

14. $28.3 \times 10^{-4} = 2.83 \times 10^{-3} = 2830 \times 10^{-6}$

15. $173 \times 10^{-2} = 1730 \times 10^{-3} = 1,730,000 \times 10^{-6}$

16. $8.3 \times 10^{-1} = 830 \times 10^{-3} = 830,000 \times 10^{-6}$

17. $173 \times 10^{-5} = 1.73 \times 10^{-3} = 1730 \times 10^{-6}$

18. $1.73 \times 10^{-1} = 173 \times 10^{-3} = 173,000 \times 10^{-6}$

19. $17.3 \times 10^{-4} = 1.73 \times 10^{-3} = 1730 \times 10^{-6}$

20. $0.0706 = 70.6 \times 10^{-3} = 70,600 \times 10^{-6}$

21. $0.0046 \times 10^{-4} = 460 \times 10^{-9} = 460,000 \times 10^{-12}$

22. $0.413 \times 10^{-7} = 41.3 \times 10^{-9} = 41,300 \times 10^{-12}$

23. $6.73 \times 10^{-10} = 0.673 \times 10^{-9} = 673 \times 10^{-12}$

24. $6.73 \times 10^{-8} = 67.3 \times 10^{-9} = 67,300 \times 10^{-12}$

25. $67.3 \times 10^{-11} = 0.673 \times 10^{-9} = 673 \times 10^{-12}$

26. $0.0173 \times 10^{-6} = 17.3 \times 10^{-9} = 17,300 \times 10^{-12}$

27. $0.0563 \times 10^{-8} = 0.563 \times 10^{-9} = 563 \times 10^{-12}$

28. $76,500 \times 10^{-15} = 76.5 \times 10^{-12} = 0.0765 \times 10^{-9}$

29. $6.43 \times 10^{-10} = 0.643 \times 10^{-9} = 643 \times 10^{-12}$

30. $906 \times 10^{-10} = 90.6 \times 10^{-9} = 90,600 \times 10^{-12}$

31. $17 \times 10^{-1} = 1.7 = 1700 \times 10^{-3}$
$\qquad = 0.0017 \times 10^{3}$

32. $0.637 \times 10^{2} = 63.7 = 63,700 \times 10^{-3}$
$\qquad = 0.0637 \times 10^{3}$

33. $716 \times 10^{-2} = 7.16 = 7160 \times 10^{-3}$
$\qquad = 0.00716 \times 10^{3}$

34. $47 \times 10 = 470 = 470,000 \times 10^{-3}$
$\qquad = 0.47 \times 10^{3}$

35. $27.3 \times 10 = 273 = 273,000 \times 10^{-3}$
$\qquad = 0.273 \times 10^{3}$

36. $3.93 \times 10^{2} = 393 = 393,000 \times 10^{-3}$
$\qquad = 0.393 \times 10^{3}$

37. $3.93 \times 10^{-2} = 0.0393 = 39.3 \times 10^{-3}$
$= 0.0000393 \times 10^{3}$

38. $0.043 \times 10^{4} = 430 = 430,000 \times 10^{-3}$
$= 0.43 \times 10^{3}$

39. $0.706 \times 10^{-1} = 0.0706 = 70.6 \times 10^{-3}$
$= 0.0000706 \times 10^{3}$

40. $89.1 \times 10^{2} = 8910 = 8,910,000 \times 10^{-3}$
$= 8.91 \times 10^{3}$

Additional practice problems are at the end of the chapter.

3.2   PREFIXES

Prefixes used with electrical units are listed in Table 3-2.

TABLE 3-2

Prefix	Symbol	Power of ten
Tera	T	10^{12}
Giga	G	10^{9}
Mega	M	10^{6}
Kilo	k	10^{3}
Milli	m	10^{-3}
Micro	μ	10^{-6}
Nano	n	10^{-9}
Pico	p	10^{-12}

Some prefixes have been omitted from the list because we would seldom, if ever, use them. Prefixes are used to make it easier to communicate with each other. For example, suppose the value of some capacitance was 0.000000003 F. This is a very clumsy number to work with. If we express it in scientific notation, we get 3×10^{-9} F. That's better, but still awkward. Looking at our table of prefixes we see that 10^{-9} is a nano unit and the symbol is n (lower-case n). We would write the value as 3 nF. "Three nanofarads" is the way we would say it. 2,700,000 Ω would be simpler to express if we used the prefix equal to 10^{6} which is *mega* and is symbolized as M.

$$2,700,000 \ \Omega = 2.7 \times 10^{6} \ \Omega = 2.7 \ \text{M}\Omega$$

Suppose we calculate a circuit current to be 5.6×10^{-4} A. We need to replace the power of ten with some prefix to make the quantity easier to read. Since our prefixes are multiples of 10^{3} or 10^{-3}, we have to change 10^{-4} to either 10^{-3} or 10^{-6}. Let's do both. First change 5.6×10^{-4} to a number times 10^{-3}.

$$5.6 \times 10^{-4} = 0.56 \times 10^{-3}$$

To change the exponent from -4 to -3, we must add 1. $10^{-4+1} = 10^{-3}$. Because we made the exponent more positive, the decimal point moved to the *left*. The prefix for 10^{-3} is milli (m). Then:

$$5.6 \times 10^{-4} \ \text{A} = 0.56 \times 10^{-3} \ \text{A} = 0.56 \ \text{mA}$$

Now let's change 5.6×10^{-4} to a number times 10^{-6}.

$$5.6 \times 10^{-4} = 560 \times 10^{-6}$$

To change the exponent from -4 to -6, we had to add -2. $10^{-4+(-2)} = 10^{-6}$. Because we made the exponent more negative, the decimal point moved to the *right*. The prefix for 10^{-6} is micro (the Greek letter μ is the symbol).

$$5.6 \times 10^{-4} = 560 \times 10^{-6} = 560 \ \mu A$$

PRACTICE PROBLEMS 3-2   Using the indicated prefixes, change the following quantities:

1. $2700 \ \Omega =$ _____ $k\Omega$ = _____ $M\Omega$

2. $12,000 \ Hz =$ _____ kHz = _____ MHz

3. $0.00076 \ A =$ _____ mA = _____ μA

4. $0.000023 \ A =$ _____ mA = _____ μA

5. $0.0000002 \ F =$ _____ μF = _____ nF

6. $0.000004 \ F =$ _____ μF = _____ nF

7. $68,000 \ \Omega =$ _____ $k\Omega$ = _____ $M\Omega$

8. $120,000 \ \Omega =$ _____ $k\Omega$ = _____ $M\Omega$

9. $3,500,000 \ Hz =$ _____ kHz = _____ MHz

10. $0.00037 \ S =$ _____ mS = _____ μS

11. $5 \times 10^{-4} \ S =$ _____ mS = _____ μS

12. $5.6 \times 10^{4} \ \Omega =$ _____ $k\Omega$ = _____ $M\Omega$

13. $21 \times 10^{-2} \ A =$ _____ mA = _____ μA

14. $68 \times 10^{-10} \ F =$ _____ nF = _____ pF

15. $45 \times 10^{-4} \ H =$ _____ mH = _____ μH

16. $0.15 \times 10^{-2} \ V =$ _____ mV = _____ μV

Solutions:

1. $2700 \ \Omega = 2.7 \ k\Omega = 0.0027 \ M\Omega$

2. $12,000 \ Hz = 12 \ kHz = 0.012 \ MHz$

3. $0.00076 \ A = 0.76 \ mA = 760 \ \mu A$

4. $0.000023 \ A = 0.023 \ mA = 23 \ \mu A$

5. $0.0000002 \ F = 0.2 \ \mu F = 200 \ nF$

6. $0.000004 \ F = 4 \ \mu F = 4000 \ nF$

7. $68,000 \ \Omega = 68 \ k\Omega = 0.068 \ M\Omega$

8. $120,000 \ \Omega = 120 \ k\Omega = 0.12 \ M\Omega$

9. $3,500,000 \ Hz = 3,500 \ kHz = 3.5 \ MHz$

10. $0.00037 \ S = 0.37 \ mS = 370 \ \mu S$

11. $5 \times 10^{-4} \ S = 0.5 \ mS = 500 \ \mu S$

12. $5.6 \times 10^{4} \ \Omega = 56 \ k\Omega = 0.056 \ M\Omega$

13. $21 \times 10^{-2} \ A = 210 \ mA = 210,000 \ \mu A$

14. $68 \times 10^{-10} \ F = 6.8 \ nF = 6800 \ pF$

15. $45 \times 10^{-4} \ H = 4.5 \ mH = 4500 \ \mu H$

16. $0.15 \times 10^{-2} \ V = 1.5 \ mV = 1500 \ \mu V$

Additional practice problems are at the end of the chapter.

The rule for changing quantities to basic units is:

Replace the prefix with the power of ten it represents. Change to a regular number according to the rules established earlier.

Here are some examples:

$$2200 \ \mu S = 2200 \times 10^{-6} \ S = 0.0022 \ S$$
$$68 \ k\Omega = 68 \times 10^3 \ \Omega = 68,000 \ \Omega$$
$$30 \ mA = 30 \times 10^{-3} \ A = 0.03 \ A$$
$$25 \ kHz = 25 \times 10^3 \ Hz = 25,000 \ Hz$$

PRACTICE PROBLEMS 3-3   Perform the indicated operations:

1. 46.7 mA = _____ A
2. 407 mA = _____ A
3. 68 kΩ = _____ Ω
4. 4.7 kΩ = _____ Ω
5. 2.73 kHz = _____ Hz
6. 300 kHz = _____ Hz
7. 670 μS = _____ S
8. 37 mS = _____ S
9. 55 mH = _____ H
10. 465 μH = _____ H
11. 120 kΩ = _____ Ω
12. 2 MΩ = _____ Ω
13. 150 mA = _____ A
14. 0.43 mA = _____ A
15. 0.68 kΩ = _____ Ω

Solutions:

1. 46.7 mA = 0.0467 A
2. 407 mA = 0.407 A
3. 68 kΩ = 68,000 Ω
4. 4.7 kΩ = 4700 Ω
5. 2.73 kHz = 2730 Hz
6. 300 kHz = 300,000 Hz
7. 670 μS = 0.00067 S
8. 37 mS = 0.037 S
9. 55 mH = 0.055 H
10. 465 μH = 0.000465 H
11. 120 kΩ = 120,000 Ω
12. 2 MΩ = 2,000,000 Ω
13. 150 mA = 0.15 A
14. 0.43 mA = 0.00043 A
15. 0.68 kΩ = 680 Ω

Additional practice problems are at the end of the chapter.

Now let's change from one prefix to another. For example, let's change a current from mA to μA.

$$20 \ mA = \text{_____} \ \mu A$$
$$20 \times 10^{-3} \ A = \text{_____} \times 10^{-6} \ A$$

In changing from mA to μA we are really changing the exponent from −3 to −6. We are making the exponent more *negative*. Remember the rule: If the exponent is made more negative, we must move the decimal point to the *right*. The change in the value of the exponent from 10^{-3} to 10^{-6} is −3 to −6 or −3. Therefore, the decimal point moves *three places* to the right.

$$20 \text{ mA} = 20,000 \text{ μA}$$

Change 450 μA to mA.

$$450 \text{ μA} = \underline{\hspace{3cm}} \text{ mA}$$

$$450 \times 10^{-6} \text{ A} = \underline{\hspace{3cm}} \times 10^{-3} \text{ A}$$

In changing from μA to mA we are really changing the exponent from 10^{-6} to 10^{-3}. We are making the exponent more *positive* by 3. Remember the rule: If the exponent is made more positive, we must move the decimal point to the *left*. The change in the value of the exponent from 10^{-6} to 10^{-3} is −6 to −3 or +3. Therefore, the decimal point moves *three places* to the left.

$$450 \text{ μA} = 0.45 \text{ mA}$$

Change 680 kΩ to MΩ.

$$680 \text{ kΩ} = \underline{\hspace{3cm}} \text{ MΩ}$$

$$680 \times 10^{3} \text{ Ω} = \underline{\hspace{3cm}} \times 10^{6} \text{ Ω}$$

In going from kΩ to MΩ we are making the exponent more positive. Therefore, the decimal point moves to the left. The change from 10^{3} to 10^{6} is from 3 to 6 or 3. Therefore, the decimal point moves *three places* to the left.

$$680 \text{ kΩ} = 0.68 \text{ MΩ}$$

Change 0.01 μF to pF.

$$0.01 \text{ μF} = \underline{\hspace{3cm}} \text{ pF}$$

$$0.01 \times 10^{-6} \text{ F} = \underline{\hspace{3cm}} \times 10^{-12} \text{ F}$$

In going from μF to pF we are making the exponent more negative. Therefore, the decimal point moves to the right. The change from 10^{-6} to 10^{-12} is from −6 to −12, or −6. Therefore, the decimal point moves *six places* to the right.

$$0.01 \text{ μF} = 10,000 \text{ pF}$$

In the following practice problems you will either be changing prefixes or be changing to basic units.

PRACTICE PROBLEMS 3-4

1. 20 mA = _____ μA = _____ A

2. 0.01 μF = _____ nF = _____ pF

3. 2 H = _____ mH = _____ μH

4. 2.5 kΩ = _____ Ω = _____ MΩ

5. 680 Ω = _____ kΩ = _____ MΩ

6. 1.8 V = _____ mV = _____ μV

7. 20 μA = _____ mA = _____ A

8. 1.5 MΩ = _____ kΩ = _____ Ω

9. 0.02 S = _____ mS = _____ μS

10. 50 nF = _____ μF = _____ pF

11. 4.7×10^4 Ω = _____ kΩ = _____ MΩ

12. 16×10^{-7} F = _____ μF = _____ nF

13. 2.4×10^{-2} A = _____ mA = _____ μA

14. 8.4×10^2 Ω = _____ kΩ = _____ MΩ

15. 56×10^5 Ω = _____ kΩ = _____ MΩ

16. 27.3 mS = _____ S = _____ μS

17. 5600 μS = _____ S = _____ mS

18. 170 mW = _____ W = _____ μW

19. 0.000642 V = _____ μV = _____ mV

20. 30 kHz = _____ MHz = _____ Hz

Solutions:

1. 20 mA = 20,000 μA = 0.02 A

2. 0.01 μF = 10 nF = 10,000 pF

3. 2 H = 2000 mH = 2,000,000 μH

4. 2.5 kΩ = 2500 Ω = 0.0025 MΩ

5. 680 Ω = 0.68 kΩ = 0.00068 MΩ

6. 1.8 V = 1800 mV = 1,800,000 μV

7. 20 μA = 0.02 mA = 0.00002 A

8. 1.5 MΩ = 1500 kΩ = 1,500,000 Ω

9. 0.02 S = 20 mS = 20,000 μS

10. 50 nF = 0.05 μF = 50,000 pF

11. 4.7×10^4 Ω = 47 kΩ = 0.047 MΩ

12. 16×10^{-7} F = 1.6 μF = 1600 nF

13. 2.4×10^{-2} A = 24 mA = 24,000 μA

14. 8.4×10^2 Ω = 0.84 kΩ = 0.00084 MΩ

15. 56×10^5 Ω = 5600 kΩ = 5.6 MΩ

16. 27.3 mS = 0.0273 S = 27,300 μS

17. 5600 μS = 0.0056 S = 5.6 mS

18. 170 mW = 0.170 W = 170,000 μW

19. 0.000642 V = 642 μV = 0.642 mV

20. 30 kHz = 0.030 MHz = 30,000 Hz

Additional practice problems are at the end of the chapter.

SELF-TEST 3-1   Express your answers in scientific notation:

1. 4760 = _____ $\times 10^3$ = _____ $\times 10^6$

2. 32.4×10^4 = _____ $\times 10^3$ = _____ $\times 10^6$

3. $2.71 \times 10^{-1} =$ _____ $\times 10^{-3} =$ _____ $\times 10^3$

4. $46.7 \times 10^{-2} =$ _____ $\times 10^{-3} =$ _____ $\times 10^3$

5. 76.3 mA = _____ A = _____ μA

6. 0.0055 μF = _____ nF = _____ pF

7. 8340 μS = _____ mS = _____ S

8. 20 kΩ = _____ Ω = _____ MΩ

9. 75,000 Hz = _____ MHz = _____ kHz

10. 146 mS = _____ S = _____ μS

Answers to Self-Test 3-1 are at the end of the chapter.

3.3 APPLICATIONS

In problem solving we perform many different operations and express answers in many different forms. In this section we will do some of these operations and express answers by using prefixes whenever practical. For example, when resistors are connected in parallel, we usually convert from resistance to conductance for problem-solving purposes. Suppose we have two parallel resistors whose values are 27 kΩ and 18 kΩ. To find the total resistance, we would use the equation:

$$R_T = \frac{1}{G_T}$$

where

$$G_T = G_1 + G_2$$

$$G_1 = \frac{1}{R_1}$$

$$G_2 = \frac{1}{R_2}$$

Since

$$G_1 = \frac{1}{R_1} \quad \text{and} \quad G_2 = \frac{1}{R_2}$$

We could write the equation like this:

$$R_T = \frac{1}{G_1 + G_2} = \frac{1}{\dfrac{1}{R_1} + \dfrac{1}{R_2}}$$

$$= \frac{1}{\dfrac{1}{27 \text{ k}\Omega} + \dfrac{1}{18 \text{ k}\Omega}} = \frac{1}{37 \ \mu S + 55.6 \ \mu S}$$

$$= \frac{1}{92.6 \ \mu S} = 10.8 \text{ k}\Omega$$

The total conductance is 92.6 μS. The total resistance is 10.8 kΩ. If three resistors are connected in parallel, the equation is:

$$R_T = \frac{1}{G_T} = \frac{1}{G_1 + G_2 + G_3} = \frac{1}{\dfrac{1}{R_1} + \dfrac{1}{R_2} + \dfrac{1}{R_3}}$$

Assume that $R_1 = 470$ Ω, $R_2 = 220$ Ω, and $R_3 = 680$ Ω. Then:

$$R_T = \frac{1}{\dfrac{1}{470\ \Omega} + \dfrac{1}{220\ \Omega} + \dfrac{1}{680\ \Omega}} = \frac{1}{2.13\ \text{mS} + 4.55\ \text{mS} + 1.47\ \text{mS}}$$

$$R_T = \frac{1}{8.14\ \text{mS}} = 123\ \Omega$$

The total conductance is 8.14 mS. The total resistance is 123 Ω.

With most calculators all operations can be performed without having to store partial answers. Only the final answer need be written down. The algorithm for this operation is in Appendix A.

To add units together when the prefixes are different, it is necessary to convert so that all prefixes are the same. For example, let's add 6300 μS and 15.3 mS. We cannot add milliunits and microunits together just as we could not add 10^{-3} and 10^{-6} together. We must convert 6300 μS to milliunits or convert 15.3 mS to microunits.

$$6300\ \mu\text{S} + 15.3\ \text{mS} =$$
$$6.30\ \text{mS} + 15.3\ \text{mS} = 21.6\ \text{mS}$$
or
$$6300\ \mu\text{S} + 15.3\ \text{mS} =$$
$$6300\ \mu\text{S} + 15,300\ \mu\text{S} = 21,600\ \mu\text{S}$$

In this case, converting to milliunits is better because it is easier to work with that number.

PRACTICE PROBLEMS 3-5   Perform the indicated operations:

1. 27 mA + 0.037 A = _____ A = _____ mA

2. 370 μS + 0.060 mS = _____ mS = _____ μS

3. $\dfrac{1}{68\ \text{k}\Omega} + \dfrac{1}{47\ \text{k}\Omega}$ = _____ mS = _____ μS

4. $\dfrac{20\ \text{V}}{27\ \text{k}\Omega}$ = _____ A = _____ mA = _____ μA

5. $\dfrac{25\ \text{V}}{2.36\ \text{mA}}$ = _____ Ω = _____ kΩ = _____ MΩ

6. $\dfrac{1}{6.28 \times 12\ \text{kHz} \times 50\ \text{nF}}$ = _____ Ω = _____ kΩ

7. $\dfrac{1}{2.7\ \text{k}\Omega} + \dfrac{1}{3.3\ \text{k}\Omega} + \dfrac{1}{4.7\ \text{k}\Omega}$ = _____ mS = _____ μS = _____ S

8. $\dfrac{1}{6340\ \mu S} =$ _____ $\Omega =$ _____ kΩ

9. 2.73 k$\Omega \times 0.43$ mA = _____ V = _____ mV

10. 1.27 V + 48 mV + 5630 μV = _____ V = _____ mV = _____ μV

Solutions:

1. 27 mA + 0.037 A = 0.064 A = 64 mA

2. 370 μS + 0.060 mS = 0.430 mS = 430 μS

3. $\dfrac{1}{68\ k\Omega} + \dfrac{1}{47\ k\Omega} = 0.036$ mS = 36 μS

4. $\dfrac{20\ V}{27\ k\Omega} = 0.000741$ A = 0.741 mA = 741 μA

5. $\dfrac{25\ V}{2.36\ mA} = 10{,}600\ \Omega = 10.6\ k\Omega = 0.0106\ M\Omega$

6. $\dfrac{1}{6.28 \times 12\ kHz \times 50\ nF} = 265\ \Omega = 0.265\ k\Omega$

7. $\dfrac{1}{2.7\ k\Omega} + \dfrac{1}{3.3\ k\Omega} + \dfrac{1}{4.7\ k\Omega} = 0.886$ mS = 886 μS = 0.000886 S

8. $\dfrac{1}{6340\ \mu S} = 158\ \Omega = 0.158\ k\Omega$

9. 2.73 k$\Omega \times 0.43$ mA = 1.17 V = 1170 mV

10. 1.27 V + 48 mV + 5630 μV = 1.32 V = 1320 mV = 1,320,000 μV

Additional practice problems are at the end of the chapter.

END OF CHAPTER PROBLEMS 3-1

1. $363 =$ _____ $\times 10^3 =$ _____ $\times 10^{-3}$

2. $76.50 =$ _____ $\times 10^3 =$ _____ $\times 10^{-3}$

3. $16.3 \times 10^{-1} =$ _____ $\times 10^3 =$ _____ $\times 10^{-3}$

4. $2.76 \times 10 =$ _____ $\times 10^3 =$ _____ $\times 10^{-3}$

5. $81{,}000 =$ _____ $\times 10^3 =$ _____ $\times 10^6$

6. $3300 =$ _____ $\times 10^3 =$ _____ $\times 10^6$

7. $1{,}000{,}000 =$ _____ $\times 10^3 =$ _____ $\times 10^6$

8. $750{,}000 =$ _____ $\times 10^3 =$ _____ $\times 10^6$

9. $0.00064 =$ _____ $\times 10^{-3} =$ _____ $\times 10^{-6}$

10. $0.00423 =$ _____ $\times 10^{-3} =$ _____ $\times 10^{-6}$

11. $0.0000706 =$ _____ $\times 10^{-3} =$ _____ $\times 10^{-6}$

12. $0.0432 =$ _____ $\times 10^{-3} =$ _____ $\times 10^{-6}$

13. $2.73 \times 10^{-9} =$ _____ $\times 10^{-6} =$ _____ $\times 10^{-12}$

14. $42.3 \times 10^{-9} =$ _____ $\times 10^{-6} =$ _____ $\times 10^{-12}$

15. $0.673 \times 10^{-9} =$ _____ $\times 10^{-6} =$ _____ $\times 10^{-12}$

16. $0.843 \times 10^{-9} =$ _____ $\times 10^{-6} =$ _____ $\times 10^{-12}$

17. $17.3 \times 10^{-6} =$ _____ $\times 10^{-9} =$ _____ $\times 10^{-3}$

18. $7.43 \times 10^{-6} =$ _____ $\times 10^{-9} =$ _____ $\times 10^{-3}$

19. $2.06 \times 10^{-6} =$ _____ $\times 10^{-9} =$ _____ $\times 10^{-3}$

20. $43.2 \times 10^{-6} =$ _____ $\times 10^{-9} =$ _____ $\times 10^{-3}$

END OF CHAPTER PROBLEMS 3-2

1. 0.00026 A = _____ mA = _____ μA

2. 0.00736 A = _____ mA = _____ μA

3. 0.000632 S = _____ mS = _____ μS

4. 0.0024 S = _____ mS = _____ μS

5. 7630 Ω = _____ kΩ = _____ MΩ

6. $470{,}000$ Ω = _____ kΩ = _____ MΩ

7. 17.3×10^{-3} A = _____ mA = _____ μA

8. 0.64×10^{-6} A = _____ mA = _____ μA

9. 71.3×10^{4} Ω = _____ kΩ = _____ MΩ

10. 6.8×10^{5} Ω = _____ kΩ = _____ MΩ

11. 5.63×10^{6} Hz- _____ kHz = _____ MHz

12. 487×10^{2} Hz = _____ kHz = _____ MHz

13. $2{,}000{,}000$ Ω = _____ kΩ = _____ MΩ

14. $470{,}000$ Ω = _____ kΩ = _____ MΩ

15. 23.7×10^{-5} S = _____ mS = _____ μS

16. 5.63×10^{-4} S = _____ mS = _____ μS

17. 30×10^{-8} F = _____ μF = _____ nF

18. 12×10^{-7} F = _____ μF = _____ nF

19. 0.000062 A = _____ mA = _____ μA

20. 0.00075 A = _____ mA = _____ μA

END OF CHAPTER PROBLEMS 3-3

1. 800 mA = _____ A

2. 70.3 mA = _____ A

3. 2.5 mS = _____ S

4. 600 μS = _____ S

5. 33 kΩ = _____ Ω

6. 3.3 kΩ = _____ Ω

7. 0.47 kΩ = _____ Ω

8. 0.56 MΩ = _____ Ω

9. 12.5 kHz = _____ Hz   10. 7.3 kHz = _____ Hz

11. 100 μF = _____ F   12. 50 μF = _____ F

13. 0.25 mA = _____ A   14. 500 mA = _____ A

15. 900 μS = _____ S   16. 25 mS = _____ S

17. 750 kΩ = _____ Ω   18. 1.2 kΩ = _____ Ω

19. 30 mA = _____ A   20. 300 mA = _____ A

END OF CHAPTER PROBLEMS 3-4

1. 0.00026 A = _____ mA = _____ μA

2. 0.00736 A = _____ mA = _____ μA

3. 0.00000632 S = _____ mS = _____ μS

4. 0.0000417 A = _____ mA = _____ μA

5. 56.2×10^{-3} S = _____ mS = _____ μS

6. 17.3×10^{-3} A = _____ mA = _____ μA

7. 613×10^{-6} s = _____ ms = _____ μs

8. 227×10^{-6} A = _____ mA = _____ μA

9. 7630 Ω = _____ kΩ = _____ MΩ

10. 12,500 Hz = _____ kHz = _____ MHz

11. 470,000 Ω = _____ kΩ = _____ MΩ

12. 1,200,000 Ω = _____ kΩ = _____ MΩ

13. 12.7×10^4 Hz = _____ kHz = _____ MHz

14. 5.63×10^6 Hz = _____ kHz = _____ MHz

15. 713×10^3 Ω = _____ kΩ = _____ MΩ

16. 8.73×10^5 Ω = _____ kΩ = _____ MΩ

17. 46.3 mA = _____ μA = _____ A

18. 213 mS = _____ S = _____ μS

19. 0.05 nF = _____ μF = _____ pF

20. 173 nF = _____ pF = _____ μF

21. 3.2 kΩ = _____ Ω = _____ MΩ

22. 17.8 kHz = _____ Hz = _____ MHz

23. 270 kΩ = _____ MΩ = _____ Ω

24. 47 kΩ = _____ MΩ = _____ Ω

25. 403 μS = _____ S = _____ mS

26. 83.6 ms = _____ s = _____ μs

27. 1.03 MHz = _____ Hz = _____ kHz

28. 423 kHz = _____ MHz = _____ Hz

29. 1.43 nF = _____ pF = _____ μF

30. 0.002 μF = _____ pF = _____ nF

31. 55×10^{-4} S = _____ mS = _____ μS

32. 4.67×10^{-3} S = _____ mS = _____ μS

33. 106×10^{-2} nF = _____ μF = _____ pF

34. 7.36×10^2 nF = _____ μF = _____ pF

35. 4.63×10^2 kΩ = _____ Ω = _____ MΩ

36. 6.67×10^2 kΩ = _____ Ω = _____ MΩ

37. 9.63×10^4 Hz = _____ kHz = _____ MHz

38. 8.23×10^2 Hz = _____ kHz = _____ MHz

39. 0.0078 A = _____ mA = _____ μA

40. 0.000104 A = _____ mA = _____ μA

41. 176 mV = _____ V = _____ μV

42. 0.00062 V = _____ mV = _____ μV

43. 1.73 W = _____ mW = _____ kW

44. 16.7 W = _____ mW = _____ kW

45. 0.025 H = _____ mH = _____ μH

46. 0.005 H = _____ mH = _____ μH

47. 173 mS = _____ S = _____ μS

48. 1460 μS = _____ mS = _____ S

49. 25×10^{-7} F = _____ μF = _____ nF

50. 2.5 mH = _____ H = _____ μH

END OF CHAPTER PROBLEMS 3-5

1. 367 μA + 1.67 mA = _____ mA = _____ μA

2. 0.417 mA + 63 μA = _____ mA = _____ μA

3. $\dfrac{1}{10\ \text{k}\Omega} + \dfrac{1}{15\ \text{k}\Omega} =$ _____ mS = _____ μS

4. $\dfrac{1}{33\ \text{k}\Omega} + \dfrac{1}{56\ \text{k}\Omega} + \dfrac{1}{100\ \text{k}\Omega} =$ _____ mS = _____ μS

5. $\dfrac{9\ \text{V}}{33\ \text{k}\Omega} =$ _____ mA = _____ μA

6. $\dfrac{12\ \text{V}}{2.7\ \text{k}\Omega} =$ _____ mA = _____ μA

7. $\dfrac{40\ \text{V}}{200\ \mu\text{A}} =$ _____ Ω = _____ kΩ

8. $\dfrac{10\ V}{25\ mA} =$ _____ $\Omega =$ _____ $k\Omega$

9. $\dfrac{1}{213\ \mu S} =$ _____ $\Omega =$ _____ $k\Omega$

10. $\dfrac{1}{1.47\ mS} =$ _____ $\Omega =$ _____ $k\Omega$

11. $\dfrac{1}{6.28 \times 75\ kHz \times 25\ pF} =$ _____ $\Omega =$ _____ $k\Omega$

12. $\dfrac{1}{6.28 \times 2.5\ kHz \times 30\ nF} =$ _____ $\Omega =$ _____ $k\Omega$

13. $68\ k\Omega \times 140\ \mu A =$ _____ $V =$ _____ mV

14. $330\ k\Omega \times 0.17\ mA =$ _____ $V =$ _____ mV

15. $500\ \mu S + 1.76\ mS + 0.000043\ S =$ _____ $mS =$ _____ μS

16. $670\ \mu S + 2\ mS + 0.0002\ S =$ _____ $mS =$ _____ μS

17. $\dfrac{1}{6.8\ k\Omega} + \dfrac{1}{5.6\ k\Omega} =$ _____ $mS =$ _____ μS

18. $\dfrac{1}{270\ \Omega} + \dfrac{1}{470\ \Omega} + \dfrac{1}{680\ \Omega} =$ _____ $mS =$ _____ μS

19. $\dfrac{50\ V}{1.2\ M\Omega} =$ _____ $mA =$ _____ μA

20. $\dfrac{40\ V}{30\ \mu A} =$ _____ $k\Omega =$ _____ $M\Omega$

ANSWERS TO SELF-TESTS

Self-Test 3-1

1. $4760 = 4.76 \times 10^3 = 0.00476 \times 10^6$

2. $32.4 \times 10^4 = 324 \times 10^3 = 0.324 \times 10^6$

3. $2.71 \times 10^{-1} = 271 \times 10^{-3} = 0.000271 \times 10^3$

4. $46.7 \times 10^{-2} = 467 \times 10^{-3} = 0.000467 \times 10^3$

5. $76.3\ mA = 0.0763\ A = 76,300\ \mu A$

6. $0.0055\ \mu F = 5.5\ nF = 5,500\ pF$

7. $8340\ \mu S = 8.34\ mS = 0.00834\ S$

8. $20\ k\Omega = 20,000\ \Omega = 0.02\ M\Omega$

9. $75,000\ Hz = 0.075\ MHz = 75\ kHz$

10. $146\ mS = 0.146\ S = 146,000\ \mu S$

CHAPTER 4

COMPUTER NUMBER SYSTEMS

The basic tools for understanding computers are an understanding of the number systems used and an understanding of basic logic functions. Number systems are covered in this chapter. Logic functions are discussed in Chapter 24.

We all know and understand the decimal number system. This is the number system we use daily, but in the computer world the *binary* number system is the number system used. All data is stored and manipulated inside the computer in binary. That is, within the computer all data is reduced to 1's and 0's. These 1's and 0's are stored in logic circuits called *registers* or they are stored in *memory*. The size of a register or memory location varies according to the kind of computer used. Microcomputers store 8 binary digits in each location. Large computers may store as many as 125 binary digits in each location.

We normally enter data into the computer in some number system other than binary because entering data in binary is too time-consuming and too prone to error. There are too many 1's and 0's. Data is entered into the computer by means of the decimal, octal or hexadecimal number system. Octal and hexadecimal are used most often because they are more closely related to binary than is the decimal system.

It is important to our understanding of these number systems that we know how they are related one to another.

4.1 BINARY NUMBER SYSTEM

In the binary number system there are two digits 0 and 1. Since digits in the units position have a value of 0 or 1, numbers greater than 1 cause a carry to the next position. Each position represents the base raised to a power. In base 10 the

MSB	4SB	3SB	2SB	LSB
2^4	2^3	2^2	2^1	2^0
16	8	4	2	1

Figure 4-1. Positional value of the first five positions for the binary number system.

units position has a power of 10^0, the next position 10^1, and so on. The same reasoning applies in binary or base 2. The units position has a power of 2^0, the next position 2^1, and so on, as illustrated in Figure 4-1. Binary digits are called *bits* (a contraction of binary digits). Therefore, the digit in the units position is called the least significant bit (LSB), and so on until the most significant bit (MSB) is reached.

4.1.1 Binary to Decimal Conversions.

Converting from base 2 to base 10 is easy. If a 1 is present in a given position, the weight of that bit is added. If a 0 is present, the weight of that bit is not added. Consider the number 11010 in Example 4.1. The decimal equivalent is 26.

EXAMPLE 4.1: $11010_2 = (?)_{10}$

Solution:

$$(1 \times 2^4) + (1 \times 2^3) + (0 \times 2^2) + (1 \times 2^1) + (0 \times 2^0) =$$
$$16 \quad + \quad 8 \quad + \quad 0 \quad + \quad 2 \quad + \quad 0 \quad = 26$$

Because there are only 1's and 0's in binary, each bit position either equals the weight of that digit position or it equals 0. In the above example the values of the bit positions are added to get the decimal equivalent. In the 2^4 or 16's position there is a 1. Therefore, that bit position converts to 16. In the 2^3 or 8's position there is a 1. That bit position converts to 8. In the 2^2 or 4's position there is a 0. That bit position converts to 0. In the 2's position there is a 1. That bit position converts to 2. There is a 0 in the units position. If we add the numbers together, we get decimal 26. *In converting from binary to decimal, find the value or weight of the MSB. Work down to the LSB adding the weight of that position if a 1 is present or a 0 if a 0 is present. Table 4-1 is a conversion table for all the computer number systems.*

EXAMPLE 4.2: Change the following binary numbers to decimal numbers:

(a) 1100 (b) 10001 (c) 101011 (d) 111101

Solution:

(a) $1100_2 =$ $(1 \times 2^3) + (1 \times 2^2) + (0 \times 2^1) + (0 \times 2^0) =$
$$8 \quad + \quad 4 \quad + \quad 0 \quad + \quad 0 = 12_{10}$$

(b) $10001_2 =$ $(1 \times 2^4) + (0 \times 2^3) + (0 \times 2^2) + (0 \times 2^1) + (1 \times 2^0)$
$$= 16 \quad + \quad 0 \quad + \quad 0 \quad + \quad 0 \quad + 1 = 17_{10}$$

TABLE 4-1.

Computer Numbering Systems

Numbering system	Decimal		Binary						Octal		Hexa-decimal	
Base	10		2						8		16	
Position weight — Power	10^1	10^0	2^5	2^4	2^3	2^2	2^1	2^0	8^1	8^0	16^1	16^0
Position weight — Value	10	1	32	16	8	4	2	1	8	1	16	1
										(LSD)		(LSD)
	0	0	0	0	0	0	0	0	0	0	0	0
	0	1	0	0	0	0	0	1		1		1
	0	2	0	0	0	0	1	0		2		2
	0	3	0	0	0	0	1	1		3		3
	0	4	0	0	0	1	0	0		4		4
	0	5	0	0	0	1	0	1		5		5
	0	6	0	0	0	1	1	0		6		6
	0	7	0	0	0	1	1	1		7		7
	0	8	0	0	1	0	0	0	1	0		8
	0	9	0	0	1	0	0	1	1	1		9
	1	0	0	0	1	0	1	0	1	2		A
	1	1	0	0	1	0	1	1	1	3		B
	1	2	0	0	1	1	0	0	1	4		C
	1	3	0	0	1	1	0	1	1	5		D
	1	4	0	0	1	1	1	0	1	6		E
	1	5	0	0	1	1	1	1	1	7		F
	1	6	0	1	0	0	0	0	2	0	1	0
	1	7	0	1	0	0	0	1	2	1	1	1
	1	8	0	1	0	0	1	0	2	2	1	2
	1	9	0	1	0	0	1	1	2	3	1	3
	2	0	0	1	0	1	0	0	2	4	1	4
	2	1	0	1	0	1	0	1	2	5	1	5
	2	2	0	1	0	1	1	0	2	6	1	6
	2	3	0	1	0	1	1	1	2	7	1	7
	2	4	0	1	1	0	0	0	3	0	1	8
	2	5	0	1	1	0	0	1	3	1	1	9
	2	6	0	1	1	0	1	0	3	2	1	A
	2	7	0	1	1	0	1	1	3	3	1	B
	2	8	0	1	1	1	0	0	3	4	1	C
	2	9	0	1	1	1	0	1	3	5	1	D
	3	0	0	1	1	1	1	0	3	6	1	E
	3	1	0	1	1	1	1	1	3	7	1	F
	3	2	1	0	0	0	0	0	4	0	2	0

Example: $1_{(10)} = 1_{(2)} = 1_{(8)} = 1_{(16)}$

Example: $10_{(10)} = 1010_{(2)} = 12_{(8)} = 0A_{(16)}$

Example: $25_{(10)} = 11001_{(2)} = 31_{(8)} = 19_{(16)}$

(c) $101011_2 = (1 \times 2^5) + (0 \times 2^4) + (1 \times 2^3) + (0 \times 2^2) + (1 \times 2^1) +$
$(1 \times 2^0) = \quad 32 \; + \quad 0 \; + \quad 8 \; + \quad 0 \; + \quad 2 \; + 1$
$= 43_{10}$

(d) $111101_2 = (1 \times 2^5) + (1 \times 2^4) + (1 \times 2^3) + (1 \times 2^2) + (0 \times 2^1) +$
$(1 \times 2^0) = \quad 32 \; + \quad 16 \; + \quad 8 \; + \quad 4 \; + \quad 0 \; + 1$
$= 61_{10}$

Let's turn the process around and find the binary equivalent of a decimal number.

EXAMPLE 4.3: Convert 11_{10} to binary:

Solution: Step 1. $\dfrac{11}{2} = 5 + 1$

Step 2. $\dfrac{5}{2} = 2 + 1$

Step 3. $\dfrac{2}{2} = 1 + 0$ LSB

Step 4. $\dfrac{1}{2} = 0 + 1$ —MSB→ 1   0   1   1

Since we are converting to base 2, we first divide the decimal number by 2 (step 1). The answer is 5 with a remainder of 1. This remainder is the LSB in the answer. The whole number left after this initial division (5) is again divided by 2 (step 2). The answer is 2 with a remainder of 1. This 1 is the bit in the 2^1 position. The whole number left (2) is divided by 2 (step 3). The result is 1 and a remainder of 0. This 0 is the bit in the 2^2 position. Step 4 is the final step because no further divisions are possible. The remainder in the final step is the MSB.

A second method of converting from decimal to binary is shown in the following example.

EXAMPLE 4.4: Convert 167 to binary.

Solution: The highest multiple of 2 contained in 167 is 128 (2^7). Therefore, we need a 128. So we start writing our binary number with a 1 in the MSB position remembering that this "1" has a value of 128. Next we determine if we need a 64. If we do, we put a 1 in the next bit position. If we don't

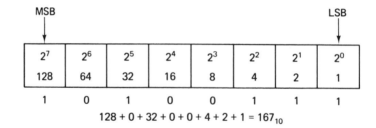

Figure 4-2. Conversion of 167_{10} to binary.

need a 64, we put a 0 in this position. Since $128+64$ is greater than 167, we must put a zero in this position. At this point we have 10 which is $128+0$.

This procedure is followed for each bit position. Can we use 32? Yes. We now have 101. Remember we started at the MSB position and are working down to the LSB position. Can we use a 16? No; a 16 is too much. Put a zero in the 16 (2^4) position. We now have 1010. Can we use an 8? No; so put a zero in the 2^3 position. Now we have 10100. Can we use a 4? Yes. Our number is now 101001. Can we use a 2? Yes. We now have 1010011. Do we need a 1 or a 0 in the LSB position? We need a 1. Our binary number is 10100111. Any decimal number can be converted to binary in this manner. With practice, this method is faster than the method shown previously. The process is detailed in Figure 4-2.

EXAMPLE 4.5: Convert the following decimal numbers to binary numbers:

(a) 7   (b) 34   (c) 89   (d) 203

Solution: (a) $\dfrac{7}{2}=3+1$

$\dfrac{3}{2}=1+1$

$\dfrac{1}{2}=0+1\longrightarrow$ 1 1 1

$7_{10}=111_2$

(b) $\dfrac{34}{2}=17+0$

$\dfrac{17}{2}=8+1$

$\dfrac{8}{2}=4+0$

$$\frac{4}{2}=2+0$$

$$\frac{2}{2}=1+0$$

$$\frac{1}{2}=0+1 \quad \rightarrow \quad 1 \quad 0 \quad 0 \quad 0 \quad 1 \quad 0$$

$$34_{10}=100010_2$$

(c) $\frac{89}{2}=44+1$

$$\frac{44}{2}=22+0$$

$$\frac{22}{2}=11+0$$

$$\frac{11}{2}=5+1$$

$$\frac{5}{2}=2+1$$

$$\frac{2}{2}=1+0$$

$$\frac{1}{2}=0+1 \quad \rightarrow \quad 1 \quad 0 \quad 1 \quad 1 \quad 0 \quad 0 \quad 1$$

$$89_{10}=1011001_2$$

(d) $\frac{203}{2}=101+1$

$$\frac{101}{2}=50+1$$

$$\frac{50}{2}=25+0$$

$$\frac{25}{2}=12+1$$

$$\frac{12}{2}=6+0$$

$$\frac{6}{2}=3+0$$

$$\frac{3}{2}=1+1$$

$$\frac{1}{2}=0+1 \quad \rightarrow \quad 1 \quad 1 \quad 0 \quad 0 \quad 1 \quad 0 \quad 1 \quad 1$$

$$203_{10}=11001011_2$$

PRACTICE PROBLEMS 4-1   Convert the following binary numbers to decimal numbers:

1. 101_2	**2.** 111_2
3. 1001_2	**4.** 1110_2
5. 10110_2	**6.** 11001_2
7. 100101_2	**8.** 111011_2
9. 1010100_2	**10.** 1111001_2

Convert the following decimal numbers to binary numbers:

11. 4	**12.** 8
13. 10	**14.** 27
15. 39	**16.** 57
17. 78	**18.** 100
19. 150	**20.** 200

Solutions:

1. 5	**2.** 7
3. 9	**4.** 14
5. 22	**6.** 25
7. 37	**8.** 59
9. 84	**10.** 121
11. 100_2	**12.** 1000_2
13. 1010_2	**14.** 11011_2
15. 100111_2	**16.** 111001_2
17. 1001110_2	**18.** 1100100_2
19. 10010110_2	**20.** 11001000_2

Additional practice problems are at the end of the chapter.

4.2 OCTAL NUMBER SYSTEM

Octal, or base 8, is an important computer number system. Since there are 8 digits in octal (0, 1, 2, 3, 4, 5, 6, 7), the digit of greatest value is 7, which is 1 less than the base. In counting, the sequence is 0 to 7. On the next count, we return to 0 in the units position (just as we do when we reach 9 in decimal) and a carry is generated. The resulting number is 10 and is read "one-zero." This number is equal to decimal 8 because the carry is equal to 8 (base 8). This process continues—the sequence goes from 0 to 7 and back to 0 again. Each time a count greater than 7 is reached, a carry is generated.

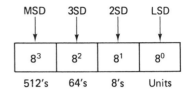

Figure 4-3. Positional value of the first four positions.

In the octal number system, digits in the units position have the indicated value. Digits in the next position have *place value*. That is, digits in this position indicate the number of 8's present. This, then, is the 8's position. The next position indicates the number of 64's present, and so on, as shown in Figure 4-3. This concept of *place value* must be understood so that we can work in any base. Recall that in the number 56_{10} the 5 has a value of 50. The 5 is in the 10's position so we have five 10's or 50. There are six 1's or 6. In the number 56_8, the 5 has a value of 40. The 5 is in the 8's position; therefore, we have five 8's or 40. There are six 1's or 6, as before.

4.2.1   *Octal to Decimal Conversions.* Let's examine the number 1025 in octal. One may be tempted to read this as "one thousand twenty-five." Such language implies base 10. The number should be read "one-zero-two-five" which does not imply any base. We normally think in decimal and decimal is the number system we use most; therefore, let's convert the number to base 10. There is a 1 in the 512's position + a 0 in the 64's position + a 2 in the 8's position + a 5 in the units position.

$$1025_8 = (1 \times 8^3) + (0 \times 8^2) + (2 \times 8^1) + (5 \times 8^0) = 533_{10}$$
$$1025_8 = (512) + (0) + (16) + (5) = 533_{10}$$

EXAMPLE 4.6: Change the following octal numbers to decimal:
(a) 73_8 (b) 432_8 (c) 600_8 (d) 1234_8

Solution:

(a) $73_8 = (7 \times 8^1) + (3 \times 8^0) = (56 + 3) = 59_{10}$
(b) $432_8 = (4 \times 8^2) + (3 \times 8^1) + (2 \times 8^0) = 256 + 24 + 2 = 282_{10}$
(c) $600_8 = (6 \times 8^2) + (0 \times 8^1) + (0 \times 8^0) = 384 + 0 + 0 = 384_{10}$
(d) $1234_8 = (1 \times 8^3) + (2 \times 8^2) + (3 \times 8^1) + (4 \times 8^0) = 512 + 128 + 24 + 4 = 668_{10}$

An algorithm for converting from octal to decimal using the calculator is in Appendix A.

Let's reverse the process. In Example 4.7 the number 189 is converted from decimal to octal.

EXAMPLE 4.7: $189_{10} = (?)_8$.

Solution:

Step 1. $\dfrac{189}{8} = 23 + 5$

Step 2. $\dfrac{23}{8} = 2 + 7$

Step 3. $\dfrac{2}{8} = 0 + 2$—MSD $\longrightarrow$ 2   7   5 $189_{10} = 275_8$

Since we are converting to base 8, we first divide the decimal number by 8 (step 1). The answer is 23 with a remainder of 5. This remainder is the LSD (least significant digit) of our base 8 number. The whole number left after this initial division (23) is again divided by 8 to determine the value of the digit in the next (8^1) position. This division leaves a remainder of 7 which becomes the digit in the 8^1's position. The whole number resulting from this division (2) is again divided by 8, and the remainder (2) is the digit in the 8^2 position. This digit is the MSD (most significant digit) since no further division is possible.

EXAMPLE 4.8: Convert the following decimal numbers to octal numbers:(a) 78   (b) 376   (c) 1463

Solution:

(a) $\dfrac{78}{8} = 9 + 6$

$\dfrac{9}{8} = 1 + 1$

$\dfrac{1}{8} = 0 + 1 \longrightarrow$ 1   1   6 $78_{10} = 116_8$

(b) $\dfrac{376}{8} = 47 + 0$

$\dfrac{47}{8} = 5 + 7$

$\dfrac{5}{8} = 0 + 5 \longrightarrow$ 5   7   0 $376_{10} = 570_8$

(c) $\dfrac{1463}{8} = 182 + 7$

$\dfrac{182}{8} = 22 + 6$

$\dfrac{22}{8} = 2 + 6$

$\dfrac{2}{8} = 0 + 2 \longrightarrow$ 2   6   6   7 $1463_{10} = 2667_8$

PRACTICE PROBLEMS 4-2

1. Convert the following numbers from base 8 to base 10:
(a) 46_8 (b) 111_8 (c) 204_8 (d) 400_8 (e) 777_8

2. Convert the following numbers from base 10 to base 8:
(a) 27_{10} (b) 89_{10} (c) 100_{10} (d) 256_{10} (e) 400_{10}

Solutions:

1. (a) 38_{10} (b) 73_{10} (c) 132_{10} (d) 256_{10} (e) 511_{10}

2. (a) 33_8 (b) 131_8 (c) 144_8 (d) 400_8 (e) 620_8

Additional practice problems are at the end of the chapter.

4.3 HEXADECIMAL NUMBER SYSTEM

Another important computer number system is the hexadecimal system (base 16). How many digits exist in hexadecimal? Sixteen, because the base is 16. What is the value of the largest digit in hexadecimal? Fifteen, one less than the base. However, 15 is a decimal number and requires *two digits* (15). In the hexadecimal number system, sixteen *different* symbols must be used. Zero through 9 are used for the first ten, and A through F are used for the remaining six, as shown in Table 4-1. Figure 4-4 indicates positional value of the first four positions of the hexadecimal number system.

4.3.1 Hexadecimal to Decimal Conversions. The hexadecimal number 3AC2 may be converted to base 10 as follows:

Three 4096's which equal	12288
+ ten 256's which equal	2560
+ twelve 16's which equal	192
+ two 1's which equal	2
SUM	15042_{10}

or:

$$(3 \times 16^3) + (10 \times 16^2) + (12 \times 16^1) + (2 \times 16^0) = 15042_{10}$$

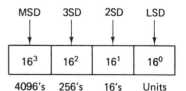

Figure 4-4. Positional value of a four-digit hexadecimal number.

EXAMPLE 4.9: Convert the following hexadecimal numbers to decimal numbers:

    (a) 72 (b) C29 (c) 12AB (d) 1A2A

Solution: (a) $72_{16} = (7 \times 16^1) + (2 \times 16^0) = 112 + 2 = 114_{10}$
     (b) $C29_{16} = (12 \times 16^2) + (2 \times 16^1) + (9 \times 16^0) = 3113_{10}$
     (c) $12AB_{16} = (1 \times 16^3) + (2 \times 16^2) + (10 \times 16^1) + (11 \times 16^0) = 4779_{10}$
     (d) $1A2A_{16} = (1 \times 16^3) + (10 \times 16^2) + (2 \times 16^1) + (10 \times 16^0) = 6698_{10}$

The conversion of decimal numbers to hexadecimal is identical to conversion using other bases. In Example 4.10 the number 1324 is converted from decimal to hexadecimal. Since the hexadecimal number system has a base of 16, division is by 16. In working with remainders greater than 9, we substitute the characters A, B, C, D, E, and F as required.

EXAMPLE 4.10: $1324_{10} = (?)_{16}$.

Solution: Step 1. $\dfrac{1324}{16} = 82 + 12$

         Step 2. $\dfrac{82}{16} = 5 + 2$ LSD

         Step 3. $\dfrac{5}{16} = 0 + 5$ —MSD→5 2 C $1324_{10} = 52C_{16}$

EXAMPLE 4.11: Convert the following decimal numbers to hexadecimal numbers:

    (a) 672 (b) 1763 (c) 12760

Solution: (a) $\dfrac{672}{16} = 42 + 0$

            $\dfrac{42}{16} = 2 + 10$

            $\dfrac{2}{16} = 0 + 2 \longrightarrow$ 2 A 0 $672_{10} = 2A0_{16}$

      (b) $\dfrac{1763}{16} = 110 + 3$

            $\dfrac{110}{16} = 6 + 14$

            $\dfrac{6}{16} = 0 + 6 \longrightarrow$ 6 E 3 $1763_{10} = 6E3_{16}$

(c) $\dfrac{12760}{16} = 797 + 8$

$\dfrac{797}{16} = 49 + 13$

$\dfrac{49}{16} = 3 + 1$

$\dfrac{3}{16} = 0 + 3 \quad \rightarrow \quad 3 \quad 1 \quad D \quad 8 \qquad 12760_{10} = 31D8_{16}$

An algorithm for converting from decimal to hexadecimal using the calculator is in Appendix A.

PRACTICE PROBLEMS 4-3

1. Convert the following hexadecimal numbers to decimal numbers:
(a) D_{16} (b) $3F_{16}$ (c) $A4_{16}$ (d) $1CD_{16}$ (e) $10BE_{16}$

2. Convert the following decimal numbers to hexadecimal numbers:
(a) 27_{10} (b) 85_{10} (c) 100_{10} (d) 256_{10} (e) 1500_{10}

Solutions:

1. (a) 13_{10} (b) 63_{10} (c) 164_{10} (d) 461_{10} (e) 4286_{10}

2. (a) $1B_{16}$ (b) 55_{16} (c) 64_{16} (d) 100_{16} (e) $5DC_{16}$

Additional practice problems are at the end of the chapter.

4.4 BINARY TO OCTAL TO HEXADECIMAL CONVERSIONS

4.4.1 Binary to Octal Conversions. Many 1's and 0's are required using binary numbers to represent large quantities. If we convert from binary to octal, the number of digits required is reduced by a factor of 3 because one octal digit equals three bits. That is, all octal digits, 0 through 7, can be represented by three bits.

$$0 = 000 \quad 4 = 100$$
$$1 = 001 \quad 5 = 101$$
$$2 = 010 \quad 6 = 110$$
$$3 = 011 \quad 7 = 111$$

For example, the binary number 110111100001 can be converted to octal by separating the bits into groups of three and substituting octal digits for each group of three bits. To separate the binary number into three-bit groups, the count begins with the LSB.

EXAMPLE 4.12: Convert 110111100001_2 to octal:

Solution:

110   111   100   001   (binary number divided into
                                 groups of three)
  6   7   4   1   (octal substitution for each
                                 three-bit group)

EXAMPLE 4.13: Convert 11101001_2 to octal:

Solution: 011   101   001
             3   5   1 $11101001_2 = 351_8$

Notice that a zero was added to the leftmost group of bits to remind us that three bits represent one octal digit.

The procedure may be reversed in order to convert from octal to binary.

EXAMPLE 4.14: Convert 417_8 to binary:

Solution: 4   1   7 $417_8 = 100001111_2$
          100   001   111

EXAMPLE 4.15: Convert 362_8 to binary:

Solution: 3   6   2 $362_8 = 11110010_2$
          011   110   010

In each example, three bits are substituted for each octal digit. Of course, a zero in the leftmost bit position has no meaning and need not be written in the answer.

4.4.2   Binary to Hexadecimal Conversions.   If we convert from binary to hexadecimal, the number of digits required is reduced by a factor of 4 because one hexadecimal digit equals four bits. That is, all hexadecimal digits (0 through F) can be represented by four bits.

$0 = 0000$	$8 = 1000$
$1 = 0001$	$9 = 1001$
$2 = 0010$	$A = 1010$
$3 = 0011$	$B = 1011$
$4 = 0100$	$C = 1100$
$5 = 0101$	$D = 1101$
$6 = 0110$	$E = 1110$
$7 = 0111$	$F = 1111$

For example, the binary number 110110011011 can be converted to hexadecimal by separating the bits into groups of four and substituting hexadecimal digits for each group of four bits. To separate the binary number into four-bit groups, the count begins with the LSB.

EXAMPLE 4.16: Convert 110110011011_2 to hexadecimal:

Solution: 1101   1001   1011   (binary number divided into four-bit groups)

     D   9   B   (hexadecimal substitution for each four-bit group)

PRACTICE PROBLEMS 4-4   Convert the following binary numbers to (a) octal numbers and then to (b) hexadecimal numbers:

1. 1101001101_2
2. 1000011001_2
3. 1100110001_2
4. 111001110101_2
5. 111101011000110_2

Convert the following numbers to binary numbers:

6. 73_8
7. 127_8
8. 617_8
9. 506_8
10. 1207_8
11. $A6_{16}$
12. $4C_{16}$
13. $C0D_{16}$
14. $BA1_{16}$
15. $F23_{16}$

Solutions:

1. (a) 1515_8 (b) $34D_{16}$
2. (a) 1031_8 (b) 219_{16}
3. (a) 1461_8 (b) 331_{16}
4. (a) 7165_8 (b) $E75_{16}$
5. (a) 75306_8 (b) $7AC6_{16}$
6. 111011_2
7. 1010111_2
8. 110001111_2
9. 101000110_2
10. 1010000111_2
11. 10100110_2
12. 1001100_2
13. 110000001101_2
14. 101110100001_2
15. 111100100011_2

Additional practice problems are at the end of the chapter.

The simplest octal to hexadecimal conversion is to first convert the number to binary and then make a conversion from binary to the other number system. We have converted from decimal to the other systems and we have converted between binary, octal, and hexadecimal. Now let's practice some conversions among all the number systems we have studied. For example, let's convert 200_{10} to binary, octal, and hexadecimal.

EXAMPLE 4.17:

$200_{10} = $ _____$_2 = $ _____$_8 = $ _____$_{16}$

> *Solution:* There are many ways to solve this problem. We either could convert from 200_{10} to each of the other systems or we could convert to base 2 and then convert to base 8 and base 16 from base 2. Another way, and probably the one requiring fewest steps, is to convert from base 10 to base 16, then from base 16 to base 2, and then from base 2 to base 8. Remember, fewer steps mean fewer chances for error.

$$\frac{200}{16} = 12 + 8 $$

$$\frac{12}{16} = 0 + 12 \longrightarrow C \quad 8$$

$$200_{10} = C8_{16}$$

Convert to binary:

$$\begin{array}{cc} C & 8 \\ 1100 & 1000 \end{array} \qquad C8_{16} = 11001000_2$$

Convert to octal:

$$\begin{array}{ccc} 011 & 001 & 000 \\ 3 & 1 & 0 \end{array} \qquad 11001000_2 = 310_8$$

$$200_{10} = 11001000_2 = 310_8 = C8_{16}$$

EXAMPLE 4.18:

$C27_{16} = $ _____$_2 = $ _____$_8 = $ _____$_{10}$

> *Solution:* In this example we have a straight conversion to binary and then to octal. Convert to binary:

$$\begin{array}{ccc} C & 2 & 7 \\ 1100 & 0010 & 0111 \end{array} \qquad C27_{16} = 110000100111_2$$

Convert to octal:

$$\begin{array}{cccc} 110 & 000 & 100 & 111 \\ 6 & 0 & 4 & 7 \end{array} \quad 110000100111_2 = 6047_8$$

The conversion to decimal can be made from any of the other systems but conversion from base 16 requires fewer steps.

$$C27_{16} = (12 \times 16^2) + (2 \times 16) + (7 \times 16^0)$$
$$= 3072 + 32 + 7$$
$$= 3111_{10}$$
$$C27_{16} = 110000100111_2 = 6047_8 = 3111_{10}$$

PRACTICE PROBLEMS 4-5

1. $73_8 =$ _____ $_{16} =$ _____ $_2 =$ _____ $_{10}$

2. $273_8 =$ _____ $_{16} =$ _____ $_2 =$ _____ $_{10}$

3. $A3_{16} =$ _____ $_8 =$ _____ $_2 =$ _____ $_{10}$

4. $12E_{16} =$ _____ $_8 =$ _____ $_2 =$ _____ $_{10}$

5. $1101001_2 =$ _____ $_8 =$ _____ $_{16} =$ _____ $_{10}$

6. $10111100_2 =$ _____ $_8 =$ _____ $_{16} =$ _____ $_{10}$

Solutions:

1. $73_8 = 3B_{16} = 111011_2 = 59_{10}$ **2.** $273_8 = BB_{16} = 10111011_2 = 187_{10}$

3. $A3_{16} = 243_8 = 10100011_2 = 163_{10}$ **4.** $12E_{16} = 456_8 = 100101110_2 = 302_{10}$

5. $1101001_2 = 151_8 = 69_{16} = 105_{10}$ **6.** $10111100_2 = 274_8 = BC_{16} = 188_{10}$

Additional practice problems are at the end of the chapter.

SELF-TEST 4-1   Convert the following numbers to decimal numbers:

1. 46_8 **2.** 276_8

3. $F6_{16}$ **4.** $C3A_{16}$

5. 1011011_2 **6.** 11001010_2

Convert the following decimal numbers to binary, octal, and hexadecimal:

7. 10   **8.** 28

9. 187   **10.** 625

11. $93_{10} =$ _____ $_2 =$ _____ $_8 =$ _____ $_{16}$

12. $E1A_{16} =$ _____ $_2 =$ _____ $_8 =$ _____ $_{10}$

13. $11010010110_2 =$ _____ $_8 =$ _____ $_{10} =$ _____ $_{16}$

14. $573_8 =$ _____ $_2 =$ _____ $_{10} =$ _____ $_{16}$

Answers to Self-Test 4-1 are at the end of the chapter.

4.6   ADDITION

    4.6.1   Adding Decimal Numbers.   Add the numbers 23 and 45. When numbers are added, they are added in columns—one number is placed below the other. Either number may be placed first. In Example 4.19 the numbers 23 and 45 are placed in the position for addition. The digits are separated to emphasize positional differences. Eight is in the units position $(3+5)$, and 6 is in the tens position $(2+4)$. The answer is 68. (Six 10's and eight 1's.)

    EXAMPLE 4.19:

$$23+45 = ?$$

2	3
4	5
6	8

    EXAMPLE 4.20: Add the numbers 64 and 87 in base 10:

    Solution:

	Carry	1	1	
			6	4
			8	7
		1	(15)	(11)
(subtract value of carry)			10	10
		1	5	1

(remainders taken as decimal sum)

    In Example 4.20, there is an 11 in the units position. This number is greater than 9, which tells us there must be a carry (10). The digit remaining in the units position is the difference of 10 and 11, or 1. The tens position now contains $6+8+$ a carry of 1, or 15. Again, since the number is greater than 9, a carry is indicated. Subtracting 10 from 15, we get 5 with a carry of 1. The third position contains only the carry. Of course, since we have been adding for years in decimal, our experience allows us to perform these operations automatically. We have taken time to discuss the arithmetic steps because this is the addition process in all bases.

EXAMPLE 4.21: Add the decimal numbers below. Follow the procedure given in Example 4.20.
(a) 738+417 (b) 9706+463 (c) 1774+7268

Solution:

(a)

```
   1       1
       7  3  8
       4  1  7
   ─────────────
   1 (11)  5 (15)
      10      10        738+417=1155
   ─────────────
   1    1  5  5
```

(b)

```
   1  1
       9  7  0  6
          4  6  3
   ─────────────────
   1(10)(11)  6  9      9706+463=10169
      10 10
   ─────────────────
   1   0   1  6  9
```

(c)

```
   1   1   1
   1   7   7  4
   7   2   6  8
   ─────────────────
   9(10)(14)(12)        1774+7268=9042
      10  10  10
   ─────────────────
   9   0   4  2
```

4.6.2 *Adding Octal Numbers*

EXAMPLE 4.22: Add the octal numbers 736 and 215:

Solution:

```
                          1        1
                          7  3  6
                          2  1  5
   (decimal sum)       1 (9)  5 (11)
(subtract value of carry)  8        8        736₈+215₈=1153₈
                       ─────────────
                       1   1  5  3
```

Again, the LSD or units position is added first. The sum of 6 and 5 is 11. The number is greater than 7 (7 is the largest digit in base 8), which tells us there must be a carry. Subtracting 8 (the value of the carry) leaves 3 with a carry of 1. The 8^1 position now contains $3+1+a$ carry of 1, or 5. Since 5 is less than 7, no carry exists. In the 8^2 position, $7+2$ is 9. Since the number is greater than 7, a carry is generated. Subtracting the carry (8) yields a remainder of 1 with a carry of 1. Since the 8^3 position contains only the carry, the answer is 1153_8.

EXAMPLE 4.23: Add the following octal numbers:
(a) 243   (b) 764   (c) 604
    172   414   777

Solution:

(a)

1		
2	4	3
1	7	2
4	(11)	5
	8	
4	3	5

$=435_8$

(b)

1	1	1	
	7	6	4
	4	1	4
1	(12)	(8)	(8)
	8	8	8
1	4	0	0

$=1400_8$

(c)

1	1	1	
	6	0	4
	7	7	7
1	(14)	(8)	(11)
	8	8	8
1	6	0	3

$=1603_8$

PRACTICE PROBLEMS 4-6   Add the following octal numbers:

1. 204   **2.** 447
    316   173

3. 605   **4.** 517
    726   567

5. 176
    761

Solutions:

1. 522_8 2. 642_8
3. 1533_8 4. 1306_8
5. 1157_8

Additional practice problems are at the end of the chapter.

4.6.3 Adding Hexadecimal Numbers

EXAMPLE 4.24: Add the hexadecimal numbers 7AF and 579.

Solution:

	$1\searrow$	$1\searrow$	
	7	A	F
	5	7	9
(decimal sum)	(13)	(18)	(24)
(subtract value of carry)		16	16
	D	2	8	$=D28_{16}$

Adding the units position yields 24. Twenty-four is greater than the largest digit (F), which is equal to decimal 15. This indicates that a carry should exist. Subtracting the value of a carry (16) leaves a remainder of 8 with a carry of 1. In the 16^1 position the sum of A (decimal 10)+7+a carry of 1 is 18. Again, a carry is generated leaving a remainder of 2 with a carry of 1. In the 16^2 position the carry+7+5 equals decimal 13. In hexadecimal, 13 is D; therefore, the answer is $D28_{16}$.

EXAMPLE 4.25: Add the following hexadecimal numbers:

(a) 4D3   (b) 789   (c) F347
     818   C47   E006

Solution:

(a)
	4	D	3
	8	1	8
	(12)	(14)	(11)
	C	E	B	$=CEB_{16}$

(b)
	$1\searrow$		$1\searrow$	
		7	8	9
		C	4	7
1	(19)	(13)	(16)
	16		16
1	3	D	0	$=13D0_{16}$

(c)

$$
\begin{array}{r}
\quad F \quad 3 \quad 4 \quad\quad 7 \\
\quad E \quad 0 \quad 0 \quad\quad 6 \\
\hline
1 \quad (29) \quad 3 \quad 4 \quad (13) \\
16
\end{array}
$$

$$1 \quad D \quad 3 \quad 4 \quad\quad D \quad = 1D34D_{16}$$

PRACTICE PROBLEMS 4-7   Add the following hexadecimal numbers:

1. C4
   1B

2. A4
   B7

3. A28
   6F9

4. 9B5
   4C8

5. FACE
   1701

Solutions:

1. DF_{16}

2. $15B_{16}$

3. 1121_{16}

4. $E7D_{16}$

5. $111CF_{16}$

Additional practice problems are found at the end of the chapter.

4.6.4   Adding Binary Numbers

EXAMPLE 4.26: Add the binary numbers 1101 and 1001.

Solution:

$$
\begin{array}{r}
1 \quad\quad\quad 1 \quad\quad (carry) \\
1 \quad 1 \quad 0 \quad 1 \\
1 \quad 0 \quad 0 \quad 1 \\
\hline
\end{array}
$$

(decimal sum) (2) 1 1 (2)

(subtract value of carry)   2   2

$$1 \quad 0 \quad 1 \quad 1 \quad 0 \quad = 10110_2$$

The method of addition is the same as that used with other bases. Notice that $1+1$ equals decimal 2. Since 2 is invalid in base 2, a carry is generated into the next position. We see from the example that a carry is generated whenever a sum greater than 1 results from the addition of numbers in a column.

EXAMPLE 4.27: Add the following binary numbers:

(a) 10 (b) 111 (c) 1001 (d) 11100
    11   101   1010   01010

Solutions:

(a)

$$
\begin{array}{cc}
1 & 0 \\
1 & 1 \\
\hline
1\ (2) & 1 \\
2 & \\
\hline
1\quad 0 & 1 = 101_2
\end{array}
$$

(b)

$$
\begin{array}{ccc}
1 & 1 & 1 \\
1 & 0 & 1 \\
\hline
1\ (3) & (2) & (2) \\
2 & 2 & 2 \\
\hline
1\quad 1 & 0 & 0 = 1100_2
\end{array}
$$

(c)

$$
\begin{array}{cccc}
1 & 0 & 0 & 1 \\
1 & 0 & 1 & 0 \\
\hline
1\ (2) & 0 & 1 & 1 \\
2 & & & \\
\hline
1\quad 0 & 0 & 1 & 1 = 10011_2
\end{array}
$$

(d)

$$
\begin{array}{ccccc}
1 & 1 & 0 & 0 \\
0 & 1 & 0 & 1 & 0 \\
\hline
1\ (2) & (2) & 1 & 1 & 0 \\
2 & 2 & & & \\
\hline
1\quad 0 & 0 & 1 & 1 & 0 = 100110_2
\end{array}
$$

From working with these and other examples of binary addition, we see that:

1. $0+0$ always equals 0
2. $0+1$ always equals 1
3. $1+1$ always equals 0 and a carry
4. $1+1+$ a carry always equals 1 and a carry

PRACTICE PROBLEMS 4-8   Add the following binary numbers:

1. 1011
   1100

2. 1001
   1111

3. 10111
   10001

4. 11001
   11110

5. 101110
   111100

Solutions:

1. 10111_2

2. 11000_2

3. 101000_2

4. 110111_2

5. 1101010_2

Additional practice problems are at the end of the chapter.

SELF-TEST 4-2   Add the following numbers:

1. AF_{16}
   36_{16}

2. $C4A_{16}$
   $17E_{16}$

3. $F4A9_{16}$
   $FE7A_{16}$

4. 73_8
   66_8

5. 63_8
   41_8

6. 73_8
   46_8

7. 101_2
   111_2

8. 1101_2
   1011_2

9. 10110_2
   11110_2

Answers to Self-Test 4-2 are at the end of the chapter.

4.7 SUBTRACTION

4.7.1 Subtracting Decimal Numbers.   Subtraction in base 10 is so automatic for us that we may do it without thinking of the rules involved. Since these rules in base 10 are identical to those for other bases, and since we are more familiar with base 10 that with other bases, let's review the process of subtraction in base 10.

EXAMPLE 4.28: Subtract 25 from 43:

Solution:

$$
\begin{array}{r}
3 \qquad\qquad \text{borrow 1} \\
\text{(minuend)} \quad \cancel{4}\!\leftarrow\!(1)\ 3 \\
\text{(subtrahend)} \quad 2 \qquad 5 \\
\hline
\text{(difference)} \quad 1 \qquad 8
\end{array}
$$

As you know, we start with the least significant position and subtract. Since we cannot subtract 5 from 3, we "borrow" from the next significant position of the minuend. We may now interpret the 3 as 13 and find the difference, which is 8. Moving to the next position, we must reduce the minuend by 1 because of the "borrow." We now subtract the number in the subtrahend from the new minuend and the difference is 1. Remember, the borrow carries with it a weight equal to that of the base.

EXAMPLE 4.29: Subtract the following decimal numbers. Show the process of borrowing.

$$
\begin{array}{llll}
\text{(a)} & 624 \quad \text{(b)} \quad 854 \quad \text{(c)} \quad 8534 \quad \text{(d)} \quad 705 \\
& -276 \qquad\quad -235 \qquad\quad -6748 \qquad\quad -528
\end{array}
$$

Solution:

(a)
$$
\begin{array}{ccc}
5 & (1)1 & \\
\cancel{6} & \cancel{2} & (1)4 \\
-2 & 7 & 6 \\
\hline
3 & 4 & 8
\end{array}
$$

(b)
$$
\begin{array}{ccc}
& 4 & \\
8 & \cancel{5} & (1)4 \\
-2 & 3 & 5 \\
\hline
6 & 1 & 9
\end{array}
$$

(c)
$$
\begin{array}{cccc}
7 & (1)4 & (1)2 & \\
\cancel{8} & \cancel{5} & \cancel{3} & (1)4 \\
-6 & 7 & 4 & 8 \\
\hline
1 & 7 & 8 & 6
\end{array}
$$

(d)
$$
\begin{array}{ccc}
6 & 9 & \\
\cancel{7} & (\cancel{1})\cancel{0} & (1)5 \\
-5 & 2 & 8 \\
\hline
1 & 7 & 7
\end{array}
$$

Notice in Example 4.29(d) that since 8 is greater than 5, a borrow is indicated. However, the next significant digit is 0 and no borrow can take place (we can't borrow something from nothing). Therefore, we must borrow from the next significant digit (7) giving the second position a value of 10. Now we can borrow from that position to yield 15 in the least significant position.

4.7.2 Subtracting Octal Numbers. In Example 4.30 the number 267_8 is subtracted from 512_8. Notice that a borrow is necessary in the first position since 7 is greater than 2. The borrow creates a new number in the minuend, 12. (This number is read "one-two"—*not* "twelve" because twelve is a decimal number. One-two in base 8 means $8+2$ in base 10.)

EXAMPLE 4.30: Subtract 267_8 from 512_8:

Solution:

$$
\begin{array}{lccc}
 & & 4 & 10 \\
\text{(minuend)} & \cancel{5} & \cancel{1}\text{-}(1)2 \\
\text{(subtrahend)} & 2 & 6 & 7 \\
\hline
\text{(difference)} & 2 & 2 & 3 = 223_8
\end{array}
$$

One-two minus 7 equals 3. If you think in base 10, then $(8+2)-7$ equals 3. In the next position, 1 was borrowed from the number in the minuend, making its value 0. A borrow is again required; it makes the value of the minuend 10 (or $8+0$ in base 10). Subtracting 6 from 10 yields a difference of 2. In the next position, 1 was borrowed from the minuend, making its value 4. The difference in this position is 2.

EXAMPLE 4.31: Solve the following subtraction problems in base 8:

(a)   43   (b)   161   (c)   2172
   $(-)16$ $(-)72$ $(-)1717$

Solution:

(a)
$$
\begin{array}{ccc}
 & 3 & \\
 & \cancel{4} & (1)3 \\
(-)1 & & 6 \\
\hline
2 & & 5 \quad = \quad 25_8
\end{array}
$$

(b)
$$
\begin{array}{ccc}
0 & (1)5 & \\
\cancel{1} & \cancel{6} & (1)1 \\
(-) & 7 & 2 \\
\hline
6 & 7 & = 67_8
\end{array}
$$

(c)
$$
\begin{array}{cccc}
1 & & 6 & \\
\cancel{2} & (1)1 & \cancel{7} & (1)2 \\
(-)1 & 7 & 1 & 7 \\
\hline
2 & 5 & 3 & = 253_8
\end{array}
$$

PRACTICE PROBLEMS 4-9   Subtract the following octal numbers:

1.   73
     $(-)56$

2.   376
     $(-)277$

3.   673
     $(-)176$

4.   2133
     $(-)1574$

5.   6014
     $(-)3247$

Solutions:

1. 15_8

2. 77_8

3. 475_8

4. 337_8

5. 2545_8

Additional practice problems are at the end of the chapter.

4.7.3   Subtracting Hexadecimal Numbers.   Applying the same rules as in base 8, let's subtract the hexadecimal number 85E from the number C37.

EXAMPLE 4.32: $C37 - 85E = (?)_{16}$.

Solution:

```
                       B    ⟋(1)2
       (minuend)       C̶  ⟋  3̶ ──→ (1)7
     (subtrahend)      8      5       E
                      ─────────────────
      (difference)     3      D       9     = 3D9₁₆
```

$$= 3D9_{16}$$

Notice that a borrow is necessary in the first position since E is greater than 7. One-seven (*not* seventeen) minus E equals 9 or, in base 10, $(16+7) - 14$ equals 9. (Remember, the borrow has a weight of 16.) In the next position, a borrow is required so that 5 is subtracted from 12 (one-two), which equals D. (In base 10, it would be $16+2-5=13$.) Finally, in the next position, subtracting 8 from B (remember that we borrowed 1, making C a B) yields 3 or, in base 10, $11-8=3$.

EXAMPLE 4.33: Subtract the following hexadecimal numbers:

	(a)	A29	(b)	3A12	(c)	F1B2
		− 7BF		− 29A0		− ABC7

Solution:

(a)

9	⟶ (1)1		
A̸	2̸ ⟶	(1)9	
7	B	F	
2	6	A	$= 26A_{16}$

(b)

	9			
3	A̸ ⟶	(1)1	2	
2	9	A	0	
1	0	7	2	$= 1072_{16}$

(c)

E	⟶ (1)0	⟶ (1)A		
F̸	1̸ ⟶	B̸ ⟶	(1)2	
A	B	C	7	
4	5	E	B	$= 45EB_{16}$

PRACTICE PROBLEMS 4-10   Subtract the following hexadecimal numbers:

1.   A17
   (−)26B

2.   B27
   (−)6A4

3.   7A29
   (−)2BC9

4.   4A9B
   (−)100F

5.   FADE
   (−)2C3F

Solutions:

1. $7AC_{16}$

2. 483_{16}

3. $4E60_{16}$

4. $3A8C_{16}$

5. $CE9F_{16}$

Additional practice problems are at the end of the chapter.

4.7.4   Subtracting Binary Numbers.   The process of subtraction in base 2 is similar to that in the other bases, as was discussed earlier in this chapter.

EXAMPLE 4.34: Subtract 0101 from 1110:

Solution:

	borrow			0		
(minuend)	1	1	$\cancel{1}$	$\longrightarrow$ (1)0	
(subtrahend)	0	1	0		1
(difference)	1	0	0		1	$= 1001_2$

EXAMPLE 4.35: Subtract 01101 from 10011:

Solution:

	borrow	0	1			
(minuend)	1 $\longrightarrow$	$(\cancel{1})\cancel{0}$ $\longrightarrow$	(1)0	1	1
(subtrahend)	0	1	1	0	1
	0	0	1	1	0	$= 00110_2$

In Example 4.34 a borrow is necessary in the least significant position since the minuend is 0. The borrow has a weight of 2 because we are working in base 2. Note that in Example 4.35 a borrow is required in the 2^2 position. However, since the 2^3 position contains a 0, the borrow must come from the 2^4 position to the 2^3 position. A borrow is now possible from the 2^3 position, and the minuend in the 2^2 position is 10 (one-zero), resulting in a difference of 1. The new minuend in the next position (2^3) is now 1, and the difference is 0. Since our original borrow came from the next position (2^4), the minuend is 0 and the difference is 0.

PRACTICE PROBLEMS 4-11   Subtract the following binary numbers:

1.   1010
   $(-)$0101

2.   10110
   $(-)$01101

3.   110010
   $(-)$100111

4.   100111
   $(-)$001101

5.   1100110
   $(-)$1011001

Solutions:

1. 101_2

2. 1001_2

3. 1011_2

4. 11010_2

5. 1101_2

Additional practice problems are at the end of the chapter.

4.7.5 Ten's Complement. In the modern computer, subtraction is usually performed by addition of the complement of the subtrahend. This is done in order to simplify the design of the arithmetic section of the computer.

EXAMPLE 4.36: Subtract 676 from 835:

$$\begin{array}{r} 835 \\ -676 \\ \hline 159 \end{array}$$

This same operation can be performed by adding the ten's complement of 676 to 835. The ten's complement of a decimal number is the difference between that number and the next higher power of ten. Therefore, the ten's complement of 676 equals 10^3 or 1000 minus 676.

$$\begin{array}{r} 1000 \\ -676 \\ \hline 324 \end{array} \quad \text{(ten's complement of 676)}$$

The next step is to add the ten's complement to the minuend and subtract 1000 from the result, as follows:

$$\begin{array}{rr} 1000 & 835 \ \text{(minuend)} \\ -676 & +\ 324 \ \text{(ten's complement)} \\ \hline 324 & 1159 \\ & -\ 1000 \\ \hline & 159 \ \text{(difference)} \end{array}$$

Notice that the last operation (subtracting 1000) is just a matter of dropping the 1 in the most significant position. Let's try another problem.

EXAMPLE 4.37: Subtract 4678 from 6392:

Solution:

$$\begin{array}{rr} 10000 & 6392 \ \text{(minuend)} \\ -\ 4678 & +\ 5322 \ \text{(ten's complement)} \\ \hline 5322 & 11714 \\ & -10000 \\ \hline & 1714 \ \text{(difference)} \end{array}$$

4.7.6 Nine's complement. Finding the ten's complement is complicated by the fact that borrowing is required. The nine's complement of a digit is the difference between 9 and that digit and can be found by a simple inspection procedure. Let's find the nine's complement of 235:

$$
\begin{array}{r}
999 \\
-\ 235 \\
\hline
764
\end{array}
\quad \text{(nine's complement of 235)}
$$

The nine's complement may be converted to the ten's complement by adding 1. Find the ten's complement of 5789:

$$
\begin{array}{r}
9999 \\
-\ 5789 \\
\hline
4210 \quad \text{(nine's complement)} \\
+\quad 1 \\
\hline
4211 \quad \text{(ten's complement)}
\end{array}
$$

4.7.7 *One's complement.* The one's complement of a binary number is similar to the nine's complement of a decimal number. The binary number to be complemented is subtracted from a binary number made up of all 1's. In the following example the one's complement of 10110010 is found.

EXAMPLE 4.38:

$$
\begin{array}{r}
1\ 1\ 1\ 1\ 1\ 1\ 1\ 1 \\
-\ 1\ 0\ 1\ 1\ 0\ 0\ 1\ 0 \\
\hline
0\ 1\ 0\ 0\ 1\ 1\ 0\ 1
\end{array}
\quad \text{(one's complement of 10110010)}
$$

Careful inspection reveals that the one's complement of a binary number may be produced by changing all of the ones in the number to zeros and changing all of the zeros to ones.

$$
\begin{array}{l}
1\ 0\ 1\ 1\ 0\ 0\ 1\ 0 \quad \text{(invert this number to get one's complement)} \\
0\ 1\ 0\ 0\ 1\ 1\ 0\ 1
\end{array}
$$

The computer is able to perform this operation easily by means of a circuit called an *inverter*. Notice that carries or borrows are not involved in this operation.

In order to subtract by using the one's complement method, the one's complement of the subtrahend is added to the minuend. The carry out of the highest bit position is then added to the *lowest bit position* (LSB). This is called *end-around carry*.

EXAMPLE 4.39: Find the difference between 10110011 and 01101101:

Solution:

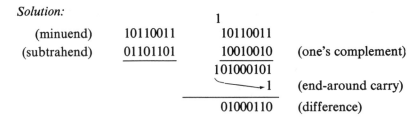

(minuend)	10110011	10110011	
(subtrahend)	01101101	10010010	(one's complement)
		101000101	
		→1	(end-around carry)
		01000110	(difference)

EXAMPLE 4.40: Find the difference between 11011000 and 10110011:

Solution:

(minuend)	11011000	11011000	
(subtrahend)	10110011	01001100	(one's complement)
		100100100	
		→1	(end-around carry)
		00100101	(difference)

4.7.8 Two's complement.   Perhaps the most popular method of subtraction used by computers is the two's complement method. The two's complement of a number may be found by adding 1 to the one's complement. Let's find the two's complement of 10011101. First, we change all 0's to 1's and all 1's to 0's (one's complement). Then we add 1 as follows:

0011101	0 1 1 0 0 0 1 0	(one's complement)
(+)	1	
	0 1 1 0 0 0 1 1	(two's complement)

In order to subtract, we obtain the two's complement of the subtrahend and then add to the minuend.

EXAMPLE 4.41: Let's subtract 01001010 from 01100111:

(minuend)	01100111	01100111	
(subtrahend)	01001010	(+)10110110	(two's complement)
	1	00011101	(difference)

(this carry is ignored)

The carry resulting from the most significant bit position is ignored. There is no end-around carry as there is with the one's complement method. The computer obtains the one's complement by using an inverter. The two's complement is then obtained by starting the addition process with a carry as shown in Example 4.42.

EXAMPLE 4.42: Subtract 00110110 from 01011011.

Solution:

(minuend)	01011011	01011011	
(subtrahend)	00110110	11001001	(one's complement)
		+ 1	(add 1 to get the two's complement)

1 00100101 (difference)

(carry is dropped)

The number of bit positions in the subtrahend must agree with the number of bit positions in the minuend. This is usually accomplished by providing a fixed number of storage cells in the computer. In the preceding examples, there were eight bit positions used; therefore, eight storage cells were used to store the numbers. Eight cells were also used to store the difference, causing the carry out of the most significant bit position to be dropped because there was no place to store it. Each of the above eight-cell combinations is called a *register*. Example 4.43 illustrates the concept of registers with fixed numbers of bit positions.

EXAMPLE 4.43: Subtract 00001100 from 01001010 (Figure 4-5).

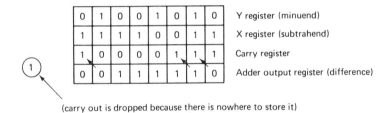

0	1	0	0	1	0	1	0	Y register (minuend)
1	1	1	1	0	0	1	1	X register (subtrahend)
1	0	0	0	0	1	1	1	Carry register
0	0	1	1	1	1	1	0	Adder output register (difference)

(carry out is dropped because there is nowhere to store it)

Figure 4-5.

Solution: The minuend is placed in the Y register; the one's complement of the subtrahend is placed in the X register. The LSB is in the rightmost bit position. A 1 is placed in the LSB position of the carry register. All the other positions of the carry register store a zero at the start. If a carry is produced with the addition of a column, a 1 is stored in the *next* bit position of the carry register. Since the output register is limited to eight bit positions, the carry out of the MSB position is dropped.

PRACTICE PROBLEMS 4-12   For the following decimal numbers find (a) the nine's complement and (b) the ten's complement:

1. 17   **2.** 36   **3.** 88

4. 276   **5.** 8976

For the following binary numbers, find the one's complement and (b) the two's complement:

6. 1101 **7.** 1001 **8.** 100110

9. 110001 **10.** 11000011

Subtract the following binary numbers by using (a) the one's complement method and (b) the two's complement method:

11.   11001   **12.**   101101   **13.**   1100011
    (−)01101   (−)001011   (−)0011001

14.   10110010   **15.**   11100100
    (−)   11101   (−)   11011

Solutions:

1. (a) 82   (b) 83   **2.** (a) 63   (b) 64

3. (a) 11   (b) 12   **4.** (a) 723   (b) 724

5. (a) 1023   (b) 1024   **6.** (a) 0010   (b) 0011

7. (a) 0110   (b) 0111   **8.** (a) 011001   (b) 011010

9. (a) 001110   (b) 001111   **10.** (a) 00111100   (b) 00111101

11. (a) 1
```
    11001
    10010   (one's complement)
  1 01011
    └──→1
    ─────
    01100
```

(b) 1
```
    11001
    10011   (two's complement)
  ⫫ 01100
```

12. (a) 1
```
    101101
    110100   (one's complement)
  1 100001
    └──→1
    ──────
    100010
```

(b) 1
```
    101101
    110101   (two's complement)
  ⫫ 100010
```

13. (a) 1   11
```
    1100011
    1100110   (one's complement)
  1 1001001
    └────→1
    ───────
    1001010
```

(b) 1   111
```
    1100011
    1100111   (two's complement)
  ⫫ 1001010
```

14. (a) 1
```
    10110010
    11100010   (one's complement)
  1 10010100
    └─────→1
    ────────
    10010101
```

(b) 1 11 1
```
    10110010
    11100011   (two's complement)
  ⫫ 10010101
```

15. (a) 1 11   1
       11100100
       11100100 (one's complement)
   1 11001000
   └────→ 1
     11001001

(b) 1 11   1
       11100100
       11100101 (two's complement)
   ̷1 11001001

Additional practice problems are at the end of the chapter.

4.7.9 Sixteen's and Eight's Complement. When large signed binary numbers are added or subtracted, it is convenient to use the sixteen's complement method. The sixteen's complement is obtained by taking the fifteen's complement and adding one. The fifteen's complement is obtained by substituting each digit with the difference between its value and 15. For example, the sixteen's complement of 3B8 is:

$$\begin{array}{rl}
\text{FFF}\quad\text{C47} & \text{(fifteen's complement)}\\
(-)\text{3B8}\quad\quad 1 & \text{(plus 1)}\\
\hline
\text{C47}\quad\text{C48} & \text{(sixteen's complement)}
\end{array}$$

In many systems, octal is used to represent large binary numbers. Therefore, it is convenient to use the eight's complement method for performing signed binary number additions and subtractions. The seven's complement is first obtained by substituting each octal digit with the difference between its value and 7. Then 1 is added to produce the eight's complement. For example, the eight's complement of 136_8 is:

$$\begin{array}{rl}
777\quad 641 & \text{(seven's complement)}\\
(-)136\quad\quad 1 & \text{(plus 1)}\\
\hline
641\quad 642 & \text{(eight's complement)}
\end{array}$$

The following subtraction is performed by using the two's complement method:

$$\begin{array}{l}
0011\quad 1100\quad 1011 \longrightarrow 0011\quad 1100\quad 1011\quad\text{(minuend)}\\
(-)0010\quad 0110\quad 0111 \longrightarrow 1101\quad 1001\quad 1001\quad\text{(two's complement)}\\
\hline
(\text{carry is dropped}) \longrightarrow \cancel{1}\ 0001\quad 0110\quad 0100\quad\text{(difference)}
\end{array}$$

The same problem may be solved by using the sixteen's complement method:

3CB	3CB	(minuend)
(−)267	D99	(subtrahend in sixteen's complement form)
	$\cancel{1}$ 164	(difference)

(carry is dropped) ⌐

If the above binary numbers are expressed in octal, the eight's complement method may be used:

1713_8	1713	(minuend)
(−)1147_8	6631	(subtrahend in eight's complement form)
	$\cancel{1}$ 0544	(difference)

(carry is dropped) ⌐

PRACTICE PROBLEMS 4-13. For the following octal numbers find (a) the seven's complement and (b) the eight's complement:

1. 6   2. 12

3. 56   4. 173

5. 726

For the following hexadecimal numbers find: the fifteen's complement and (b) the sixteen's complement.

6. A   7. 2C

8. 87   9. 4A3

10. F209

Perform the following subtractions: (a) in binary by using the two's complement method; (b) in octal by using the eight's complement method, and (c) in hexadecimal by using the sixteen's complement method.

11.	11010011	12.	10011101
	(−)00111101		(−)00110011
13.	10111100	14.	11110000
	(−)01001110		(−)00111100
15.	11001111		
	(−)10000011		

Solutions:

1. (a) 1	(b) 2	2. (a) 65	(b) 66
3. (a) 21	(b) 22	4. (a) 604	(b) 605

5. (a) 051   (b) 052   **6.** (a) 5   (b) 6

7. (a) D3   (b) D4   **8.** (a) 78   (b) 79

9. (a) B5C   (b) B5D   **10.** (a) 0DF6   (b) 0DF7

11. (a)   11010011
          11000011   (two's complement)
    $\cancel{1}$ 10010110$_2$

   (b) 1101   0011 =   D3
          1100   0011 =   C3   (sixteen's complement)
              $\cancel{1}$ 96$_{16}$

   (c) 011   010   011 =   323
          111   000   011 =   703
                  $\cancel{1}$ 226$_8$

The subtrahend is 00111101. The two's complement is 11000011. Since we have already found the two's complement of 00111101 to be 11000011, we separate the two's complement into 2 four-bit groups. Then 1100   0011 = C3 which is the sixteen's complement. To find the eight's complement, we must separate the subtrahend into 3 three-bit groups. Since we have only eight bits, we add a zero to the subtrahend in the original problem to get 000   111   101. The two's complement is 111   000   011. The eight's complement is 703$_8$.

12. (a)   10011101
          11001101   (two's complement)
    $\cancel{1}$ 01101010$_2$

   (b) 1001   1101 =   9D
          1100   1101 =   CD   (sixteen's complement)
              $\cancel{1}$ 6A$_{16}$

   (c) 010   011   101 =   235
          111   001   101 =   715   (eight's complement)
                  $\cancel{1}$ 152$_8$

13. (a)   10111100
          10110010   (two's complement)
    $\cancel{1}$ 01101110$_2$

   (b) 1011   1100 =   BC
          1011   0010 =   B2   (sixteen's complement)
              $\cancel{1}$ 6E$_{16}$

   (c) 010   111   100 =   274
          110   110   010 =   662
                  $\cancel{1}$ 156$_8$

14. (a)   11110000
          11000100   (two's complement)
    $\cancel{1}$ 10110100$_2$

   (b) 1111   0000 =   F0
          1100   0100 =   C4
              $\cancel{1}$ B4$_{16}$

(c) 011   110   000 =   360
     111   000   100 =   704
                  ̸Х 264₈

15. (a)   11001111
          01111101   (two's complement)
        ̸Х 01001100

(b) 1100   1111 =   CF
     0111   1101 =   7D   (sixteen's complement)
              ̸Х 4C₁₆

(c) 011   001   111 =   317
     101   111   101 =   575   (eight's complement)
                  ̸Х 114₈

Additional practice problems are at the end of the chapter.

SELF-TEST 4-3   Work the following problems by using (a) the straight subtraction method and (b) the complement method:

1. 11101011_2
     $(-)01001101_2$

2. 11000110_2
     $(-)10011011_2$

3. 73_8
     $(-)46_8$

4. 603_8
     $(-)364_8$

5. $F6_{16}$
     $(-)3A_{16}$

6. $B03_{16}$
     $(-)8AE_{16}$

Answers to Self-Test 4-3 are at the end of the chapter.

END OF CHAPTER PROBLEMS 4-1

Convert the following binary numbers to decimal numbers:

1. 111_2

2. 110_2

3. 1010_2

4. 1011_2

5. 1101_2

6. 1000_2

7. 10111_2

8. 11100_2

9. 100110_2

10. 110011_2

11. 111000_2

12. 101011_2

13. 1100101_2

14. 1000111_2

15. 1110111_2

16. 1111111_2

17. 11110000_2

18. 10101010_2

19. 11000011_2

20. 10011001_2

Convert the following decimal numbers to binary numbers:

21. 5

22. 6

23. 12

24. 15

25. 21	**26.** 31
27. 45	**28.** 53
29. 59	**30.** 62
31. 68	**32.** 80
33. 96	**34.** 115
35. 135	**36.** 160
37. 210	**38.** 245

END OF CHAPTER PROBLEMS 4-2

Convert the following numbers from octal to decimal:

1. 14_8	**2.** 27_8
3. 77_8	**4.** 100_8
5. 276_8	**6.** 667_8
7. 4176_8	**8.** 11714_8

Convert the following numbers from decimal to octal:

9. 20_{10}	**10.** 42_{10}
11. 80_{10}	**12.** 98_{10}
13. 360_{10}	**14.** 517_{10}
15. 1417_{10}	**16.** 5716_{10}

END OF CHAPTER PROBLEMS 4-3

Convert the following numbers from hexadecimal to decimal:

1. $1A_{16}$	**2.** $A1_{16}$
3. $4C_{16}$	**4.** $1BC_{16}$
5. 200_{16}	**6.** 400_{16}
7. $11AA_{16}$	**8.** $FACE_{16}$

Convert the following numbers from decimal to hexadecimal:

9. 22_{10}	**10.** 40_{10}
11. 97_{10}	**12.** 127_{10}
13. 512_{10}	**14.** 2700_{10}
15. 6075_{10}	**16.** 8080_{10}

END OF CHAPTER PROBLEMS 4-4

1. $75_8 =$ _____ $_2 =$ _____ $_{10} =$ _____ $_{16}$

2. $103_8 =$ _____ $_2 =$ _____ $_{10} =$ _____ $_{16}$

3. $140_8 =$ _____ $_2 =$ _____ $_{10} =$ _____ $_{16}$

4. $175_8 =$ _____ $_2 =$ _____ $_{10} =$ _____ $_{16}$

5. $2A_{16} =$ _____ $_2 =$ _____ $_8 =$ _____ $_{10}$

6. $6C_{16} =$ _____ $_2 =$ _____ $_8 =$ _____ $_{10}$

7. $1A7_{16} =$ _____ $_2 =$ _____ $_8 =$ _____ $_{10}$

8. $2FB_{16} =$ _____ $_2 =$ _____ $_8 =$ _____ $_{10}$

9. $100110_2 =$ _____ $_8 =$ _____ $_{10} =$ _____ $_{16}$

10. $110101_2 =$ _____ $_8 =$ _____ $_{10} =$ _____ $_{16}$

11. $111000101_2 =$ _____ $_8 =$ _____ $_{10} =$ _____ $_{16}$

12. $101111011111_2 =$ _____ $_8 =$ _____ $_{10} =$ _____ $_{16}$

13. $10_{10} =$ _____ $_2 =$ _____ $_8 =$ _____ $_{16}$

14. $100_{10} =$ _____ $_2 =$ _____ $_8 =$ _____ $_{16}$

15. $290_{10} =$ _____ $_2 =$ _____ $_8 =$ _____ $_{16}$

16. $413_{10} =$ _____ $_2$ _____ $_8 =$ _____ $_{16}$

END OF CHAPTER PROBLEMS 4-5

Add the following octal numbers:

1. 123
   234

2. 451
   116

3. 456
   317

4. 517
   116

5. 667
   107

6. 273
   606

7. 604
   617

8. 176
   412

9. 726
   161

10. 471
    707

Add the following hexadecimal numbers:

11. 4A
    A3

12. B2
    1C

13. A07
    29E

14. B18
    398

15. 4AB
    8AB

16. C17
    8B9

17. ABF
    CDE

✎18. FED
    CBA

19. F0C6
    E9A6

20. BA09
    E6E4

Add the following binary numbers:

21. 1100
    1001

22. 1001
    1111

23. 11100
    10111

24. 10111
    11101

✓25. 10011
    10110

✐26. 10101
    11101

27. 110011
    101111

28. 111000
    011100

29. 101110
    111011

30. 111101
    100111

END OF CHAPTER PROBLEMS 4-6

Subtract the following octal numbers.

1.   73
   (−)26

2.   64
   (−)40

3.   616
   (−)554

4.   416
   (−)267

5.   540
   (−)273

✐6.   405
   (−)167

7. 2004
   (−)1445

✓8. 3001
   (−)2563

9. 6334
   (−)4617

10. 5503
   (−)3473

Subtract the following hexadecimal numbers:

11.   A6
   (−)8C

12.   CA
   (−)8B

13.   E06
   (−)A25

✐14.   C0A
   (−)149

✓15.   9A0
   (−)3AF

16.   8B4
   (−)3C6

17.   C00
      (−)103

18.   B00
      (−)608

19.   B029
      (−)A04D

20.   C24B
      (−)174C

Subtract the following binary numbers:

21.   1011
      (−)0110

22.   1101
      (−)0110

23.   1101
      (−)1011

24.   1001
      (−)0110

25.   11001
      (−)10111

26.   111011
      (−)011101

27.   10010
      (−)01111

28.   110010
      (−)100111

29.   110011
      (−)011101

30.   111001
      (−)010110

END OF CHAPTER PROBLEMS 4-7

Find the two's complement of:

1. 1010_2

2. 11001_2

3. 101101_2

4. 11000010_2

Find the eight's complement of:

5. 64_8

6. 526_8

7. 635_8

8. 2723_8

Find the sixteen's complement of:

9. $A3_{16}$

10. $3B4_{16}$

11. $C1E_{16}$

12. $3F03_{16}$

Use the one's complement method to find:

13. 11001_2
      $(−)01100_2$

14. 11100_2
      $(−)10011_2$

15. 101101_2
      $(−)001010_2$

16. 1101001_2
      $(−)0010110_2$

Use the two's complement method to find:

17. 10110_2
      $(−)01011_2$

18. 10011_2
      $(−)00100_2$

19. 1011001_2
      $(−)0010110_2$

20. 11100101_2
      $(−)10011010_2$

Use the eight's complement method to find:

21. 63_8
$(-)25_8$

22. 503_8
$(-)127_8$

23. 626_8
$(-)377_8$

24. 6026_8
$(-)4562_8$

Use the sixteen's complement method to find:

25. $A2_{16}$
$(-)8C_{16}$

26. $E4D_{16}$
$(-)BCA_{16}$

27. $6A2_{16}$
$(-)4FF_{16}$

28. $F40B_{16}$
$(-)2AEC_{16}$

ANSWERS TO SELF-TESTS

Self-Test 4-1

1. 38

2. 190

3. 246

4. 3130

5. 91

6. 202

7. (a) 1010_2
(b) 12_8
(c) A_{16}

8. (a) 11100_2
(b) 34_8
(c) $1C_{16}$

9. (a) 10111011_2
(b) 273_8
(c) BB_{16}

10. (a) 1001110001_2
(b) 1161_8
(c) 271_{16}

11. (a) 1011101_2
(b) 135_8
(c) $5D_{16}$

12. (a) 111000011010_2
(b) 7032_8
(c) 3610_{10}

13. (a) 3226_8
(b) 1686_{10}
(c) 696_{16}

14. (a) 101111011_2
(b) 379_{10}
(c) $17B_{16}$

Self-Test 4-2

1. $E5_{16}$

2. $DC8_{16}$

3. $1F323_{16}$

4. 161_8

5. 124_8

6. 141_8

7. 1100_2

8. 11000_2

9. 110100_2

Self-Test 4-3

1. 10011110

2. 101011_2

3. 25_8

4. 217_8

5. BC_{16}

6. 255_{16}

5

ALGEBRAIC TERMS. ROOTS AND POWERS

In order to communicate with another person, we must be able to speak the same language. Further, we must have the same words in our vocabulary so that when these words are used, we understand their meaning. When we move into areas new to us, whether that area is math or electronics or some other technology, we must learn new words. In this chapter we are going to define some of the words used in mathematics to describe or define mathematical operations or processes.

5.1 LITERAL AND REAL NUMBERS

We may divide numbers into two groups. The first group we can define as *real* numbers. These are the numbers we use all the time: 7, 468, 3.25, -6, and $2\frac{1}{2}$ are all examples of real numbers. Real numbers can be either positive numbers or negative numbers. *Literal* numbers are letters that are used to represent numbers or quantities. Literal numbers are used in electronics equations to represent quantities such as current, voltage, and resistance. $E = IR$ is Ohm's law written as an equation where E represents the energy source, I represents circuit current, and R represents circuit resistance. Real numbers always have the same value. The number 6 always has a value of 6. It can never mean some other amount. Literal numbers can represent different values. The letter E can stand for some value in one circuit and an entirely different value in another circuit. E might equal 10 volts in one circuit, 25 volts in a second circuit, and so on. Because literal numbers have no constant value, they are also referred to as *variable* numbers.

5.2 ALGEBRAIC EXPRESSIONS AND TERMS

An algebraic expression is a number or a group of numbers. The numbers can be real numbers, literal numbers, or both. $6A, 3+2x, 7(x-2), 1.6+7x, 3ab+4c-5ac$ are all examples of algebraic expressions. If we knew the value of the literal numbers, then we could determine the numerical value of the expression.

When numbers are separated into groups with $+$ or $-$ signs, each group in the expression is called a *term*. The algebraic expression $3ab+4c-5ac$ contains three terms: $3ab$, $4c$, and $5ac$. $6A$ is one term. The expression $1.6+7x$ contains two terms: 1.6 and $7x$.

When an algebraic expression contains one term, it is called a *monomial*. Where an expression contains more than one term, it is called a *polynomial*. Two-term expressions are also called *binomials*. Three-term expressions are also called *trinomials*. These words are more descriptive than just *polynomials*.

5.3 NUMERICAL COEFFICIENTS

The *numerical coefficient* of an algebraic term is the real number part of the term. If there is no numerical value given, it is understood to be 1. In the term $6xy$, 6 is the numerical coefficient. In the term ax, 1 is the coefficient.

PRACTICE PROBLEMS 5-1   List the number of terms in the following expressions. Also indicate the kind of expression (monomial, binomial, etc.).

1. $2a+3y$

2. $2x+y+1$

3. $4(x+1)-3$

4. $\dfrac{3a}{8}+26$

5. $\dfrac{4x}{5}+1$

6. $3(a-1)+2b$

7. $\dfrac{3a-1}{4}+6$

8. $\dfrac{4c+3}{2d}-b$

9. $\dfrac{7y}{8}+6-4(x-3)$

10. $3a+6b-4c-3$

11. $4c-3a+4$

12. $\dfrac{4x-3y}{4}-6x+3$

Solutions:

1. 2, binomial

2. 3, trinomial

3. 2, binomial

4. 2, binomial

5. 2, binomial

6. 2, binomial

7. 2, binomial

8. 2, binomial

9. 3, trinomial

10. 4, polynomial

11. 3, trinomial

12. 3, trinomial

The binomials and trinomials may also be called polynomials.

5.4 EXPONENTS

In Chapter 2 we discussed the exponent whose base was 10. The exponent told us how many times 10 was multiplied by itself. This is the function of an exponent in any base. The base may be a real number or a literal number. $5^3 = 5 \cdot 5 \cdot 5$. $x^2 = x \cdot x$. We can find the answer to 5^3 directly: $5^3 = 125$. We can find x^2 only if we are given the value of x. If $x = 4$, $x^2 = 4^2 = 16$. If $x = 6$, $x^2 = 6^2 = 36$.

PRACTICE PROBLEMS 5-2   Solve the following problems. Round your answers to three significant figures.

1. 7^2

2. 6^3

3. 4.5^2

4. 8.73^2

5. 24^2

Solve the following problems where $x = 3, y = 4$:

6. x^3

7. y^2

8. y^3

9. x^2

10. x^4

Solutions:

1. 49.0

2. 216

3. 20.3

4. 76.2

5. 576

6. 27.0

7. 16.0

8. 64.0

9. 9.00

10. 81.0

Additional practice problems are at the end of the chapter.

When real numbers and literal numbers appear in the same term of an algebraic expression, we can determine the value of the term if we know the value of the literal numbers. For example, in the term $6x$, if $x = 3$, then $6x^2 = 6 \cdot x^2 = 6 \cdot 3^2 = 6 \cdot 9 = 54$. When one or more of the numbers in a term is raised to a power, only that number is affected. When 3 is substituted for x in our example, we get $6 \cdot 3^2$. Our first step is to find 3^2. Then we complete the multiplication. We can find the value of any term by multiplying the real numbers times the known values of the literal numbers.

EXAMPLE 5.1: Let $x=3$ and $y=4$. Find: (a) xy (b) x^2y (c) x^2y^2 (d) $4x^3y^2$.

Solution:

(a) $xy = 3 \cdot 4 = 12$
(b) $x^2y = 3^2 \cdot 4 = 9 \cdot 4 = 36$
(c) $x^2y^2 = 3^2 \cdot 4^2 = 9 \cdot 16 = 144$
(d) $4x^3y^2 = 4 \cdot 3^3 \cdot 4^2 = 4 \cdot 27 \cdot 16 = 1{,}728$

If a sign of grouping is used, then all numbers within that group are acted on by the exponent. For example, in the term $(ab)^2$, both a and b are raised to the second power: $(ab)^2 = a^2 \cdot b^2$.

EXAMPLE 5.2: Find $(ab)^2$ where $a=3$ and $b=4$.

Solution:

$(ab)^2 = a^2 \cdot b^2 = 3^2 \cdot 4^2 = 9 \cdot 16 = 144$
It would also be correct to find the product of a and b and then raise that value to the second power.

$$(ab)^2 = (3 \cdot 4)^2 = 12^2 = 144$$

Either method is correct and either method may be used in problem solving.

If there is more than one term in an algebraic expression, we find the value of each term and then perform the indicated addition or subtraction.

EXAMPLE 5.3: Find $3a^2b - ab^2$ where $a=5$ and $b=6$.

Solution:

$$3a^2b - ab^2 = 3 \cdot 5^2 \cdot 6 - 5 \cdot 6^2$$
$$= 3 \cdot 25 \cdot 6 - 5 \cdot 36$$
$$= 450 - 180 = 270$$

PRACTICE PROBLEMS 5-3   Find the values of the following algebraic expressions where $x=3, y=4, z=5$:

1. xy

2. x^2y

3. $(xy)^2$

4. $2y^2z$

5. $3(xz)^2$

6. $x^3y - z$

7. $y^3 - z^2$

8. $(x^2y - z^2)^2$

Solutions:

1. 12

2. 36

3. 144

4. 160

5. 675

6. 103

7. 39

8. 121

Additional practice problems are at the end of the chapter.

5.5 ROOTS

The *square root* of a number is one of its two equal factors. $6 \cdot 6 = 36$. The two equal factors are 6; therefore, 6 is the square root of 36. Actually, there are *two* square roots of 36 because $(-6)(-6)$ also equals 36. All positive numbers have two square roots: one negative and one positive. Usually, in solving electrical problems, the negative root is not considered.

The symbol used to denote the square root of a number is $\sqrt{}$ and is called the *radical*. $\sqrt{36} = 6$. Raising a number to the one-half power is another way of denoting square root. $36^{1/2} = 6$. Both notations will be used in this text.

We can extract from memory the square root of numbers like 25, 36, 64, and 81. If we go much beyond $\sqrt{100}$, most of us have to use the calculator. To find the square root of a number by using the calculator, we punch in the number and then the $\sqrt{}$ key. We can check our answer by squaring it.

EXAMPLE 5.4: Find $\sqrt{40}$.

Solution:

$$\sqrt{40} = 6.32 \qquad \text{(accurate to three places)}$$
$$\text{Check: } 6.32 \cdot 6.32 \simeq 40$$

Because we rounded to three places we don't get exactly 40 when we check.

The cube root of a number is one of its three equal parts. The symbol is $\sqrt[3]{}$. Raising a number to the one-third power is another way of indicating cube root. $\sqrt[3]{x} = x^{1/3}$. Because we seldom work with cube roots or higher roots in electronics, we will limit problem solving in this chapter to square roots.

PRACTICE PROBLEMS 5-4   Solve the following problems accurate to three places:

1. $5^{1/2}$

2. $7^{1/2}$

3. $27^{1/2}$

4. $256^{1/2}$

5. $512^{1/2}$

Solutions:

1. 2.24   **2.** 2.65

3. 5.20   **4.** 16

5. 22.6

Additional practice problems are at the end of the chapter.

We can find the square root of literal numbers if we know their values. When we extract the square root of an expression, we perform all mathematical functions under the radical *first*. Then we find the square root.

> EXAMPLE 5.5: Find the value of the following algebraic expressions where $x = 4$ and $y = 5$:
>
> (a) $\sqrt{2x + 3y}$ (b) $\sqrt{3xy^2}$ (c) $\sqrt{x^2y^2} + 4$
>
> *Solution:* (a) $\sqrt{2x + 3y} = \sqrt{8 + 15} = \sqrt{23} = 4.80$
>
> (b) $\sqrt{3xy^2} = \sqrt{3 \cdot 4 \cdot 5^2} = \sqrt{12 \cdot 25} = \sqrt{300} = 17.3$
>
> (c) $\sqrt{x^2y^2} + 4 = \sqrt{4^2 \cdot 5^2} + 4 = \sqrt{16 \cdot 25} + 4$
> $= \sqrt{400} + 4 = 20 + 4 = 24$

PRACTICE PROBLEMS 5-5   Find the value of the following algebraic expressions where $x = 3, y = 4, z = 5$.

1. $\sqrt{xy}$ **2.** $\sqrt{z^2}$

3. $\sqrt{y^2z^2}$ **4.** $\sqrt{5x^2z}$

5. $\sqrt{x^2y^2}$ **6.** $\sqrt{4y^2 + z^2}$

7. $\sqrt{x^4 - 8y}$

Solutions:

1. 3.46   **2.** 5

3. 20   **4.** 15

5. 12   **6.** 9.43

7. 7

Additional practice problems are at the end of the chapter.

SELF-TEST 5-1   Find the value of the following expressions where $x = 3, y = 5, z = 4$. Consider only the positive root.

1. $x^2 y$

2. $\sqrt{xz}$

3. $(2xy)^2$

4. $\sqrt{3xy^2}$

5. $xy + 2z$

6. $3y^2 - x^2 z$

7. $\sqrt{3x^3 + 2yz}$

8. $\sqrt{3xy + yz^2}$

9. $\sqrt{2xz^2 + 3y^2 z}$

10. $\sqrt{4y^2 z - 2xz^2}$

   Answers to Self-Test 5-1 are at the end of the chapter.

5.6   PRACTICAL APPLICATIONS

Literal numbers are used in equations to represent the relationship between variables in electronics circuits. In this section we will use some of these equations and solve for circuit variables. Where square roots are found, only the positive root is considered.

One form of Ohm's law is $E = IR$. E is the applied voltage and is measured in volts; I is the circuit current and is measured in amperes; and R is the circuit resistance and is measured in ohms.

EXAMPLE 5.6: Find E when $I = 5$ mA and $R = 2.7$ kΩ.

Solution: Recall from Chapter 3 that the prefix milli (m) is 10^{-3} and kilo (k) is 10^3. Then

$$E = IR = 5 \text{ mA} \times 2.7 \text{ k}\Omega = 5 \times 10^{-3} \times 2.7 \times 10^3 = 13.5 \text{ V}$$

The power dissipated in a circuit can be found by using the equation $P = I^2 R$. Power is measured in watts.

EXAMPLE 5.7: Find P when $I = 50$ mA and $R = 220$ Ω.

Solution:

$$P = I^2 R = (50 \text{ mA})^2 \times 220 \ \Omega = (50 \times 10^{-3})^2 \times 220 = 0.55 \text{ W}$$

When the power dissipated and the resistance are known, circuit current may be found by using the equation $I = \sqrt{\dfrac{P}{R}}$

EXAMPLE 5.8: Find I when $P = 700$ mW and $R = 3.3$ kΩ.

Solution:

$$I = \sqrt{\frac{P}{R}} = \sqrt{\frac{700 \text{ mW}}{3.3 \text{ k}\Omega}} = \sqrt{\frac{700 \times 10^{-3}}{3.3 \times 10^3}} = 1.46 \times 10^{-2} \text{ A}$$

$$= 14.6 \times 10^{-3} \text{ A} = 14.6 \text{ mA}$$

Prefixes are changed to their powers of 10 values to solve problems but it is standard practice to use prefixes instead of powers of ten in answers.

The applied voltage may be found when P and R are known by using the equation $E = \sqrt{PR}$.

EXAMPLE 5.9: Find E when $P = 100$ mW and $R = 10$ kΩ.

Solution:

$$E = \sqrt{PR} = \sqrt{100 \text{ mW} \times 10 \text{ k}\Omega}$$
$$= \sqrt{100 \times 10^{-3} \times 10 \times 10^3} = 31.6 \text{ V}$$

P can be found when E and R are known by using the equation $P = \dfrac{E^2}{R}$.

EXAMPLE 5.10: Find P when $E = 15$ V and $R = 2$ kΩ.

Solution:

$$P = \frac{E^2}{R} = \frac{15^2}{2 \text{ k}\Omega} = \frac{15^2}{2 \times 10^3} = 0.113 \text{ W}$$

Algorithms for solving problems dealing with powers of 10 are shown in Appendix A.

PRACTICE PROBLEMS 5-6 $E = IR$. Find E when:

1. $I = 2$ mA, $R = 2.7$ kΩ   **2.** $I = 4.3$ mA, $R = 3.3$ kΩ

3. $I = 400$ μA, $R = 22$ kΩ   **4.** $I = 40$ mA, $R = 680$ Ω

5. $I = 5.6$ mA, $R = 4.7$ kΩ

$P = I^2 R$. Find P when:

6. $I = 40$ mA, $R = 1$ kΩ   **7.** $I = 170$ mA, $R = 470$ Ω

8. $I = 2.3$ A, $R = 33$ Ω   **9.** $I = 1.76$ mA, $R = 27$ kΩ

10. $I = 400$ μA, $R = 56$ kΩ

$I = \sqrt{\dfrac{P}{R}}$. Find I when:

11. $P = 250$ mW, $R = 120$ Ω   **12.** $P = 1.2$ W, $R = 2.2$ kΩ

13. $P = 37.3$ mW, $R = 680$ Ω   **14.** $P = 4.73$ mW, $R = 12$ kΩ

15. $P = 680$ mW, $R = 1.8$ kΩ

$E = \sqrt{PR}$. Find E when:

16. $P = 56$ mW, $R = 2$ kΩ   **17.** $P = 475$ mW, $R = 100$ Ω

18. $P = 2.7$ W, $R = 150$ Ω   **19.** $P = 780$ mW, $R = 2.7$ kΩ

20. $P = 30$ mW, $R = 3.9$ kΩ

$P = \dfrac{E^2}{R}$. Find P when:

21. $E = 12$ V, $R = 470$ Ω   **22.** $E = 9$ V, $R = 1.8$ kΩ

23. $E = 20$ V, $R = 4.7$ kΩ   **24.** $E = 30$ V, $R = 120$ Ω

25. $E = 15$ V, $R = 12$ kΩ

Solutions:

1. 5.4 V	**2.** 14.2 V
3. 8.8 V	**4.** 27.2 V
5. 26.3 V	**6.** 1.6 W
7. 13.6 W	**8.** 175 W
9. 83.6 mW	**10.** 8.96 mW
11. 45.6 mA	**12.** 23.4 mA
13. 7.41 mA	**14.** 628 μA
15. 19.4 mA	**16.** 10.6 V
17. 6.89 V	**18.** 20.1 V
19. 45.9 V	**20.** 10.8 V
21. 306 mW	**22.** 45 mW
23. 85.1 mW	**24.** 7.5 W
25. 18.8 mW	

Additional practice problems are at the end of the chapter.

Capacitive reactance in an AC circuit is measured in ohms. The equation is:

$$X_C = \frac{1}{2\pi f C}$$

where π (the Greek letter pi) is a constant and equals 3.14 accurate to three places, f is the frequency in hertz (Hz), and C is the capacitance.

EXAMPLE 5.11: Find the capacitive reactance (X_C) when $f = 7.5$ kHz and $C = 0.5$ μF.

Solution:

$$X_C = \frac{1}{2\pi f C} = \frac{1}{2 \times \pi \times 7.5 \text{ kHz} \times 0.5 \text{ μF}} = 42.4 \ \Omega$$

An algorithm for solving this kind of problem by using a calculator is in Appendix A. Most calculators have a π key so that it is not necessary to key in the value.

An important parameter in AC circuits is the resonant frequency. This frequency is found by using the following equation:

$$f_r = \frac{1}{2\pi\sqrt{LC}}$$

where L is the circuit inductance and C is the circuit capacitance.

EXAMPLE 5.12: Find the resonant frequency when $L = 200$ mH and $C = 300$ pF.

Solution:

$$f_0 = \frac{1}{2\pi\sqrt{LC}} = \frac{1}{2 \times \pi \times \sqrt{200 \text{ mH} \times 300 \text{ pF}}}$$

$$= \frac{1}{6.28 \times \sqrt{6 \times 10^{-11}}} = \frac{1}{6.28 \times 7.75 \times 10^{-6}}$$

$$= \frac{1}{4.87 \times 10^{-5}} = 2.05 \times 10^4 \text{ Hz} = 20.5 \text{ kHz}$$

Always use prefixes instead of powers of ten form in your final answers. An algorithm for solving this kind of problem is in Appendix A.

Another important parameter in AC circuits is *impedance*. Impedance is measured in ohms. One method used to find impedance when both resistance and reactance are known is:

$$Z = \sqrt{R^2 + X^2}$$

where R is the circuit resistance and X is the circuit reactance.

EXAMPLE 5.13. Find Z when $R = 5.6$ kΩ and $X = 4$ kΩ.

Solution:

$$Z = \sqrt{R^2 + X^2} = \sqrt{(5.6 \text{ k}\Omega)^2 + (4 \text{ k}\Omega)^2} = \sqrt{(5.6 \times 10^3)^2 + (4 \times 10^3)^2}$$

$$= \sqrt{3.14 \times 10^7 + 1.6 \times 10^7} = \sqrt{4.74 \times 10^7} = 6.88 \times 10^3 \ \Omega = 6.88 \text{ k}\Omega$$

Remember, when adding numbers the exponents to the base 10 must be the same. (Of course, when using the calculator to perform the addition, no changes need to be made.) An algorithm for solving this kind of problem is in Appendix A.

PRACTICE PROBLEMS 5-7 $X_C = \dfrac{1}{2\pi f C}$. Find X_C when:

1. $f = 120$ Hz, $C = 2 \ \mu$F

2. $f = 2.5$ kHz, $C = 300$ nF

3. $f = 420$ Hz, $C = 0.5 \ \mu$F

4. $f = 60$ Hz, $C = 10 \ \mu$F

5. $f = 1.2$ kHz, $C = 50$ nF

$f_r = \dfrac{1}{2\pi\sqrt{LC}}$. Find f_r when:

6. $L = 300$ mH, $C = 100$ nF

7. $L = 500$ mH, $C = 470$ pF

8. $L=200$ mH, $C=2$ μF   **9.** $L=1.7$ H, $C=20$ nF

10. $L=37$ mH, $C=75$ nF

$Z=\sqrt{R^2+X^2}$. Find Z when:

11. $R=1.8$ kΩ, $X=3$ kΩ **12.** $R=27$ kΩ, $X=50$ kΩ

13. $R=1$ kΩ, $X=600$ Ω **14.** $R=4.7$ kΩ, $X=3.1$ kΩ

15. $R=12$ kΩ, $X=16$ kΩ

Solutions:

1. 663 Ω **2.** 212 Ω

3. 758 Ω **4.** 265 Ω

5. 2.65 kΩ **6.** 919 Hz

7. 10.4 kHz   **8.** 252 Hz

9. 863 Hz   **10.** 3.02 kHz

11. 3.5 kΩ **12.** 56.8 kΩ

13. 1.17 kΩ **14.** 5.63 kΩ

15. 20 kΩ

SELF-TEST 5-2

1. $E=IR$. Find E if $I=7.5$ mA and $R=2.7$ kΩ.   **2.** $P=I^2R$. Find P if $I=27$ mA and $R=150$ Ω.

3. $I=\sqrt{\dfrac{P}{R}}$. Find I when $P=750$ mW and $R=3.3$ kΩ.   **4.** $E=\sqrt{PR}$. Find E when $P=60$ mW and $R=6.8$ kΩ.

5. $P=\dfrac{E^2}{R}$. Find P when $E=15$ V and $R=1.8$ kΩ.   **6.** $X_C=\dfrac{1}{2\pi fC}$. Find X_C when $f=3.2$ kHz and $C=150$ nF.

7. $f_r=\dfrac{1}{2\pi\sqrt{LC}}$. Find f_r when $L=700$ mH and $C=470$ pF.   **8.** $Z=\sqrt{R^2+X^2}$. Find Z when $R=5.6$ kΩ and $X=8$ kΩ.

Answers to Self-Test 5-2 are at the end of the chapter.

END OF CHAPTER PROBLEMS 5-1

Solve the following problems:

1. 2^5 **2.** 3^3

3. 8^3 **4.** 3.6^3

5. 18^2 **6.** 16^2

Solve the following problems where $a = 5$, $b = 3$, $c = 6$:

7. a^2 **8.** b^2

9. b^3 **10.** a^3

11. ac **12.** $b^2 c$

13. $a^2 c$ **14.** $(bc)^2$

15. $(ac)^2$ **16.** $a^3 - b^3$

17. $2(ab)^2$ **18.** $3(bc)^2$

19. $a^2 b^2 - 3c^2$ **20.** $abc^2 - b^2 c^2$

21. $ac^2 - 2abc$ **22.** $b^2 c + a^2$

23. $(a^2 b - 2c^2)^2$ **24.** $(2a^2 b - 4c^2)^2$

END OF CHAPTER PROBLEMS 5-2

Solve the following problems:

1. $30^{1/2}$ **2.** $18^{1/2}$

3. $40^{1/2}$ **4.** $\sqrt{80}$

5. $\sqrt{200}$ **6.** $\sqrt{625}$

Solve the following problems where $x = 3$, $y = 4$, $z = 5$.

7. $\sqrt{yz}$ **8.** $\sqrt{xz}$

9. $\sqrt{y^2}$ **10.** $\sqrt{3y^2 z}$

11. $\sqrt{6xz^2}$ **12.** $\sqrt{3x^2 + y^2}$

13. $\sqrt{3y^2 + x^2}$ **14.** $\sqrt{x^2 z^2} + y$

15. $\sqrt{y^2 z^2} + x$ **16.** $\sqrt{x^3 + 2z}$

END OF CHAPTER PROBLEMS 5-3

$E = IR$. Find E when:

1. $I = 660$ μA, $R = 6.8$ kΩ   **2.** $I = 1.73$ mA, $R = 12$ kΩ

3. $I = 3.7$ mA, $R = 18$ kΩ   **4.** $I = 3.7$ mA, $R = 1.8$ kΩ

5. $I = 50$ mA, $R = 2.7$ kΩ

$P = I^2 R$. Find P when:

6. $I = 760$ μA, $R = 1.8$ kΩ   **7.** $I = 7.5$ mA, $R = 680$ Ω

8. $I = 2$ mA, $R = 120$ Ω   **9.** $I = 36.5$ mA, $R = 3.3$ kΩ

10. $I = 620$ μA, $R = 1.5$ kΩ   **11.** $I = 1.2$ A, $R = 5$ Ω

$I = \sqrt{\dfrac{P}{R}}$. Find I when:

12. $P = 65.4$ mW, $R = 5.6$ kΩ

13. $P = 100$ mW, $R = 1$ kΩ

14. $P = 50$ mW, $R = 18$ kΩ

15. $P = 500$ mW, $R = 330$ Ω

16. $P = 4.6$ W, $R = 10$ Ω

17. $P = 25$ W, $R = 8$ Ω

$E = \sqrt{PR}$. Find E when:

18. $P = 8.3$ mW, $R = 27$ kΩ

19. $P = 78.3$ mW, $R = 5.6$ kΩ

20. $P = 130$ mW, $R = 750$ Ω

21. $P = 500$ mW, $R = 1.8$ kΩ

22. $P = 18$ mW, $R = 6.8$ kΩ

23. $P = 1.2$ W, $R = 220$ Ω

$P = \dfrac{E^2}{R}$. Find P when:

24. $E = 16$ V, $R = 6.8$ kΩ

25. $E = 4.7$ V, $R = 560$ Ω

26. $E = 13.3$ V, $R = 33$ kΩ

27. $E = 8.73$ V, $R = 2.7$ kΩ

28. $E = 18.2$ V, $R = 10$ kΩ

29. $E = 25$ V, $R = 1.5$ kΩ

END OF CHAPTER PROBLEMS 5-4

$X_C = \dfrac{1}{2\pi fC}$. Find X_C when:

1. $f = 5$ kHz, $C = 100$ nF

2. $f = 10$ kHz, $C = 1$ μF

3. $f = 200$ Hz, $C = 200$ nF

4. $f = 15$ kHz, $C = 470$ pF

5. $f = 6.3$ kHz, $C = 2$ μF

6. $f = 500$ Hz, $C = 500$ nF

$f_r = \dfrac{1}{2\pi\sqrt{LC}}$. Find f_r when:

7. $L = 80$ mH, $C = 150$ nF

8. $L = 1$ mH, $C = 10$ μF

9. $L = 150$ μH, $C = 50$ nF

10. $L = 6.5$ mH, $C = 600$ nF

11. $L = 50$ mH, $C = 250$ nF

12. $L = 300$ mH, $C = 50$ nF

$Z = \sqrt{R^2 + X^2}$. Find Z when:

13. $R = 120$ kΩ, $X = 80$ kΩ

14. $R = 100$ Ω, $X = 100$ Ω

15. $R = 200$ Ω, $X = 100$ Ω

16. $R = 3$ kΩ, $X = 4$ kΩ

17. $R = 33$ kΩ, $X = 20$ kΩ

18. $R = 27$ kΩ, $X = 20$ kΩ

ANSWERS TO SELF-TESTS

Self-Test 5-1

1. 45

2. 3.46

3. 900

4. 15

5. 23

7. 11

9. 19.9

6. 39

8. 11.2

10. 17.4

Self-Test 5-2

1. 20.3 V

3. 15.1 mA

5. 125 mW

7. 8.77 kHz

2. 109 mW

4. 20.2 V

6. 332 Ω

8. 9.77 kΩ

FRACTIONS

6.1 PRIME NUMBERS

A number that has no whole-number factors except 1 and itself is called a prime number. A number that is not prime is divisible by more than one prime number. The prime numbers that make up a non-prime number are called *prime factors*. In determining prime factors, we ignore 1 as a factor.

Some examples of prime numbers are:

$$1, 2, 3, 5, 7, 11, 13, 17, 19, 23, 29, 31, 37, 41, 43, 47$$

Let's look at some non-prime numbers and their prime factors:

$$10 = 2 \cdot 5$$
$$15 = 3 \cdot 5$$
$$21 = 3 \cdot 7$$
$$26 = 2 \cdot 13$$

Quite often a prime factor appears more than once as a factor:

$$8 = 2 \cdot 2 \cdot 2 = 2^3$$
$$9 = 3 \cdot 3 = 3^2$$
$$36 = 2 \cdot 2 \cdot 3 \cdot 3 = 2^2 \cdot 3^2$$
$$150 = 2 \cdot 3 \cdot 5 \cdot 5 = 2 \cdot 3 \cdot 5^2$$

When we found the prime factors of 8, the answer was $2 \cdot 2 \cdot 2$ or 2^3. Either form is acceptable.

PRACTICE PROBLEMS 6-1   Find the prime factors of:

1. 12	**2.** 24
3. 39	**4.** 45
5. 70	**6.** 72
7. 100	**8.** 120
9. 210	**10.** 300

Solutions:

1. $2 \cdot 2 \cdot 3$	**2.** $2 \cdot 2 \cdot 2 \cdot 3$
3. $3 \cdot 13$	**4.** $3 \cdot 3 \cdot 5$
5. $2 \cdot 5 \cdot 7$	**6.** $2^3 \cdot 3^2$
7. $2^2 \cdot 5^2$	**8.** $2^3 \cdot 3 \cdot 5$
9. $2 \cdot 3 \cdot 5 \cdot 7$	**10.** $2^2 \cdot 3 \cdot 5^2$

Additional practice problems are at the end of the chapter.

Literal numbers are prime when they are raised to the first power. x is prime. x^2 is not prime because $x^2 = x \cdot x$. The prime factors of x^3y^2 are $x \cdot x \cdot x \cdot y \cdot y \cdot$. Real numbers and literal numbers often appear in the same term. Let's factor some of these numbers.

EXAMPLE 6.1: Factor $40a^2b^3$.

Solution:

$$40a^2b^3 = 2 \cdot 2 \cdot 2 \cdot 5 \cdot a \cdot a \cdot b \cdot b \cdot b$$

EXAMPLE 6.2: Factor $33x^2y^2z$.

Solution:

$$33x^2y^2z = 3 \cdot 11 \cdot x \cdot x \cdot y \cdot y \cdot z$$

PRACTICE PROBLEMS 6-2   Find the prime factors of the following:

1. $6ab^2$	**2.** $8a^2bc^2$
3. $22x^3$	**4.** $38x^2y^2$
5. $30x^2y^3$	**6.** $115z^2$
7. $136x^2z^3$	**8.** $168a^2c^2$
9. $175b^2c^3$	**10.** $200y^3$

Solutions:

1. $2 \cdot 3 \cdot a \cdot b \cdot b$
2. $2 \cdot 2 \cdot 2 \cdot a \cdot a \cdot b \cdot c \cdot c$
3. $2 \cdot 11 \cdot x \cdot x \cdot x$
4. $2 \cdot 19 \cdot x \cdot x \cdot y \cdot y$
5. $2 \cdot 3 \cdot 5 \cdot x \cdot x \cdot y \cdot y \cdot y$
6. $5 \cdot 23 \cdot z \cdot z$
7. $2 \cdot 2 \cdot 2 \cdot 17 \cdot x \cdot x \cdot z \cdot z \cdot z$
8. $2 \cdot 2 \cdot 2 \cdot 3 \cdot 7 \cdot a \cdot a \cdot c \cdot c$
9. $5 \cdot 5 \cdot 7 \cdot b \cdot b \cdot c \cdot c \cdot c$
10. $2 \cdot 2 \cdot 2 \cdot 5 \cdot 5 \cdot y \cdot y \cdot y$

Additional practice problems are at the end of the chapter.

6.2 LOWEST COMMON MULTIPLE

In dividing monomials or polynomials, or in working with fractions, it is often necessary for us to find the prime factors of two or more terms. Once the prime factors are found, we can determine the *lowest common multiple* (LCM). Knowing the LCM, we can reduce the problem to its simplest form.

To find the LCM, the rule is:

1. Factor each term.
2. Determine the *maximum* number of times a prime factor appears in any one term.
3. Multiply the resulting prime factors found in step 2. This number is the LCM.

The lowest common multiple is the *lowest* multiple of each term that is common to all terms. For example, the LCM of $6a$ and $18a^2$ is found by finding the prime factors of each term.

$$6a = 2 \cdot 3 \cdot a$$

$$18a^2 = 2 \cdot 3 \cdot 3 \cdot a \cdot a$$

The maximum number of times 2 appears as a factor is once; therefore, 2 is part of the LCM. Three appears a maximum of two times; therefore, $3 \cdot 3$ or 3^2 is part of the LCM. Since a appears a maximum of two times, $a \cdot a$ or a^2 is part of the LCM. This takes care of all the prime factors. Our next step is to multiply these factors together to get the LCM.

$$2 \cdot 3 \cdot 3 \cdot a \cdot a = 18a^2$$

$18a^2$ is the lowest number that contains both $6a$ and $18a^2$ as factors. That is, $6a$ is a factor of $18a^2$ ($6a \cdot 3a = 18a^2$) and $18a^2$ is a factor ($18a^2 \cdot 1 = 18a^2$). Of course, there are other common multiples: $36a^2$, $36a^3$, and $54a^3$ to name a few, but $18a^2$ is the *lowest*.

EXAMPLE 6.3: Find the LCM of $10xy^2$ and $15x^3y$.

Solution:

$$10xy^2 = 2 \cdot 5 \cdot x \cdot y \cdot y$$

$$15x^3y = 3 \cdot 5 \cdot x \cdot x \cdot x \cdot y$$

Two, three, and five, each appear as factors once. Therefore, 2 and 3 and 5 are part of the LCM. Since x appears a maximum of three times, x^3 is part of the LCM. Since y appears a maximum of two times, y^2 is part of the LCM.

$$LCM = 2 \cdot 3 \cdot 5 \cdot x^3 \cdot y^2$$

$$= 30x^3y^2$$

$30x^3y^2$ is the lowest number that has both $10xy^2$ and $15x^3y$ as factors.

EXAMPLE 6.4: Find the LCM of $14a^2b$, $42abc^2$, and $12b$.

Solution:

$$14a^2b = 2 \cdot 7 \cdot a \cdot a \cdot b$$

$$42abc^2 = 2 \cdot 3 \cdot 7 \cdot a \cdot b \cdot c \cdot c$$

$$12b = 2 \cdot 2 \cdot 3 \cdot b$$

$$LCM = 2 \cdot 2 \cdot 3 \cdot 7 \cdot a \cdot a \cdot b \cdot c \cdot c = 84a^2bc^2$$

$84a^2bc^2$ is the smallest number that has $14a^2b$, $42abc^2$, and $12b$ as factors. That is, $84a^2bc^2$ is the smallest number that is divisible by all three factors.

PRACTICE PROBLEMS 6-3   Find the LCM of the following numbers:

1. $3, 8$
2. $9, 15$
3. $3, 4, 8$
4. $4, 8, 12$
5. $12, 15, 25$
6. a^2b^3, a^3b^2
7. a^2b^4, a^2b^3
8. a^2b^2, a^3b, ab
9. a^3b^2, a^2b^3, a^3b^3
10. $a^3b^3c^3, abc, a^2b^3c^2$
11. $12a^2b, 18ab^2$
12. $3ab^2, 8a^2b^2$
13. $12xyz^2, 30x^3yz^2, 45xyz$
14. $11xz^2, 44x^3y, y^2z$
15. $108xy^4, 72y^4, x^2z$

Solutions:

1. 24
2. 45
3. 24
4. 24

5. 300   **6.** a^3b^3

7. a^2b^4 **8.** a^3b^2

9. a^3b^3 **10.** $a^3b^3c^3$

11. $36a^2b^2$ **12.** $24a^2b^2$

13. $180x^2yz^2$ **14.** $44x^2y^2z^2$

15. $216x^2y^4z$

Additional practice problems are at the end of the chapter.

SELF-TEST 6-1   Find the prime factors of the following numbers:

1. 40   **2.** 48

3. 84   **4.** 180

5. $28xy^2$ **6.** $54a^3b^2$

7. $120x^2z^2$ **8.** $315a^2b^2$

Find the LCM of the following numbers:

9. 12, 15   **10.** 20, 24

11. 10, 21, 30   **12.** 14, 35, 105

13. a^2b, ab^3 **14.** $8xz^2, 9xz$

15. $18a^2b, 24ac^2, 36b^2c$

Answers to Self-Test 6-1 are at the end of the chapter.

6.3   MULTIPLICATION AND DIVISION OF MONOMIALS

6.3.1   Multiplication.   To multiply monomials, we multiply the real numbers and then multiply the literal numbers by using the laws of exponents learned in previous chapters. Recall from arithmetic that each part of a multiplication problem has a name.

$$
\begin{array}{rl}
6a & \text{(multiplicand)} \\
\times 7a & \text{(multiplier)} \\
\hline
42a^2 & \text{(product)}
\end{array}
$$

The product cannot be simplified further because we have not been given a value for a. Remember that we can indicate the operation to be performed, in this case multiplication, in a number of ways. We indicate multiplication with "×" or "·" or "( )" or other signs of grouping.

In the problem $4a^2b \cdot 3ab^3$ we multiply 4×3 and get 12. $a^2 \cdot a = a^{2+1} = a^3$ and $b \times b^3 = b^{1+3} = b^4$. The answer is $12a^3b^4$. With practice, we can do the addition of exponents of like bases in our heads.

In multiplication all the literal numbers contained in the multiplicand and the multiplier must appear in the product unless the exponent is zero.

EXAMPLE 6.5: $3x^3y^{-2}z \times 5x^2y^2z^2 = 15x^5z^3$. If we put like bases together, we can more easily see the result:

$$3 \cdot 5 \cdot x^3 \cdot x^2 \cdot y^{-2} \cdot y^2 \cdot z \cdot z^2 = 15 \cdot x^{3+2} \cdot y^{-2+2} \cdot z^{1+2}$$
$$= 15x^5y^0z^3$$
$$= 15x^5z^3$$

Remember, any number raised to the zero power equals 1.

EXAMPLE 6.6:

$$4x^{-3}yz^{-2} \cdot 2xyz^2 \cdot 3x^{-1}y^{-1}z^2 = 24x^{-3}yz^2.$$
$$4 \times 2 \times 3 = 24$$
$$x^{-3+1-1} = x^{-3}$$
$$y^{1+1-1} = y^1 = y$$
$$z^{-2+2+2} = z^2$$

Multiplying the resulting numbers together gives the product $24x^{-3}yz^2$. Separating the numbers and adding the exponents of like bases may help you see the solution. Don't forget to multiply the parts together to get the final product!

PRACTICE PROBLEMS 6-4   Perform the indicated operations:

1. $2ab \cdot 3a^2b$

2. $4a^3b \cdot 3ab^2$

3. $3a^4b \cdot 3a^{-3}b^{-4}$

4. $6x^{-2}y^4 \cdot 8x^4y^{-1}$

5. $2x^3y^2 \cdot 3x^{-2}y^{-1} \cdot 3x^4y^3$

6. $5x^6y^2z^2 \cdot x^{-2}y^{-4}z^{-3} \cdot x^{-1}y^3$

7. $3x^3y^{-3}z^{-2} \cdot 3x^{-3}y^3 \cdot 3xyz^4$

8. $7x^5y^{-2}z^2 \cdot 3x^{-1}y^5z^4 \cdot 2x^{-1}y^{-1}z^{-1}$

Solutions:

1. $6a^3b^2$

2. $12a^4b^3$

3. $9ab^{-3}$

4. $48x^2y^3$

5. $18x^5y^4$

6. $5x^3yz^{-1}$

7. $27xyz^2$

8. $42x^3y^2z^5$

Additional practice problems are at the end of the chapter.

6.3.2 Division.   Recall from arithmetic that the parts of a division problem are as follows:

(a) $\dfrac{18a}{6a} = 3$ dividend / quotient / divisor

(b) $6a\overline{)18a}$ — 3 quotient / 18a dividend / divisor

When written as a fraction as in (a), the dividend is the numerator and the divisor is the denominator.

In arithmetic we learned how to do "long division." Here, our long division is done by means of a calculator, but we must do the division of literal numbers ourselves because calculators can only deal with real numbers.

In dividing monomials we find the prime factors of the dividend and the divisor. Then we cancel each factor that appears in both the dividend and the divisor. The resultant number is the quotient.

EXAMPLE 6.7: $\dfrac{10a^3}{2a^2}$.

Solution: Find the prime factors and cancel those appearing in both numerator and denominator.

$$\frac{2 \cdot 5 \cdot \cancel{a} \cdot \cancel{a} \cdot a}{2 \cdot \cancel{a} \cdot \cancel{a}} = \frac{5a}{1} = 5a$$

Another way to solve the problem would be to move all literal numbers into the numerator and multiply. Recall that we can move numbers from the numerator to the denominator or from the denominator to the numerator by changing the sign of the exponent. Using this method in Example 6.7, we get:

$$\frac{10a^3}{2a^2} = \frac{10a^3 \cdot a^{-2}}{2} = 5a^{3-2} = 5a$$

EXAMPLE 6.8: $\dfrac{6a^2b^3}{18a^4b}$.

Solution:

$$\frac{2 \cdot 3 \cdot \cancel{a} \cdot \cancel{a} \cdot \cancel{b} \cdot b \cdot b}{2 \cdot 3 \cdot 3 \cdot \cancel{a} \cdot \cancel{a} \cdot a \cdot a \cdot \cancel{b}} = \frac{b^2}{3a^2} = \frac{a^{-2}b^2}{3}$$

or $$\frac{6a^2a^{-4}b^3b^{-1}}{18} = \frac{6a^{-2}b^2}{18} = \frac{a^{-2}b^2}{3}$$

EXAMPLE 6.9: $\dfrac{40x^2y^{-1}}{10x^{-1}y^{-3}}$.

Solution:

$$\frac{40x\cdot x\cdot y^{-1}}{10x^{-1}\cdot y^{-1}\cdot y^{-1}\cdot y^{-1}} = \frac{4x^2}{x^{-1}y^{-2}}$$

$$\frac{4x^2\cdot x}{y^{-2}} = \frac{4x^3}{y^{-2}} = 4x^3y^2$$

and also $\quad \dfrac{40x^2y^{-1}}{10x^{-1}y^{-3}} = \dfrac{40x^2\cdot x\cdot y^{-1}\cdot y^3}{10} = 4x^3y^2$

The second method is usually quicker. Both methods of solution are important in that they help develop important math concepts and ideas. The first method helps us develop an understanding of factoring so that we may deal more easily with fractions. The second method helps us strengthen our understanding of exponents and bases.

PRACTICE PROBLEMS 6-5   Perform the following divisions by (a) canceling like factors and (b) moving all literal numbers to the numerator and multiplying:

1. $\dfrac{15a^4}{5a}$

2. $\dfrac{36a^2}{6a^3}$

3. $\dfrac{42a^3b^2}{7ab}$

4. $\dfrac{56ab^3}{7a^3b}$

5. $\dfrac{20a^4b^2c}{3a^2b^2c^2}$

6. $\dfrac{39a^4bc^4}{3a^2b^3c}$

7. $\dfrac{7x^{-2}y}{63xy^2}$

8. $\dfrac{72xy^2}{9x^{-3}y^{-4}}$

9. $\dfrac{84x^{-1}y^2z^{-3}}{6x^3y^{-1}z^{-2}}$

10. $\dfrac{46x^{-1}y^2z^{-3}}{23xy^{-1}z^2}$

Solutions:

1. (a) $\dfrac{3\cdot 5\cdot a\cdot a\cdot a\cdot a}{5a} = 3a^3$

   (b) $\dfrac{15a^4\cdot a^{-1}}{5} = 3a^3$

2. (a) $\dfrac{2\cdot 2\cdot 3\cdot 3\cdot a\cdot a}{2\cdot 3\cdot a\cdot a\cdot a} = \dfrac{6}{a} = 6a^{-1}$

   (b) $\dfrac{36a^2\cdot a^{-3}}{6} = 6a^{-1}$

3. (a) $\dfrac{2\cdot 3\cdot 7\cdot a\cdot a\cdot a\cdot b\cdot b}{7\cdot a\cdot b} = 6a^2b$

   (b) $\dfrac{42a^3\cdot a^{-1}\cdot b^2\cdot b^{-1}}{7} = 6a^2b$

4. (a) $\dfrac{2\cdot 2\cdot 2\cdot 7\cdot a\cdot b\cdot b\cdot b}{7\cdot a\cdot a\cdot a\cdot b} = \dfrac{8b^2}{a^2} = 8a^{-2}b^2$

   (b) $\dfrac{56a\cdot a^{-3}\cdot b^3\cdot b^{-1}}{7} = 8a^{-2}b^2$

5. (a) $\dfrac{2\cdot2\cdot5\cdot\cancel{a}\cdot\cancel{a}\cdot a\cdot a\cdot b\cdot\cancel{b}\cdot\cancel{c}}{3\cdot\cancel{a}\cdot\cancel{a}\cdot\cancel{b}\cdot\cancel{b}\cdot\cancel{c}\cdot c} = \dfrac{20a^2}{3c} = \dfrac{20a^2c^{-1}}{3}$

(b) $\dfrac{20a^4a^{-2}b^2b^{-2}cc^{-2}}{3} = \dfrac{20a^2b^0c^{-1}}{3} = \dfrac{20a^2c^{-1}}{3}$

6. (a) $\dfrac{3\cdot13\cdot\cancel{a}\cdot\cancel{a}\cdot a\cdot a\cdot b\cdot\cancel{c}\cdot c\cdot c\cdot c}{3\cdot\cancel{a}\cdot\cancel{a}\cdot b\cdot b\cdot b\cdot\cancel{c}} = \dfrac{13a^2c^3}{b^2} = 13a^2b^{-2}c^3$

(b) $\dfrac{39a^4a^{-2}bb^{-3}c^4c^{-1}}{3} = 13a^2b^{-2}c^3$

7. (a) $\dfrac{7\cdot x^{-1}\cdot x^{-1}\cdot y}{3\cdot3\cdot7\cdot x\cdot y\cdot y} = \dfrac{x^{-2}}{9xy} = \dfrac{x^{-2}\cdot x^{-1}\cdot y^{-1}}{9} = \dfrac{x^{-3}y^{-1}}{9}$

(b) $\dfrac{7x^{-2}\cdot x^{-1}\cdot y\cdot y^{-2}}{63} = \dfrac{x^{-3}y^{-1}}{9}$

8. (a) $\dfrac{2\cdot2\cdot2\cdot3\cdot3\cdot x\cdot y\cdot y}{3\cdot3\cdot x^{-1}\cdot x^{-1}\cdot x^{-1}\cdot y^{-1}\cdot y^{-1}\cdot y^{-1}\cdot y^{-1}} = \dfrac{8xy^2}{x^{-3}\cdot y^{-4}} = 8x\cdot x^3\cdot y^2\cdot y^4 = 8x^4y^6$

(b) $\dfrac{72x\cdot x^3\cdot y^2\cdot y^4}{9} = 8x^4y^6$

9. (a) $\dfrac{2\cdot2\cdot3\cdot7\cdot x^{-1}\cdot y\cdot y\cdot z^{-1}\cdot z^{-1}\cdot z^{-1}}{2\cdot3\cdot x\cdot x\cdot x\cdot y^{-1}\cdot z^{-1}\cdot z^{-1}} = \dfrac{14x^{-1}y^2z^{-1}}{x^3y^{-1}} = 14x^{-1}x^{-3}y^2y^1z^{-1} = 14x^{-4}y^3z^{-1}$

(b) $\dfrac{84x^{-1}\cdot x^{-3}\cdot y^2\cdot y\cdot z^{-3}\cdot z^2}{6} = 14x^{-4}y^3z^{-1}$

10. (a) $\dfrac{2\cdot23\cdot x^{-1}\cdot y\cdot y\cdot z^{-1}\cdot z^{-1}\cdot z^{-1}}{23x\cdot y^{-1}\cdot z\cdot z} = 2x^{-1}\cdot x^{-1}\cdot y^2\cdot y\cdot z^{-3}\cdot z^{-2} = 2x^{-2}\cdot y^3\cdot z^{-5}$

(b) $\dfrac{46x^{-1}\cdot x^{-1}\cdot y^2\cdot y\cdot z^{-3}\cdot z^{-2}}{23} = 2x^{-2}y^3\cdot z^{-5}$

Additional practice problems are at the end of the chapter.

SELF-TEST 6-2   Perform the indicated operations:

1. $3x\cdot5y$

2. $3x^2y\cdot4x^2y^3$

3. $5a^{-1}b^2\cdot6a^3b^{-1}$

4. $3a^2b^{-1}\cdot5ab^{-2}\cdot2a^{-4}b$

5. $6ab^{-3}c^2\cdot7a^{-4}b^2c\cdot a^{-1}$

6. $\dfrac{48x^3y}{6y^2}$

7. $\dfrac{24x^{-2}y}{3xy^2}$

8. $\dfrac{56a^2bc^{-3}}{8b}$

9. $\dfrac{63a^{-1}b^2c^{-4}}{7a^3c^{-3}}$

10. $\dfrac{48xyz^{-3}}{3x^{-1}y^3z}$

Answers to Self-Test 6-2 are at the end of the chapter.

We learned how to multiply and divide fractions in arithmetic. At that time we learned the following rules:

To multiply fractions we (1) multiply the numerators (this product is the numerator in the answer); (2) multiply the denominators (this product is the denominator in the answer); and (3) cancel like factors to reduce to lowest terms.

Let's try a few together:

EXAMPLE 6.10:

(a) $\dfrac{2}{3} \times \dfrac{4}{5} = \dfrac{2 \times 4}{3 \times 5} = \dfrac{8}{15}$

(b) $\dfrac{1}{4} \times \dfrac{3}{4} \times \dfrac{3}{8} = \dfrac{1 \times 3 \times 3}{4 \times 4 \times 8} = \dfrac{9}{128}$

(c) $\dfrac{5}{20} \times \dfrac{4}{10} = \dfrac{5 \times 4}{20 \times 10} = \dfrac{20}{200}$

This time we recognize like factors in both numerator and denominator; therefore, we find the prime factors and cancel all like factors to reduce to lowest terms.

$$\frac{20}{200} = \frac{2 \cdot 2 \cdot \cancel{5}}{2 \cdot 2 \cdot 2 \cdot \cancel{5} \cdot 5} = \frac{1}{10}$$

(d) $\dfrac{3a}{4} \times \dfrac{6a^2}{10} = \dfrac{3a \cdot 6a^2}{4 \cdot 10} = \dfrac{18a^3}{40} = \dfrac{2 \cdot 3 \cdot 3 \cdot a^3}{2 \cdot 2 \cdot 2 \cdot 5} = \dfrac{9a^3}{20}$

(e) $\dfrac{7x^2y}{2x} \times \dfrac{3x^3y^3}{14x^2} = \dfrac{7x^2y \cdot 3x^3y^3}{2x \cdot 14x^2} = \dfrac{21x^5y^4}{28x^3}$

$= \dfrac{3 \cdot 7 \cdot x^5 \cdot y^4}{2 \cdot 2 \cdot 7 \cdot x^3} = \dfrac{3 \cdot \cancel{7} \cdot x^5 \cdot x^{-3} \cdot y^4}{2 \cdot 2 \cdot \cancel{7}} = \dfrac{3x^2y^4}{4}$

(f) $\dfrac{3a^{-1}b^2}{4a^2b^3} \times \dfrac{4a^3b^{-3}}{9a^{-1}b^{-1}} \times \dfrac{2a^2b^4}{a^3b^2} = \dfrac{3a^{-1}b^2 \cdot 4a^3b^{-3} \cdot 2a^2b^4}{4a^2b^3 \cdot 9a^{-1}b^{-1} \cdot a^3b^2}$

At this point we can save some time by examining both numerator and denominator for like factors. The factors don't have to be prime; they only have to be *like* factors. $4, a^2, a^3, b^2, a^{-1}$ are found in both numerator and denominator and can be canceled at this time. We are now left with:

$$\frac{3a^{-1}b^2 \cdot 4a^2b^{-3} \cdot 2a^2b^4}{4a^2b^3 \cdot 9a^{-1}b^{-1} \cdot a^3b^2} = \frac{3 \cdot b^{-3} \cdot 2b^4}{b^3 \cdot 9b^{-1}} = \frac{6b}{9b^2}$$

$$= \frac{2 \cdot 3 \cdot b}{3 \cdot 3 \cdot b \cdot b} = \frac{2}{3b} = \frac{2b^{-1}}{3}$$

In each case, we reduce to lowest terms by canceling like terms. In Example 6.10(f) the literal number, b, was moved to the numerator in the final answer. Although it is mathematically correct to leave literal numbers in the denominator, we normally give answers with all literal numbers in the numerator.

PRACTICE PROBLEMS 6-6   Find the product of the following numbers. Reduce your answers to lowest terms.

1. $\frac{3}{4} \times \frac{7}{8}$

2. $\frac{3}{8} \times \frac{2}{3}$

3. $\frac{1}{5} \times \frac{4}{7} \times \frac{1}{2}$

4. $\frac{1}{3} \times \frac{1}{4} \times \frac{4}{5}$

5. $\frac{3x}{5} \times \frac{2x}{4}$

6. $\frac{4x^2}{7} \times \frac{3x}{4}$

7. $\frac{5x^2}{8y} \times \frac{6xy}{5}$

8. $\frac{2x^3}{3y^2} \times \frac{3y^3}{4x}$

9. $\frac{3xy^{-1}}{5z} \times \frac{5y^3z^2}{9x^{-2}}$

10. $\frac{4a^{-1}b^2}{5c^2} \times \frac{b^{-3}c}{3a^2} \times \frac{3a}{4b^{-1}c^3}$

11. $\frac{2x^2y^{-1}}{5z^2} \times \frac{20y^3z^3}{30xz^{-1}} \times \frac{xz^2}{y^3}$

Solutions:

1. $\frac{21}{32}$

2. $\frac{6}{24} = \frac{1}{4}$

3. $\frac{4}{70} = \frac{2}{35}$

4. $\frac{4}{60} = \frac{1}{15}$

5. $\frac{6x^2}{20} = \frac{3x^2}{10}$

6. $\frac{12x^3}{28} = \frac{3x^3}{7}$

7. $\frac{30x^3y}{40y} = \frac{3x^3}{4}$

8. $\frac{6x^3y^3}{12xy^2} = \frac{x^2y}{2}$

9. $\frac{15xy^2z^2}{45x^{-2}z} = \frac{x^3y^2z}{3}$

10. $\frac{12b^{-1}c}{60a^2b^{-1}c^5} = \frac{a^{-2}c^{-4}}{5}$

11. $\frac{40x^3y^2z^5}{150xy^3z} = \frac{4x^2y^{-1}z^4}{15}$

Additional practice problems are at the end of the chapter.

In arithmetic we learned that when we divide fractions we simply invert the divisor and then multiply. Here are some examples:

EXAMPLE 6.11:

(a) $\dfrac{1}{3} \div \dfrac{1}{5} = \dfrac{1}{3} \times \dfrac{5}{1} = \dfrac{5}{3}$

(b) $\dfrac{5}{8} \div 3 = \dfrac{5}{8} \times \dfrac{1}{3} = \dfrac{5}{24}$

(c) $\dfrac{3a}{4} \div \dfrac{4a}{5} = \dfrac{3a}{4} \times \dfrac{5}{4a} = \dfrac{15a}{16a} = \dfrac{15}{16}$

(d) $\dfrac{15x^2y}{8z^2} \div \dfrac{12x^2z}{21y^3} = \dfrac{15x^2y}{8z^2} \times \dfrac{21y^3}{12x^2z}$

$= \dfrac{315x^2y^4}{96x^2z^3} = \dfrac{3 \cdot 3 \cdot 5 \cdot 7 x^2 y^4}{3 \cdot 2 \cdot 2 \cdot 2 \cdot 2 \cdot 2 \cdot x^2 \cdot z^3} = \dfrac{105y^4z^{-3}}{32}$

Notice that in each example we inverted the divisor and then multiplied. Like factors in the numerator and denominator of the answers were canceled as before.

PRACTICE PROBLEMS 6-7   Perform the following divisions. Reduce your answers to lowest terms.

1. $\dfrac{7}{8} \div \dfrac{3}{5}$

2. $\dfrac{5}{7} \div \dfrac{5}{3}$

3. $\dfrac{4a}{5} \div \dfrac{6a^2}{7}$

4. $\dfrac{5a^2}{6} \div \dfrac{15a}{18}$

5. $\dfrac{7xy^2}{9z} \div \dfrac{14x^2y}{3z^2}$

6. $\dfrac{42x^{-2}y}{15y^{-1}z^2} \div \dfrac{7xz}{3y}$

7. $\dfrac{20a^2b^{-1}c}{9} \div \dfrac{5ab^3c^{-2}}{27}$

8. $\dfrac{10a^{-2}b^3}{7c^2} \div \dfrac{5b^{-1}}{28ac^4}$

Solutions:

1. $\dfrac{35}{24}$

2. $\dfrac{15}{35} = \dfrac{3}{7}$

3. $\dfrac{28a}{30a^2} = \dfrac{14a^{-1}}{15}$

4. $\dfrac{90a^2}{90a} = a$

5. $\dfrac{21xy^2z^2}{126x^2yz} = \dfrac{x^{-1}yz}{6}$

6. $\dfrac{126x^{-2}y^2}{105xy^{-1}z^3} = \dfrac{6x^{-3}y^3z^{-3}}{5}$

7. $\dfrac{540a^2b^{-1}c}{45ab^3c^{-2}} = 12ab^{-4}c^3$

8. $\dfrac{280a^{-1}b^3c^4}{35b^{-1}c^2} = 8a^{-1}b^4c^2$

Additional practice problems are at the end of the chapter.

SELF-TEST 6-3   Perform the indicated operations:

1. $\dfrac{2}{5} \times \dfrac{7}{8}$

2. $\dfrac{3x^2}{7} \times \dfrac{2x^2}{5}$

3. $\dfrac{2}{3} \times \dfrac{1}{5} \times \dfrac{1}{4}$

4. $\dfrac{3ab^2}{5c} \times \dfrac{2bc^{-1}}{3a^2}$

5. $\dfrac{2a^2bc^{-1}}{3b^{-1}} \times \dfrac{4b^3}{2ac^2} \times \dfrac{7a^{-3}}{8}$

6. $\dfrac{1}{5} \div \dfrac{2}{3}$

7. $\dfrac{5a}{6} \div \dfrac{3a^2}{4}$

8. $\dfrac{36a^2b}{42c^{-3}} \div \dfrac{9b^3}{21a^3c}$

9. $\dfrac{12b}{21a^2c} \div \dfrac{3b^2}{7a^{-1}c^2}$

10. $\dfrac{15x^{-1}y^{-1}}{27z^3} \div \dfrac{5x^{-1}y}{9z}$

Answers to Self-Test 6-3 are at the end of the chapter.

6.5   ADDITION AND SUBTRACTION OF FRACTIONS

6.5.1   Adding fractions with common denominators.   Two or more fractions can be combined into one when the denominators are identical. If fractions to be added have the same denominator, the denominator in the answer is that number. The numerator in the answer is found by adding all the numerators together.

EXAMPLE 6.12:

(a) $\dfrac{1}{5} + \dfrac{2}{5} = \dfrac{1+2}{5} = \dfrac{3}{5}$

(b) $\dfrac{2a}{7} + \dfrac{3a}{7} = \dfrac{2a+3a}{7} = \dfrac{5a}{7}$

(c) $\dfrac{2x}{9} + \dfrac{3x}{9} = \dfrac{2x+3x}{9} = \dfrac{5x}{9}$

(d) $\dfrac{4}{11} + \dfrac{3}{11} + \dfrac{2}{11} = \dfrac{4+3+2}{11} = \dfrac{9}{11}$

In each of the examples above the denominators are the same. That is, each fraction had a denominator of 5 in Example 6.12(a); a denominator of 7 in (b), and so on. Thus, 5 is the *common denominator* in (a) and 7 is the common denominator in (b). Said another way, when the denominators are identical, that number is the common denominator.

When adding fractions, then, the first step is to determine if the denominators are identical. If they are, the next step is to add the numerators. We learned in a previous chapter that we can only combine like terms. Therefore, the numerator in the answer may contain more than one term.

EXAMPLE 6.13:

(a) $\dfrac{2a}{5} + \dfrac{3b}{5} = \dfrac{2a+3b}{5}$

(The answer cannot be simplified further because $2a$ and $3b$ are unlike terms.)

(b) $\dfrac{3xy}{7} + \dfrac{2x}{7} + \dfrac{xy}{7} = \dfrac{3xy+2x+xy}{7} = \dfrac{4xy+2x}{7}$

(c) $\dfrac{3}{8x} + \dfrac{5a}{8x} + \dfrac{2}{8x} = \dfrac{3+5a+2}{8x} = \dfrac{5+5a}{8x}$

(d) $\dfrac{a}{12} + \dfrac{5a}{12} + \dfrac{7a}{12} = \dfrac{13a}{12}$

6.5.2 Subtracting fractions with common denominators. Up to this point we have been working with positive fractions. Let's consider a fraction in which a part or all of the fraction is negative. Some examples are shown below.

EXAMPLE 6.14:

(a) $-\dfrac{3}{8}$ (b) $\dfrac{-3}{8}$ (c) $\dfrac{3}{-8}$

(d) $\dfrac{-3}{-8}$ (e) $-\dfrac{-3}{8}$ (f) $-\dfrac{-3}{-8}$

Consider that there are three signs associated with any fraction: the sign of the numerator, the sign of the denominator, and the sign of the entire fraction. In the fraction $\frac{3}{8}$ we know the signs are positive because no sign is given. In Example 6.14(a), the sign of the fraction is negative but the numerator and denominator are positive. In Example 6.14(b), the numerator is negative, the denominator and the fraction are positive, and so on for each fraction.

We don't leave fractions with negative numerators or denominators if we can avoid it. We change the signs so that the fraction in the answer is either positive or negative and both numerator and denominator are positive. To change the signs, the rule is:

Any two signs of a fraction may be changed without changing the value of the fraction.

Consider Example 6.14(a), applying the above rule:

$$-\frac{+3}{+8} = +\frac{-3}{+8} = -\frac{-3}{-8} = +\frac{+3}{-8}$$

(The plus signs were included here to help us change signs.) Notice that in each case *two* signs were changed.

Suppose we want to subtract one fraction from another as in Example 6.15.

EXAMPLE 6.15: Perform the indicated operation:

$$\frac{7}{8} - \frac{3}{8}$$

Solution: The subtraction is accomplished by changing the sign of the fraction and adding. However, if we change the sign of the fraction, we must also change *one more sign*. Remember, we must change *two* signs. The other sign we will change is the sign of the numerator. We don't want to change the sign of the denominator because in order to add the fractions the denominators must be identical. Eight and -8 are not identical.

To solve the problem in Example 6.15 then, we must first change the sign of the second fraction and add (remembering to also change the sign of the numerator).

$$\frac{7}{8} - \frac{3}{8} = \frac{7}{8} + \frac{-3}{8} = \frac{7-3}{8} = \frac{4}{8} = \frac{1}{2}$$

We perform the addition as before. When we add the numerators, we are adding $+7$ and -3. The result is 4. The resultant fraction $\frac{4}{8}$ is reduced to lowest terms or $\frac{1}{2}$. Here are some more examples:

EXAMPLE 6.16:

(a) $\dfrac{7x}{9} - \dfrac{3x}{9} = \dfrac{7x}{9} + \dfrac{-3x}{9} = \dfrac{7x-3x}{9} = \dfrac{4x}{9}$

(b) $\dfrac{4}{15} + \dfrac{3}{15} - \dfrac{2}{15} = \dfrac{4}{15} + \dfrac{3}{15} + \dfrac{-2}{15}$

$\qquad = \dfrac{4+3-2}{15} = \dfrac{5}{15} = \dfrac{1}{3}$

Of course, we soon recognize that whenever a subtraction is indicated we merely subtract the numerator instead of adding, so when we see a problem like (b) above, we can skip a step.

(b) $\dfrac{4}{15} + \dfrac{3}{15} - \dfrac{2}{15} = \dfrac{4+3-2}{15} = \dfrac{5}{15} = \dfrac{1}{3}$

(c) $\dfrac{3}{8} - \dfrac{7}{8} + \dfrac{1}{8} = \dfrac{3-7+1}{8} = \dfrac{-3}{8} = -\dfrac{3}{8}$

Instead of leaving the answer as a positive fraction with a negative numerator, it is standard practice to change the signs so that the numerator is positive and the fraction is negative.

$$(d) \quad \frac{11a}{12} + \frac{5a}{12} - \frac{7a}{12} = \frac{11a+5a-7a}{12} = \frac{16a-7a}{12}$$
$$= \frac{9a}{12} = \frac{3a}{4}$$

PRACTICE PROBLEMS 6-8   Perform the indicated operations:

1. $\dfrac{1}{3} + \dfrac{2}{3}$

2. $\dfrac{1}{5} + \dfrac{2}{5}$

3. $\dfrac{5}{7} - \dfrac{2}{7}$

4. $\dfrac{9}{16} - \dfrac{5}{16}$

5. $\dfrac{1}{24} + \dfrac{3}{24} + \dfrac{11}{24}$

6. $\dfrac{2}{15} + \dfrac{3}{15} - \dfrac{4}{15}$

7. $\dfrac{3}{5a} + \dfrac{1}{5a}$

8. $\dfrac{3}{8ab} + \dfrac{1}{8ab} + \dfrac{5}{8ab}$

9. $\dfrac{2}{9x} + \dfrac{5}{9x} - \dfrac{2}{9x}$

10. $\dfrac{5x}{11} + \dfrac{2x}{11}$

11. $\dfrac{7x}{20} - \dfrac{3x}{20}$

12. $\dfrac{3xy}{16} + \dfrac{xy}{16} + \dfrac{5xy}{16}$

13. $\dfrac{9ab}{20} - \dfrac{5ab}{20} - \dfrac{3ab}{20}$

14. $\dfrac{3a}{8b} + \dfrac{5}{8b} + \dfrac{2a}{8b}$

15. $\dfrac{2xy}{7} - \dfrac{3y}{7} + \dfrac{2y}{7}$

16. $\dfrac{3}{4y} - \dfrac{5}{4y}$

17. $\dfrac{8a}{21} - \dfrac{14a}{21} + \dfrac{2a}{21}$

Solutions:

1. $\dfrac{1+2}{3} = \dfrac{3}{3} = 1$

2. $\dfrac{1+2}{5} = \dfrac{3}{5}$

3. $\dfrac{5-2}{7} = \dfrac{3}{7}$

4. $\dfrac{9-5}{16} = \dfrac{4}{16} = \dfrac{1}{4}$

5. $\dfrac{1+3+11}{24} = \dfrac{15}{24} = \dfrac{5}{8}$

6. $\dfrac{2+3-4}{15} = \dfrac{1}{15}$

7. $\dfrac{3+1}{5a} = \dfrac{4}{5a}$

8. $\dfrac{3+1+5}{8ab} = \dfrac{9}{8ab}$

9. $\dfrac{2+5-2}{9x} = \dfrac{5}{9x}$

10. $\dfrac{5x+2x}{11} = \dfrac{7x}{11}$

11. $\dfrac{7x-3x}{20} = \dfrac{4x}{20} = \dfrac{x}{5}$

12. $\dfrac{3xy+xy+5xy}{16} = \dfrac{9xy}{16}$

13. $\dfrac{9ab-5ab-3ab}{20} = \dfrac{ab}{20}$

14. $\dfrac{3a+5+2a}{8b} = \dfrac{5a+5}{8b}$

15. $\dfrac{2xy-3y+2y}{7} = \dfrac{2xy-y}{7}$

16. $\dfrac{3-5}{4y} = \dfrac{-2}{4y} = \dfrac{-1}{2y} = -\dfrac{1}{2y}$

17. $\dfrac{8a-14a+2a}{21} = \dfrac{10a-14a}{21} = \dfrac{-4a}{21} = -\dfrac{4a}{21}$

Additional practice problems are at the end of the chapter.

6.5.3 Adding and Subtracting Fractions with Unlike Denominators. When the fractions we are adding or subtracting have unlike denominators we cannot add them directly as before. We must first find a common denominator. We do this by finding the least common multiple (LCM) of all the denominators. Then we change each fraction so that all fractions have this LCM as its denominator. The LCM is the common denominator. Once we have a common denominator, we can add and subtract as before.

EXAMPLE 6.17: Perform the indicated operation:

$$\frac{1}{4} + \frac{1}{6}$$

Solution: Since the denominators in this example are different, we find the LCM.

$$4 = 2 \cdot 2$$
$$6 = 2 \cdot 3$$
$$\text{LCM} = 2 \cdot 2 \cdot 3 = 12$$

Next we change each fraction so that its denominator is 12 (the LCM).

$$\frac{1}{4} = \frac{?}{12}$$

If we change the denominator of a fraction, we must also change the numerator; otherwise, the fractions are not equal.

$$\frac{1}{4} \neq \frac{1}{12}$$

If we divide the new denominator (12) by the old denominator (4), we get 3: $12 \div 4 = 3$. This tells us that the new denominator is 3 times greater than the original denominator. If the new denominator is 3 times greater, then the new numerator must also be 3 times greater. Look at it this way:

$$\frac{1}{4} = \frac{1 \times 3}{4 \times 3} = \frac{3}{12}$$

If we reduce $\frac{3}{12}$, we get the original fraction $\frac{1}{4}$. The fractions are equal.

The second fraction is changed in the same manner.

$$\frac{1}{6} = \frac{?}{12}$$

$$\frac{1}{6} = \frac{1 \times 2}{6 \times 2} = \frac{2}{12}$$

$$\frac{1}{4} + \frac{1}{6} = \frac{3}{12} + \frac{2}{12} = \frac{5}{12}$$

An easy rule is:

1. Divide the LCM by the old denominator.
2. Multiply this number by the old numerator to find the new numerator.

With practice we can usually do this by inspection.

EXAMPLE 6.18: Perform the indicated operation:

$$\frac{1}{6} + \frac{3}{8}$$

Solution: Step 1. Find the common denominator:

$$6 = 2 \cdot 3$$
$$8 = 2 \cdot 2 \cdot 2$$
$$CD = 2 \cdot 2 \cdot 2 \cdot 3 = 24$$

Step 2. Change each fraction so that its denominator is the CD.

$$\frac{1}{6} = \frac{1 \times 4}{24} = \frac{4}{24}$$

$$\frac{3}{8} = \frac{3 \times 3}{24} = \frac{9}{24}$$

Step 3. Add the new fractions together. Reduce to lowest terms where possible.

$$\frac{4}{24} + \frac{9}{24} = \frac{4+9}{24} = \frac{13}{24}$$

EXAMPLE 6.19: $\dfrac{3a}{20} + \dfrac{2a}{15} + \dfrac{3a}{10}$.

Solution: Step 1. Find the common denominator:

$$20 = 2 \cdot 2 \cdot 5$$
$$15 = 3 \cdot 5$$
$$10 = 2 \cdot 5$$
$$CD = 2 \cdot 2 \cdot 3 \cdot 5 = 60$$

Step 2. Change each fraction so that its denominator is the CD.

$$\frac{3a}{20} = \frac{3a \times 3}{60} = \frac{9a}{60}$$

$$\frac{2a}{15} = \frac{2a \times 4}{60} = \frac{8a}{60}$$

$$\frac{3a}{10} = \frac{3a \times 6}{60} = \frac{18a}{60}$$

Step 3. Add the new fractions together. Reduce to lowest terms where possible.

$$\frac{9a}{60} + \frac{8a}{60} + \frac{18a}{60} = \frac{9a+8a+18a}{60} = \frac{35a}{60} = \frac{7a}{12}$$

EXAMPLE 6.20: $\dfrac{3}{10a^2} + \dfrac{4}{15a} + \dfrac{b}{6a^2}$.

Solution:

$$10a^2 = 2\cdot5\cdot a\cdot a$$
$$15a = 3\cdot5\cdot a$$
$$6a^2 = 2\cdot3\cdot a\cdot a$$
$$CD = 2\cdot3\cdot5\cdot a\cdot a = 30a^2$$
$$\frac{3}{10a^2} = \frac{3\times3}{30a^2} = \frac{9}{30a^2}$$
$$\frac{4}{15a} = \frac{4\times2a}{30a^2} = \frac{8a}{30a^2}$$
$$\frac{b}{6a^2} = \frac{b\times5}{30a^2} = \frac{5b}{30a^2}$$
$$\frac{9}{30a^2} + \frac{8a}{30a^2} + \frac{5b}{30a^2} = \frac{9+8a+5b}{30a^2}$$

(No reduction of the numerator is possible here. We have no like terms.)

PRACTICE PROBLEMS 6-9 Perform the indicated operations. Reduce to lowest terms where possible.

1. $\dfrac{1}{3} + \dfrac{1}{4} + \dfrac{1}{5}$

2. $\dfrac{7}{20} + \dfrac{5}{24}$

3. $\dfrac{3a}{4} + \dfrac{2b}{3}$

4. $\dfrac{2a}{3} + \dfrac{3a}{2}$

5. $\dfrac{b}{3a} + \dfrac{2b}{5}$

6. $\dfrac{x}{5y} + \dfrac{3x}{4}$

7. $\dfrac{x}{2y} + \dfrac{y}{5x}$

8. $\dfrac{3x}{y} + \dfrac{2y}{x}$

9. $\dfrac{5}{12} + \dfrac{5}{18} - \dfrac{5}{36}$

10. $\dfrac{5}{6a} + \dfrac{4}{15a^3} + \dfrac{1}{4a^2}$

11. $\dfrac{2xy}{15z} + \dfrac{3xz}{5y}$

12. $\dfrac{5}{12} + \dfrac{5}{8} - \dfrac{1}{3}$

13. $\dfrac{7a}{8} - \dfrac{2a}{5} + \dfrac{3a}{20}$

14. $\dfrac{3bc}{16a} - \dfrac{5ab}{8c}$

15. $\dfrac{11a}{15} - \dfrac{3a}{5} + \dfrac{2a}{3}$

16. $\dfrac{2a^2c}{3b} - \dfrac{4a^2c}{15b} + \dfrac{2a^2c}{5b}$

Solutions:

1. $\dfrac{20}{60} + \dfrac{15}{60} + \dfrac{12}{60} = \dfrac{20+15+12}{60} = \dfrac{47}{60}$

2. $\dfrac{42}{120} + \dfrac{25}{120} = \dfrac{42+25}{120} = \dfrac{67}{120}$

3. $\dfrac{9a}{12} + \dfrac{8b}{12} = \dfrac{9a+8b}{12}$

4. $\dfrac{4a}{6} + \dfrac{9a}{6} = \dfrac{4a+9a}{6} = \dfrac{13a}{6}$

5. $\dfrac{5b}{15a} + \dfrac{6ab}{15a} = \dfrac{5b+6ab}{15a}$

6. $\dfrac{4x}{20y} + \dfrac{15xy}{20y} = \dfrac{4x+15xy}{20y}$

7. $\dfrac{5x^2}{10xy} + \dfrac{2y^2}{10xy} = \dfrac{5x^2+2y^2}{10xy}$

8. $\dfrac{3x^2}{xy} + \dfrac{2y^2}{xy} = \dfrac{3x^2+2y^2}{xy}$

9. $\dfrac{15}{36} + \dfrac{10}{36} + \dfrac{-5}{36} = \dfrac{15+10-5}{36} = \dfrac{20}{36} = \dfrac{5}{9}$

10. $\dfrac{50a^2}{60a^3} + \dfrac{16}{60a^3} + \dfrac{15a}{60a^3} = \dfrac{50a^2+16+15a}{60a^3}$

11. $\dfrac{2xy^2}{15yz} + \dfrac{9xz^2}{15yz} = \dfrac{2xy^2+9xz^2}{15yz}$

12. $\dfrac{10}{24} + \dfrac{15}{24} + \dfrac{-8}{24} = \dfrac{10+15-8}{24} = \dfrac{17}{24}$

13. $\dfrac{35a}{40} + \dfrac{-16a}{40} + \dfrac{6a}{40} = \dfrac{35a-16a+6a}{40} = \dfrac{25a}{40}$
$= \dfrac{5a}{8}$

14. $\dfrac{3bc^2}{16ac} + \dfrac{-10a^2b}{16ac} = \dfrac{3bc^2-10a^2b}{16ac}$

15. $\dfrac{11a}{15} + \dfrac{-9a}{15} + \dfrac{10a}{15} = \dfrac{11a-9a+10a}{15} = \dfrac{12a}{15}$

$= \dfrac{4a}{5}$

16. $\dfrac{10a^2c}{15b} + \dfrac{-4a^2c}{15b} + \dfrac{6a^2c}{15b} = \dfrac{10a^2c-4a^2c+6a^2c}{15b}$

$= \dfrac{12a^2c}{15b} = \dfrac{4a^2c}{5b}$

Additional practice problems are at the end of the chapter.

SELF-TEST 6-4   Perform the indicated operations:

1. $\dfrac{1}{8} + \dfrac{3}{8}$

2. $\dfrac{2}{7a} + \dfrac{3}{7a}$

3. $\dfrac{8}{9x} - \dfrac{2}{9x}$

4. $\dfrac{15a}{12} - \dfrac{a}{12}$

5. $\dfrac{2a}{9c} + \dfrac{4a}{9c} - \dfrac{a}{9c}$

6. $\dfrac{3}{8} + \dfrac{1}{6}$

7. $\dfrac{5}{12} - \dfrac{2}{15}$

8. $\dfrac{4a}{5b} + \dfrac{2}{3}$

9. $\dfrac{3}{5x} + \dfrac{2}{7x^2} + \dfrac{3}{10x^3}$

10. $\dfrac{4}{15b} - \dfrac{5b}{18a^2c} + \dfrac{3}{10}$

Answers to Self-Test 6-4 are at the end of the chapter.

When the numerator is greater than the denominator, the fraction is called an *improper* fraction. Some examples are $\frac{4}{3}$, $\frac{6}{5}$, $\frac{17}{12}$. Improper fractions always have a value greater than one.

Mixed numbers are numbers which have a whole number part and a fractional part. Some examples of mixed numbers are $2\frac{2}{3}$, $5\frac{1}{4}$, $8\frac{7}{8}$. When we write the number $2\frac{2}{3}$, we are really saying that we have 2 plus $\frac{2}{3}$: $2+\frac{2}{3}$. Even though we don't write the "$+$" between the whole number and fractional parts, we know it is implied.

We can express non-whole numbers as either mixed numbers or as improper fractions. In problem solving it is easier to work with improper fractions. Answers, however, are usually written as mixed numbers. The conversion from one form to the other is shown in the following examples.

EXAMPLE 6.21: Change $\frac{4}{3}$ to a mixed number.

Solution: Step 1. Perform the indicated division. Write the remainder as a fraction.

$$\begin{array}{r} 1+\dfrac{1}{3} \\ \hline 3\overline{)4} \\ 3 \\ \hline 1 \end{array}$$

Step 2. Write the answer as a mixed number.

$$1+\frac{1}{3}=1\frac{1}{3}$$

Remember, $1\frac{1}{3}$ means $1+\frac{1}{3}$

EXAMPLE 6.22: Change $\frac{17}{12}$ to a mixed number.

Solution:

$$\begin{array}{r} 1+\dfrac{5}{12} \\ \hline 12\overline{)17} \\ 12 \\ \hline 5 \end{array} \qquad 1+\frac{5}{12}=1\frac{5}{12}$$

EXAMPLE 6.23: Change $\frac{16}{3}$ to a mixed number.

Solution:

$$5 + \frac{1}{3}$$

$$3\overline{)16}$$
$$\underline{15}$$
$$1$$

$$5 + \frac{1}{3} = 5\frac{1}{3}$$

EXAMPLE 6.24: Change $2\frac{2}{3}$ to an improper fraction.

Solution: Step 1. Separate the whole number and fractional parts.

$$2\frac{2}{3} = 2 + \frac{2}{3}$$

Step 2. Find the CD and add.
The denominators are 1 $\left(2 = \frac{2}{1}\right)$ and 3.

$$2 = \frac{6}{3}$$

Therefore, $\frac{6}{3} + \frac{2}{3} = \frac{6+2}{3} = \frac{8}{3}$.

$$2\frac{2}{3} = \frac{8}{3}$$

A shortcut method is:
Multiply the whole number and the denominator of the fraction.
$2 \times 3 = 6$
Add the numerator to this number.
$6 + 2 = 8$
This number is the numerator in the answer.
The denominator is unchanged.

$$2\frac{2}{3} = \frac{6+2}{3} = \frac{8}{3}$$

EXAMPLE 6.25: Using the shortcut method, change $5\frac{1}{4}$ to an improper fraction.

Solution: $5\frac{1}{4} = \frac{5 \times 4 + 1}{4} = \frac{20+1}{4} = \frac{21}{4}$

EXAMPLE 6.26: Change $8\frac{7}{8}$ to an improper fraction.

Solution: $8\frac{7}{8} = \frac{8 \times 8 + 7}{8} = \frac{64+7}{8} = \frac{71}{8}$

PRACTICE PROBLEMS 6-10   Change the following improper fractions to mixed numbers:

1. $\dfrac{27}{4}$ 2. $\dfrac{23}{8}$

3. $\dfrac{15}{3}$ 4. $\dfrac{17}{3}$

5. $\dfrac{17}{6}$ 6. $\dfrac{21}{4}$

7. $\dfrac{28}{3}$ 8. $\dfrac{13}{4}$

Change the following mixed numbers to improper fractions:

9. $3\dfrac{5}{8}$ 10. $5\dfrac{3}{4}$

11. $7\dfrac{1}{3}$ 12. $2\dfrac{1}{2}$

13. $4\dfrac{1}{8}$ 14. $6\dfrac{1}{3}$

15. $3\dfrac{1}{12}$ 16. $5\dfrac{3}{8}$

Solutions:

1. $6\dfrac{3}{4}$ 2. $2\dfrac{7}{8}$

3. 5 4. $5\dfrac{2}{3}$

5. $2\dfrac{5}{6}$ 6. $5\dfrac{1}{4}$

7. $9\dfrac{1}{3}$ 8. $3\dfrac{1}{4}$

9. $\dfrac{29}{8}$ 10. $\dfrac{23}{4}$

11. $\dfrac{22}{3}$ 12. $\dfrac{5}{2}$

13. $\dfrac{33}{8}$ 14. $\dfrac{19}{3}$

15. $\dfrac{37}{12}$ 16. $\dfrac{43}{8}$

Additional practice problems are at the end of the chapter.

6.6.1 Multiplication and Division.   Multiplying and dividing improper fractions is no different from multiplying and dividing proper fractions. When the answer is an improper fraction, we change it to a mixed number.

EXAMPLE 6.27:

(a) $\dfrac{4}{3} \times \dfrac{8}{7} = \dfrac{32}{21} = 1\dfrac{11}{21}$

(b) $\dfrac{5}{3} \times 2 = \dfrac{10}{3} = 3\dfrac{1}{3}$

(c) $\dfrac{5}{2} \div \dfrac{6}{5} = \dfrac{5}{2} \times \dfrac{5}{6} = \dfrac{25}{12} = 2\dfrac{1}{12}$

(d) $\dfrac{8}{3} \div \dfrac{1}{4} = \dfrac{8}{3} \times 4 = \dfrac{32}{3} = 10\dfrac{2}{3}$

When we multiply or divide mixed numbers, we change the mixed numbers to improper fractions and perform the multiplication or division as before.

EXAMPLE 6.28:

(a) $2\dfrac{1}{3} \times 3\dfrac{7}{8} = \dfrac{7}{3} \times \dfrac{31}{8} = \dfrac{217}{24} = 9\dfrac{1}{24}$

(b) $3\dfrac{1}{4} \times 4\dfrac{1}{3} = \dfrac{13}{4} \times \dfrac{13}{3} = \dfrac{169}{12} = 14\dfrac{1}{12}$

(c) $5\dfrac{1}{5} \div 3\dfrac{2}{3} = \dfrac{26}{5} \div \dfrac{11}{3} = \dfrac{26}{5} \times \dfrac{3}{11} = \dfrac{78}{55} = 1\dfrac{23}{55}$

(d) $2\dfrac{1}{3} \div 6\dfrac{1}{4} = \dfrac{7}{3} \div \dfrac{25}{4} = \dfrac{7}{3} \times \dfrac{4}{25} = \dfrac{28}{75}$

No change is necessary here since the answer is a proper fraction.

PRACTICE PROBLEMS 6-11   Perform the indicated operation. If the answer is an improper fraction, change it to a mixed number.

1. $1\dfrac{7}{8} \times 3\dfrac{3}{4}$

2. $\dfrac{7}{8} \times \dfrac{16}{9}$

3. $3 \times \dfrac{8}{3}$

4. $\dfrac{5}{3} \times 3\dfrac{1}{2}$

5. $\dfrac{9}{5} \times 2\dfrac{2}{5}$

6. $\dfrac{9}{7} \div \dfrac{4}{3}$

7. $2\dfrac{3}{4} \div \dfrac{4}{3}$

8. $4 \div 1\dfrac{2}{3}$

9. $5\dfrac{1}{6} \div 4\dfrac{5}{12}$

10. $5\dfrac{1}{2} \div \dfrac{3}{4}$

Solutions:

1. $7\dfrac{1}{32}$

2. $1\dfrac{5}{9}$

3. 8

4. $5\dfrac{5}{6}$

5. $4\frac{8}{25}$ **6.** $\frac{27}{28}$

7. $2\frac{1}{16}$ **8.** $2\frac{2}{5}$

9. $1\frac{9}{53}$ **10.** $7\frac{1}{3}$

Additional practice problems are at the end of the chapter.

6.6.2 Addition and Subtraction.   To add and subtract we change all mixed numbers to improper fractions and perform the additions and subtractions as we did with proper fractions. Let's work some together.

EXAMPLE 6.29:

(a) $\dfrac{9}{8} + \dfrac{5}{4}$

Find the common denominator and add. The CD is 8.

$$\frac{9}{8} + \frac{5}{4} = \frac{9}{8} + \frac{10}{8} = \frac{9+10}{8} = \frac{19}{8} = 2\frac{3}{8}$$

(b) $\dfrac{7}{3} + 2\dfrac{5}{6}$

Change $2\frac{5}{6}$ to an improper fraction:

$$2\frac{5}{6} = \frac{17}{6}$$

Find the CD. The CD is 6.

$$\frac{7}{3} + \frac{17}{6} = \frac{14}{6} + \frac{17}{6} = \frac{14+17}{6} = \frac{31}{6} = 5\frac{1}{6}$$

(c) $3\dfrac{3}{4} - 1\dfrac{7}{16}$

Change to improper fractions:

$$3\frac{3}{4} = \frac{15}{4} \text{ and } 1\frac{7}{16} = \frac{23}{16}$$

Find the CD. The CD is 16.

$$\frac{15}{4} - \frac{23}{16} = \frac{60}{16} - \frac{23}{16} = \frac{60-23}{16} = \frac{37}{16} = 2\frac{5}{16}$$

(d) $\dfrac{9}{4} - 7\dfrac{2}{3}$

Change $7\frac{2}{3}$ to an improper fraction:

$$7\frac{2}{3} = \frac{23}{3}$$

Find the CD. The CD is 12.

$$\frac{9}{4} - \frac{23}{3} = \frac{27}{12} - \frac{92}{12} = \frac{27-92}{12} = \frac{-65}{12} = -\frac{65}{12} = -5\frac{5}{12}$$

Remember, when the numerator or denominator is negative, we change *two* signs. In this case we change the sign of the numerator and the sign of the fraction.

PRACTICE PROBLEMS 6-12   Perform the indicated operations. Express your answers as mixed numbers or as proper fractions.

1. $\dfrac{16}{9} + \dfrac{4}{3}$

2. $3\dfrac{1}{3} + \dfrac{8}{15}$

3. $4\dfrac{7}{8} + \dfrac{9}{4}$

4. $5\dfrac{1}{3} + 6\dfrac{5}{8}$

5. $\dfrac{9}{8} - \dfrac{1}{2}$

6. $2\dfrac{1}{7} - \dfrac{8}{3}$

7. $4\dfrac{2}{5} - 2\dfrac{1}{10}$

8. $2\dfrac{9}{10} - 4\dfrac{1}{15}$

9. $\dfrac{9}{7} + \dfrac{8}{3} - 1\dfrac{2}{3}$

10. $5\dfrac{5}{6} - 1\dfrac{2}{3} + 3\dfrac{3}{4}$

Solutions:

1. $\dfrac{28}{9} = 3\dfrac{1}{9}$

2. $\dfrac{58}{15} = 3\dfrac{13}{15}$

3. $\dfrac{57}{8} = 7\dfrac{1}{8}$

4. $\dfrac{287}{24} = 11\dfrac{23}{24}$

5. $\dfrac{5}{8}$

6. $\dfrac{-11}{21}$

7. $\dfrac{23}{10} = 2\dfrac{3}{10}$

8. $-\dfrac{35}{30} = -\dfrac{7}{6} = -1\dfrac{1}{6}$

9. $\dfrac{48}{21} = 2\dfrac{6}{21} = 2\dfrac{2}{7}$

10. $\dfrac{190}{24} = \dfrac{95}{12} = 7\dfrac{11}{12}$

Additional practice problems are at the end of the chapter.

SELF-TEST 6-5   Change to mixed numbers:

1. $\dfrac{29}{8}$

2. $\dfrac{33}{5}$

Change to improper fractions:

3. $3\frac{3}{16}$

4. $8\frac{2}{3}$

Perform the indicated operation:

5. $2\frac{5}{8} \times \frac{9}{4}$

6. $3\frac{1}{3} \times 4\frac{1}{2}$

7. $\frac{9}{8} + 1\frac{7}{8}$

8. $5\frac{1}{3} \div 2\frac{5}{6}$

9. $\frac{8}{3} + 1\frac{1}{2}$

10. $2\frac{7}{8} + 5\frac{1}{4}$

11. $6\frac{1}{2} - 1\frac{3}{32}$

12. $8\frac{1}{3} - 4\frac{5}{12}$

13. $1\frac{1}{4} + \frac{9}{8} - \frac{10}{3}$

14. $6\frac{1}{2} + 4\frac{1}{3} - 5\frac{1}{5}$

Answers to Self-Test 6-5 are at the end of the chapter.

6.7 DECIMAL FRACTIONS

It is often desirable or necessary to express answers as decimal fractions rather than as mixed numbers when working with fractions. We worked with decimal fractions in Chapter 1. Let's perform all four functions on two fractions and give the answers as decimal fractions.

EXAMPLE 6.30: Add, subtract, multiply, and divide the following two fractions: $\frac{3}{8}$ and $\frac{2}{3}$.

$$\frac{3}{8} + \frac{2}{3}$$

$$\frac{3}{8} - \frac{2}{3}$$

$$\frac{3}{8} \times \frac{2}{3}$$

$$\frac{3}{8} \div \frac{2}{3}$$

Solution:

$$\frac{3}{8} + \frac{2}{3} = 3 \div 8 + 2 \div 3 = 0.375 + 0.667 = 1.04$$

$$\frac{3}{8} - \frac{2}{3} = 0.375 - 0.667 = -0.292$$

$$\frac{3}{8} \times \frac{2}{3} = 0.375 \times 0.667 = 0.250$$

$$\frac{3}{8} \div \frac{2}{3} = 0.375 \div 0.667 = 0.563$$

This problem is easily solved using your calculator, but don't forget the heirarchy: **M**ultiply, **D**ivide, **A**dd, and **S**ubtract (remember **My Dear Aunt Sally**). Problem solving of fractional numbers using a calculator is shown in Appendix A.

EXAMPLE 6.31: Add, subtract, multiply, and divide the following two mixed numbers: $3\frac{1}{4}$ and $2\frac{1}{6}$.

$$3\frac{1}{4} + 2\frac{1}{6}$$

$$3\frac{1}{4} - 2\frac{1}{6}$$

$$3\frac{1}{4} \times 2\frac{1}{6}$$

$$3\frac{1}{4} \div 2\frac{1}{6}$$

Solution:

$$3\frac{1}{4} + 2\frac{1}{6} = 3 + 1 \div 4 + 2 + 1 \div 6 = 3.25 + 2.17 = 5.42$$

$$3\frac{1}{4} - 2\frac{1}{6} = 3.25 - 2.17 = 1.08$$

$$3\frac{1}{4} \times 2\frac{1}{6} = 3.25 \times 2.17 = 7.04$$

$$3\frac{1}{4} \div 2\frac{1}{6} = 3.25 \div 2.17 = 1.50$$

PRACTICE PROBLEMS 6-13   Add, subtract, multiply, and divide the following numbers. Express answers as decimal numbers.

1. $\frac{1}{3}$ and $\frac{5}{8}$

2. $\frac{6}{7}$ and $\frac{3}{5}$

3. $\frac{9}{16}$ and $\frac{3}{32}$

4. $\frac{7}{64}$ and $\frac{2}{3}$

5. $1\frac{3}{5}$ and $7\frac{1}{8}$

6. $9\frac{3}{16}$ and $2\frac{3}{4}$

7. $3\frac{1}{3}$ and $1\frac{5}{32}$

8. $4\frac{2}{9}$ and $2\frac{1}{3}$

9. $2\frac{11}{16}$ and $8\frac{1}{8}$

10. $6\frac{1}{2}$ and $4\frac{7}{8}$

Solutions:

1.
$$\frac{1}{3} + \frac{5}{8} = 0.958$$
$$\frac{1}{3} - \frac{5}{8} = -0.292$$
$$\frac{1}{3} \times \frac{5}{8} = 0.208$$
$$\frac{1}{3} \div \frac{5}{8} = 0.533$$

2.
$$\frac{6}{7} + \frac{3}{5} = 1.46$$
$$\frac{6}{7} - \frac{3}{5} = 0.257$$
$$\frac{6}{7} \times \frac{3}{5} = 0.514$$
$$\frac{6}{7} \div \frac{3}{5} = 1.43$$

3.
$$\frac{9}{16} + \frac{3}{32} = 0.656$$
$$\frac{9}{16} - \frac{3}{32} = 0.469$$
$$\frac{9}{16} \times \frac{3}{32} = 0.0527$$
$$\frac{9}{16} \div \frac{3}{32} = 6.00$$

4.
$$\frac{7}{16} + \frac{2}{3} = 1.10$$
$$\frac{7}{16} - \frac{2}{3} = -0.229$$
$$\frac{7}{16} \times \frac{2}{3} = 0.292$$
$$\frac{7}{16} \div \frac{2}{3} = 0.656$$

5. $1\frac{3}{5}+7\frac{1}{8}=8.73$

$1\frac{3}{5}-7\frac{1}{8}=-5.53$

$1\frac{3}{5}\times7\frac{1}{8}=11.4$

$1\frac{3}{5}\div7\frac{1}{8}=0.225$

6. $9\frac{3}{16}+2\frac{3}{4}=11.9$

$9\frac{3}{16}-2\frac{3}{4}=6.43$

$9\frac{3}{16}\times2\frac{3}{4}=25.3$

$9\frac{3}{16}\div2\frac{3}{4}=3.34$

7. $3\frac{1}{3}+1\frac{5}{32}=4.49$

$3\frac{1}{3}-1\frac{5}{32}=2.18$

$3\frac{1}{3}\times1\frac{5}{32}=3.85$

$3\frac{1}{3}\div1\frac{5}{32}=2.88$

8. $4\frac{2}{9}+2\frac{1}{3}=6.56$

$4\frac{2}{9}-2\frac{1}{3}=1.89$

$4\frac{2}{9}\times2\frac{1}{3}=9.85$

$4\frac{2}{9}\div2\frac{1}{3}=1.81$

9. $2\frac{11}{16}+8\frac{1}{8}=10.8$

$2\frac{11}{16}-8\frac{1}{8}=-5.44$

$2\frac{11}{16}\times8\frac{1}{8}=21.8$

$2\frac{11}{16}\div8\frac{1}{8}=0.331$

10. $6\frac{1}{2}+4\frac{7}{8}=11.4$

$6\frac{1}{2}-4\frac{7}{8}=1.63$

$6\frac{1}{2}\times4\frac{7}{8}=31.7$

$6\frac{1}{2}\div4\frac{7}{8}=1.33$

Additional practice problems are at the end of the chapter.

SELF-TEST 6-6   Add, subtract, multiply, and divide the following numbers. Express answers as decimal numbers.

1. $\frac{9}{16}$ and $\frac{3}{8}$

2. $\frac{2}{3}$ and $\frac{1}{6}$

3. $5\frac{7}{8}$ and $2\frac{1}{4}$

4. $3\frac{3}{16}$ and $6\frac{1}{3}$

Answers to Self-Test 6-6 are at the end of the chapter.

END OF CHAPTER PROBLEMS 6-1

Find the prime factors of the following:

1. 18

2. 27

3. 44

4. 50

5. 63

6. 45

7. 92

8. 105

9. 231

10. 195

11. $8a^2b^2$

12. $12x^2$

13. $24x^2y$

14. $32ab^2$

15. $42c^3$

16. $56yz^3$

17. $88y^2z$

18. $132a^3b$

END OF CHAPTER PROBLEMS 6-2

Find the LCM of the following numbers:

1. $6, 15$

2. $24, 36$

3. $8, 24, 36$

4. $15, 18, 24$

5. $12, 30, 36$

6. $25, 75, 100$

7. ab^3, a^3b

8. a^3b^3, a^4b

9. ab^2, a^3b^3, a^3b^2

10. a^4b^2, a^4b^3, ab^3

11. $a^2bc^3, a^4b^2c^2$

12. $ab^2c^2, a^3bc^2, a^2b^2c^3$

13. a^4b^2c, a^4b^3c, abc^2

14. $15x^2y^2, 27x^3y$

15. $26xy^3, 39x^2y^2$

16. $8x^4y^2z, 32xy$

17. $16x^2y^2z, 32xy, 8x^3z^2$

18. $64x^2y, 16x^3y^3, 128xy^2$

END OF CHAPTER PROBLEMS 6-3

Perform the indicated operations:

1. $3a^2b \cdot 7a^2b^2$

2. $4a^3b \cdot 4ab$

3. $5a^{-1}b^{-3} \cdot 4a^{-3}b^{-2}$

4. $6a^{-2}b^{-2} \cdot 5a^{-3}b^{-4}$

5. $10x^2yz \cdot x^3y^2z \cdot 2xz^2$

6. $8x^3y^2z \cdot x^2z^2 \cdot 3xyz$

7. $a^{-1}bc^2 \cdot 3a^2b^{-2}c^{-4} \cdot 2a^2b^{-1}c$

8. $x^{-2}y^{-1}z^2 \cdot 4x^2y^{-1}z \cdot 3x^4y^{-2}z^3$

9. $3x^2y^{-1}z \cdot 4x^4y^{-2}z \cdot 5x^{-1}y^{-1}z^2$

10. $4x^{-3}yz^2 \cdot 5x^{-3}y^{-3}z \cdot x^2y^2z^{-3}$

END OF CHAPTER PROBLEMS 6-4

Reduce to lowest terms:

1. $\dfrac{18a^3}{2a}$

2. $\dfrac{27a^5}{3a^2}$

3. $\dfrac{48a^2b^3}{6ab}$

4. $\dfrac{54a^4b^5}{9a^3b^2}$

5. $\dfrac{25a^4b^2c}{5abc^2}$

6. $\dfrac{8ab^3c^2}{16a^3bc^3}$

7. $\dfrac{9x^{-2}y^2}{45x^{-3}y}$

8. $\dfrac{63x^3y^{-2}}{7x^2y^{-4}}$

9. $\dfrac{44x^{-3}yz^{-2}}{4x^2y^2z^2}$

10. $\dfrac{96x^2y^{-3}z}{12x^{-3}y^2z^{-2}}$

END OF CHAPTER PROBLEMS 6-5

Perform the indicated multiplications. Reduce to lowest terms where possible.

1. $\dfrac{4}{5} \times \dfrac{3}{8}$

2. $\dfrac{2}{7} \times \dfrac{3}{4}$

3. $\dfrac{5}{12} \times \dfrac{3}{4} \times \dfrac{2}{3}$

4. $\dfrac{5}{8} \times \dfrac{3}{5} \times \dfrac{2}{3}$

5. $\dfrac{5a}{9} \times \dfrac{6a^2}{7}$

6. $\dfrac{7a^2}{8} \times \dfrac{6a^3}{7}$

7. $\dfrac{3x^2}{10y} \times \dfrac{5x}{8y}$

8. $\dfrac{10a}{21b^2} \times \dfrac{3a^2}{5b^2}$

9. $\dfrac{3}{4y^2} \times \dfrac{8xy}{9}$

10. $\dfrac{10x^2y}{12} \times \dfrac{5}{x}$

11. $\dfrac{7x^{-2}y}{8z} \times \dfrac{4y^2z^{-1}}{21x}$

12. $\dfrac{12x^3y^{-1}}{21z^2} \times \dfrac{7xz}{9y^3z^{-1}}$

13. $\dfrac{6x^2y^3}{7x^{-1}z} \times \dfrac{3x^2z}{4y^{-1}z^{-1}} \times \dfrac{2y^{-1}z^3}{3x^2y^2}$

14. $\dfrac{3y^{-1}z^2}{5x^{-2}} \times \dfrac{10x^3z^{-1}}{21y^2z^2} \times \dfrac{x^{-1}y^{-1}}{2}$

END OF CHAPTER PROBLEMS 6-6

Perform the following divisions. Reduce your answers to lowest terms.

1. $\dfrac{3}{4} \div \dfrac{9}{16}$

2. $\dfrac{4}{5} \div \dfrac{10}{3}$

3. $\dfrac{3}{16} \div \dfrac{7}{8}$

4. $\dfrac{20}{3} \div \dfrac{5}{6}$

5. $\dfrac{3x}{8} \div \dfrac{7x^3}{3}$

6. $\dfrac{7x^3}{9} \div \dfrac{28x}{21}$

7. $\dfrac{14ab^2}{9c^3} \div \dfrac{35b}{c^2}$

8. $\dfrac{35a^3b^{-1}}{24c^2} \div \dfrac{7ab^2}{12c^3}$

9. $\dfrac{3xy^3}{10x^{-1}} \div \dfrac{3xy^3}{10x^{-1}}$

10. $\dfrac{10a}{21b^2c^3} \div \dfrac{20a^{-2}}{28b^{-1}c^2}$

END OF CHAPTER PROBLEMS 6-7

Add the following fractions. Reduce to lowest terms where possible.

1. $\dfrac{1}{8} + \dfrac{3}{8}$

2. $\dfrac{5}{16} + \dfrac{7}{16}$

3. $\dfrac{11}{12} - \dfrac{5}{12}$

4. $\dfrac{5}{8} - \dfrac{3}{8}$

5. $\dfrac{5}{12} + \dfrac{7}{12} + \dfrac{11}{12}$

6. $\dfrac{3}{18} + \dfrac{5}{18} + \dfrac{1}{18}$

7. $\dfrac{3}{16} - \dfrac{5}{16} + \dfrac{7}{16}$

8. $\dfrac{5}{12} + \dfrac{7}{12} - \dfrac{1}{12}$

9. $\dfrac{2}{7x} + \dfrac{3}{7x}$

10. $\dfrac{2}{9a} + \dfrac{4}{9a}$

11. $\dfrac{6}{7x} - \dfrac{2}{7x}$

12. $\dfrac{5}{3x} - \dfrac{4}{3x}$

13. $\dfrac{1}{12ab} + \dfrac{5}{12ab} + \dfrac{3}{12ab}$

14. $\dfrac{1}{14ab} + \dfrac{3}{14ab} + \dfrac{5}{14ab}$

15. $\dfrac{3}{14a} + \dfrac{9}{14a} - \dfrac{5}{14a}$

16. $\dfrac{9}{16x} - \dfrac{5}{16x} + \dfrac{3}{16x}$

17. $\dfrac{5x}{24} + \dfrac{7x}{24}$

18. $\dfrac{3a}{20} + \dfrac{5a}{20}$

19. $\dfrac{8x}{15} - \dfrac{2x}{15}$

20. $\dfrac{7ab}{9} - \dfrac{4ab}{9}$

21. $\dfrac{7ab}{25} + \dfrac{6ab}{25} + \dfrac{2ab}{25}$

22. $\dfrac{2xy}{15} + \dfrac{4xy}{15} + \dfrac{6xy}{15}$

23. $\dfrac{9ab}{25} - \dfrac{2ab}{25} + \dfrac{8ab}{25}$

24. $\dfrac{3xy}{20} + \dfrac{7xy}{20} - \dfrac{9xy}{20}$

25. $\dfrac{2x}{9} + \dfrac{4x}{9} + \dfrac{4x}{9}$

26. $\dfrac{3ab}{14} + \dfrac{3ab}{14} + \dfrac{5ab}{14}$

27. $\dfrac{3x}{7} - \dfrac{5x}{7}$

28. $\dfrac{7}{11a} - \dfrac{9}{11a}$

29. $\dfrac{9x}{24} - \dfrac{15x}{24} + \dfrac{x}{24}$

30. $\dfrac{6a}{15} - \dfrac{7a}{15} - \dfrac{2a}{15}$

END OF CHAPTER PROBLEMS 6-8

Perform the indicated operations. Reduce to lowest terms where possible.

1. $\dfrac{3x}{5} + \dfrac{2x}{15}$

2. $\dfrac{3x}{8} + \dfrac{3x}{16}$

3. $\dfrac{2a}{3} + \dfrac{3a}{2}$

4. $\dfrac{b}{3} + \dfrac{2b}{5}$

5. $\dfrac{y}{2x} + \dfrac{5y}{8}$

6. $\dfrac{2b}{3a} + \dfrac{3a}{2b}$

7. $\dfrac{3x}{4y} + \dfrac{2y}{3x}$

8. $\dfrac{13ab}{16c} - \dfrac{ac}{4b}$

9. $\dfrac{3b}{4} + \dfrac{3b}{8} - \dfrac{15b}{16}$

10. $\dfrac{x}{10y^2} - \dfrac{2x}{5y^2} + \dfrac{x}{3y^2}$

11. $\dfrac{4x}{3yz} + \dfrac{2y}{5x} - \dfrac{z}{6y}$

12. $\dfrac{5y^2}{6x^2y} + \dfrac{4xy}{15y^2} - \dfrac{2x^2}{9y}$

END OF CHAPTER PROBLEMS 6-9

Change the following improper fractions to mixed numbers:

1. $\dfrac{17}{2}$

2. $\dfrac{9}{4}$

3. $\dfrac{37}{4}$

4. $\dfrac{24}{5}$

5. $\dfrac{19}{5}$

6. $\dfrac{17}{5}$

7. $\dfrac{43}{7}$

8. $\dfrac{36}{8}$

9. $\dfrac{27}{4}$

10. $\dfrac{19}{6}$

Change the following mixed numbers to improper fractions:

11. $10\dfrac{1}{2}$

12. $8\dfrac{3}{4}$

13. $4\dfrac{3}{16}$

14. $3\dfrac{2}{5}$

15. $3\dfrac{5}{8}$

16. $4\dfrac{1}{3}$

17. $9\dfrac{1}{3}$

18. $8\dfrac{1}{4}$

19. $5\dfrac{1}{6}$

20. $6\dfrac{2}{3}$

END OF CHAPTER PROBLEMS 6-10

Perform the indicated operations. If the answer is an improper fraction, change to a mixed number:

1. $\dfrac{3}{8} \times \dfrac{10}{3}$

2. $\dfrac{8}{5} \times \dfrac{5}{3}$

3. $2\dfrac{1}{3} \times 3\dfrac{1}{4}$

4. $1\dfrac{5}{6} \times 2\dfrac{7}{8}$

5. $\dfrac{9}{4} \times 2\dfrac{1}{5}$

6. $\dfrac{8}{5} \times 2\dfrac{5}{8}$

7. $3 \times 3\dfrac{1}{4}$

8. $2 \times 5\dfrac{5}{6}$

9. $\dfrac{7}{2} \div \dfrac{4}{3}$

10. $\dfrac{9}{5} \div \dfrac{5}{3}$

11. $3 \div \dfrac{7}{5}$

12. $4 \div \dfrac{8}{3}$

13. $\dfrac{9}{4} \div 1\dfrac{2}{3}$

14. $\dfrac{8}{7} \div 2\dfrac{1}{4}$

15. $3\dfrac{1}{2} \div 4\dfrac{1}{3}$

16. $5\dfrac{2}{3} \div 3\dfrac{2}{3}$

17. $2\dfrac{1}{3} \div 6\dfrac{1}{2}$

18. $1\dfrac{7}{16} \div 3\dfrac{3}{8}$

END OF CHAPTER PROBLEMS 6-11

Perform the indicated operations. If the answer is an improper fraction, change to a mixed number:

1. $\dfrac{16}{5} + \dfrac{10}{3}$

2. $\dfrac{15}{2} + \dfrac{9}{7}$

3. $4\dfrac{5}{6} + \dfrac{5}{7}$

4. $6\dfrac{1}{3} + \dfrac{2}{5}$

5. $2\dfrac{1}{3} + \dfrac{9}{5}$

6. $3\dfrac{3}{8} + \dfrac{10}{9}$

7. $3\dfrac{7}{8} + 4\dfrac{1}{2}$

8. $4\dfrac{2}{5} + 3\dfrac{3}{10}$

9. $\dfrac{10}{3} - \dfrac{1}{4}$

10. $\dfrac{12}{5} - \dfrac{1}{3}$

11. $3\dfrac{5}{6} - \dfrac{9}{4}$

12. $7\dfrac{1}{2} - \dfrac{9}{8}$

13. $3\frac{1}{8} - 1\frac{1}{4}$

14. $3\frac{7}{12} - 5\frac{3}{5}$

15. $\frac{9}{4} + \frac{7}{2} - 2\frac{1}{3}$

16. $\frac{8}{3} + \frac{11}{5} - 1\frac{1}{2}$

17. $7\frac{1}{2} - 2\frac{1}{3} + 3\frac{1}{4}$

18. $6\frac{2}{3} + 3\frac{1}{5} - 2\frac{1}{4}$

END OF CHAPTER PROBLEMS 6-12

Add, subtract, multiply, and divide the following numbers. Express answers as decimal numbers.

1. $\frac{3}{8}$ and $\frac{3}{32}$

2. $\frac{3}{5}$ and $\frac{1}{3}$

3. $\frac{7}{16}$ and $\frac{2}{7}$

4. $\frac{13}{32}$ and $\frac{5}{64}$

5. $\frac{5}{8}$ and $\frac{37}{64}$

6. $\frac{5}{16}$ and $\frac{13}{32}$

7. $\frac{3}{5}$ and $\frac{3}{10}$

8. $\frac{5}{12}$ and $\frac{7}{10}$

9. $3\frac{3}{8}$ and $2\frac{1}{3}$

10. $7\frac{5}{32}$ and $4\frac{1}{12}$

11. $4\frac{2}{5}$ and $2\frac{15}{64}$

12. $9\frac{7}{16}$ and $4\frac{3}{32}$

13. $7\frac{1}{2}$ and $3\frac{5}{9}$

14. $8\frac{5}{16}$ and $5\frac{23}{64}$

15. $4\frac{7}{8}$ and $8\frac{7}{16}$

16. $5\frac{4}{7}$ and $9\frac{1}{12}$

17. $3\frac{2}{3}$ and $6\frac{2}{3}$

18. $6\frac{3}{64}$ and $9\frac{1}{2}$

19. $1\frac{7}{12}$ and $3\frac{2}{3}$

20. $2\frac{9}{16}$ and $4\frac{9}{32}$

ANSWERS TO SELF-TESTS

Self-Test 6-1

1. $2 \cdot 2 \cdot 2 \cdot 5$

2. $2 \cdot 2 \cdot 2 \cdot 2 \cdot 3$

3. $2 \cdot 2 \cdot 3 \cdot 7$

4. $2 \cdot 2 \cdot 3 \cdot 3 \cdot 5$

5. $2 \cdot 2 \cdot 7 \cdot x \cdot y^2$

6. $2 \cdot 3 \cdot 3 \cdot 3 \cdot a^3 \cdot b^2$

7. $2 \cdot 2 \cdot 2 \cdot 3 \cdot 5 x^2 z^2$

8. $3 \cdot 3 \cdot 5 \cdot 7 \cdot a^2 \cdot b^2$

9. 60

10. 120

11. 210

12. 210

13. $a^2 b^3$

14. $72xz^2$

15. $72a^2b^2c^2$

Self-Test 6-2

1. $15xy$

2. $12x^4y^4$

3. $30a^2b$

4. $30a^{-1}b^{-2}$

5. $42a^{-4}b^{-1}c^3$

6. $8x^2y^{-1}$

7. $8x^{-3}y^{-1}$

8. $7a^2c^{-3}$

9. $9a^{-4}b^2c^{-1}$

10. $16x^2y^{-2}z^{-4}$

Self-Test 6-3

1. $\dfrac{7}{20}$

2. $\dfrac{6x^4}{35}$

3. $\dfrac{1}{30}$

4. $\dfrac{2a^{-1}b^3c^{-2}}{5}$

5. $\dfrac{7a^{-2}b^5c^{-3}}{6}$

6. $\dfrac{3}{10}$

7. $\dfrac{10a^{-1}}{9}$

8. $2a^5b^{-2}c^4$

9. $\dfrac{4a^{-3}b^{-1}c}{3}$

10. $y^{-2}z^{-2}$

Self-Test 6-4

1. $\dfrac{1}{2}$

2. $\dfrac{5}{7a}$

3. $\dfrac{2}{3x}$

4. $\dfrac{7a}{6}$

5. $\dfrac{5a}{9c}$

6. $\dfrac{13}{24}$

7. $\dfrac{17}{60}$

8. $\dfrac{12a+10b}{15b}$

9. $\dfrac{42x^2+20x+21}{70x^3}$

10. $\dfrac{24a^2c-25b^2+27a^2bc}{90a^2bc}$

Self-Test 6-5

1. $3\frac{5}{8}$

2. $6\frac{3}{5}$

3. $\dfrac{51}{16}$

4. $\dfrac{26}{3}$

5. $5\frac{29}{32}$

6. 15

7. $\dfrac{3}{5}$

8. $1\frac{15}{17}$

9. $4\frac{1}{6}$

10. $8\frac{1}{8}$

11. $5\frac{13}{32}$

12. $3\frac{11}{12}$

13. $-\dfrac{23}{24}$

14. $5\frac{19}{30}$

Self-Test 6-6

1. $\dfrac{9}{16} + \dfrac{3}{8} = 0.938$

   $\dfrac{9}{16} - \dfrac{3}{8} = 0.188$

   $\dfrac{9}{16} \times \dfrac{3}{8} = 0.211$

   $\dfrac{9}{16} \div \dfrac{3}{8} = 1.50$

2. $\dfrac{2}{3} + \dfrac{1}{6} = 0.833$

   $\dfrac{2}{3} - \dfrac{1}{6} = 0.500$

   $\dfrac{2}{3} \times \dfrac{1}{6} = 0.111$

   $\dfrac{2}{3} \div \dfrac{1}{6} = 4.00$

3. $5\dfrac{7}{8} + 2\dfrac{1}{4} = 8.13$

   $5\dfrac{7}{8} - 2\dfrac{1}{4} = 3.63$

   $5\dfrac{7}{8} \times 2\dfrac{1}{4} = 13.2$

   $5\dfrac{7}{8} \div 2\dfrac{1}{4} = 2.61$

4. $3\dfrac{3}{16} + 6\dfrac{1}{3} = 9.52$

   $3\dfrac{3}{16} - 6\dfrac{1}{3} = -3.15$

   $3\dfrac{3}{16} \times 6\dfrac{1}{3} = 20.2$

   $3\dfrac{3}{16} \div 6\dfrac{1}{3} = 0.503$

CHAPTER 7

LINEAR EQUATIONS

7.1 IDENTIFYING EQUATIONS

Algebraic expressions separated by an equals sign (=) are called equations. The equals sign implies that the expressions are equal. Let's call the expression to the left of the equals sign the *left side* and the expression to the right of the equals sign the *right side*. All of the following algebraic statements are equations.

EXAMPLE 7.1

(a) $16 + 4 = 20$ (b) $16 + x = 20$

(c) $x + 4 = 20$ (d) $x^3 + 4 = 20$

(e) $x^2 + x = 20$ (f) $x + 4 = 20$

(g) $4x = 20$ (h) $\dfrac{x}{4} = 20$

(i) $x - 4 = 20$

In Example 7.1(a) all numbers are real numbers. Both sides have a value of 20. In all the other examples a part of the left side is unknown. The value of the unknown (x) must be of such a value as to make the left side equal to 20. Sometimes the value of x can be found by inspection. This should be the case in Example 7.1(b) and (c). In (b) $x = 4$,

$$16 + x = 20$$
$$16 + 4 = 20$$
$$20 = 20$$

and in (c), x must equal 16.

$$x+4=20$$
$$16+4=20$$
$$20=20$$

In other examples we must work a little harder. Notice in Examples 7.1(d) and (e) we have x^3 and x^2 terms. In all other equations x is raised to the first power. Equations in which the unknown is raised to the first power are called *linear* equations. Equations containing x^2 terms are called *second-degree* equations. Equations containing x^3 terms are called *third-degree* equations.

7.2 LINEAR EQUATIONS

Linear equations are easy to solve if we remember these simple rules:

1. The value of the left side equals the value of the right side.
2. We may change or manipulate the equation provided we do not destroy this equality.
3. We may perform any mathematical operation on one side provided we perform the same operation on the other side.

Let's examine rule 3 very closely. Rule 3 means that we can make any change we want in the value of one side so long as we make the same change in the other side. That is, we must maintain the equality between the sides. This means we can add, subtract, multiply, divide, square, etc. We can do anything necessary to solve the equation so long as we maintain the equality of sides.

In solving equations we typically solve for some unknown value. In simple linear equations there is only one unknown and only one value for that unknown.

Consider Example 7.1(c). We know from inspection that $x=16$, but let's work it mathematically. In solving equations we must somehow get the unknown on one side by itself. That is, we have to move all terms that are real numbers to one side and leave the term containing the unknown on the other side. It is standard practice to put the term containing the unknown on the left side. We would solve the equation in the following manner.

EXAMPLE 7.1(c): $x+4=20$.

Solution: We must move the 4 from the left side to the right side. This can be done by subtracting 4. If we subtract 4 from the left side, we must not forget to subtract 4 from the right side.

$$x+4-4=20-4$$

$$x=16$$

Then we replace the unknown with its value and check for equality:

$$x + 4 = 20$$
$$16 + 4 = 20$$
$$20 = 20$$

When we want to move a term from one side to the other, we do it by adding or subtracting. In the example above we subtracted. In the example below we will add.

EXAMPLE 7.1(i): $x - 4 = 20$.

Solution:

$$x - 4 + 4 = 20 + 4 \qquad \text{(add 4 to both sides)}$$
$$x = 24$$

Check: $\quad 24 - 4 = 20$
$$20 = 20$$

From these examples we can make up the following rule:

Terms may be moved from one side to another by changing their signs.

Notice that in Example 7.1(c) that 4 was positive when it was on the left side and became negative when it was moved to the right side. In Example 7.1(i), 4 was negative when it was on the left side and it became positive when it was moved to the right side.

EXAMPLE 7.1(g): $4x = 20$.

Solution:

$$4x = 20$$
$$\frac{4x}{4} = \frac{20}{4} \qquad \text{(divide \textit{both sides} by 4)}$$
$$x = 5$$

Check: $\quad 4(5) = 20$
$$20 = 20$$

In this example, the left-side term is $4x$. Since we have to find the value of x, we must somehow get rid of the 4. That is, we want the left side to equal x. We can make the left side equal to x by dividing by 4.

Let's look at Example 7.1(h):

$$\frac{x}{4} = 20$$

How can we make the left side equal to x? Multiplication by 4 is the only way it can be done.

$$\frac{x}{4} \times 4 = \frac{4x}{4} = x$$

Solving the equation for x, we get:

$$\frac{x}{4} = 20$$

$$\frac{x}{4}(4) = 20(4)$$

$$x = 80$$

Check: $\quad \dfrac{80}{4} = 20$

$$20 = 20$$

These two examples (g and h) show how we can move parts of terms from one side to another by multiplying or dividing.

EXAMPLE 7.2: $\dfrac{20}{x} = 5$. Find x.

Solution: We must get x out of the denominator. There are two ways we can do that. The first way is to take the reciprocal of both sides:

$$\frac{20}{x} = 5$$

$$\frac{x}{20} = \frac{1}{5}$$

Next we multiply both sides by 20. This is the only way we can move 20 to the right side.

$$\frac{x}{20}(20) = \frac{1}{5}(20)$$

$$\frac{20x}{20} = \frac{20}{5}$$

$$x = 4$$

Check: $\quad \dfrac{20}{4} = 5$

$$5 = 5$$

The other way to solve the problem is to multiply both sides by x.

$$\frac{20}{x}(x)=5x$$

$$\frac{20x}{x}=5x$$

$$20=5x$$

Next we divide both sides by 5.

$$\frac{20}{5}=\frac{5x}{5}$$

$$4=x$$

or $$x=4$$

EXAMPLE 7.3: $\frac{x}{3}=-5$. Find x.

Solution:

$$\frac{x}{3}(3)=-5(3) \qquad \text{(multiply both sides by 3)}$$

$$x=-15$$

Check: $$\frac{-15}{3}=-5$$

$$-5=-5$$

EXAMPLE 7.4: $\frac{24}{-x}=-2$. Find x.

Solution:

$$\frac{-x}{24}=-\frac{1}{2} \qquad \text{(take the reciprocal of both sides)}$$

$$\frac{-x}{24}(24)=-\frac{1}{2}(24)$$

$$-x=-12$$

$$x=12 \qquad \text{(change the sign of both sides)}$$

Check: $$\frac{24}{-12}=-2$$

$$-2=-2$$

PRACTICE PROBLEMS 7-1   Solve for the unknown in the following equations. Check your answers by plugging the answer back into the original problem.

1. $x - 3 = 14$

2. $a + 5 = 6$

3. $4 - a = 6$

4. $5 + y = 6$

5. $3x = 12$

6. $5a = 30$

7. $6a = -42$

8. $-3a = 18$

9. $\dfrac{x}{3} = 4$

10. $\dfrac{a}{7} = 3$

11. $\dfrac{a}{5} = -3$

12. $\dfrac{-x}{6} = 7$

13. $\dfrac{24}{x} = 6$

14. $\dfrac{28}{a} = 7$

15. $\dfrac{-32}{b} = 4$

16. $\dfrac{-48}{y} = -6$

17. $\dfrac{2x}{3} = 4$

18. $\dfrac{4}{3x} = 8$

Solutions:

1. 17

2. 1

3. -2

4. 1

5. 4

6. 6

7. -7

8. -6

9. 12

10. 21

11. -15

12. -42

13. 4

14. 4

15. -8

16. 8

17. 6

18. $\dfrac{1}{6}$

Additional practice problems are at the end of the chapter.

7.3 SECOND-DEGREE EQUATIONS

Up to this point we have solved equations by means of addition, subtraction, multiplication, or division. Some equations we use in electronics require additional operations. Let's consider some general case equations.

EXAMPLE 7.5: $x^2 = 81$. Find x.

Solution: Whenever a number is squared, we can find that number by finding its square root. $\sqrt{6^2} = 6$, $\sqrt{7^2} = 7$, and $\sqrt{x^2} = x$.

Of course, if we take the square root of the left side, we must also take the square root of the right side.

$$x^2 = 81$$
$$\sqrt{x^2} = \sqrt{81}$$
$$x = 9$$

A good rule to remember is:

The square root of the square of a number is that number.

We can turn this rule around and say:

The square of the square root of a number is that number.

An example of how to apply this rule follows.

EXAMPLE 7.6: $\sqrt{x} = 5$. Find x.

Solution: We need x where we have $\sqrt{x}$. We must square the term in order to get rid of the square root.

$$(\sqrt{x})^2 = x \quad \text{or} \quad (x^{1/2})^2 = x^1 = x$$

Of course, we must also square the right side.

$$\sqrt{x} = 5$$
$$(\sqrt{x})^2 = 5^2$$
$$x = 25$$

Check: $\sqrt{25} = 5$ $5 = 5$

Let's look at some more examples of problems involving squares and square roots.

EXAMPLE 7.7: $\dfrac{x^2}{4} = 6$. Find x.

Solution:

$$\frac{x^2}{4}(4) = 6(4) \qquad \text{(multiply both sides by 4)}$$

$$x^2 = 24$$

$$\sqrt{x^2} = \sqrt{24} \qquad \text{(take the square root of both sides)}$$

$$x = 4.9$$

Check: $\dfrac{4.9^2}{4} = 6$ $6 = 6$

EXAMPLE 7.8: $\dfrac{\sqrt{y}}{2} = 4$. Find y.

Solution:

$$\frac{\sqrt{y}}{2}(2) = 4(2) \qquad \text{(multiply both sides by 2)}$$

$$\sqrt{y} = 8$$

$$y = 64 \qquad \text{(square both sides)}$$

Check:
$$\frac{\sqrt{64}}{2} = 4$$

$$\frac{8}{2} = 4$$

$$4 = 4$$

EXAMPLE 7.9: $15 = 3xy$. Find x.

Solution:

$$\frac{15}{3} = \frac{\cancel{3}xy}{\cancel{3}} \qquad \text{(divide by 3)}$$

$$5 = xy$$

$$\frac{5}{y} = \frac{x\cancel{y}}{\cancel{y}} \qquad \text{(divide by } y\text{)}$$

$$\frac{5}{y} = x$$

$$\text{or} \qquad x = \frac{5}{y}$$

Check:
$$15 = 3\left(\frac{5}{y}\right)y$$

$$15 = \frac{3 \cdot 5 \cdot \cancel{y}}{\cancel{y}}$$

$$15 = 15$$

EXAMPLE 7.10: $4 = \dfrac{1}{3a\sqrt{b}}$. Find a and then b.

Solution:

$$4(a) = \frac{1}{3a\sqrt{b}}(a) \qquad \text{(multiply by } a\text{)}$$

$$4a = \frac{1}{3\sqrt{b}}$$

$$\frac{4a}{4} = \frac{1}{3\sqrt{b}}\left(\frac{1}{4}\right) \qquad \text{(divide by 4)}$$

$$a = \frac{1}{12\sqrt{b}}$$

Check: $\quad 4 = \dfrac{1}{3\left(\dfrac{1}{12\sqrt{b}}\right)(\sqrt{b})} = \dfrac{1}{3\left(\dfrac{1}{12}\right)} = \dfrac{1}{\dfrac{3}{12}} = \dfrac{12}{3} = 4$

$4 = \dfrac{1}{3a\sqrt{b}}$. Find b.

$$4(\sqrt{b}) = \frac{1}{3a\sqrt{b}}(\sqrt{b}) \qquad \text{(multiply by } \sqrt{b}\text{)}$$

$$4\sqrt{b} = \frac{1}{3a}$$

$$\frac{4\sqrt{b}}{4} = \frac{1}{3a}\left(\frac{1}{4}\right) \qquad \text{(divide by 4)}$$

$$\sqrt{b} = \frac{1}{12a}$$

$$(\sqrt{b})^2 = \left(\frac{1}{12}a\right)^2 \qquad \text{(square both sides)}$$

$$b = \frac{1}{144a^2}$$

$$\text{or} \quad b = \frac{a^{-2}}{144}$$

Check: $\quad 4 = \dfrac{1}{3a\sqrt{\dfrac{1}{144a^2}}}$

$$4 = \frac{1}{3a\left(\dfrac{1}{12}a\right)} = \frac{1}{\dfrac{3a}{12a}} = \frac{12}{3} = 4$$

We can't cover all the kinds of equations that might be encountered, but the examples above should provide enough problem-solving situations so that any similar equations can be solved.

PRACTICE PROBLEMS 7-2   Solve for the unknown in the following equations:

1. $x^2 = 36$

2. $\dfrac{x^2}{5} = 7$

3. $\sqrt{x} = 6$

4. $\dfrac{\sqrt{b}}{6} = 2$

5. $4 = \dfrac{1}{3\sqrt{x}}$

Solve for each literal number in the following equations:

6. $6 = 3ab$

7. $2a = 3b$

8. $4\ k\Omega = R_1 + R_2$

9. $\dfrac{P}{R} = I^2$

10. $X_L = 2\pi fL$

11. $f_r = \dfrac{1}{2\pi\sqrt{LC}}$

(Do not solve for π. π is a constant, not a variable. π always equals 3.14.)

12. $4 = \dfrac{1}{3a\sqrt{b}}$

13. $6 = \dfrac{2}{3\sqrt{ab}}$

14. $X_C = \dfrac{1}{2\pi fC}$

Solutions:

1. 6

2. 5.92

3. 36

4. 144

5. $\dfrac{1}{144}$

6. $a = \dfrac{2}{b}$, $b = \dfrac{2}{a}$

7. $a = \dfrac{3b}{2}$, $b = \dfrac{2a}{3}$

8. $R_1 = 4\ k\Omega - R_2$, $R_2 = 4\ k\Omega - R_1$

9. $I = \sqrt{\dfrac{P}{R}}$, $R = \dfrac{P}{I^2}$

10. $f = \dfrac{X_L}{2\pi L}$, $L = \dfrac{X_L}{2\pi f}$

11. $L = \dfrac{1}{4\pi^2 f_r^2 C}$, $C = \dfrac{1}{4\pi^2 f_r^2 L}$

12. $a = \dfrac{1}{12\sqrt{b}}$, $b = \dfrac{1}{144a^2}$

13. $a = \dfrac{1}{81b}$, $b = \dfrac{1}{81a}$

14. $f = \dfrac{1}{2\pi CX_C}$, $C = \dfrac{1}{2\pi fX_C}$

Additional practice problems are at the end of the chapter.

SELF-TEST 7-1 Solve for the unknown in the following equations:

1. $y - 6 = 4$ **2.** $4x = 20$

3. $\dfrac{-a}{4} = 2$ **4.** $\dfrac{5}{2y} = 2$

5. $a^2 = 49$ **6.** $\dfrac{a^2}{3} = 3$

7. $\sqrt{a} = 5$ **8.** $\dfrac{\sqrt{y}}{5} = 6$

9. $2 = \dfrac{1}{4\sqrt{x}}$

Solve for each literal number in the following equations:

10. $3xy = 7$ **11.** $R = \dfrac{E^2}{P}$

12. $3 = \dfrac{1}{4x\sqrt{y}}$

Answers to Self-Test 7-1 are at the end of the chapter.

7.4 APPLICATIONS

Quite often in electronics problem solving, we are solving an equation—solving for the unknown. In this section we will change equations so that the unknown is on one side and all known values are on the other. The following problems are examples of problems that have to be solved when working with electrical circuits.

EXAMPLE 7.11: $E = IR$. Find I when $E = 12$ V and $R = 12$ kΩ.

Solution: First solve the equation for I.

$$E = IR$$

$$\frac{E}{R} = \frac{IR}{R} \qquad \text{(divide by } R\text{)}$$

$$\frac{E}{R} = I$$

$$I = \frac{E}{R} \qquad \begin{array}{l}\text{(change sides so that the unknown} \\ \text{is on the left side)}\end{array}$$

Now plug in values of E and R and solve.

$$I = \frac{12 \text{ V}}{12 \text{ k}\Omega} = 1 \text{ mA}$$

EXAMPLE 7.12: $P = \dfrac{E^2}{R}$. Find E when $P = 300$ mW and $R = 1.5$ kΩ.

Solution: Solve for E.

$$PR = \frac{E^2}{R}(R) \quad \text{(multiply by } R)$$

$$PR = E^2$$

$$\sqrt{PR} = \sqrt{E^2} \quad \text{(take the square root of both sides)}$$

$$\sqrt{PR} = E$$

$$E = \sqrt{PR}$$

Plug in the known values.

$$E = \sqrt{300 \text{ mW} \times 1.5 \text{ k}\Omega} = \sqrt{450}$$
$$E = 21.2 \text{ V}$$

EXAMPLE 7.13: $\dfrac{N_P}{N_S} = \sqrt{\dfrac{Z_P}{Z_S}}$. Find Z_P when $Z_S = 8$ Ω, $N_P = 500$, and $N_S = 25$.

Solution: Solve for Z_P.

$$\left(\frac{N_P}{N_S}\right)^2 = \sqrt{\left(\frac{Z_P}{Z_S}\right)^2} \quad \text{(square both sides)}$$

$$\left(\frac{N_P}{N_S}\right)^2 = \frac{Z_P}{Z_S}$$

$$\left(\frac{N_P}{N_S}\right)^2 \times Z_S = \frac{Z_P}{Z_S}(Z_S) \quad \text{(multiply by } Z_S)$$

$$\left(\frac{N_P}{N_S}\right)^2 Z_S = Z_P$$

$$Z_P = \left(\frac{N_P}{N_S}\right)^2 Z_S$$

Plug in the known values.

$$Z_P = \left(\frac{500}{25}\right)^2 \times 8 \ \Omega$$
$$= 3200 \ \Omega = 3.2 \text{ k}\Omega$$

EXAMPLE 7.14: $f_r = \dfrac{1}{2\pi\sqrt{LC}}$. Find L if $C=400pF$ and $f_r = 15$ kHz.

Solution: Solve for L.

$$f_r\sqrt{LC} = \frac{1}{2\pi} \qquad \text{(multiply by } \sqrt{LC}\text{)}$$

$$\sqrt{LC} = \frac{1}{2\pi f_r} \qquad \text{(divide by } f_r\text{)}$$

$$LC = \left(\frac{1}{2\pi f_r}\right)^2 \qquad \text{(square both sides)}$$

$$LC = \frac{1}{4\pi^2 f_r^2}$$

$$L = \frac{1}{4\pi^2 f_r^2 C} \qquad \text{(divide by } C\text{)}$$

Plug in the known values.

$$L = \frac{1}{4\pi^2 \times (15 \text{ kHz})^2 \times 400 \text{ pF}} = \frac{1}{3.55}$$

$$= 2.81 \times 10^{-1} \text{ H} = 281 \text{ mH}$$

PRACTICE PROBLEMS 7-3

1. $R_T = R_1 + R_2$. Find R_2 when $R_T = 678$ Ω and $R_1 = 238$ Ω.

2. $G_T = G_1 + G_2$. Find G_2 when $G_T = 600$ μS and $G_1 = 170$ μS.

3. $A_V = \dfrac{V_o}{V_{in}}$. Find V_o if $A_V = 48$ and $V_{in} = 25$ mV.

4. $P = \dfrac{E^2}{R}$. Find R if $E = 30$ V and $P = 780$ mW.

5. $P = I^2 R$. Find I if $P = 170$ mW and $R = 1.2$ kΩ.

6. $\dfrac{N_P^2}{N_S^2} = \dfrac{Z_P}{Z_S}$. Find N_S if $N_P = 200$, $Z_P = 2$ kΩ, and $Z_S = 16$ Ω.

7. $X_C = \dfrac{1}{2\pi f C}$. Find C if $f = 75$ Hz and $X_C = 13.7$ kΩ.

8. $X_L = 2\pi f L$. Find L if $f = 10$ kHz and $X_L = 2.73$ kΩ.

9. $B_C = 2\pi f C$. Find C if $f = 2.7$ kHz and $B_C = 200$ μS.

10. $f_r = \dfrac{1}{2\pi\sqrt{LC}}$. Find L if $f_r = 7.85$ kHz and $C = 20$ nF.

Solutions:

1. $R_2 = 440 \ \Omega$

2. $G_2 = 430 \ \mu S$

3. $V_o = 1.2$ V

4. $R = 1.15$ kΩ

5. $I = 11.9$ mA

6. $N_S = 17.9$

7. $C = 155$ nF

8. $L = 43.4$ mH

9. $C = 11.8$ nF

10. $L = 20.6$ mH

END OF CHAPTER PROBLEMS 7-1

Solve for the unknown in the following problems:

1. $a - 7 = 3$

2. $a - 5 = 3$

3. $5 - x = 3$

4. $8 - x = 5$

5. $b + 3 = 10$

6. $b + 2 = 6$

7. $3 - x = 7$

8. $2 - x = 7$

9. $5y = 20$

10. $6y = 24$

11. $2b = -18$

12. $7b = -28$

13. $-6x = 42$

14. $-3x = 21$

15. $\dfrac{c}{3} = 4$

16. $\dfrac{c}{6} = 2$

17. $\dfrac{a}{5} = -2$

18. $\dfrac{a}{4} = -6$

19. $\dfrac{x}{-2} = -8$

20. $\dfrac{x}{-4} = -3$

21. $\dfrac{3x}{4} = 6$

22. $\dfrac{2x}{5} = 4$

23. $\dfrac{7}{2y} = 14$

24. $\dfrac{5}{3a} = 4$

25. $\dfrac{5y}{-3} = 10$

26. $\dfrac{6y}{-5} = 8$

27. $\dfrac{-6}{7y} = 3$

28. $\dfrac{-3}{2a} = 9$

END OF CHAPTER PROBLEMS 7-2

Solve for the unknown in the following equations:

1. $x^2 = 25$

2. $x^2 = 16$

3. $\dfrac{a^2}{3} = 18$

4. $\dfrac{x^2}{6} = 5$

5. $\sqrt{a} = 7$

6. $\sqrt{a} = 4$

7. $\dfrac{\sqrt{x}}{3}=2$ **8.** $\dfrac{\sqrt{y}}{4}=3$

9. $2=\dfrac{1}{2\sqrt{x}}$ **10.** $3=\dfrac{1}{3\sqrt{x}}$

Solve for each literal number in the following equations:

11. $2ax=12$ **12.** $6bc=18$

13. $8bc=14$ **14.** $2\mathrm{mS}=G_1+G_2$

15. $P=\dfrac{E^2}{R}$ **16.** $B_L=\dfrac{1}{2\pi fL}$

17. $B_C=2\pi fC$ **18.** $6=\dfrac{1}{2a\sqrt{x}}$

19. $3=\dfrac{1}{2a\sqrt{b}}$ **20.** $3=\dfrac{4}{\sqrt{bc}}$

21. $5=\dfrac{6}{\sqrt{xy}}$

END OF CHAPTER PROBLEMS 7-3

1. $R_T=R_1+R_2$. Find R_1 if $R_T=12.4$ kΩ and $R_2=5.6$ kΩ.

2. $R_T=R_1+R_2$. Find R_2 if $R_T=27$ kΩ and $R_1=12$ kΩ.

3. $G_T=G_1+G_2$. Find G_1 if $G_T=54.9$ μS and $G_2=37$ μS.

4. $G_T=G_1+G_2$. Find G_2 if $G_T=516$ μS and $G_1=303$ μS.

5. $A_v=\dfrac{V_0}{V_{\text{in}}}$. Find V_0 if $A_V=150$ and $V_{\text{in}}=55$ mV.

6. $A_V=\dfrac{V_0}{V_{\text{in}}}$. Find V_{in} if $A_V=63$ and $V_0=3.15$ mV.

7. $P=\dfrac{E^2}{R}$. Find R if $P=2$ W and $E=15$ V.

8. $P=\dfrac{E^2}{R}$. Find E if $P=300$ mW and $R=270$ Ω.

9. $P=I^2R$. Find I if $P=600$ mW and $R=680$ Ω.

10. $P=I^2R$. Find R if $P=500$ mW and $I=10$ mA.

11. $\dfrac{N_P^2}{N_S^2}=\dfrac{Z_P}{Z_S}$. Find N_P if $N_S=100$, $Z_P=2.5$ kΩ, and $Z_S=8$ Ω.

12. $(\dfrac{N_P}{N_S})^2=\dfrac{Z_P}{Z_S}$. Find Z_S if $N_P=1200$, $N_S=100$, and $Z_P=1$ kΩ.

13. $X_C=\dfrac{1}{2\pi fC}$. Find C if $f=2.7$ kHz and $X_C=700$ Ω.

14. $X_C=\dfrac{1}{2\pi fC}$. Find f if $C=10$ nF and $X_C=5$ kΩ.

15. $X_L=2\pi fL$. Find L if $f=4.8$ kHz and $X_L=3.85$ kΩ.

16. $X_L=2\pi fL$. Find f if $L=54.1$ mH and $X_L=170$ Ω.

17. $B_C=2\pi fC$. Find C if $f=400$ Hz and $B_C=330$ μS.

18. $B_C=2\pi fC$. Find f if $B_C=2.3$ mS and $C=2.15$ μF.

19. $f_r=\dfrac{1}{2\pi\sqrt{LC}}$. Find L if $f_r=30$ kHz and $C=100$ pF.

20. $f_r=\dfrac{1}{2\pi\sqrt{LC}}$. Find L if $f_r=3.5$ kHz and $C=250$ nF.

Self-Test 7-1

1. $y = 10$

2. $x = 5$

3. $a = -8$

4. $y = \dfrac{5}{4}$

5. $a = 7$

6. $a = 3$

7. $a = 25$

8. $y = 900$

9. $x = \dfrac{1}{64}$

10. $x = \dfrac{7}{3y}$, $y = \dfrac{7}{3x}$

11. $P = \dfrac{E^2}{R}$, $E = \sqrt{PR}$

12. $x = \dfrac{1}{12\sqrt{y}}$, $y = \dfrac{1}{144x^2}$

CHAPTER 8

DC CIRCUIT ANALYSIS—KIRCHHOFF'S LAWS

There are three basic laws dealing with circuit analysis. These are Kirchhoff's current and voltage laws and Ohm's law. Theorems have been developed from these basic laws to help in solving the more complex problems. These theorems will be discussed in Chapter 10. Since Kirchhoff's laws and Ohm's law are basic to the understanding of all electrical and electronics circuits, these laws will be discussed in detail in this and the following chapter.

All electrical circuits consist of three basic parts: an *electrical energy source*, a *transfer network*, and a *load*. Some typical energy loads are lamps, motors, and radios. If a load is an energy user, it is said to have the electrical property of resistance. In this kind of load, electrical energy is converted to some other form —usually heat and light. If the load is an energy storer, it has the property of capacitance or inductance. In this chapter we will limit ourselves to loads that are energy users, that is, those loads that are resistive.

The transfer network in an electrical circuit is usually a copper conductor that provides a path for energy to get from the source to the load. An example of such a simple circuit is shown in Figure 8-1.

8.1 KIRCHHOFF'S CURRENT LAW

The physical duality of Kirchhoff's current law can be visualized by using the following analogy. Consider water flowing in a pipe. If we were to choose a point anywhere along the length of the pipe, the rate of water flow (say gallons per minute) *into* that point would have to equal the rate of water flow *leaving* it, as illustrated in Figure 8-2.

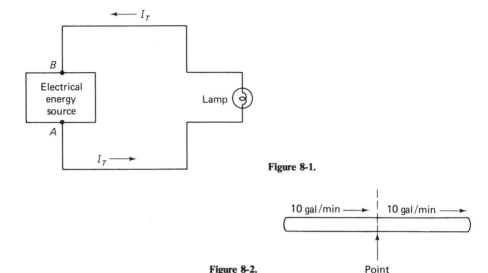

Figure 8-1.

Figure 8-2.

Point

Charges moving in a given direction in an electrical conductor are analogous to water flowing in a pipe. When a charge in motion approaches a point already containing a charge, there is a repelling force between them. If there is a continuous supply of charges, incoming charges cause other charges to move because like charges repel each other. If charges, then, are removed from a point, other charges take their place.

Kirchhoff's current law essentially expresses the law of conservation of charges and further defines their behavior at a point in an electrical circuit. It can be stated as follows:

> The number of charges per second (current) flowing *to* a point in a conductor must equal the number of charges flowing *from* that point.

Stated another way:

> The algebraic sum of currents *into* and *out* of a point must equal zero if currents toward the point are given an arbitrary plus sign (+) and if currents leaving the point are given an arbitrary negative sign (−). (Figure 8-3.)

8.1.1 Current in Series Circuits. Practical applications of Kirchhoff's current law may be further visualized by the following: Note that the current (I in Figure 8-1) has but one path. Therefore, according to Kirchhoff's current law, the magnitude of the current must be the same at any point in the closed circuit. If, for example, 1 A leaves point A in the diagram, 1 A must return to point B at that same instant. The current through the energy source, the lamp, and the

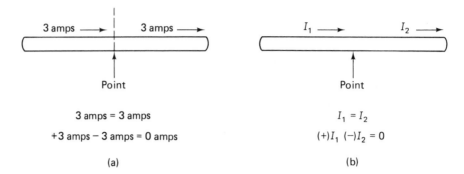

$$3 \text{ amps} = 3 \text{ amps}$$
$$+3 \text{ amps} - 3 \text{ amps} = 0 \text{ amps}$$

(a)

$$I_1 = I_2$$
$$(+)I_1 \ (-)I_2 = 0$$

(b)

Figure 8-3.

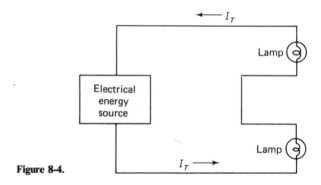

Figure 8-4.

conductor must be the same. Such an electrical circuit is called a series circuit. A series circuit is a circuit in which there is only one path for current.

The diagram in Figure 8-4 consists of two lamps tied end to end. Again, note that the current has but one path and its magnitude must be the same at any point in that path.

8.1.2   Current in Parallel Circuits.   Let us now go back to our water pipe analogy. This time we will add a "T" to our pipe. At a given instant, if 10 gal/min enter the "T," then 10 gal/min must leave at that same instant, as shown in Figure 8-5. The rate of water flowing *into* the "T" is equal to the rate of water flowing *out* of the "T."

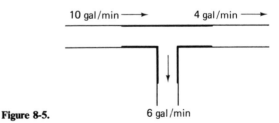

Figure 8-5.

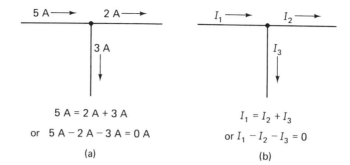

Figure 8-6.

The electrical analogy of the "T" is called a *junction*. A junction is a point in an electrical circuit where three or more conductors meet. Kirchhoff's current law for a junction can be stated as follows:

> The sum of currents toward a junction must equal the sum of currents leaving that junction (as illustrated in Figure 8-6).

Stated another way:

> The algebraic sum of currents into and out of a junction must equal zero if currents toward the junction are given an arbitrary (+) sign and if currents leaving the point are given an arbitrary (−) sign.

In Figure 8-7, the current into the junction is:

$$I_1 + I_2 = 2\text{ A} + 3\text{ A} = 5\text{ A}$$

Then, according to Kirchhoff's current law, the current out of the junction (I_3) must equal 5 A. Algebraically:

$$I_1 + I_2 = I_3$$

or

$$I_1 + I_2 - I_3 = 0$$

$I_1 = 2$ A ⟶

$I_3 = 5$ A ⟶

$I_2 = 3$ A ⟶   **Figure 8-7.**

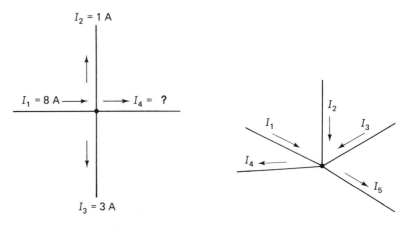

Figure 8-8.   **Figure 8-9.**

In Figure 8-8 the current into the junction, I_1, equals 8 A. Therefore, the current out of the junction must equal 8 A, or:

$$I_2 + I_3 + I_4 = 8 \text{ A}$$

$$1 \text{ A} + 3 \text{ A} + I_4 = 8 \text{ A}$$

Therefore, $I_4 = 4 \text{ A}$

Algebraically:

$$I_1 = I_2 + I_3 + I_4$$

or $I_1 - I_2 - I_3 - I_4 = 0$

Solve for I_1 and I_3 in Figure 8-9.

$$I_1 + I_2 + I_3 = I_4 + I_5$$

Then $I_1 = I_4 + I_5 - I_3 - I_2$

and $I_3 = I_4 + I_5 - I_2 - I_1$

The diagram in Figure 8-10 consists of two lamps connected *across* an energy source. Note that the current, I_T, into junction A must equal the currents out of that junction $(I_1 + I_2)$. Therefore, from Kirchhoff's current law:

$$I_T = I_1 + I_2$$

If the current, I_T, in Figure 8-10 is 5 A and I_1 is measured to be 2 A, then I_2 must equal 3 A.

$$I_T = I_1 + I_2$$

$$5 \text{ A} = 2 \text{ A} + I_2$$

$$5 \text{ A} - 2 \text{ A} = I_2$$

$$3 \text{ A} = I_2$$

The branch currents into and out of junction B must also be equal to 5 A. Circuits that have multiple current paths are called *parallel circuits*.

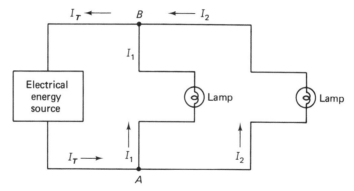

Figure 8-10.

8.1.3 Series-Parallel Circuits. An electrical circuit that has both series- and parallel-circuit characteristics is called a *series-parallel circuit*. Any number of energy sources and loads may make up a series-parallel circuit. For example, Figure 8-11 shows three lamps and a battery connected in a series-parallel arrangement. The part of the transfer network between points *A* and *B* carries the total current. At point *B* the current divides (parallel circuit)—one part flows through lamp L_2 and the remainder flows through lamp L_3. The current recombines at point *C* and total current flows in the transfer network between point *C* and lamp L_1 and on to point *D* (series circuit). So we have one circuit component (L_1) in which all the current flows (series portion) and two circuit components in which part of the total current flows (parallel portion).

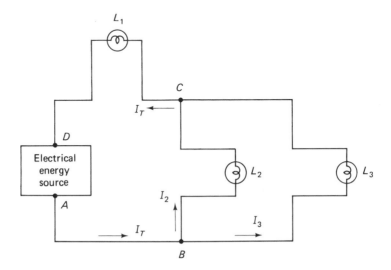

Figure 8-11.

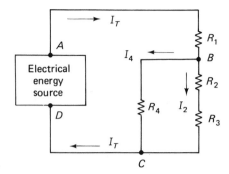

Figure 8-12.

Figure 8-12 shows another circuit configuration. Application of Kirchhoff's current law yields these current relationships:

$$I_T = I_1$$

$$I_T = I_4 + I_3$$

$$I_T = I_4 + I_2$$

$$I_2 = I_3$$

$$I_1 = I_2 + I_4$$

Inspection of the circuit shows that R_2 and R_3 are in series. Together they are in parallel with R_4. Finally, this whole combination is in series with R_1. In the circuit, then, total current flows through R_1 and a division takes place at point B resulting in two currents, I_4 and I_2 ($I_T = I_2 + I_4$). Since R_2 and R_3 are in series, the current through them is the same ($I_2 = I_3$). A recombination of currents takes place at the junction of R_3 and R_4 (point C). Then total current flows from point C to the electrical energy source (point D).

EXAMPLE 8.1: In Figure 8-13, if $I_T = 1$ A, $I_3 = 0.4$ A, and $I_4 = 0.5$ A, find I_1 and I_2.

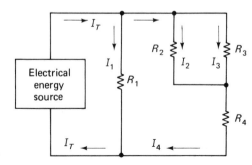

Figure 8-13.

Solution: Using Kirchhoff's current law, we can write equations for various currents:

$$I_T = I_1 + I_4$$

$$I_4 = I_2 + I_3$$

$$I_T = I_1 + I_2 + I_3$$

Choosing an equation containing only one unknown, we get:

$$I_T = I_1 + I_4$$

or $$I_1 = I_T - I_4 = 1 \text{ A} - 0.5 \text{ A} = 0.5 \text{ A}$$

and $$I_4 = I_2 + I_3$$

or $$I_2 = I_4 - I_3 = 0.5 \text{ A} - 0.4 \text{ A} = 0.1 \text{ A}$$

PRACTICE PROBLEMS 8-1   Refer to Figure 8-14.

1. Find I_T if $I_1 = 1$ A, $I_2 = 2$ A, and $I_3 = 3$ A.
2. Find I_3 if $I_T = 12$ mA, $I_1 = 1$ mA, and $I_2 = 7$ mA.
3. Find I_2 if $I_T = 10$ μA, $I_1 = 4$ μA, and $I_3 = 2$ μA.
4. Find I_1 if $I_T = 2$ A, $I_2 = 1.2$ A, and $I_3 = 300$ mA.

Refer to Figure 8-15.

5. $I_1 = 4$ mA and $I_2 = 3$ mA. Find I_T and I_3.
6. $I_T = 2.3$ mA and $I_3 = 600$ μA. Find I_1 and I_2.
7. $I_T = 400$ mA and $I_1 = 300$ mA. Find I_2 and I_3.

Refer to Figure 8-12.

8. $I_1 = 400$ mA, $I_2 = 150$ mA. Find I_T, I_3, and I_4.
9. $I_3 = 600$ μA, $I_4 = 800$ μA. Find I_T, I_1, and I_2.

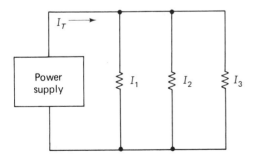

Figure 8-14.

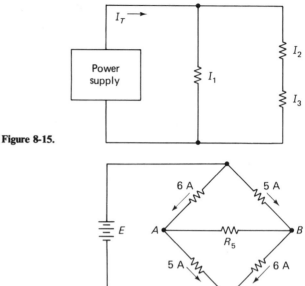

Figure 8-15.

Figure 8-16.

Refer to Figure 8-13.

10. $I_T = 17$ mA, $I_2 = 6$ mA, $I_4 = 13$ mA. Find I_1 and I_3.

11. $I_1 = 500$ μA, $I_2 = 1.3$ mA, $I_3 = 850$ μA. Find I_T and I_4.

12. Refer to Figure 8-16. Does current flow through R_5 from A to B or from B to A? What is the magnitude of this current?

Solutions:

1. 6 A

3. 4 μA

5. $I_T = 7$ mA, $I_3 = 3$ mA

7. $I_2 = 100$ mA, $I_3 = 100$ mA

9. $I_T = 1.4$ mA, $I_1 = 1.4$ mA, $I_2 = 600$ μA

11. $I_T = 2.65$ mA, $I_4 = 2.15$ mA

2. 4 mA

4. 500 mA

6. $I_1 = 1.7$ mA, $I_2 = 600$ μA

8. $I_T = 400$ mA, $I_3 = 150$ mA, $I_4 = 250$ mA

10. $I_1 = 4$ mA, $I_3 = 7$ mA

12. From A to B. $I_5 = 1$ A

Additional practice problems are at the end of the chapter.

8.2 KIRCHHOFF'S VOLTAGE LAW

If a charge travels around any closed path in an electrical network and returns to its starting point, its net potential energy change must be zero. That is, it must go through equal potential energy gains and potential energy losses.

Kirchhoff's voltage law expresses the law of conservation of energy and can be stated as follows:

> Around any closed path the sum of potential rises must equal the sum of potential drops ($\Sigma E = \Sigma V$).

Stated another way:

> The algebraic sum of emf's (E) and voltage drops (V) around a closed path is equal to zero ($\Sigma E - \Sigma V = 0$).

In the circuit of Figure 8-17 both L_1 and L_2 are connected *across* the electrical energy source. Kirchhoff's voltage law tells us that in any closed path the emf's and potential drops must be equal. For loop 1, $E = V_1$ and for loop 2, $E = V_2$. Therefore, $V_1 = V_2$.

Consider a single charge at the reference point in the circuit. The electrical energy source raises the potential energy of the charge to 10 V at point *A*. In order for the charge to return to the reference point, it must travel through lamp 1 or lamp 2. No matter which path it takes, it must give up the potential energy which the electrical energy source provided it.

Therefore, in parallel circuits the potential drop is the same for each branch and equals the emf.

In Figure 8-18 we have an electrical energy source connected in series with two lamps. Once again let's consider a single charge at the reference point. The electrical energy source raises the potential energy of the charge to 10 V (point *A*). In this circuit there is only one path the charge may take. It must pass through lamp 1 and then through lamp 2. When it has passed through lamp one, it has converted some, but not all, of the energy provided by the electrical energy source. The remainder of its energy is converted when it travels through lamp 2. The charge drops across a potential difference of 6 V for lamp 1. Since it

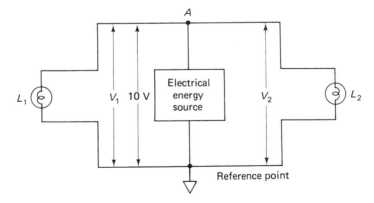

Figure 8-17.

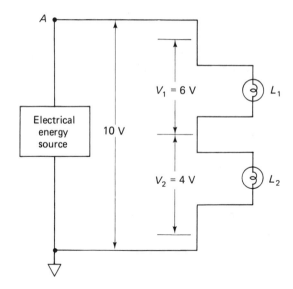

Figure 8-18.

drops across a total potential difference of 10 V, there remains 4 V to drop across lamp 2. When the charge finally passes through lamp 2, all of the potential energy provided to it by the electrical energy source is converted and the charge has made a complete trip around the circuit and returned to the reference point.

From the above example we may conclude that (a) in a series circuit the sum of the potential drops must equal the emf and (b) the current through each element is the same.

According to Kirchhoff's potential law, the total rise in potential (E) in Figure 8-18 must equal the total drop in potential:

$$E = V_1 + V_2$$

If, for the above example, the potential rise is equal to 10 V, then the sum of potential drops $(V_1 + V_2)$ must also equal 10 V.

> EXAMPLE 8.2: In Figure 8-18, find the magnitude of V_2 if $E = 30$ V and V_1 equals 10 V.

Solution:

$$E = V_1 + V_2$$

Then
$$V_2 = E - V_1$$

$$V_2 = 30 \text{ V} - 10 \text{ V} = 20 \text{ V}$$

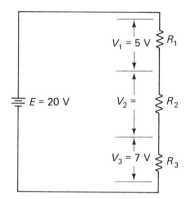

Figure 8-19.

EXAMPLE 8.3: In Figure 8-19, determine V_2.

Solution:

$$E = V_1 + V_2 + V_3$$

Then $\quad V_2 = E - V_1 - V_3 = 20\text{ V} - 5\text{ V} - 7\text{ V} = 8\text{ V}$

EXAMPLE 8.4: In Figure 8-20, determine V_4.

Solution:

$$E = V_1 + V_2 + V_3 + V_4$$

Then $\quad V_4 = E - V_1 - V_2 - V_3 = 25\text{ V} - 12\text{ V} - 8\text{ V} - 1\text{ V} = 4\text{ V}$

In Figure 8-21 Kirchhoff's voltage law is satisfied as part of the EMF is dropped across the circuit between points A and B (across R_2 and R_3) and part is

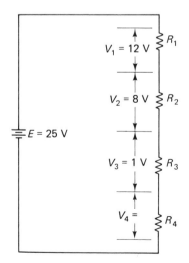

Figure 8-20.

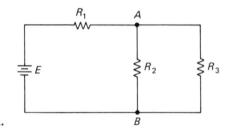

Figure 8-21.

dropped across R_1. The total potential drops equal V_1 plus V_2 (the drop between points A and B) and equal the EMF.

$$E = V_1 + V_2$$

and $E = V_1 + V_3$ (since there are two paths, either may be chosen
                          in writing the equation)

In Figure 8-22 application of Kirchhoff's potential law yields:

$$E = V_1 + V_2 + V_3$$ (because the sum of the potential drops
                              must equal the EMF)

and $E = V_1 + V_4$

or $E - V_1 - V_4 = 0$

That is, we may choose *any* closed path (or loop) and the sum of the drops and EMF's in that loop must equal zero.

EXAMPLE 8.5: In Figure 8-22 let $E = 10$ V, $V_2 = 5$ V, and $V_3 = 2$ V. Find V_1 and V_4.

Solution:

$$V_4 = V_2 + V_3$$
$$V_4 = 5\ V + 2\ V = 7\ V$$
$$E = V_1 + V_2 + V_3$$
$$V_1 = E - V_2 - V_3 = 10\ V - 5\ V - 2\ V = 3\ V$$

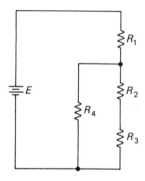

Figure 8-22.

PRACTICE PROBLEMS 8-2

1. In Figure 8-23, if $E = 25$ V and $V_1 = 10$ V, find V_2.

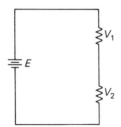

Figure 8-23.

2. In Figure 8-19, if $E = 20$ V, $V_1 = 5$ V, and $V_2 = 7$ V, find V_3.

3. In Figure 8-19, if $V_1 = 3$ V, $V_2 = 30$ V, and $V_3 = 10$ V, find E.

Refer to Figure 8-21.

4. $E = 50$ V, $V_3 = 20$ V. Find V_1 and V_2.

5. $E = 50$ V, $V_1 = 15$ V. Find V_2 and V_3.

6. $V_1 = 7$ V, $V_3 = 14$ V. Find E and V_2.

Refer to Figure 8-22.

7. $E = 10$ V, $V_2 = 5$ V, $V_3 = 2$ V. Find V_1 and V_4.

8. $E = 150$ mV, $V_2 = 30$ mV, $V_4 = 80$ mV. Find V_1 and V_3.

9. $V_1 = 7$ V, $V_2 = 5$ V, $V_4 = 8$ V. Find E and V_3.

Refer to Figure 8-24.

10. $E = 35$ V, $V_1 = 12$ V, $V_3 = 10$ V, $V_5 = 5$ V. Find V_2, V_4, and V_6.

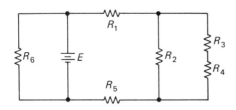

Figure 8-24.

Solutions:

1. 15 V **2.** 8 V

3. 43 V **4.** $V_1 = 30$ V, $V_2 = 20$ V

5. $V_2 = 35$ V, $V_3 = 35$ V **6.** $E = 21$ V, $V_2 = 14$ V

7. $V_1 = 3$ V, $V_4 = 7$ V **8.** $V_1 = 70$ mV, $V_3 = 50$ mV

9. $E = 15$ V, $V_3 = 2$ V **10.** $V_2 = 18$ V, $V_4 = 8$ V, $V_6 = 35$ V

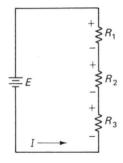

Figure 8-25.

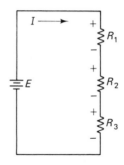

Figure 8-26.

8.3 POLARITY

The polarity of the potential drops across the individual resistances depends on the direction of current. If we use electron current, as in Figure 8-25, the polarities are as shown. The end of the resistor where current enters is assigned a negative $(-)$ sign. This makes the other end positive $(+)$. If we use conventional current, as in Figure 8-26, the end where current enters is assigned a positive $(+)$ sign. This makes the other end negative $(-)$. We see in Figures 8-25 and 8-26 that the resultant polarities are the same regardless of whether we use electron current or conventional current. In our circuits polarities are assigned, but usually one chooses the direction of current with which one is most comfortable.

In Figure 8-27, polarities are shown as dictated by the direction of current. Suppose that $V_1=5$ V, $V_2=7$ V, and $V_3=8$ V. The general equation states that:

$$E+V_1+V_2+V_3=0$$

Observing polarity, let's start at point A and go around the loop in a counterclockwise direction. Using the sign we encounter first as we go through each source and resistance, we get:

$$+20\text{ V}-8\text{ V}-7\text{ V}-5\text{ V}=0$$
$$20\text{ V}-20\text{ V}=0$$

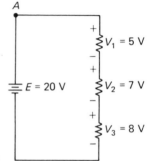

Figure 8-27.

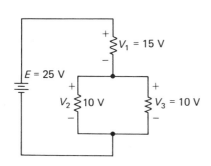

Figure 8-28.

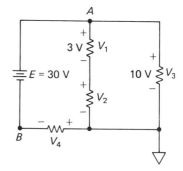

Figure 8-29.

In Figure 8-28, R_2 and R_3 are in parallel and they are in series with R_1. Because R_2 and R_3 are in parallel, $V_2 = V_3$. One loop includes E, R_1, and R_2. Therefore, $E = V_1 + V_2$. A second loop includes E, R_1, and R_3. Therefore, $E = V_1 + V_3$. In each case 25 V = 15 V + 10 V.

Let's find V_2 and V_4 in Figure 8-29. One loop includes E, R_1, R_2 and R_4. Since we have two unknown potential drops in this loop, we can't use it. Another loop includes E, R_3, and R_4. In this loop the only unknown is V_4.

$$E = V_3 + V_4$$
$$V_4 = E - V_3 = 30 \text{ V} - 10 \text{ V} = 20 \text{ V}$$

Now that we know V_4, we can use the loop which includes E, R_1, and R_4 to find V_2.

$$E = V_1 + V_2 + V_4$$
$$V_2 = E - V_1 - V_4$$
$$V_2 = 30 \text{ V} - 3 \text{ V} - 20 \text{ V} = 30 \text{ V} - 23 \text{ V} = 7 \text{ V}$$

We could also use the loop consisting of R_1, R_2, and R_3 to find V_2. Observing polarity and moving around the loop clockwise from R_2, we get:

$$-V_2 - V_1 + V_3 = 0$$
$$V_2 = V_3 - V_1$$
$$V_2 = 10 \text{ V} - 3 \text{ V} = 7 \text{ V}$$

The loop we use may or may not include a source.

Suppose we were interested in the potential at one point in a circuit with reference to another point. For example, in Figure 8-29 we might ask the question, "What is the potential at point A with respect to common?" "With respect to" means "with reference to." Said another way, "If common is our reference point, what is the potential at point A?" The answer is 10 V. Point A is

10 V more positive than common. This can be determined three different ways:

1. The circuit from point A to common through R_3 shows that point A is the positive side of R_3 and common is the negative side. The drop across R_3 is 10 V; so point A is 10 V positive with respect to common.
2. Starting at point A and going to common through the path consisting of R_1 and R_2 yields $+3\ V+7\ V=10\ V$.
3. Starting at point A and going to common through the path consisting of E and R_4, we get $+30\ V-20\ V=10\ V$.

In each case we observed polarity and added (algebraically) the emf's and voltage drops.

What is the potential at point B with respect to common? The answer is -20 V. The easiest way to determine this is to observe the polarity of the drop across R_4. We could arrive at the same answer by using the path consisting of E and R_3 or by using the path consisting of E, R_1, and R_2. Using E and R_3, we get $-30\ V+10\ V=-20\ V$. Using E, R_1, and R_2, we get $-30\ V+3\ V+7\ V=-20$ V.

EXAMPLE 8.6: In Figure 8-30; a) What is the drop across R_2 and what is its polarity? b) What is the drop across R_3 and what is its polarity? c) What is the potential at point A with respect to common?

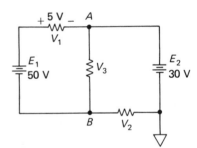

Figure 8-30.

Solution: (a) The drop across R_2 (V_2) can be found by going around the outer loop. The other choice is the loop consisting of E_2, R_2, and R_3, but we don't know V_3; therefore we can't use that loop. The outer loop consists of E_1, E_2, R_1, and R_2.

Let's start at the left side of R_2 which is point B and go around the loop in a clockwise direction. The general equation for the loop is:

$$E_1 + V_1 + E_2 + V_2 = 0$$

Observing polarity we get

$$-50 \text{ V} + 5 \text{ V} + 30 \text{ V} + V_2 = 0$$

Notice that a positive polarity was assigned to V_2. Because we don't know the polarity of V_2 yet, we give it a "+" sign until we solve the equation. Solving, we get

$$-15 \text{ V} + V_2 = 0$$
$$V_2 = 15 \text{ V}$$

Notice that V_2 is positive. This tells us that the first sign we encounter when we get to R_2 is a "+" sign. We went in a clockwise direction so the + sign must be assigned to the right side of R_2 (which makes the left side negative), as shown in Figure 8-31. If we had gone in a counterclockwise direction from point B, we would have this equation

$$V_2 + E_2 + V_1 + E_1 = 0$$

Assigning polarities and values, we get:

$$V_2 - 30 \text{ V} - 5 \text{ V} + 50 \text{ V} = 0$$
$$V_2 + 15 \text{ V} = 0$$
$$V_2 = -15 \text{ V}$$

Going counterclockwise we see that V_2 is *negative* 15 V. That is, the left side of R_2 is negative (which makes the right side positive) and agrees with our first method.

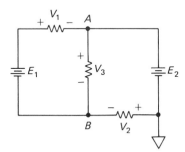

Figure 8-31.

Solution: (b) Knowing both V_1 and V_2 allows us to choose either the loop that includes R_3, E_1 and R_1 or the loop that includes R_3, R_2 and E_2. Let's use the loop that includes R_3, E_1, and R_1. Starting at point A and moving clockwise (remember we can go in either direction) we get

$$V_3 + E_1 + V_1 = 0 \qquad \text{(general equation)}$$

$$V_3 - 50 \text{ V} + 5 \text{ V} = 0 \qquad \text{(plug in known values and observe polarity)}$$

$$V_3 - 45 \text{ V} = 0$$

$$V_3 = 45 \text{ V}$$

We started at point A. Since the first sign we encountered was a + sign ($V_3 = 45$ V), the polarity is as shown in Figure 8-31.

Solution: (c) Notice that there are three paths from point A to common. One is through E_2. Another is through R_2 and R_3. A third is through R_1, E_1, and R_2. No matter which one is chosen, the result should show that point A is 30 V positive with respect to common.

$$P_t A \text{ to common} = E_2 = 30 \text{ V}$$

$$= V_3 + V_2 = 45 \text{ V} - 15 \text{ V} = 30 \text{ V}$$

$$= V_1 + E_1 + V_2 = -5 \text{ V} + 50 \text{ V} - 15 \text{ V}$$

$$= 30 \text{ V}$$

PRACTICE PROBLEMS 8-3

1. Refer to Figure 8-29. Let $V_1 = 7$ V and $V_3 = 15$ V. Find (a) V_2, (b) V_4, (c) potential at point A with respect to common, (d) potential at point A with respect to point B, and (e) potential at point B with respect to common.

2. Refer to Figure 8-32. Find (a) V_1, (b) V_3, (c) potential at point A with respect to common, (d) potential at point B with respect to common, and (e) potential at point A with respect to point B.

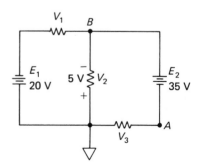

Figure 8-32.

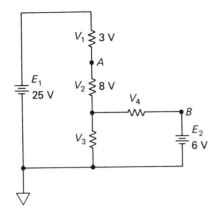

Figure 8-33.

3. Refer to Figure 8-33. Find (a) V_3, (b) V_4, (c) potential at point A with respect to common, and (d) potential at point B with respect to point A.

Solutions:

1. (a) 8 V   (b) 15 V   (c) 15 V   (d) 30 V   (e) -15 V

2. (a) 15 V   (b) 40 V   (c) -40 V   (d) -5 V   (e) -35 V

3. 14 V   (b) 8 V   (c) 22 V   (d) -16 V

Additional practice problems are at the end of the chapter.

SELF-TEST 8-1

1. Refer to Figure 8-14. Find I_3 if $I_T = 16$ mA, $I_1 = 4$ mA, and $I_2 = 2$ mA.

2. Refer to Figure 8-15. $I_1 = 40$ μA and $I_2 = 60$ μA. Find I_T and I_3.

3. Refer to Figure 8-15. $I_T = 3$ mA and $I_2 = 2.6$ mA. Find I_1 and I_3.

4. Refer to Figure 8-12. $I_T = 800$ μA, $I_2 = 308$ μA. Find I_1, I_3, and I_4.

5. Refer to Figure 8-13. $I_1 = 15$ mA, $I_2 = 5$ mA, $I_3 = 7$ mA. Find I_T and I_4.

6. Refer to Figure 8-19. $E = 30$ V, $V_1 = 7$ V, $V_2 = 10$ V. Find V_3.

7. Refer to Figure 8-19. $V_1 = 200$ mV, $V_2 = 1.3$ V, $V_3 = 400$ mV. Find E.

8. Refer to Figure 8-21. $E = 20$ V, $V_1 = 7$ V. Find V_2 and V_3.

9. Refer to Figure 8-22. $V_1 = 3$ V, $V_2 = 10$ V, $V_4 = 12$ V. Find E and V_3.

10. Refer to Figure 8-34. Find V_2 and V_3.

11. Refer to Figure 8-29. Let $V_1 = 13$ V and $V_3 = 21$ V. Find (a) V_2, (b) V_4, (c) potential at point B with respect to point A, and (d) potential at point A with respect to common.

12. Refer to Figure 8-33. Let $V_1 = 4$ V and $V_2 = 12$ V. Find (a) V_3, (b) V_4, (c) potential at common with respect to point A, and (d) potential at point B with respect to point A.

Answers to Self-Test 8-1 are at the end of the chapter.

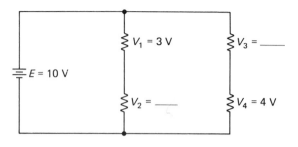

Figure 8-34.

END OF CHAPTER PROBLEMS 8-1

Refer to Figure 8-14.

1. Find I_T if $I_1 = 3$ A, $I_2 = 2$ A, and $I_3 = 4$ A.

2. Find I_T if $I_1 = 10$ mA, $I_2 = 15$ mA, and $I_3 = 2$ mA.

3. Find I_3 if $I_T = 20$ mA, $I_1 = 6$ mA, and $I_2 = 11$ mA.

4. Find I_3 if $I_T = 400$ mA, $I_1 = 100$ mA, and $I_2 = 150$ mA.

5. Find I_2 if $I_T = 65$ μA, $I_1 = 10$ μA, and $I_3 = 10$ μA.

6. Find I_2 if $I_T = 7$ mA, $I_1 = 500$ μA, and $I_3 = 3.2$ mA.

7. Find I_1 if $I_T = 1.3$ A, $I_2 = 300$ mA, and $I_3 = 40$ mA.

8. Find I_1 if $I_T = 400$ mA, $I_2 = 120$ mA, $I_3 = 150$ mA.

Refer to Figure 8-15.

9. $I_1 = 10$ mA and $I_2 = 15$ mA. Find I_T and I_3.

10. $I_1 = 600$ μA and $I_2 = 1.2$ mA. Find I_T and I_3.

11. $I_T = 1.6$ A and $I_3 = 800$ mA. Find I_1 and I_2.

12. $I_T = 350$ mA and $I_3 = 150$ mA. Find I_1 and I_2.

13. $I_T = 600$ μA and $I_1 = 400$ μA. Find I_2 and I_3.

14. $I_T = 2.1$ mA and $I_1 = 700$ μA. Find I_2 and I_3.

Refer to Figure 8-12.

15. $I_1 = 1.6$ A, $I_2 = 800$ mA. Find I_T, I_3, and I_4.

16. $I_1 = 750$ mA, $I_2 = 300$ mA. Find I_T, I_3, and I_4.

17. $I_3 = 500$ μA, $I_4 = 2.7$ mA. Find I_T, I_1, and I_2.

18. $I_3 = 3$ mA, $I_4 = 3.6$ mA. Find I_T, I_1, and I_2.

Refer to Figure 8-13.

19. $I_T = 6.7$ mA, $I_2 = 1.2$ mA, $I_4 = 5$ mA. Find I_1 and I_3.

20. $I_T = 206$ mA, $I_2 = 65$ mA, $I_4 = 160$ mA. Find I_1 and I_3.

21. $I_1 = 6$ mA, $I_3 = 3.4$ mA, $I_4 = 8.2$ mA. Find I_T and I_2.

22. $I_1 = 760$ mA, $I_3 = 300$ mA, $I_4 = 465$ mA. Find I_T and I_2.

23. Refer to Figure 8-24. $I_T = 100$ mA, $I_2 = 20$ mA, and $I_5 = 30$ mA. Find $I_1, I_3, I_4,$ and I_6.

END OF CHAPTER PROBLEMS 8-2

Refer to Figure 8-23.

1. $E = 30$ V and $V_1 = 12$ V. Find V_2.

2. $E = 17$ V and $V_1 = 4$ V. Find V_2.

3. $V_1 = 7$ V and $V_2 = 6$ V. Find E.

4. $V_1 = 500$ mV and $V_2 = 1.2$ V. Find E.

Refer to Figure 8-19.

5. $E = 70$ V, $V_1 = 22$ V, $V_2 = 37$ V. Find V_3.

6. $E = 12$ V, $V_1 = 2$ V, $V_2 = 6$ V. Find V_3.

7. $V_1 = 3$ V, $V_2 = 8$ V, $V_3 = 7$ V. Find E.

8. $V_1 = 40$ V, $V_2 = 22$ V, $V_3 = 15$ V. Find E.

9. $E = 37$ V, $V_2 = 20$ V, $V_3 = 3$ V. Find V_1.

10. $E = 45$ V, $V_2 = 15$ V, $V_3 = 20$ V. Find V_1.

Refer to Figure 8-21.

11. $E = 20$ V, $V_3 = 8$ V. Find V_1 and V_2.

12. $E = 7.8$ V, $V_3 = 3.6$ V. Find V_1 and V_2.

13. $E = 18$ V, $V_1 = 5$ V. Find V_2 and V_3.

14. $E = 40$ V, $V_1 = 18$ V. Find V_2 and V_3.

15. $V_1 = 12$ V, $V_3 = 5$ V. Find E and V_2.

16. $V_1 = 300$ mV, $V_3 = 1.1$ V. Find E and V_2.

Refer to Figure 8-22.

17. $E = 12$ V, $V_2 = 3$ V, $V_3 = 6$ V. Find V_1 and V_4.

18. $E = 25$ V, $V_2 = 8$ V, $V_3 = 7$ V. Find V_1 and V_4.

19. $E = 3$ V, $V_2 = 500$ mV, $V_4 = 1.3$ V. Find V_1 and V_3.

20. $E = 35$ V, $V_2 = 10$ V, $V_4 = 12$ V. Find V_1 and V_3.

21. $V_1 = 20$ V, $V_2 = 30$ V, $V_4 = 40$ V. Find E and V_3.

22. $V_1 = 3.8$ V, $V_2 = 1.6$ V, $V_4 = 4$ V. Find E and V_3.

END OF CHAPTER PROBLEMS 8-3

1. Refer to Figure 8-29. Let $V_1 = 15$ V and $V_3 = 25$ V. Find (a) V_2, (b) V_4, (c) potential at point A with respect to common, (d) potential at point A with respect to point B, and (e) potential at point B with respect to common.

2. Refer to Figure 8-29. Let $V_1 = 6$ V and $V_3 = 15$ V. Find (a) V_2, (b) V_4, (c) potential at point A with respect to common, (d) potential at point A with respect to point B, and (e) potential at point B with respect to common.

3. Refer to Figure 8-32. Let $V_2 = 8$ V. Find (a) V_1, (b) V_3, (c) potential at point A with respect to common, (d) potential at point A with respect to point B, and (e) potential at point B with respect to common.

4. Refer to Figure 8-32. Let $V_2 = 14$ V. Find (a) V_1, (b) V_3, (c) potential at point A with respect to common, (d) potential at point A with respect to point B, and (e) potential at point B with respect to common.

5. Refer to Figure 8-33. Let $V_1 = 6$ V and $V_2 = 8$ V. Find (a) V_3, (b) V_4, (c) potential at point A with respect to common, and (d) potential at point B with respect to point A.

6. Refer to Figure 8-33. Let $V_1 = 3$ V and $V_2 = 3$ V. Find (a) V_3, (b) V_4, (c) potential at point A with respect to common and, (d) potential at point B with respect to point A.

ANSWERS TO SELF-TESTS

Self-Test 8-1

1. 10 mA

2. $I_T = 100\ \mu A$, $I_3 = 60\ \mu A$

3. $I_1 = 400\ \mu A$, $I_3 = 2.6$ mA

4. $I_1 = 800\ \mu A$, $I_3 = 308\ \mu A$, $I_4 = 492\ \mu A$

5. $I_T = 27$ mA, $I_4 = 12$ mA

6. 13 V

7. 1.9 V

8. $V_2 = 13$ V, $V_3 = 13$ V

9. $E = 15$ V, $V_3 = 2$ V

10. $V_2 = 7$ V, $V_3 = 6$ V

11. (a) 8 V   (b) 9 V   (c) -30 V   (d) 21 V

12. (a) 9 V   (b) 3 V   (c) -21 V   (d) -15 V

CHAPTER

DC CIRCUIT ANALYSIS—OHM'S LAW

In Chapter 8 we learned how voltage drops and currents divide in series and parallel circuits. In this chapter we will learn how to compute those voltage drops and currents.

Ohm's law expresses the relationship between voltage, current, and resistance in an electrical circuit. Ohm's law states that circuit current is directly proportional to voltage and inversely proportional to resistance. Expressed as an equation, Ohm's law is:

$$I = \frac{E}{R}$$

If we know the applied voltage and the circuit resistance, we can calculate the circuit current. If E and I are known, we can rearrange the equation to solve for R.

$$R = \frac{E}{I}$$

If I and R are known, we can solve for E.

$$E = IR$$

Of course, we must always know two of the variables in order to solve for the third.

9.1 CIRCUIT RESISTANCE

Before we get into solving circuit problems in terms of voltage drops and currents, let's learn how to calculate circuit resistance. If there is only one resistance in an electrical circuit, the total circuit resistance is equal to that

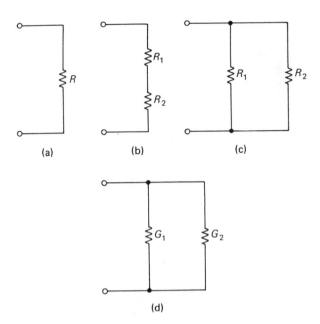

Figure 9-1.

resistance. Such a circuit is shown in Figure 9-1(a). $R_T = R$. In Figure 9-1(b) there are two resistances in series. In order to identify which resistance we are talking about, they are labeled R_1 and R_2.

In a series circuit the total resistance equals the sum of the individual resistances.

$$R_T = R_1 + R_2 + R_3 + \cdots + R_N$$

Therefore, in Figure 9-1(b), $R_T = R_1 + R_2$

EXAMPLE 9.1: In Figure 9-2(a) let $R_1 = 2.7$ kΩ and $R_2 = 4.7$ kΩ. Find R_T.

Solution: $R_T = R_1 + R_2$
$R_T = 2.7$ k$\Omega + 4.7$ k$\Omega = 7.4$ kΩ

EXAMPLE 9.2: In Figure 9-2(b) let $R_1 = 68$ kΩ, $R_2 = 100$ kΩ and $R_3 = 39$ kΩ. Find R_T

Solution: $R_T = R_1 + R_2 + R_3$
$R_T = 68$ k$\Omega + 100$ k$\Omega + 39$ k$\Omega = 207$ kΩ

EXAMPLE 9.3: In Figure 9-2(c) let $R_1 = 120$ Ω, $R_2 = 330$ Ω, $R_3 = 100$ Ω and $R_4 = 560$ Ω. Find R_T.

Solution: $R_T = R_1 + R_2 + R_3 + R_4$

$R_T = 120\ \Omega + 330\ \Omega + 100\ \Omega + 560\ \Omega = 1.11\ k\Omega$

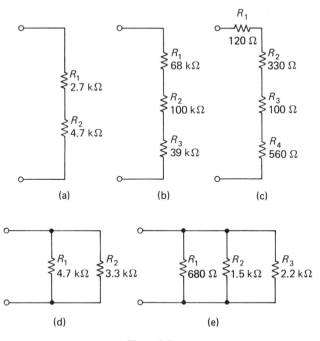

Figure 9-2.

In Figure 9-1(c), R_1 and R_2 are connected in parallel. Connecting resistances in parallel *reduces* total circuit resistance. This is so because as parallel paths are added, more circuit current flows. If total circuit current is increased, then total circuit resistance must decrease.

It is more logical to think of the parallel circuit in terms of circuit conductances. Conductance and resistance are reciprocals.

$$G = \frac{1}{R} \qquad R = \frac{1}{G}$$

Figure 9-1(d) is the same as Figure 9-1(c) except that we are using the symbol for conductance (G) to show that we will work with that unit in the parallel circuit.

In a parallel circuit the total conductance equals the sum of the individual branch conductances.

$$G_T = G_1 + G_2 + G_3 + \cdots + G_N$$

EXAMPLE 9.4: In Figure 9-2(d) let $R_1 = 4.7\ k\Omega$ and $R_2 = 3.3\ k\Omega$. Find R_T.

Solution: Because we are dealing with a parallel circuit, we must first find G_1 and G_2.

$$G_1 = \frac{1}{R_1} = \frac{1}{4.7\text{ k}} = 213\ \mu S$$

$$G_2 = \frac{1}{R_2} = \frac{1}{3.3\text{ k}} = 303\ \mu S$$

Now we can find G_T

$$G_T = G_1 + G_2 = 213\ \mu S + 303\ \mu S = 516 \mu S$$

R_T is the reciprocal of G_T.

$$R_T = \frac{1}{G_T} = \frac{1}{516\ \mu S} = 1.94\text{ k}\Omega$$

We can put the whole solution into one equation:

$$R_T = \frac{1}{\dfrac{1}{R_1} + \dfrac{1}{R_2}}$$

or

$$R_T = \frac{1}{G_1 + G_2}$$

The algorithm for solving this kind of problem with the calculator is in Appendix A.

EXAMPLE 9.5: In Figure 9-2(e), let $R_1 = 680\ \Omega$, $R_2 = 1.5$ kΩ and $R_3 = 2.2$ kΩ. Find R_T.

Solution:

$$G_T = G_1 + G_2 + G_3$$

$$G_T = 1.47\text{ mS} + 667\ \mu S + 445\ \mu S = 2.59\text{ mS}$$

$$R_T = \frac{1}{G_T} = \frac{1}{2.59\text{ mS}} = 386\ \Omega$$

It is always true that *in a parallel circuit,* R_T *is always less than the smallest branch resistance.*

PRACTICE PROBLEMS 9-1

1. Refer to Figure 9-2(a). Find R_T when (a) $R_1 = 82$ kΩ, $R_2 = 100$ kΩ and (b) $R_1 = 33$ kΩ, $R_2 = 15$ kΩ.

2. Refer to Figure 9-2(b). Find R_T when (a) $R_1 = 120\ \Omega$, $R_2 = 220\ \Omega$, $R_3 = 1$ kΩ and (b) $R_1 = 68$ kΩ, $R_2 = 120$ kΩ, $R_3 = 39$ kΩ.

3. Refer to Figure 9-2(d). Find R_T when (a) $R_1=6.8$ kΩ, $R_2=2.7$ kΩ, (b) $R_1=7.5$ kΩ, $R_2=15$ kΩ, (c) $R_1=47$ kΩ, $R_2=120$ kΩ, and (d) $R_1=68$ Ω, $R_2=220$ Ω.

4. Refer to Figure 9-2(e). Find R_T when (a) $R_1=1.5$ kΩ, $R_2=4.7$ kΩ, $R_3=3.3$ kΩ, (b) $R_1=680$ Ω, $R_2=1.8$ kΩ, $R_3=470$ Ω, (c) $R_1=120$ kΩ, $R_2=120$ kΩ, $R_3=180$ kΩ, and (d) $R_1=22$ kΩ, $R_2=100$ kΩ, $R_3=10$ kΩ.

Solutions:

1. (a) $R_T=R_1+R_2=82$ k$\Omega+100$ k$\Omega=182$ kΩ
   (b) $R_T=R_1+R_2=33$ k$\Omega+15$ k$\Omega=48$ kΩ

2. (a) $R_T=R_1+R_2+R_3=120$ $\Omega+220$ $\Omega+1$ k$\Omega=1.34$ kΩ
   (b) $R_T=R_1+R_2+R_3=68$ k$\Omega+120$ k$\Omega+39$ k$\Omega=227$ kΩ

3. (a) $G_T=G_1+G_2=147$ μS$+370$ μS$=517$ μS
   $R_T=\dfrac{1}{G_T}=\dfrac{1}{517\ \mu S}=1.93$ kΩ
   (b) $G_T=G_1+G_2=133$ μS$+66.7$ μS$=200$ μS
   $R_T=\dfrac{1}{G_T}=\dfrac{1}{200\ \mu S}=5$ kΩ
   (c) $G_T=G_1+G_2=21.3$ μS$+8.33$ μS$=29.6$ μS
   $R_T=\dfrac{1}{G_T}=\dfrac{1}{29.6\ \mu S}=33.8$ kΩ
   (d) $G_T=G_1+G_2=14.7$ mS$+4.55$ mS$=19.3$ mS
   $R_T=\dfrac{1}{G_T}=\dfrac{1}{19.3\ mS}=51.9$ Ω

4. (a) $G_T=G_1+G_2+G_3=667$ μS$+213$ μS$+303$ μS
   $R_T=\dfrac{1}{G_T}=\dfrac{1}{1.18\ mS}=846$ Ω
   (b) $G_T+G_1+G_2+G_3=1.47$ mS$+556$ μS$+2.13$ mS$=4.15$ mS
   $R_T=\dfrac{1}{G_T}=\dfrac{1}{4.15\ mS}=241$ Ω
   (c) $G_T=G_1+G_2+G_3=8.33$ μS$+8.33$ μS$+5.56$ μS$=22.2$ μS
   $R_T=\dfrac{1}{G_T}=\dfrac{1}{22.2\ \mu S}=45$ kΩ
   (d) $G_T=G_1+G_2+G_3=45.5$ μS$+10$ μS$+100$ μS$=155$ μS
   $R_T=\dfrac{1}{G_T}=\dfrac{1}{155\ \mu S}=6.43$ kΩ

Additional practice problems are at the end of the chapter.

Now let's look at some series-parallel circuits. Let's find the total resistance of the circuit in Figure 9-3(a) where $R_1=4.7$ kΩ, $R_2=6.8$ kΩ, and $R_3=12$ kΩ. Notice that R_2 and R_3 are in parallel and they are in series with R_1. The equation is:

$$R_T=R_1+R_2\|R_3$$

(The symbol "$\|$" means "in parallel with"). Let's first reduce R_2 and R_3 to an

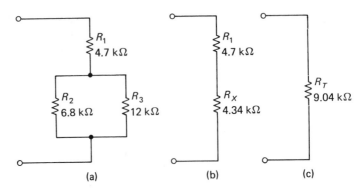

Figure 9-3.

equivalent series resistance which we will call R_X. We will solve for R_X the same way we solved for R_T in Figure 9-2(d).

$$G_X = G_2 + G_3 = 147 \ \mu S + 83.3 \ \mu S = 230 \ \mu S$$

$$R_X = \frac{1}{G_X} = \frac{1}{230 \ \mu S} = 4.34 \ k\Omega$$

We now have a circuit that looks like Figure 9-3(b). Then

$$R_T = R_1 + R_X = 4.7 \ k\Omega + 4.34 \ k\Omega = 9.04 \ k\Omega$$

We have reduced a complex circuit to a single resistance as shown in Figure 9-3(c).

Let's try another one. In Figure 9-4(a) let $R_1 = 20 \ k\Omega$, $R_2 = 30 \ k\Omega$, and $R_3 = 68 \ k\Omega$. Notice that R_1 and R_2 are in series. The equivalent resistance (R_X) is equal to $R_1 + R_2$.

$$R_X = R_1 + R_2 = 20 \ k\Omega + 30 \ k\Omega = 50 \ k\Omega$$

The circuit now looks like the one in Figure 9-4(b). The final step is to reduce the circuit to a single resistance, R_T.

$$G_T = G_X + G_3 = 20 \ \mu S + 14.7 \ \mu S = 34.7 \ \mu S$$

$$R_T = \frac{1}{G_T} = \frac{1}{34.7 \ \mu S} = 28.8 \ k\Omega$$

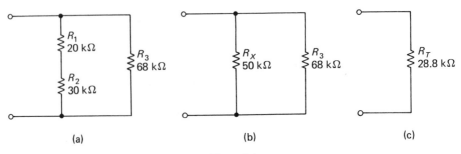

Figure 9-4.

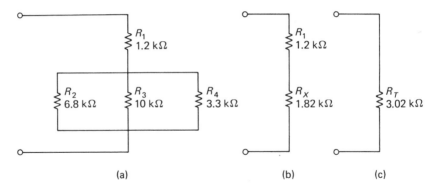

Figure 9-5.

Let's try one more. In Figure 9-5(a) let $R_1 = 1.2$ kΩ, $R_2 = 6.8$ kΩ, $R_3 = 10$ kΩ, and $R_4 = 3.3$ kΩ. Notice that R_2, R_3, and R_4 are connected in parallel. To find R_X, we must first find G_X.

$$G_X = G_2 + G_3 + G_4 = 147 \ \mu S + 100 \ \mu S + 303 \ \mu S = 550 \ \mu S$$

$$R_X = \frac{1}{G_X} = \frac{1}{550 \ \mu S} = 1.82 \text{ k}\Omega$$

This equivalent resistance, R_X, is in series with R_1, as shown in Figure 9-5(b). R_T then is equal to R_1 and R_X in series.

$$R_T = R_X + R_1 = 1.82 \text{ k}\Omega + 1.2 \text{ k}\Omega = 3.02 \text{ k}\Omega$$

The circuit reduces to the equivalent resistance shown in Figure 9-5(c).

PRACTICE PROBLEMS 9-2

1. Refer to Figure 9-3(a) and find R_T when (a) $R_1 = 330$ Ω, $R_2 = 680$ Ω, $R_3 = 910$ Ω and (b) $R_1 = 82$ kΩ, $R_2 = 270$ kΩ, $R_3 = 180$ kΩ.

2. Refer to Figure 9-4(a) and find R_T when (a) $R_1 = 150$ Ω, $R_2 = 560$ Ω, $R_3 = 1.5$ kΩ and (b) $R_1 = 22$ kΩ, $R_2 = 56$ kΩ, $R_3 = 39$ kΩ.

3. Refer to Figure 9-5(a) and find R_T when (a) $R_1 = 750$ kΩ, $R_2 = 1.2$ MΩ, $R_3 = 2.2$ MΩ, $R_4 = 910$ kΩ and (b) $R_1 = 2.7$ kΩ, $R_2 = 1$ kΩ, $R_3 = 10$ kΩ, $R_4 = 8.2$ kΩ.

Solutions:

1. (a) $R_T = R_1 + R_2 \| R_3 = 330 \ \Omega + 389 \ \Omega = 719 \ \Omega$
   (b) $R_T = R_1 + R_2 \| R_3 = 82 \ \text{k}\Omega + 108 \ \text{k}\Omega = 190 \text{ k}\Omega$

2. (a) $R_T = (R_1 + R_2) \| R_3 = 710 \ \Omega \| 1.5 \text{ k}\Omega = 482 \ \Omega$
   (b) $R_T = (R_1 + R_2) \| R_3 = 15.8 \ \text{k}\Omega \| 39 \ \text{k}\Omega = 11.2 \text{ k}\Omega$

3. (a) $R_1 + R_2 \| R_3 \| R_4 = 750 \ \text{k}\Omega + 419 \ \text{k}\Omega =$
1.17 MΩ
(b) $R_1 + R_2 \| R_3 \| R_4 = 2.7 \ \text{k}\Omega + 818 \ \Omega = 3.52 \ \text{k}\Omega$

Additional practice problems are at the end of the chapter.

Of course, there are many kinds of series-parallel circuits. We have examined only a few. In all problem solving we must recognize which resistors are in parallel and reduce that part of the circuit to an equivalent resistance. We then add series resistances and continue both operations until we have reduced the circuit to one resistance (R_T).

SELF-TEST 9-1

1. Refer to Figure 9-2(b). Find R_T when $R_1 =$ 39 kΩ, $R_2 = 56$ kΩ, and $R_3 = 27$ kΩ.

2. Refer to Figure 9-2(d). Find R_T when $R_1 =$ 820 Ω and $R_2 = 1.2$ kΩ.

3. Refer to Figure 9-2(e). Find R_T when $R_1 =$ 33 kΩ, $R_2 = 100$ kΩ, and $R_3 = 82$ kΩ.

4. Refer to Figure 9-3(a). Find R_T when $R_1 =$ 12 kΩ, $R_2 = 27$ kΩ, and $R_3 = 47$ kΩ.

5. Refer to Figure 9-4(a). Find R_T when $R_1 =$ 330 Ω, $R_2 = 1.2$ kΩ, and $R_3 = 910$ Ω.

6. Refer to Figure 9-5(a). Find R_T when $R_1 =$ 270 kΩ, $R_2 = 1.2$ MΩ, $R_3 = 750$ kΩ, and $R_4 =$ 470 kΩ.

Answers to Self-Test 9-1 are at the end of the chapter.

9.2 SERIES CIRCUITS

Consider the circuit in Figure 9-6. Let's find the current, I, and the voltage drops across R_1 and R_2. The general Ohm's law equation, as mentioned previously, is:

$$I = \frac{E}{R}$$

When dealing with a specific situation we must be more specific in labeling the

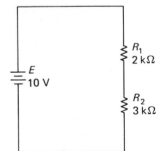

Figure 9-6.

variables in our equation. For example, in a series circuit we know from Kirchhoff's law that current is constant. $I_T = I_1 = I_2$. Therefore, we can label the current "I" and not use a subscript. There are three resistances though—R_1, R_2, and R_T— and there are three voltages—E, V_1, and V_2. If we use R_T in the equation, we must use E. If we use R_1, we must use V_1, and so on. To find I, we could write the equation three ways then, depending on known values:

$$I = \frac{E}{R_T}$$

$$I = \frac{V_1}{R_1}$$

$$I = \frac{V_2}{R_2}$$

In our circuit we know E and can find R_T:

$$R_T = R_1 + R_2 = 5 \text{ k}\Omega$$

Then

$$I = \frac{E}{R_T} = \frac{10 \text{ V}}{5 \text{ k}\Omega} = 2 \text{ mA}$$

Knowing I, we can solve for V_1 and V_2 by rearranging the equation and solving for the unknown:

$$V_1 = IR_1 = 2 \text{ mA} \times 2 \text{ k}\Omega = 4 \text{ V}$$

$$V_2 = IR_2 = 2 \text{ mA} \times 3 \text{ k}\Omega = 6 \text{ V}$$

Using Kirchhoff's voltage law to check our answers, we get:

$$E = V_1 + V_2$$

$$10 \text{ V} = 4 \text{ V} + 6 \text{ V}$$

Let's look at the problem from another angle. Because the current is constant, and $IR = V$, by inspection we can see that the greater voltage drops across R_2.

> *In any series circuit the greater voltage drops across the greater resistance.*

Further, the voltage drops are always in proportion to the resistances. In this problem the resistances are in the ratio of 2 to 3. Expressed as a proportion:

$$\frac{R_1}{R_2} = \frac{V_1}{V_2} \tag{9-1}$$

We could also say:

$$\frac{R_1}{R_T} = \frac{V_1}{E} \tag{9-2}$$

$$\frac{R_2}{R_T} = \frac{V_2}{E} \tag{9-3}$$

R_1 is 0.4 or 40% of total resistance. Therefore, 40% of the total voltage

$$\frac{R_1}{R_T} = \frac{2 \text{ k}\Omega}{5 \text{ k}\Omega} = 0.4$$

must drop across it. We can rearrange our proportion (Equation 9-2):

$$\frac{R_1}{R_T} = \frac{V_1}{E} \tag{9-2}$$

$$V_1 = E\left(\frac{R_1}{R_T}\right)$$

or

$$V_1 = \frac{ER_1}{R_T} \tag{9-4}$$

Then it follows that

$$V_2 = \frac{ER_2}{R_T} \tag{9-5}$$

We will use these equations often in problem solving. Looking at Equation 9-4 again, we see:

$$V_1 = \left(\frac{E}{R_T}\right)R_1$$

or

$$V_1 = IR_1 \qquad \left(I = \frac{E}{R_T}\right)$$

EXAMPLE 9.6: In Figure 9-6 let $E = 12$ V, $R_2 = 1.5$ kΩ and $V_1 = 4$ V. Find R_1, V_2, I, and R_T.

Solution: In this problem V_2 is the only unknown variable that can be found from the information given. We can't find R_1 or V_2 because we don't know I. We can't find R_T because we don't know I or R_1. We can find V_2 by using Kirchhoff's voltage law.

$$E = V_1 + V_2$$

$$12 \text{ V} = 4 \text{ V} + V_2$$

$$8 \text{ V} = V_2$$

Once we have determined the value of V_2 we can find I.

$$I = \frac{V_2}{R_2} = \frac{8 \text{ V}}{1.5 \text{ k}\Omega} = 5.33 \text{ mA}$$

Now that we know the value of I we can find R_1 and R_T.

$$R_1 = \frac{V_1}{I} = \frac{4 \text{ V}}{5.33 \text{ mA}} = 750 \text{ }\Omega$$

$$R_T = R_1 + R_2 = 750 \text{ }\Omega + 1.5 \text{ k}\Omega = 2.25 \text{ k}\Omega$$

also $\qquad R_T = \frac{E}{I} = \frac{12 \text{ V}}{5.33 \text{ mA}} = 2.25 \text{ k}\Omega$

EXAMPLE 9.7: In Figure 9-7 let $E = 12$ V, $V_1 = 2.73$ V, $I = 38.7$ mA, and $R_2 = 100 \text{ }\Omega$. Find V_2, V_3, R_1, R_3 and R_T.

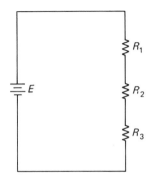

Figure 9-7.

Solution: A check of given values shows that with the information given we can find R_T because we know E and I.

$$R_T = \frac{E}{I} = \frac{12 \text{ V}}{38.7 \text{ mA}} = 310 \text{ }\Omega$$

We can also find R_1 because we know I and V_1.

$$R_1 = \frac{V_1}{I} = \frac{2.73 \text{ V}}{38.7 \text{ mA}} = 70.5 \text{ }\Omega$$

We can find V_2 because we know I and R_2

$$V_2 = IR_2 = 38.7 \text{ mA} \times 100 \text{ }\Omega = 3.87 \text{ V}$$

The given values plus the unknowns we have found provides us with enough information to solve for the rest of the unknowns which are V_3 and R_3. Although other solutions may be possible, we will solve for these variables as follows:

$$V_3 = E - V_1 - V_2 = 12 \text{ V} - 2.73 \text{ V} - 3.87 \text{ V} = 5.4 \text{ V}$$

Then

$$R_3 = \frac{V_3}{I} = \frac{5.4 \text{ V}}{38.7 \text{ mA}} = 140 \text{ }\Omega$$

Check R_T:

$$R_T = R_1 + R_2 + R_3 = 70.5 \text{ }\Omega + 100 \text{ }\Omega + 140 \text{ }\Omega = 311 \text{ }\Omega$$

(rounding causes a difference of one
in the LSD position)

Check E:

$$E = V_1 + V_2 + V_3 = 2.73 \text{ V} + 3.87 \text{ V} + 5.4 \text{ V} = 12 \text{ V}$$

PRACTICE PROBLEMS 9-3

1. Refer to Figure 9-6. Find I, R_T, V_1, and V_2 when (a) $E = 25$ V, $R_1 = 2.7$ kΩ, and $R_2 = 4.7$ kΩ, (b) $E = 30$ V, $R_1 = 10$ kΩ, and $R_2 = 15$ kΩ, and (c) $E = 12$ V, $R_1 = 470$ Ω, and $R_2 = 1.2$ kΩ.

2. Refer to Figure 9-6. Find R_T, R_1, V_1, and V_2 when $E = 10$ V, $I = 150$ μA, and $R_2 = 20$ kΩ.

3. Refer to Figure 9-6. Find E, R_T, V_2, and R_1 when $I = 12.6$ mA, $V_1 = 6.7$ V, and $R_2 = 810$ Ω.

4. Refer to Figure 9-7. Find I, R_T, V_1, V_2, and V_3 when (a) $E = 10$ V, $R_1 = 3.3$ kΩ, $R_2 = 6.8$ kΩ, and $R_3 = 2.2$ kΩ, (b) $E = 20$ V, $R_1 = 4.7$ kΩ, $R_2 = 10$ kΩ, and $R_3 = 12$ kΩ, and (c) $E = 15$ V, $R_1 = 12$ kΩ, $R_2 = 33$ kΩ, $R_3 = 18$ kΩ.

5. Refer to Figure 9-7. Find V_2, V_3, R_1, R_3, and R_T when $E = 18$ V, $V_1 = 8.2$ V, $I = 40$ μA, and $R_2 = 180$ kΩ.

6. Refer to Figure 9-7. Find E, R_2, R_3, V_1, and V_3 when $I = 37.6$ mA, $R_1 = 47$ Ω, $V_2 = 1.92$ V, and $R_T = 137$ Ω.

Solutions:

1. (a) $R_T = R_1 + R_2 = 2.7$ k$\Omega + 4.7$ k$\Omega = 7.4$ kΩ
   $I = \dfrac{E}{R_T} = \dfrac{25 \text{ V}}{7.4 \text{ k}\Omega} = 3.38$ mA
   $V_1 = IR_1 = 3.38$ mA $\times 2.7$ k$\Omega = 9.12$ V
   $V_2 = IR_2 = 3.38$ mA $\times 4.7$ k$\Omega = 15.9$ V

   (b) $R_T = 25$ kΩ
   $I = 1.2$ mA
   $V_1 = 12$ V
   $V_2 = 18$ V

   (c) $R_T = 1.67$ kΩ
   $I = 7.19$ mA
   $V_1 = 3.38$ V
   $V_2 = 8.62$ V

2. $R_T = 66.7$ kΩ, $R_1 = 46.7$ kΩ, $V_1 = 7$ V, $V_2 = 3$ V

3. $E = 16.9$ V, $V_2 = 10.2$ V, $R_1 = 532$ Ω, $R_T = 1.34$ kΩ

4. (a) $R_T = R_1 + R_2 + R_3 = 3.3 \text{ k}\Omega + 6.8 \text{ k}\Omega +$ (b) $R_T = 26.7 \text{ k}\Omega$
$2.2 \text{ k}\Omega = 12.3 \text{ k}\Omega$

$\qquad I = \dfrac{E}{R_T} = \dfrac{10 \text{ V}}{12.3 \text{ k}\Omega} = 813 \ \mu\text{A}$

$\qquad V_1 = IR_1 = 813 \ \mu\text{A} \times 3.3 \text{ k}\Omega = 2.68 \text{ V}$
$\qquad V_2 = IR_2 = 813 \ \mu\text{A} \times 6.8 \text{ k}\Omega = 5.53 \text{ V}$
$\qquad V_3 = IR_3 = 813 \ \mu\text{A} \times 2.2 \text{ k}\Omega = 1.79 \text{ V}$

(b) $R_T = 26.7 \text{ k}\Omega$
$\quad I = 749 \ \mu\text{A}$
$\quad V_1 = 3.52 \text{ V}$
$\quad V_2 = 7.49 \text{ V}$
$\quad V_3 = 8.99 \text{ V}$

(c) $R_T = 63 \text{ k}\Omega$
$\quad I = 238 \ \mu\text{A}$
$\quad V_1 = 2.86 \text{ V}$
$\quad V_2 = 7.86 \text{ V}$
$\quad V_3 = 4.29 \text{ V}$

5. $R_T = 450 \text{ k}\Omega$, $V_2 = 7.2 \text{ V}$, $V_3 = 2.6 \text{ V}$, $R_3 =$ **6.** $E = 5.15 \text{ V}$, $V_1 = 1.77 \text{ V}$, $V_3 = 1.46 \text{ V}$, $R_2 =$
$65 \text{ k}\Omega$, $R_1 = 205 \text{ k}\Omega$ $51.1 \ \Omega$, $R_3 = 38.8 \ \Omega$

Additional practice problems are at the end of the chapter.

9.3 PARALLEL CIRCUITS

Consider the circuit in Figure 9-8. Let's find V, R_T, I_T, and I_2. The general equation is:

$$V = IR = I\left(\frac{1}{G}\right) = \frac{I}{G}$$

Specifically,

$$V = \frac{I_T}{G_T} = \frac{I_1}{G_1} = \frac{I_2}{G_2}$$

Since we are given I_T, we will find V by using:

$$V = \frac{I_T}{G_T}$$

$$V = \frac{10 \text{ mA}}{2 \text{ mS} + 3 \text{ mS}} = \frac{10 \text{ mA}}{5 \text{ mS}} = 2 \text{ V}$$

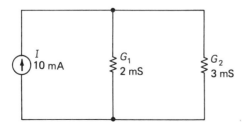

Figure 9-8.

Now we can find I_1 and I_2:

$$I = \frac{V}{R} = VG$$

$$I_1 = VG_1 = 2 \text{ V} \times 2 \text{ mS} = 4 \text{ mA}$$

$$I_2 = VG_2 = 2 \text{ V} \times 3 \text{ mS} = 6 \text{ mA}$$

We must keep in mind that V is the constant in a parallel circuit. Since we know G_T,

$$R_T = \frac{1}{G_T} = \frac{1}{5 \text{ mS}} = 200 \text{ }\Omega$$

We see that I_2 was the greater current. That was because G_2 was the greater conductance. *In a parallel circuit the greater current flows through the greater conductance.* The currents are always in proportion to the conductance. In this problem the conductances are in the ratio of 2 to 3. Expressed as a proportion:

$$\frac{G_1}{G_2} = \frac{I_1}{I_2} \tag{9-6}$$

We could also say:

$$\frac{G_1}{G_T} = \frac{I_1}{I_T} \tag{9-7}$$

and:

$$\frac{G_2}{G_T} = \frac{I_2}{I_T} \tag{9-8}$$

G_1 is 0.4 or 40% of the total conductance. Therefore, 40% of the total current flows through it. We can rearrange our proportion (Equation 9-7)

$$\frac{G_1}{G_T} = \frac{I_1}{I_T} \tag{9-7}$$

$$I_1 = I_T \left(\frac{G_1}{G_T} \right)$$

or

$$I_1 = \frac{I_T G_1}{G_T} \tag{9-9}$$

Then it follows that

$$I_2 = \frac{I_T G_2}{G_T} \tag{9-10}$$

We will use these equations very often in problem solving.

Notice the duality or similarity between Equations 9-4 and 9-9. The relationship between voltage drops and resistance in a series circuit is the same as the relationship between current and conductance in a parallel circuit.

EXAMPLE 9.8: In Figure 9-8 let $I_T = 400$ μA, $R_1 = 8.1$ kΩ, and $I_2 = 150$ μA. Solve for V, I_1, R_2, G_T, and R_T.

Solution: An examination of known variables shows that we have one unknown current, I_1. We can find this current by applying Kirchhoff's current law. We don't have enough information to find any of the other unknowns at this time.

$$I_T = I_1 + I_2$$
$$400 \text{ μA} = I_1 + 150 \text{ μA}$$
$$250 \text{ μA} = I_1$$

Now that we know I_1 we can find V.

$$V = I_1 R_1 = 250 \text{ μA} \times 8.1 \text{ kΩ} = 2.03 \text{ V}$$

With V we can find R_T, G_T, and R_2.

$$R_2 = \frac{V}{I_2} = \frac{2.03 \text{ V}}{150 \text{ μA}} = 13.5 \text{ kΩ}$$

$$G_T = \frac{I_T}{V} = \frac{400 \text{ μA}}{2.03 \text{ V}} = 198 \text{ μS}$$

$$R_T = \frac{1}{G_T} = \frac{1}{198 \text{ μS}} = 5.06 \text{ kΩ}$$

There may be alternate solutions which result in the same answers. Typically, there is more than one valid approach that can be used to solve problems like these.

EXAMPLE 9.9: Refer to Figure 9-9. Let $V = 50$ V, $G_1 = 90$ μS, $I_2 = 400$ μA, and $G_T = 250$ μS.

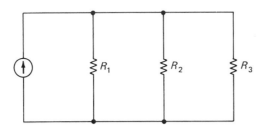

Figure 9-9.

Solution: Find R_1, R_2, R_3, R_T, I_1, I_3, and I_T. From given data we can find I_T.

$$I_T = VG_T = 50 \text{ V} \times 250 \text{ μS} = 12.5 \text{ mA}$$

We can also find R_2:

$$R_2 = \frac{V}{I_2} = \frac{50 \text{ V}}{400 \text{ μA}} = 125 \text{ kΩ}$$

Since we know G_1, we can find R_1:

$$R_1 = \frac{1}{G_1} = \frac{1}{90 \ \mu\text{S}} = 11.1 \ \text{k}\Omega$$

Since we know G_1 and V, we can find I_1:

$$I_1 = VG_1 = 50 \ \text{V} \times 90 \ \mu\text{S} = 4.5 \ \text{mA}$$

Since we know G_T, we can find R_T:

$$R_T = \frac{1}{G_T} = \frac{1}{250 \ \mu\text{S}} = 4 \ \text{k}\Omega$$

The rest of the unknowns can now be found:

$$I_3 = I_T - I_1 - I_2 = 12.5 \ \text{mA} - 4.5 \ \text{mA} - 400 \ \mu\text{A} = 7.6 \ \text{mA}$$

$$R_3 = \frac{V}{I_3} = \frac{50 \ \text{V}}{7.6 \ \text{mA}} = 6.58 \ \text{k}\Omega$$

PRACTICE PROBLEMS 9-4

1. Refer to Figure 9-8. Find G_T, R_T, V, I_1, and I_2 when (a) $I_T = 20$ mA, $R_1 = 2.2$ kΩ, and $R_2 = 4.7$ kΩ and (b) $I_T = 200 \ \mu$A, $R_1 = 15$ kΩ, $R_2 = 27$ kΩ.

2. Refer to Figure 9-8. Find I_1, V, R_2, G_T, and R_T when $I_T = 1.5$ mA, $R_1 = 10$ kΩ, and $I_2 = 800 \ \mu$A.

3. Refer to Figure 9-8. Find I_1, I_2, R_2, G_T, and R_T when $I_T = 1$ mA, $V = 2.7$ V, and $R_1 = 5.6$ kΩ.

4. Refer to Figure 9-9. Find G_T, R_T, V, I_1, I_2, and I_3 when (a) $I_T = 10$ mA, $R_1 = 330 \ \Omega$, $R_2 = 470 \ \Omega$, and $R_3 = 680 \ \Omega$ and (b) $I_T = 250 \ \mu$A, $R_1 = 5.6$ kΩ, $R_2 = 2.2$ kΩ, and $R_3 = 3.9$ kΩ.

5. Refer to Figure 9-9. Given $I_T = 50$ mA, $I_1 = 20$ mA, $V = 9$ V, and $R_2 = 1$ kΩ, find R_1, R_3, I_2, I_3, R_T, and G_T.

6. Refer to Figure 9-9. Given $V = 12$ V, $G_1 = 17.9 \ \mu$S, $I_2 = 255 \ \mu$A, and $G_T = 69.4 \ \mu$S, find R_1, R_2, R_3, I_1, I_3, and I_T.

Solutions:

1. (a) $G_T = G_1 + G_2 = 455 \ \mu\text{S} + 213 \ \mu\text{S} = 667 \ \mu\text{S}$

$$R_T = \frac{1}{G_T} = \frac{1}{667 \ \mu\text{S}} = 1.5 \ \text{k}\Omega$$

$$V = \frac{I_T}{G_T} = I_T R_T = 30 \ \text{V} \qquad \text{(either expression yields 30 V)}$$

$$I_1 = VG_1 = 30 \ \text{V} \times 455 \ \mu\text{S} = 13.6 \ \text{mA} \qquad \left(\text{we could have used } \frac{V}{R_1} \right)$$

$$I_2 = VG_2 = 30 \ \text{V} \times 213 \ \mu\text{S} = 6.38 \ \text{mA}$$

(b) $G_T = 104\ \mu S$
$\quad R_T = 9.64\ k\Omega$
$\quad V = 1.93\ V$
$\quad I_1 = 129\ \mu A$
$\quad I_2 = 71.4\ \mu A$

2. $I_1 = 700\ \mu A$, $V = 7\ V$, $R_2 = 8.75\ k\Omega$, $R_T = 4.67\ k\Omega$, $G_T = 214\ \mu S$

3. $I_1 = 482\ \mu A$, $I_2 = 518\ \mu A$, $R_2 = 5.21\ k\Omega$, $G_T = 370\ \mu S$, $R_T = 2.7\ k\Omega$

4. (a) $G_T = G_1 + G_2 + G_3 = 3.03\ mS + 2.13\ mS + 1.47\ mS = 6.63\ mS$
$$R_T = \frac{1}{G_T} = \frac{1}{6.63\ mS} = 151\ \Omega$$
$\quad V = I_T R_T = 10\ mA \times 151\ \Omega = 1.51\ V$
$\quad I_1 = VG_1 = 1.51\ V \times 3.03\ mS = 4.57\ mA$
$\quad I_2 = VG_2 = 1.51\ V \times 2.13\ mS = 3.21\ mS$
$\quad I_3 = VG_3 = 1.51\ V \times 1.47\ mS = 2.22\ mS$
(b) $G_T = 890\ \mu S$
$\quad R_T = 1.12\ k\Omega$
$\quad V = 281\ mV$
$\quad I_1 = 50.2\ \mu A$
$\quad I_2 = 128\ \mu A$
$\quad I_3 = 72.1\ \mu A$

5. $I_2 = 9\ mA$, $I_3 = 21\ mA$, $R_1 = 450\ \Omega$, $R_3 = 429\ \Omega$, $R_T = 180\ \Omega$, $G_T = 5.56\ mS$

6. $R_T = 14.4\ k\Omega$, $R_1 = 55.9\ k\Omega$, $R_2 = 47.1\ k\Omega$, $R_3 = 33.1\ k\Omega$, $I_1 = 215\ \mu A$, $I_3 = 363\ \mu A$, $I_T = 833\ \mu A$

Additional practice problems are at the end of the chapter.

SELF-TEST 9-2

1. Refer to Figure 9-6. Let $E = 25\ V$, $R_1 = 18\ k\Omega$, and $V_2 = 6.5\ V$. Find I, R_T, R_2, and V_1.

2. Refer to Figure 9-7. Let $I = 800\ \mu A$, $R_1 = 68\ k\Omega$, $V_2 = 50\ V$, and $R_T = 200\ k\Omega$. Find E, R_2, R_3, V_1, and V_3.

3. Refer to Figure 9-8. Let $I_T = 30\ mA$, $V = 1\ V$, and $R_1 = 100\ \Omega$. Find I_1, I_2, R_2, G_T, and R_T.

4. Refer to Figure 9-9. Let $I_T = 1\ mA$, $I_1 = 200\ \mu A$, $V = 10\ V$, and $R_2 = 27\ k\Omega$. Find R_1, R_3, I_2, I_3, G_T, and R_T.

Answers to Self-Test 9-2 are at the end of the chapter.

9.4 SERIES-PARALLEL CIRCUITS

Consider the circuit in Figure 9-10. Let's find the unknown variables which are I_T, I_1, I_2, I_3, R_T, V_1, V_2, and V_3. Since we know all values of R and we know E, we will first find R_T. Then we can find I_T.
$$R_T = R_1 + R_2 \| R_3$$
Let
$$R_X = R_2 \| R_3$$

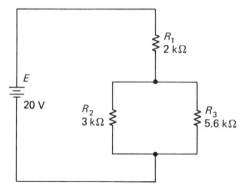

Figure 9-10.

Then
$$G_X = G_2 + G_3 = 333 \ \mu S + 179 \ \mu S = 512 \ \mu S$$

$$R_X = \frac{1}{G_X} = \frac{1}{512 \ \mu S} = 1.95 \ k\Omega$$

$$R_T = R_1 + R_X = 2 \ k\Omega + 1.95 \ k\Omega = 3.95 \ k\Omega$$

$$I_T = \frac{E}{R_T} = \frac{20 \ V}{3.95 \ k\Omega} = 5.06 \ mA$$

Since R_1 is a series resistor, $I_T = I_1$.

$$I_1 = I_T = 5.06 \ mA$$

Now we can find V_1.

$$V_1 = I_1 R_1 = 5.06 \ mA \times 2 \ k\Omega = 10.1 \ V$$

Applying Kirchhoff's voltage law, we get:

$$V_2 = V_3 = E - V_1 = 20 \ V - 10.1 \ V = 9.9 \ V$$

We could also find V_2 and V_3 by using

$$V_2 = V_3 = \frac{ER_X}{R_T} = \frac{20 \ V \times 1.95 \ k\Omega}{3.95 \ k\Omega} = 9.87 \ V$$

[The difference in the two methods (9.9 V versus 9.87 V) resulted from rounding V_1. V_1 was actually 10.12 V. Either answer is acceptable.] Now that we know V_2 and V_3, we can find I_2 and I_3.

$$I_2 = \frac{V_2}{R_2} = \frac{9.9 \ V}{3 \ k\Omega} = 3.3 \ mA$$

$$I_3 = \frac{V_2}{R_2} = \frac{9.9 \ V}{5.6 \ k\Omega} = 1.77 \ mA$$

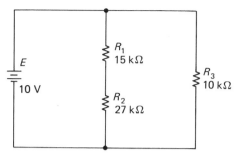

Figure 9-11.

EXAMPLE 9.10: Solve for R_T, I_T, I_3, I_1, V_1, and V_2 in Figure 9-11.

Solution: R_1 and R_2 are in series. This makes the branch resistance equal to $R_1 + R_2$. This branch is in parallel with R_3.

$$R_T = (R_1 + R_2) \| R_3$$

Let $R_X = R_1 + R_2$

Then $R_X = 15\ k\Omega + 27\ k\Omega = 42\ k\Omega$

$$G_T = G_X + G_3 = 23.8\ \mu S + 100\ \mu S = 12.4\ \mu S$$

$$R_T = \frac{1}{G_T} = \frac{1}{124\ \mu S} = 8.08\ k\Omega$$

Now we can find I_T:

$$I_T = \frac{E}{R_T} = \frac{10\ V}{8.08\ k\Omega} = 1.24\ mA$$

Knowing I_T, we can find I_3.

$$I_3 = \frac{I_T G_3}{G_T} = \frac{1.24\ mA \times 100\ \mu S}{124\ \mu S} = 1\ mA$$

or $$I_3 = \frac{V_3}{R_3} = \frac{10\ V}{10\ k\Omega} = 1\ mA$$

(We knew $V_3 = 10$ V because $V_3 = E$. The second method was easier. We must be aware that very often there is more than one path to take in finding the unknowns in complex circuits.) Since $I_3 = 1$ mA,

$$I_1 = I_2 = I_T - I_3 = 1.24\ mA - 1\ mA = 240\ \mu S$$

$$V_1 = \frac{E R_1}{R_X} = \frac{10\ V \times 15\ k\Omega}{42\ k\Omega} = 3.57\ V$$

$$V_2 = E - V_1 = 10\ V - 3.57\ V = 6.43\ V$$

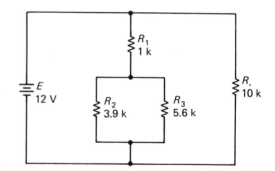

Figure 9-12.

EXAMPLE 9.11: Consider the circuit in Figure 9-12. Let's find all the currents and voltage drops and the total resistance.

Solution: Since we are given all the circuit resistances, we can find R_T. A look at the circuit shows that R_2 and R_3 are in parallel. Let

$$R_A = R_2 \| R_3$$

$$R_A = \frac{1}{G_2 + G_3} = \frac{1}{435 \ \mu S} = 2.3 \ k\Omega$$

This resistance, R_A, is in series with R_1. The total branch resistance is $R_A + R_1$. Let's call this resistance R_B.

$$R_B = R_1 + R_A = 1 \ k\Omega + 2.3 \ k\Omega = 3.3 \ k\Omega$$

R_B is in parallel with R_4.

$$R_T = R_B \| R_4 = \frac{1}{G_B + G_4} = \frac{1}{403 \ \mu S} = 2.48 \ k\Omega$$

Now we can find I_T.

$$I_T = \frac{E}{R_T} = \frac{12 \ V}{2.48 \ k\Omega} = 4.84 \ mA$$

We can see that R_4 is connected across the source. Therefore, $V_4 = E$. Solving for I_4, we get:

$$I_4 = \frac{V_4}{R_4} = \frac{12 \ V}{10 \ k\Omega} = 1.2 \ mA$$

Applying Kirchhoff's current law, we get:

$$I_T = I_1 + I_4$$

$$I_1 = I_T - I_4 = 4.84 \ mA - 1.2 \ mA = 3.64 \ mA$$

Knowing I_1, we can find I_2 and I_3.

$$I_2 = \frac{I_1 G_2}{G_2 + G_3} = 2.14 \text{ mA}$$

$$I_3 = I_T - I_2 = 1.49 \text{ mA}$$

We can find the remaining voltage drops now.

$$V_1 = I_1 R_1 = 3.64 \text{ mA} \times 1 \text{ k}\Omega = 3.64 \text{ V}$$

$$V_2 = I_2 R_2 = 2.14 \text{ mA} \times 3.9 \text{ k}\Omega = 8.36 \text{ V}$$

$$V_3 = V_2 = 8.36 \text{ V}$$

PRACTICE PROBLEMS 9-5

1. Refer to Figure 9-10. Let $E = 25$ V, $R_1 = 2.7$ kΩ, $R_2 = 6.8$ kΩ, and $R_3 = 4.7$ kΩ. Find V_1, V_2, V_3, I_1, I_2, I_3, I_T, and R_T.

2. Refer to Figure 9-11. $E = 40$ V, $R_1 = 680$ Ω, $R_2 = 1.2$ kΩ, and $R_3 = 2.7$ kΩ. Find V_1, V_2, V_3, I_1, I_2, I_3, I_T, and R_T.

3. Refer to Figure 9-10. Let $E = 20$ V, $I_T = 5$ mA, $R_1 = 1.5$ kΩ, and $R_2 = 6.8$ kΩ. Find R_3, R_T, V_1, V_2, V_3, I_2, and I_3.

4. Refer to Figure 9-13. Let $E = 10$ V, $I_T = 4$ mA, $V_2 = 3$ V, $R_3 = 1.5$ kΩ, and $R_4 = 6$ kΩ. Find I_2, I_3, I_4, R_1, R_2, and R_T.

5. Refer to Figure 9-12. Let $E = 9$ V, $R_1 = 9.1$ kΩ, $R_2 = 27$ kΩ, $R_3 = 56$ kΩ, and $R_4 = 120$ kΩ. Find V_1, V_2, V_3, V_4, I_T, I_1, I_2, I_3, I_4, and R_T.

6. Refer to Figure 9-14. Let $E = 9$ V, $R_1 = 330$ Ω, $R_2 = 810$ Ω, $R_3 = 1$ kΩ, and $R_4 = 560$ Ω. Find V_1, V_2, V_3, V_4, I_1, I_2, I_3, I_4, I_T, and R_T.

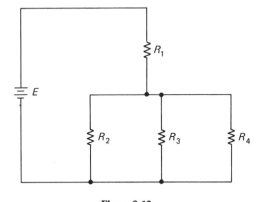

Figure 9-13.

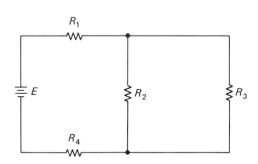

Figure 9-14.

Solutions:

1. $R_T = 5.48$ kΩ, $I_T = 4.56$ mA, $I_1 = 4.56$ mA, $I_2 = 1.86$ mA, $I_3 = 2.7$ mA, $V_1 = 12.3$ V, $V_2 = 12.7$ V, $V_3 = 12.7$ V.

2. $R_T = 1.11$ kΩ, $I_T = 36.1$ mA, $I_1 = 21.3$ mA, $I_2 = 21.3$ mA, $I_3 = 14.8$ mA, $V_1 = 14.5$ V, $V_2 = 25.5$ V, $V_3 = 40$ V

3. $R_T = 4$ kΩ, $V_1 = 7.5$ V, $V_2 = 12.5$ V, $V_3 = 12.5$ V, $I_2 = 1.84$ mA, $I_3 = 3.16$ mA, $R_3 = 3.96$ kΩ

4. $R_T = 2.5$ kΩ, $R_1 = 1.75$ kΩ, $R_2 = 2$ kΩ, $I_2 = 1.5$ mA, $I_3 = 2$ mA, $I_4 = 500$ μA

5. $R_T = 22.3$ kΩ, $I_T = 404$ μA, $I_1 = 329$ μA, $I_2 = 223$ μA, $I_3 = 107$ μA, $I_4 = 75$ μA, $V_1 = 2.99$ V, $V_2 = 6.01$ V, $V_3 = 6.01$ V, $V_4 = 9$ V

6. $R_T = 1.34$ kΩ, $I_T = I_1 = I_4 = 6.73$ mA, $I_2 = 3.72$ mA, $I_3 = 3.01$ mA, $V_1 = 2.22$ V, $V_2 = V_3 = 3.01$ V, $V_4 = 3.77$ V

SELF-TEST 9-3

1. Refer to Figure 9-10. Let $E = 3$ V, $I_T = 70$ μA, $R_2 = 33$ kΩ, and $R_3 = 47$ kΩ. Find V_1, V_2, V_3, I_2, I_3, R_T, and R_1.

2. Refer to Figure 9-11. Let $E = 20$ V, $R_1 = 20$ kΩ, $R_3 = 10$ kΩ, $V_2 = 6.57$ V. Find I_T, I_1, I_2, I_3, V_1, V_3, R_T, and R_2.

3. Refer to Figure 9-13. Let $E = 15$ V, $R_1 = 12$ kΩ, $R_2 = 100$ kΩ, $R_3 = 22$ kΩ, and $R_4 = 68$ kΩ. Find I_T, I_1, I_2, I_3, I_4, V_1, V_2, V_3, V_4, and R_T.

4. Refer to Figure 9-14. Let $E = 25$ V, $R_1 = 2.7$ kΩ, $R_2 = 3.9$ kΩ, $R_3 = 6.8$ kΩ, and $R_4 = 2.2$ kΩ. Find V_1, V_2, V_3, V_4, I_T, I_1, I_2, I_3, I_4, and R_T.

Answers to Self-Test 9-3 are at the end of the chapter.

END OF CHAPTER PROBLEMS 9-1

1. Refer to Figure 9-2(a). Find R_T when (a) $R_1 = 30$ kΩ, $R_2 = 15$ kΩ, (b) $R_1 = 750$ Ω, $R_2 = 1.2$ kΩ, (c) $R_1 = 6.8$ kΩ, $R_2 = 10$ kΩ, and (d) $R_1 = 120$ kΩ, $R_2 = 180$ kΩ.

2. Refer to Figure 9-2(b). Find R_T when (a) $R_1 = 470$ Ω, $R_2 = 820$ Ω, $R_3 = 1$ kΩ, (b) $R_1 = 15$ kΩ, $R_2 = 47$ kΩ, $R_3 = 68$ kΩ, (c) $R_1 = 47$ Ω, $R_2 = 100$ Ω, $R_3 = 150$ Ω, and (d) $R_1 = 330$ kΩ, $R_2 = 470$ kΩ, $R_3 = 560$ kΩ.

3. Refer to Figure 9-2(d). Find G_T and R_T when (a) $R_1 = 10$ kΩ, $R_2 = 20$ kΩ, (b) $R_1 = 1.2$ kΩ, $R_2 = 2.7$ kΩ, (c) $R_1 = 47$ kΩ, $R_2 = 18$ kΩ, (d) $R_1 = 75$ kΩ, $R_2 = 150$ kΩ, (e) $R_1 = 68$ Ω, $R_2 = 120$ Ω, and (f) $R_1 = 560$ Ω, $R_2 = 1.8$ kΩ.

4. Refer to Figure 9-2(e). Find G_T and R_T when (a) $R_1 = 2.2$ kΩ, $R_2 = 6.8$ kΩ, $R_3 = 12$ kΩ, (b) $R_1 = 68$ kΩ, $R_2 = 200$ kΩ, $R_3 = 82$ kΩ, (c) $R_1 = 910$ Ω, $R_2 = 2.2$ kΩ, $R_3 = 1.2$ kΩ, (d) $R_1 = 120$ kΩ, $R_2 = 68$ kΩ, $R_3 = 270$ kΩ, (e) $R_1 = 18$ kΩ, $R_2 = 18$ kΩ, $R_3 = 18$ kΩ, and (f) $R_1 = 330$ Ω, $R_2 = 820$ Ω, $R_3 = 910$ Ω.

5. Refer to Figure 9-3(a) and find R_T when (a) $R_1 = 560$ Ω, $R_2 = 1$ kΩ, $R_3 = 820$ Ω, (b) $R_1 = 7.5$ kΩ, $R_2 = 18$ kΩ, $R_3 = 27$ kΩ, (c) $R_1 = 27$ kΩ, $R_2 = 12$ kΩ, $R_3 = 47$ kΩ, and (d) $R_1 = 330$ kΩ, $R_2 = 680$ kΩ, $R_3 = 1$ MΩ.

6. Refer to Figure 9-4(a) and find R_T when (a) $R_1 = 4.7$ kΩ, $R_2 = 6.8$ kΩ, $R_3 = 9.1$ kΩ, (b) $R_1 = 1.8$ kΩ, $R_2 = 1.5$ kΩ, $R_3 = 820$ Ω, (c) $R_1 = 27$ kΩ, $R_2 = 18$ kΩ, $R_3 = 56$ kΩ, and (d) $R_1 = 680$ Ω, $R_2 = 470$ Ω, $R_3 = 2.2$ kΩ.

7. Refer to Figure 9-5(a) and find R_T when (a) $R_1 = 12$ kΩ, $R_2 = 47$ kΩ, $R_3 = 33$ kΩ, $R_4 = 27$ kΩ, (b) $R_1 = 120$ Ω, $R_2 = 470$ Ω, $R_3 = 1$ kΩ, $R_4 = 560$ Ω, (c) $R_1 = 390$ kΩ, $R_2 = 1.2$ M, $R_3 = 820$ kΩ, $R_4 = 750$ kΩ, and (d) $R_1 = 5.1$ kΩ, $R_2 = 6.8$ kΩ, $R_3 = 10$ kΩ, $R_4 = 10$ kΩ.

1. Refer to Figure 9-6. Find I, R_T, V_1, and V_2 when (a) $E=9$ V, $R_1=270$ Ω, $R_2=680$ Ω and (b) $E=12$ V, $R_1=1.2$ kΩ, $R_2=3.3$ kΩ.

2. Refer to Figure 9-6. Find I, R_T, V_1, and R_2 when (a) $E=10$ V, $R_1=1$ kΩ, $V_2=3$ V and (b) $E=40$ V, $R_1=6.8$ kΩ, $V_2=23.5$ V.

3. Refer to Figure 9-6. Find R_1, R_T, V_1, and V_2 when (a) $E=20$ V, $I=213$ μA, $R_2=33$ kΩ and (b) $E=25$ V, $I=1$ mA, $R_2=10$ kΩ.

4. Refer to Figure 9-6. Find E, R_1, R_T, and V_2 when (a) $I=75$ μA, $V_1=6.2$ mV, $R_2=3$ kΩ and (b) $I=2$ mA, $V_1=14.3$ V, $R_2=12$ kΩ.

5. Refer to Figure 9-7. Find I, R_T, V_1, V_2, and V_3 when (a) $E=9$ V, $R_1=33$ kΩ, $R_2=10$ kΩ, $R_3=18$ kΩ and (b) $E=40$ V, $R_1=220$ kΩ, $R_2=100$ kΩ, $R_3=68$ kΩ.

6. Refer to Figure 9-7. Find V_2, V_3, R_1, R_3, and R_T when (a) $E=25$ V, $V_1=5$ V, $I=1$ mA, $R_2=10$ kΩ and (b) $E=20$ V, $V_1=6.7$ V, $I=150$ μA, $R_2=47$ kΩ.

7. Refer to Figure 9-7. Find E, R_2, R_3, V_1, and V_3 when (a) $I=3$ mA, $R_1=4.7$ kΩ, $V_2=7.3$ V, $R_T=21$ kΩ and (b) $I=270$ μA, $R_1=15$ kΩ, $V_2=3.7$ V, $R_T=43.2$ kΩ.

8. Refer to Figure 9-8. Find G_T, R_T, V, I_1, and I_2 when (a) $I_T=3$ mA, $R_1=270$ Ω, $R_2=560$ Ω and (b) $I_T=700$ μA, $R_1=68$ kΩ, $R_2=39$ kΩ.

9. Refer to Figure 9-8. Find V, R_2, I_1, G_T, and R_T when (a) $I_T=25$ mA, $R_1=680$ Ω, $I_2=15$ mA and (b) $I_T=5$ mA, $R_1=1.8$ kΩ, $I_2=1$ mA.

10. Refer to Figure 9-8. Find I_1, I_2, R_2, G_T, and R_T when (a) $I_T=10$ mA, $V=20$ V, $R_1=4.7$ kΩ and (b) $I_T=350$ μA, $V=25$ V, $R_1=180$ kΩ.

11. Refer to Figure 9-8. Find I_2, I, G_T, R_T, and R_1 when (a) $V=12$ V, $I_1=530$ μA, $R_2=20$ kΩ and (b) $V=5$ V, $I_1=1$ mA, $R_2=15$ kΩ.

12. Refer to Figure 9-9. Find G_T, R_T, V, I_1, I_2, and I_3 when (a) $I_T=1.73$ mA, $R_1=2.7$ kΩ, $R_2=1.2$ kΩ, $R_3=1$ kΩ and (b) $I_T=500$ μA, $R_1=8.1$ kΩ, $R_2=7.5$ kΩ, $R_3=4.7$ kΩ.

13. Refer to Figure 9-9. Find R_1, R_3, R_T, G_T, I_2, and I_3 when (a) $I_T=4$ mA, $I_1=700$ μA, $V=12.6$ V, $R_2=12$ kΩ and (b) $I_T=300$ μA, $I_1=75$ μA, $V=20$ V, $R_2=270$ kΩ.

14. Refer to Figure 9-9. Find I_1, I_3, I_T, R_1, R_2, R_3, and R_T when (a) $V=15$ V, $G_1=213$ μS, $I_2=3$ mA, $G_T=1$ mS and (b) $V=30$ V, $G_1=2$ mS, $I_2=60$ mA, $G_T=8$ mS.

1. Refer to Figure 9-10. Find V_1, V_2, V_3, I_1, I_2, I_3, I_T, and R_T when (a) $E=15$ V, $R_1=10$ kΩ, $R_2=33$ kΩ, $R_3=47$ kΩ and (b) $E=9$ V, $R_1=180$ Ω, $R_2=560$ Ω, $R_3=680$ Ω.

2. Refer to Figure 9-11. Find V_1, V_2, V_3, I_1, I_2, I_3, I_T, and R_T when (a) $E=60$ V, $R_1=12$ kΩ, $R_2=22$ kΩ, $R_3=18$ kΩ and (b) $E=3$ V, $R_1=7.5$ kΩ, $R_2=27$ kΩ, $R_3=47$ kΩ.

3. Refer to Figure 9-10. (a) Let $E=10$ V, $I_T=500$ μA, $R_1=12$ kΩ, $R_2=47$ kΩ. Find R_3, R_T, V_1, V_2, V_3, I_2, and I_3. (b) Let $E=12$ V, $I_T=600$ μA, $R_2=39$ kΩ, $R_3=27$ kΩ. Find R_1, R_T, V_1, V_2, V_3, I_2, I_3.

4. Refer to Figure 9-13. (a) Let $E=25$ V, $R_1=3$ kΩ, $R_2=15$ kΩ, $R_3=22$ kΩ, $R_4=18$ kΩ. Find I_T, I_1, I_2, I_3, I_4, R_T, V_1, V_2, V_3, and V_4. (b) Let $E=25$ V, $I_T=250$ μA, $V_2=10$ V, $R_3=150$ kΩ, $R_4=200$ kΩ. Find I_1, I_2, I_3, I_4, R_1, R_2, and R_T.

5. Refer to Figure 9-12. (a) Let $E=40$ V, $R_1=$ 680 Ω, $R_2=1.5$ kΩ, $R_3=2.7$ kΩ, $R_4=4.7$ kΩ. Find V_1, V_2, V_3, V_4, I_T, I_1, I_2, I_3, I_4, and R_T. (b) Let $E=30$ V, $R_1=150$ kΩ, $R_2=470$ kΩ, $R_3=680$ kΩ, $R_4=680$ kΩ. Find V_1, V_2, V_3, V_4, I_T, I_1, I_2, I_3, I_4, and R_T.

6. Refer to Figure 9-14. (a) Let $E=40$ V, $R_1=$ 10 kΩ, $R_2=27$ kΩ, $R_3=56$ kΩ, $R_4=33$ kΩ. Find V_1, V_2, V_3, V_4, I_1, I_2, I_3, I_4, I_T, and R_T. (b) Let $E=15$ V, $R_1=150$ kΩ, $R_2=470$ kΩ, $R_3=680$ kΩ, $R_4=150$ kΩ. Find V_1, V_2, V_3, V_4, I_1, I_2, I_3, I_4, I_T, and R_T.

ANSWERS TO SELF-TESTS

Self-Test 9-1

1. 122 kΩ

2. 487 Ω

3. 19 kΩ

4. 29.1 kΩ

5. 571 Ω

6. 503 kΩ

Self-Test 9-2

1. $R_T=24.3$ kΩ, $I=1.03$ mA, $R_2=6.31$ kΩ, $V_1=$ 18.5 V

2. $E=160$ V, $V_1=54.4$ V, $V_3=55.6$ V, $R_3=$ 69.5 kΩ, $R_2=62.5$ kΩ

3. $R_T=33.3$ Ω, $G_T=30$ mS, $I_1=10$ mA, $I_2=$ 20 mA, $R_2=50$ Ω

4. $G_T=100$ μS, $R_T=10$ kΩ, $I_2=370$ μA, $I_3=$ 430 μA, $R_1=50$ kΩ, $R_3=23.3$ kΩ

Self-Test 9-3

1. $R_T=42.9$ kΩ, $V_1=1.64$ V, $V_2=V_3=1.36$ V, $I_2=41.2$ μA, $I_3=28.9$ μA, $R_1=23.4$ kΩ

2. $V_1=13.4$ V, $V_3=20$ V, $I_1=I_2=672$ μA, $I_3=$ 2 mA, $I_T=2.67$ mA, $R_T=7.49$ kΩ, $R_2=$ 9.78 kΩ

3. $I_T=I_1=570$ μA, $I_2=81.4$ μA, $I_3=370$ μA, $I_4=120$ μA, $V_1=6.86$ V, $V_2=V_3=V_4=8.14$ V

4. $V_1=9.15$ V, $V_2=V_3=8.4$ V, $V_4=7.45$ V, $I_T=$ 3.39 mA, $I_1=I_4=3.39$ mA, $I_2=2.15$ mA, $I_3=1.24$ mA, $R_T=7.38$ kΩ

CHAPTER 10

DC CIRCUIT ANALYSIS—CIRCUIT THEOREMS

It is often necessary to determine the internal resistance (or conductance) of an electrical energy source, amplifier, or system. We are also frequently asked to determine current and voltage distribution in electrical circuits in which Ohm's law and Kirchhoff's laws are not adequate. A variety of theorems have been developed over the years to assist the technician and engineer in solving these complex circuit problems. Thévenin's theorem, Norton's theorem and the Superposition theorem are some of the more widely used theorems. We will explain and use each method in this unit.

10.1 SUPERPOSITION THEOREM

A method often used to solve complex circuit problems utilizes the Superposition theorem. Using this method we will find the current through, and the drop across, R_2 in Figure 10-1. This is done by first replacing one source with a short circuit and determining the magnitude and direction of I_2. If we replace E_2 with a short circuit, the circuit shown in Figure 10-2 results.

$$I_T = \frac{E_1}{R_1 + R_X}$$

where $R_X = R_2 \| R_3$.

$$I_T = \frac{20 \text{ V}}{3 \text{ k}\Omega + 667 \text{ }\Omega} = 5.45 \text{ mA}$$

$$I_2 = \frac{I_T R_X}{R_2} = \frac{5.45 \text{ mA} \times 667 \text{ }\Omega}{2 \text{ k}\Omega} = 1.82 \text{ mA}$$

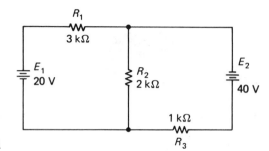

Figure 10-1

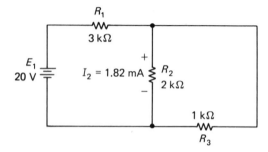

Figure 10-2

The current is 1.82 mA and has the direction shown. Now let's replace E_1 with a short circuit and compute the current due to E_2. This is done in Figure 10-3.

$$I_T = \frac{E_2}{R_3 + R_X}$$

where $R_X = R_1 \| R_2$

$$I_T = \frac{40\ V}{1\ k\Omega + 1.2\ k\Omega} = 18.2\ mA$$

$$I_2 = \frac{I_T R_X}{R_2} = \frac{18.2\ mA \times 1.2\ k\Omega}{2\ k\Omega} = 10.9\ mA$$

The current due to E_2 alone is 10.9 mA and the polarity is as shown in Figure 10-3. Now if we go back to the original circuit and superimpose the currents through R_2 as in Figure 10-4, we get one current whose value is

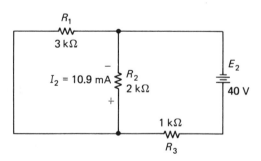

Figure 10-3

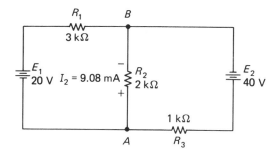

Figure 10-4

1.82 mA causing a polarity at point A which is negative with respect to point B, and we get a second current of 10.9 mA causing a polarity at point A which is positive with respect to point B. The currents are opposing. If we add the currents algebraically, we get a resulting current which is the difference between the two: 10.9 mA − 1.82 mA = 9.08 mA. That is, the actual current through R_2 is 9.08 mA and point A is positive with respect to point B. The voltage drop across R_2 is 9.08 mA × 2 kΩ which equals 18.2 V.

Knowing V_2, we can determine V_1 and V_3 by using Kirchhoff's voltage law. Using the loop containing E_1, R_1, and R_2 in Figure 10-5(a), we get the general equation $E_1 + V_1 + V_2 = 0$. Starting at point A and moving clockwise, we get:

$$-E_1 + V_1 - V_2 = 0$$

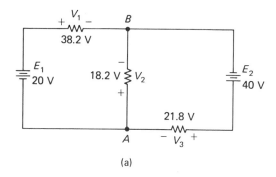

(a)

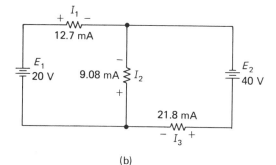

(b)

Figure 10-5

At this point we don't know the polarity of the drop across V_1; therefore, we assigned positive on the left-hand side. If our assumption is wrong, when we solve for V_1 its magnitude will be correct but its value will be negative. If that turns out to be the case, we simply reverse the assigned polarity. Plugging in known values:

$$-20 \text{ V} + V_1 - 18.2 \text{ V} = 0$$
$$V_1 - 38.2 \text{ V} = 0$$
$$V_1 = 38.2 \text{ V}$$

Because the computed value for V_1 is positive, we have assigned the correct polarity and no change is necessary. Then:

$$I_1 = \frac{V_1}{R_1} = \frac{38.2 \text{ V}}{3 \text{ k}\Omega} = 12.7 \text{ mA}$$

Now let's look at the loop containing E_2, R_2, and R_3. Referring again to Figure 10-5(a), the general equation is $E_2 + V_2 + V_3 = 0$. Starting at point B and moving counter-clockwise, we get $-V_2 + V_3 + E_2 = 0$. Again, the polarity of the drop across V_3 is not known and we have assumed it to be positive. Plugging in known values and solving for V_3, we get:

$$-18.2 \text{ V} + V_3 + 40 \text{ V} = 0$$
$$V_3 + 21.8 \text{ V} = 0$$
$$V_3 = -21.8 \text{ V}$$

V_3 equals 21.8 V, but the polarity assigned must be reversed since the computed value was negative. Solving for I_3, we get:

$$I_3 = \frac{V_3}{R_3} = \frac{21.8 \text{ V}}{1 \text{ k}\Omega} = 21.8 \text{ mA}$$

The complete circuit showing currents is in Figure 10-5(b).

Let's try another one. Find the currents and voltage drops in the circuit shown in Figure 10-6.

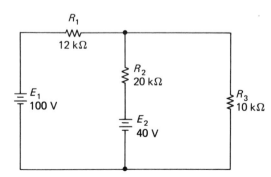

Figure 10-6

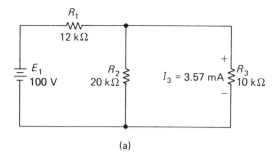

(a)

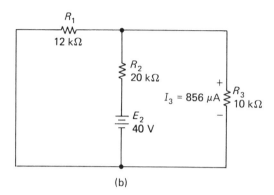

(b)   **Figure 10-7**

In Figure 10-7(a) the circuit is redrawn with E_2 replaced by a short circuit. The current through R_3 is calculated:

$$I_T = \frac{E_1}{R_1 + R_X}$$

where $R_X = R_2 \| R_3$.

$$I_T = \frac{100 \text{ V}}{12 \text{ k}\Omega + 6.67 \text{ k}\Omega} = 5.36 \text{ mA}$$

$$I_3 = \frac{I_T R_X}{R_3} = \frac{5.36 \text{ mA} \times 6.67 \text{ k}\Omega}{10 \text{ k}\Omega} = 3.57 \text{ mA}$$

In Figure 10-7(b) the circuit is again redrawn, but this time E_1 is replaced with a short circuit. I_3 in this circuit is calculated:

$$I_T = \frac{E_2}{R_2 + R_X}$$

where $R_X = R_1 \| R_3$.

$$I_T = \frac{40 \text{ V}}{20 \text{ k}\Omega + 5.45 \text{ k}\Omega} = 1.57 \text{ mA}$$

$$I_3 = \frac{I_T R_X}{R_3} = \frac{1.57 \text{ mA} \times 5.45 \text{ k}\Omega}{10 \text{ k}\Omega} = 856 \text{ }\mu\text{A}$$

The current through R_3 in the original circuit (Figure 10-6) is the algebraic sum of these two currents. Since the two currents are in the same direction, the resulting current is the sum of the two individual currents: $I_3 = 3.57$ mA + 856 μA = 4.43 mA. The voltage drop across R_3 can now be determined:

$$V_3 = I_3 R_3 = 4.43 \text{ mA} \times 10 \text{ k}\Omega = 44.3 \text{ V}$$

In Figure 10-8, polarities have been assigned. The polarity of the voltage drop across R_3 had to be as shown. This was dictated by the resulting direction of I_3. The drops across R_1 and R_2 were chosen arbitrarily. Remember, if we assigned the wrong polarity to V_1 and V_2, the calculated values will be correct but they will be negative. Using Kirchhoff's voltage law, we get:

$$E_1 + V_1 + V_3 = 0$$
$$100 \text{ V} - V_1 - 44.3 \text{ V} = 0$$
$$- V_1 + 55.7 \text{ V} = 0$$
$$V_1 = 55.7 \text{ V}$$
$$E_2 - V_3 - V_2 = 0$$
$$40 \text{ V} - 44.3 \text{ V} - V_2 = 0$$
$$V_2 = - 4.3 \text{ V}$$

V_2 is negative which tells us that the assigned polarity is wrong. The polarity is corrected in Figure 10-9. Now we can find I_1 and I_2 by using Ohm's law:

$$I_1 = \frac{V_1}{R_1} = \frac{55.7 \text{ V}}{12 \text{ k}\Omega} = 4.63 \text{ mA}$$

$$I_2 = \frac{V_2}{R_2} = \frac{4.3 \text{ V}}{20 \text{ k}\Omega} = 215 \text{ }\mu\text{A}$$

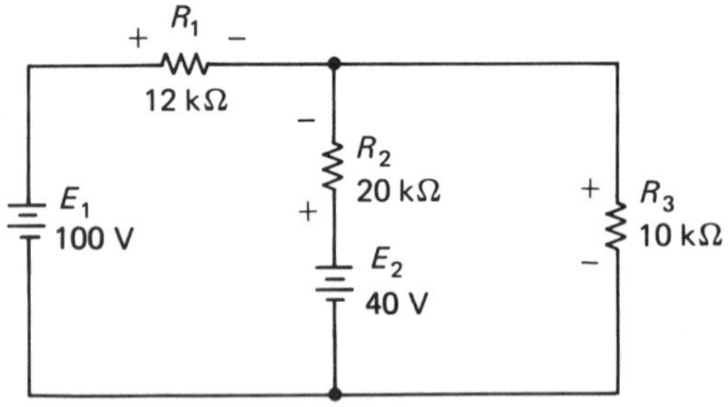

Figure 10-8

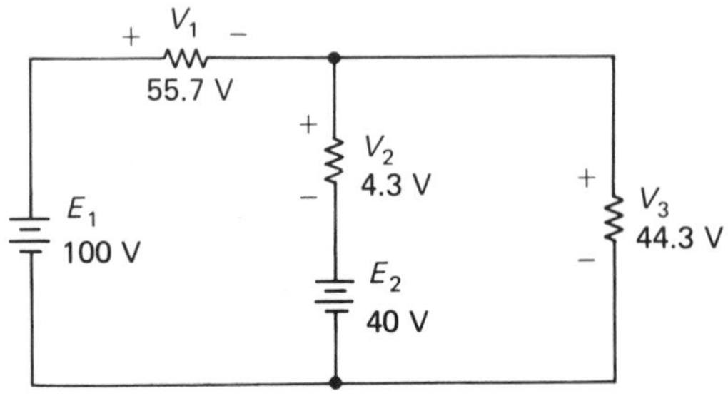

Figure 10-9

Kirchhoff's laws are now used to check our work. I_1 is total current. Therefore:

$$I_1 = I_2 + I_3$$

$$4.64 \text{ mA} = 222 \text{ } \mu\text{A} + 4.43 \text{ mA}$$

Also,

$$E_1 - V_3 - V_1 = 0$$

$$100 \text{ V} - 44.3 \text{ V} - 55.7 \text{ V} = 0$$

And

$$E_1 - E_2 - V_2 - V_1 = 0$$

$$100 \text{ V} - 40 \text{ V} - 4.3 \text{ V} - 55.7 \text{ V} = 0$$

PRACTICE PROBLEMS 10-1   Find the various voltage drops and currents in the circuits of Figure 10-10.

Solutions: In Figure 10-10(a) the current through R_2 would be 45.3 mA. Replacing E_2 with a short circuit, we would get:

$$R_T = R_1 + R_2 \| R_3 = 100 \text{ } \Omega + 132 \text{ } \Omega = 232 \text{ } \Omega$$

$$I_T = \frac{E_1}{R_T} = \frac{40 \text{ V}}{232 \text{ } \Omega} = 172 \text{ mA}$$

$$I_2 = \frac{I_T R_X}{R_2} = \frac{172 \text{ mA} \times 132 \text{ } \Omega}{330 \text{ } \Omega} = 68.8 \text{ mA}$$

I_2 due to E_1 is 68.8 mA.

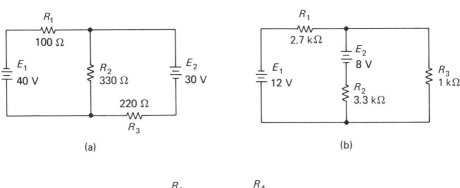

(a)   (b)

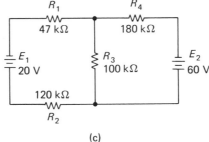

(c)

Figure 10-10

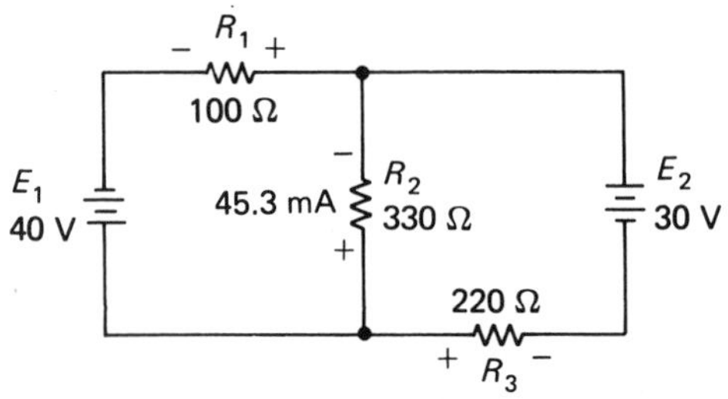

Figure 10-11

Replacing E_1 with a short circuit, we get:

$$R_T = R_3 + R_1 \| R_2 = 220\ \Omega + 76.7\ \Omega = 297\ \Omega$$

$$I_T = \frac{E_2}{R_T} = \frac{30\ \text{V}}{297\ \Omega} = 101\ \text{mA}$$

$$I_2 = \frac{I_T R_X}{R_2} = \frac{101\ \text{mA} \times 76.7\ \Omega}{330\ \Omega} = 23.5\ \text{mA}$$

I_2 due to E_2 is 23.5 mA.

The currents are opposing. Therefore, the resulting current is the difference between the two currents, or 45.3 mA. The polarities are shown in Figure 10-11. Using Ohm's law and Kirchhoff's laws, we can find the rest of the currents and voltage drops.

$$V_2 = I_2 R_2 = 45.3\ \text{mA} \times 330\ \Omega = 14.9\ \text{V}$$

$$V_1 = E_1 - V_2 = 40\ \text{V} - 14.9\ \text{V} = 25.1\ \text{V}$$

$$V_3 = E_2 + V_2 = 30\ \text{V} + 14.9\ \text{V} = 44.9\ \text{V}$$

$$I_1 = \frac{V_1}{R_1} = \frac{25.1\ \text{V}}{100\ \Omega} = 251\ \text{mA}$$

$$I_3 = \frac{V_3}{R_3} = \frac{44.9\ \text{V}}{220\ \Omega} = 204\ \text{mA}$$

In Figure 10-10(b) the current through R_3 would be 1.2 mA. Replacing E_1 with a short circuit, we would get:

$$R_T = R_2 + R_1 \| R_3 = 3.3\ \text{k}\Omega + 730\ \Omega = 4.03\ \text{k}\Omega$$

$$I_T = \frac{E_2}{R_T} + \frac{8\ \text{V}}{4.03\ \text{k}\Omega} = 1.99\ \text{mA}$$

$$I_3 = \frac{I_T R_X}{R_3} = \frac{1.99\ \text{mA} \times 730\ \Omega}{1\ \text{k}\Omega} = 1.45\ \text{mA}$$

I_3 due to E_2 is 1.45 mA.

Replacing E_2 with a short circuit, we would get:

$$R_T = R_1 + R_2 \| R_3 = 2.7\ \text{k}\Omega + 767\ \Omega = 3.47\ \text{k}\Omega$$

$$I_T = \frac{E_1}{R_T} = \frac{12\ \text{V}}{3.47\ \text{k}\Omega} = 3.46\ \text{mA}$$

$$I_3 = \frac{I_T R_X}{R_3} = \frac{3.46\ \text{mA} \times 767\ \Omega}{1\ \text{k}\Omega} = 2.65\ \text{mA}$$

I_3 due to E_1 is 2.65 mA.

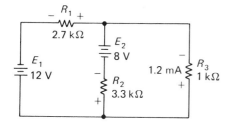

Figure 10-12

The two currents are opposing and the resulting current through R_3 is 1.2 mA. The polarities are shown in Figure 10-12. Using Ohm's law and Kirchhoff's laws, we can find the rest of the currents and voltage drops.

$$V_3 = I_3 R_3 = 1.2 \text{ mA} \times 1 \text{ k}\Omega = 1.2 \text{ V}$$

$$V_2 = E_2 + V_3 = 8 \text{ V} + 1.2 \text{ V} = 9.2 \text{ V}$$

$$V_1 = E_1 - V_3 = 12 \text{ V} - 1.2 \text{ V} = 10.8 \text{ V}$$

$$I_2 = \frac{V_2}{R_2} = \frac{9.2 \text{ V}}{3.3 \text{ k}\Omega} = 2.79 \text{ mA}$$

$$I_1 = \frac{V_1}{R_1} = \frac{10.8 \text{ V}}{2.7 \text{ k}\Omega} = 4 \text{ mA}$$

In Figure 10-10(c) the current through R_3 would be $211\,\mu\text{A}$. Replacing E_1 with a short circuit, we would get:

$$R_T = R_4 + R_3 \| (R_1 + R_2) = 180 \text{ k}\Omega + 62.5 \text{ k}\Omega = 243 \text{ k}\Omega$$

$$I_T = \frac{E_2}{R_T} = \frac{60 \text{ V}}{243 \text{ k}\Omega} = 247 \text{ }\mu\text{A}$$

$$I_3 = \frac{I_T R_X}{R_3} = \frac{247 \text{ }\mu\text{A} \times 62.5 \text{ k}\Omega}{100 \text{ k}\Omega} = 155 \text{ }\mu\text{A}$$

I_3 due to E_2 is 155 μA. Replacing E_2 with a short circuit:

$$R_T = R_1 + R_2 + R_3 \| R_4 = 47 \text{ k}\Omega + 120 \text{ k}\Omega + 64.3 \text{ k}\Omega = 231 \text{ k}\Omega$$

$$I_T = \frac{E_1}{R_T} = \frac{20 \text{ V}}{231 \text{ k}\Omega} = 86.6 \text{ }\mu\text{A}$$

$$I_3 = \frac{I_T R_X}{R_3} = \frac{86.6 \text{ }\mu\text{A} \times 64.3 \text{ k}\Omega}{100 \text{ k}\Omega} = 55.7 \text{ }\mu\text{A}$$

I_3 due to E_1 is 55.7 μA.

The two currents are aiding in making the resulting current through R_3 the sum of the two currents or 211 μA. The polarities are shown in Figure 10-13. Using Ohm's law and Kirchhoff's laws,

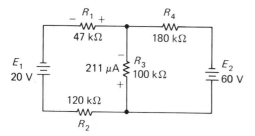

Figure 10-13

the rest of the currents and voltage drops may be found.

$$V_3 = I_3 R_3 = 211 \ \mu A \times 100 \ k\Omega = 21.1 \ V$$

$$V_4 = E_2 - V_3 = 60 \ V - 21.1 \ V = 38.9 \ V$$

$$I_4 = \frac{V_4}{R_4} = \frac{38.9 \ V}{180 \ k\Omega} = 216 \ \mu A$$

$$V_1 + V_2 = E_2 - V_4 - E_1 = 60 \ V - 38.9 \ V - 20 \ V = 1.1 \ V$$

$$I_1 = I_2 = \frac{V_1 + V_2}{R_1 + R_2} = \frac{1.1 \ V}{167 \ k\Omega} = 6.59 \ \mu A$$

$$V_1 = I_1 R_1 = 6.59 \ \mu A \times 47 \ k\Omega = 0.31 \ V$$

$$V_2 = I_2 R_2 = 6.59 \ \mu A \times 120 \ k\Omega = 0.791 \ V$$

Additional practice problems are at the end of the chapter.

SELF-TEST 10-1

Use the superposition theorem to find the various currents and voltage drops in the circuits in Figure 10-14.

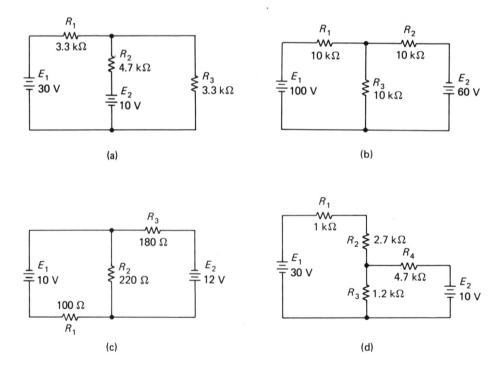

(a)

(b)

(c)

(d)

Figure 10-14

Answers to Self-Test 10-1 are at the end of the chapter.

10.2 THÉVENIN'S THEOREM

Thévenin's theorem states that any network, no matter how complex, can be reduced to an equivalent voltage source and series resistance. The voltage source is labeled V_{OC} and the series resistance is R_{TH} as illustrated in Figure 10-15. Consider the circuit in Figure 10-16. Because this is a simple series-parallel circuit, the various currents and voltage drops can be found easily by using Ohm's law and Kirchhoff's laws.

$$R_T = R_1 + R_2 \| R_3 = 1 \text{ k}\Omega + 500 \ \Omega = 1.5 \text{ k}\Omega$$

$$I_T = \frac{E}{R_T} = \frac{10 \text{ V}}{1.5 \text{ k}\Omega} = 6.67 \text{ mA}$$

$$I_1 = I_T = 6.67 \text{ mA}$$

$$V_1 = I_1 R_1 = 6.67 \text{ mA} \times 1 \text{ k}\Omega = 6.67 \text{ V}$$

$$V_2 = V_3 = E - V_1 = 10 \text{ V} - 6.67 \text{ V} = 3.33 \text{ V}$$

$$I_2 = \frac{V_2}{R_2} = \frac{3.33 \text{ V}}{1 \text{ k}\Omega} = 3.33 \text{ mA}$$

$$I_3 = \frac{V_3}{R_3} = \frac{3.33 \text{ V}}{1 \text{ k}\Omega} = 3.33 \text{ mA}$$

In circuits such as this, circuit theorems are not needed. Let's go ahead and Thévenize this circuit anyway just to develop a basic understanding of Thévenin's theorem. Remember that we said Thévenin's circuit is an *equivalent* circuit. This means that the Thévenin's equivalent circuit and the circuit it replaces furnish the same energy to the load. Suppose in Figure 10-16 we consider that R_3 is the load. We may consider that E, R_1, and R_2 is the circuit that supplies energy to R_3, just as V_{OC} and R_{TH}, as indicated in Figure 10-17.

In Thévenizing, we first determine from which points we wish to examine the circuit. In Figure 10-17 we are considering that R_3 is the load. Therefore, the circuit to the left of terminals x–y will be Thévenized.

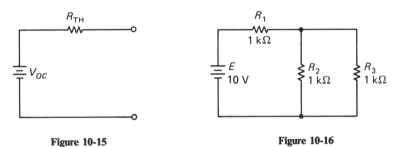

Figure 10-15   Figure 10-16

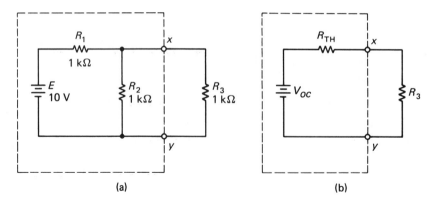

Figure 10-17

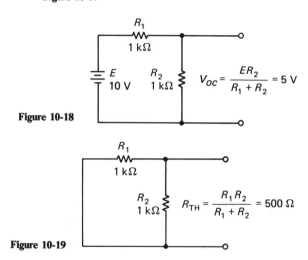

Figure 10-18

Figure 10-19

To find Thévenin's equivalent circuit, we use the following procedure:

1. *Remove the load and calculate the difference in potential between the open circuit terminals* (x–y). *This is the voltage source* V_{OC} *in our equivalent circuit.* In Figure 10-18 the load (R_3) has been removed. Looking back from the open-circuit terminals, we see that the circuit has been reduced to a simple series circuit and $V_{OC} = V_2$. Applying Ohm's law, we get $V_{OC} = 5$ V.

2. *With the load removed as in step 1,* replace the source with a short circuit and determine the resistance at the open circuit terminals.* This resistance is R_{TH} in our equivalent circuit. Replacing the source with a short circuit as in Figure 10-19 results in a circuit with R_1 and R_2 in

---

*We have assumed that the source is an ideal voltage source and $R_{int} = 0$ Ω. If the source resistance were some finite value, then the source would be replaced with that resistance instead of a short circuit.

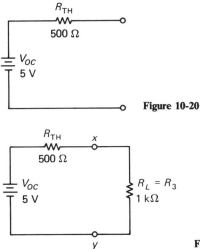

Figure 10-20

Figure 10-21.

parallel. This parallel circuit is seen looking back into the circuit from the open-circuit terminals. The equivalent resistance (R_{TH}) is 500 Ω. V_{OC} and R_{TH} are connected in series to form Thévenin's equivalent circuit in Figure 10-20.

3. *Connect the load across the output terminals of the Thévenin's equivalent circuit. Calculate* V_L *and* I_L, *the voltage drop across and the current through the load.* Using Ohm's law in Figure 10.21, we get:

$$I_L = \frac{V_{OC}}{R_{TH} + R_L} \tag{10-1}$$

$$I_L = \frac{5\text{ V}}{1.5\text{ k}\Omega} = 3.33\text{ mA}$$

$$V_L = \frac{V_{OC}R_L}{R_{TH} + R_L} \tag{10-2}$$

$$V_L = \frac{5\text{ V} \times 1\text{ k}\Omega}{1.5\text{ k}\Omega} = 3.33\text{ V}$$

Thévenin's equivalent circuit causes the same current through R_3 and the same voltage drop across it, as does the original circuit. Further, the polarity associated with V_{OC} in Figure 10-18 is the polarity of the voltage drop across R_3 in the original circuit. (Of course, in Figure 10-16 it was obvious that the point y was negative with respect to point x, but in some complex circuits it is not so obvious.)

Knowing V_3 and I_3 (remember we called R_3 the load), we could find the rest of the voltage drops and currents just as we did in Section 10.1 (Superposition theorem).

Let's Thévenize some rather simple series-parallel circuits. We will find only V_L and I_L.

PRACTICE PROBLEMS 10-2 For the problems in Figure 10-22:

1. Develop Thévenin's equivalent circuit.

2. Find V_L and I_L.

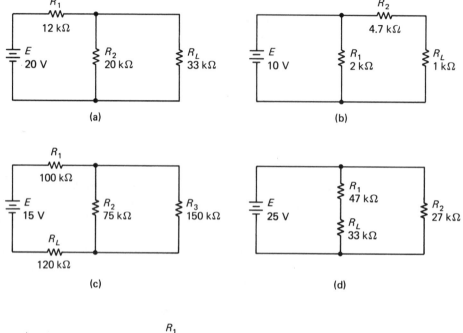

(a)

(b)

(c)

(d)

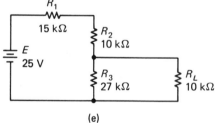

(e)

Figure 10-22

Solutions:

In Figure 10-22(a) if we open circuit the load, $V_{OC} = V_2$.

$$V_{OC} = V_2 = \frac{ER_2}{R_1 + R_2} = \frac{20\ \text{V} \times 20\ \text{k}\Omega}{32\ \text{k}\Omega} = 12.5\ \text{V}$$

$$R_{TH} = R_1 \| R_2 = 7.5\ \text{k}\Omega$$

Thévenin's equivalent circuit with the load connected is shown in Figure 10-23. Applying Ohm's law, we get:

$$I_L = \frac{V_{OC}}{R_{TH} + R_L} = \frac{12.5 \text{ V}}{40.5 \text{ k}\Omega} = 309 \text{ }\mu\text{A}$$

$$V_L = \frac{V_{OC}R_L}{R_{TH} + R_L} = \frac{12.5 \text{ V} \times 33 \text{ K}\Omega}{40.5 \text{ k}\Omega} = 10.2 \text{ V}$$

or

$$V_L = I_L R_L = 309 \text{ }\mu\text{A} \times 33 \text{ k}\Omega = 10.2 \text{ V}$$

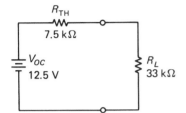

Figure 10-23

In Figure 10-22(b) if we open circuit the load, $V_{OC} = V_1 = E$. Shorting the source places a short circuit across R_1 so that $R_{TH} = R_2 = 4.7$ kΩ. Thévenin's equivalent circuit with the load connected is shown in Figure 10-24. Applying Ohm's law, we get:

$$I_L = \frac{V_{OC}}{R_{TH} + R_L} = \frac{10 \text{ V}}{5.7 \text{ k}\Omega} = 1.75 \text{ mA}$$

$$V_L = I_L R_L = 1.75 \text{ mA} \times 1 \text{ k}\Omega = 1.75 \text{ V}$$

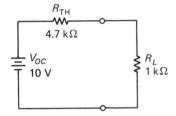

Figure 10-24

In Figure 10-22(c) if we open circuit the load, there is no circuit current. Therefore, $V_{OC} = E = 15$ V.

$$R_{TH} = R_1 + R_2 \| R_3 = 100 \text{ k}\Omega + 50 \text{ k}\Omega = 150 \text{ k}\Omega$$

Thévenin's equivalent circuit with the load connected is shown in Figure 10-25.

$$I_L = \frac{V_{OC}}{R_{TH} + R_L} = \frac{15 \text{ V}}{270 \text{ k}\Omega} = 55.6 \text{ }\mu\text{A}$$

$$V_L = I_L R_L = 55.6 \text{ }\mu\text{A} \times 120 \text{ k}\Omega = 6.67 \text{ V}$$

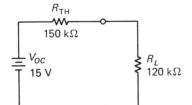

Figure 10-25

In Figure 10-22(d) if we open circuit the load, V_{OC} again equals E since V_1 would equal 0 V. Shorting the source to find R_{TH} causes a short circuit across R_2. Therefore, $R_{TH} = R_1 = 47$ kΩ. Thévenin's equivalent circuit with the load connected is shown in Figure 10-26.

$$I_L = \frac{V_{OC}}{R_{TH} + R_L} = \frac{25 \text{ V}}{80 \text{ k}\Omega} = 313 \text{ } \mu\text{A}$$

$$V_L = I_L R_L = 313 \text{ } \mu\text{A} \times 33 \text{ k}\Omega = 10.3 \text{ V}$$

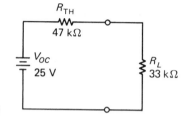

Figure 10-26

In Figure 10-22(e) if we open circuit the load, $V_{OC} = V_3$.

$$V_{OC} = V_3 = \frac{ER_3}{R_1 + R_2 + R_3} = \frac{25 \text{ V} \times 27 \text{ k}\Omega}{52 \text{ k}\Omega} = 13 \text{ V}$$

$$R_{TH} = R_3 \| (R_1 + R_2) = 13 \text{ k}\Omega$$

Thévenin's equivalent circuit with the load connected is shown in Figure 10-27.

$$I_L = \frac{V_{OC}}{R_{TH} + R_L} = \frac{13 \text{ V}}{23 \text{ k}\Omega} = 565 \text{ } \mu\text{A}$$

$$V_L = I_L R_L = 565 \text{ } \mu\text{A} \times 10 \text{ k}\Omega = 5.65 \text{ V}$$

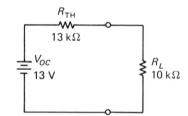

Figure 10-27

Additional practice problems are at the end of the chapter.

Now let's Thévenize some circuits that are more complex. Consider the circuit in Figure 10-1. We have already found the various currents and voltage drops by using the Superposition theorem. We will consider that R_2 is the load and find I_2 and V_2 by using Thévenin's theorem.

First, let's open circuit the load, as in Figure 10-28(a), and calculate V_{OC}. Because E_1 and E_2 are connected series aiding, $E_T = E_1 + E_2 = 60$ V. Therefore,

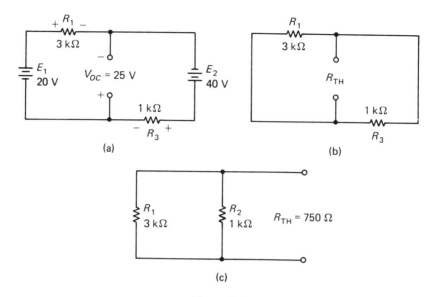

Figure 10-28

we may consider that there is an EMF of 60 V in series with R_1 and R_3.

$$V_1 = \frac{E_T R_1}{R_1 + R_3} = \frac{60 \text{ V} \times 3 \text{ k}\Omega}{4 \text{ k}\Omega} = 45 \text{ V}$$

$$V_3 = E_T - V_1 = 60 \text{ V} - 45 \text{ V} = 15 \text{ V}$$

Using Kirchhoff's voltage law, we get $V_{OC} = E_2 - V_3 = 25$ V (or $V_{OC} = V_1 - E_1 = 25$ V) and the polarity is as shown.

Next, we replace E_1 and E_2 with short circuits as in Figure 10-28(b) and calculate R_{TH}. In Figure 10-28(c) the circuit has been redrawn so that we can see that R_1 and R_3 are in parallel.

$$R_{TH} = R_1 \| R_3 = 750 \ \Omega$$

Now let's draw Thévenin's equivalent circuit and connect the load across it as in Figure 10-29.

$$I_L = \frac{V_{OC}}{R_{TH} + R_L} = \frac{25 \text{ V}}{2.75 \text{ k}\Omega} = 9.09 \text{ mA}$$

$$V_L = I_L R_L = 9.09 \text{ mA} \times 2 \text{ k}\Omega = 18.2 \text{ V}$$

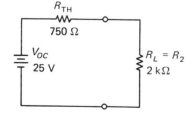

Figure 10-29

Knowing V_L and I_L (V_2 and I_2), we could find the rest of the voltage drops and currents as before. Notice that when we found V_{OC} we also found the polarity of the voltage drop across the load in the original circuit.

Let's find I_3 and V_3 in Figure 10-6 by using Thévenin's theorem. In Figure 10-30 the load has been removed resulting in a series circuit. The sources are connected series opposing so that $E_T = E_1 - E_2 = 60$ V.

$$V_2 = \frac{E_T R_2}{R_1 + R_2} = \frac{60 \text{ V} \times 20 \text{ k}\Omega}{32 \text{ k}\Omega} = 37.5 \text{ V}$$

Using Kirchhoff's voltage law, we get: $V_{OC} = E_2 + V_2 = 77.5$ V.

In Figure 10-30(b) E_1 and E_2 are replaced with short circuits and R_{TH} is calculated.

$$R_{TH} = R_1 \| R_2 = 7.5 \text{ k}\Omega$$

R_L is connected across Thévenin's equivalent circuit in Figure 10-31.

$$I_L = \frac{V_{OC}}{R_{TH} + R_L} = \frac{77.5 \text{ V}}{17.5 \text{ k}\Omega} = 4.43 \text{ mA}$$

$$V_L = I_L R_L = 4.43 \text{ mA} \times 10 \text{ k}\Omega = 44.3 \text{ V}$$

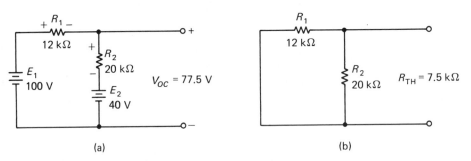

(a)   (b)

Figure 10-30

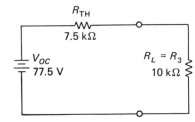

Figure 10-31

Consider the bridge circuit in Figure 10-32. Let's remove the load and redraw it slightly to help us see the circuit. With the load removed (Figure 10-33)

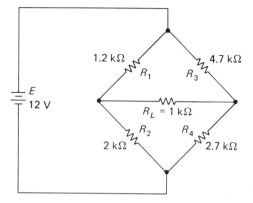

Figure 10-32

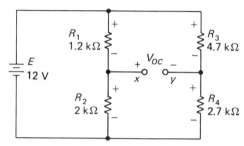

Figure 10-33

we can treat the circuit as a series-parallel circuit where $E = V_1 + V_2 = V_3 + V_4$.

$$V_1 = \frac{ER_1}{R_1 + R_2} = 4.5 \text{ V}$$

$$V_2 = E - V_1 = 7.5 \text{ V}$$

$$V_3 = \frac{ER_3}{R_3 + R_4} = 7.62 \text{ V}$$

$$V_4 = E - V_3 = 4.38 \text{ V}$$

Applying Kirchhoff's voltage law starting at point x and moving clockwise, we get:

$$V_{OC} + V_4 - V_2 = 0$$

$$V_{OC} + 4.38 \text{ V} - 7.5 \text{ V} = 0$$

$$V_{OC} = 3.12 \text{ V}$$

We assumed that point x was positive with respect to point y. The fact that V_{OC} is positive in our calculation proves that we were right. If V_{OC} had been -3.12 V in our equation, it would simply have meant that point x was *negative* with respect to point y.

Now let's short E and calculate R_{TH}. Figure 10-34(a) shows the circuit with E shorted. Since it may be difficult to see the series-parallel circuit, let's redraw the circuit as in Figure 10-34(b).

$$R_{TH} = R_1 \| R_2 + R_3 \| R_4 = 750 \text{ }\Omega + 1.71 \text{ K}\Omega = 2.46 \text{ k}\Omega$$

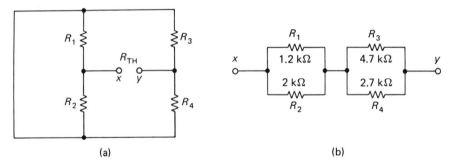

(a)   (b)

Figure 10-34

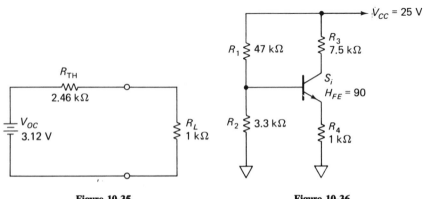

Figure 10-35   **Figure 10-36**

Thévenin's equivalent circuit with the load connected is shown in Figure 10-35.

$$I_L = \frac{V_{OC}}{R_{TH} + R_L} = \frac{3.12 \text{ V}}{3.46 \text{ k}\Omega} = 0.902 \text{ mA}$$

$$V_L = I_L R_L = 0.902 \text{ mA} \times 1 \text{ k}\Omega = 0.902 \text{ V}$$

Let's look at a typical bias arrangement for a bipolar transistor. Such a circuit is drawn in Figure 10-36. Let's find V_C and I_C. If we assume that $I_B = 0 \ \mu\text{A}$, then $I_C = I_E$.

$$I_E = \frac{\dfrac{V_{CC} R_2}{R_1 + R_2} - V_{BE}}{R_4} = \frac{\dfrac{25 \text{ V} \times 3.3 \text{ k}\Omega}{47 \text{ k}\Omega + 3.3 \text{ k}\Omega} - 0.6 \text{ V}}{1 \text{ k}\Omega} = 1.04 \text{ mA}$$

$$I_C = I_E = 1.04 \text{ mA}$$

$$V_C = V_{CC} - I_C R_3 = 25 \text{ V} - 7.8 \text{ V} = 17.2 \text{ V}$$

However, $I_B = 0 \ \mu\text{A}$ only if H_{FE} is infinitely large. Since H_{FE} is 90, I_B is some finite value and $I_C \neq I_E$. Using Thévenin's theorem, we can find the exact value of I_C for this transistor. Figure 10-37 shows the circuit with the base lead

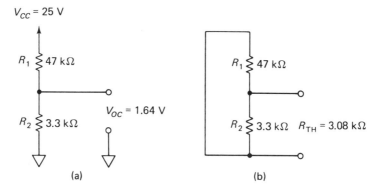

Figure 10-37

disconnected. That is, the circuit looking into the base is the load.

$$V_{OC} = \frac{V_{CC}R_2}{R_1 + R_2} = \frac{25 \text{ V} \times 3.3 \text{ k}\Omega}{47 \text{ k}\Omega + 3.3 \text{ k}\Omega} = 1.64 \text{ V}$$

In Figure 10-37(b) the source is shorted and:

$$R_{TH} = R_1 \| R_2 = 3.08 \text{ k}\Omega$$

Next we will connect the load to the equivalent circuit as in Figure 10-38. We needn't consider the collector since the collector-base junction is reverse biased. Therefore, the load consists of the emitter-base junction in series with some resistance. This resistance, labeled R, is the resistance we see looking across the junction from base to emitter. The base current is the current in this circuit. Therefore:

$$I_B = \frac{V_{OC} - V_{BE}}{R_{TH} + R_L} = \frac{1.64 \text{ V} - 0.6 \text{ V}}{3.08 \text{ k}\Omega + 91 \text{ k}\Omega} = 11.1 \ \mu\text{A}$$

$$I_C = H_{FE}I_B = 90 \times 11.1 \ \mu\text{A} = 995 \ \mu\text{A}$$

$$V_C = V_{CC} - I_C R_3 = 25 \text{ V} - 7.46 \text{ V} = 17.5 \text{ V}$$

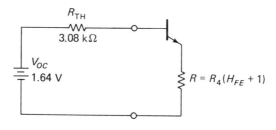

Figure 10-38

PRACTICE PROBLEMS 10-3   Develop Thévenin's equivalent circuit for the following problems and find V_L and I_L:

1. Figure 10-10(a). Assume that $R_2 = R_L$.
2. Figure 10-10(b). Assume that $R_3 = R_L$.
3. Figure 10-10(c). Assume that $R_3 = R_L$.
4. Figure 10-39.
5. Figure 10-40. Find I_B, I_C, and V_C.

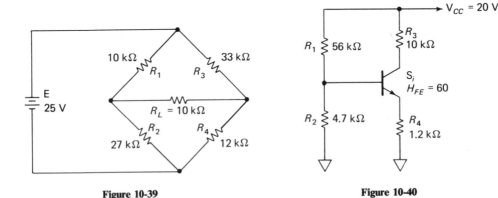

Figure 10-39   **Figure 10-40**

Solutions:

1. With R_2 removed, as in Figure 10-41(a), a series circuit results. E_1 and E_2 are connected series aiding. Therefore, 70 V will drop across R_1 and R_3.

$$V_1 = \frac{70 \text{ V} \times R_1}{R_1 + R_3} = \frac{70 \text{ V} \times 100 \text{ }\Omega}{320 \text{ }\Omega} = 21.9 \text{ V}$$

Applying Kirchhoff's voltage law to find V_{OC}, we see that the loop equation is $V_{OC} + V_1 + E = 0$. Moving counter-clockwise from point A, we get:

$$V_{OC} + 21.9 \text{ V} - 40 \text{ V} = 0$$

$$V_{OC} = 18.1 \text{ V}$$

We assumed that the potential at point A was positive with respect to point B. The fact that V_{OC} is positive indicates that we chose the correct polarity. In Figure 10-41(b) the sources have been shorted. Looking into the circuit from the open-circuit terminals, we see that R_1 and R_3 are in parallel and the equivalent resistance (R_{TH}) is 68.7 Ω. The Thévenin's equivalent circuit is shown in series with the load in Figure 10-41(c).

$$I_L = \frac{V_{OC}}{R_{TH} + R_L} = \frac{18.1 \text{ V}}{399 \text{ }\Omega} = 45.4 \text{ mA}$$

$$V_L = I_L R_L = 45.4 \text{ mA} \times 330 \text{ }\Omega = 15 \text{ V}$$

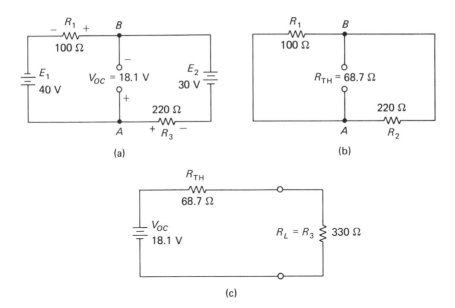

Figure 10-41

2. With R_3 removed, a series circuit results. E_1 and E_2 are connected series aiding and 20 V will drop across R_1 and R_2.

$$V_2 = \frac{E_T R_2}{R_1 + R_2} = \frac{20 \text{ V} \times 3.3 \text{ k}\Omega}{6 \text{ k}\Omega} = 11 \text{ V}$$

Applying Kirchhoff's voltage law to find V_{OC} in Figure 10-42(a), we see that the loop equation is $V_{OC} + E_2 + V_2 = 0$. Moving counterclockwise from point A, we get:

$$8 \text{ V} - 11 \text{ V} + V_{OC} = 0$$
$$V_{OC} = 3 \text{ V}$$

Shorting E_1 and E_2 results in R_1 in parallel with R_2 looking back from the open-circuit terminals as shown in Figure 10-42(b). This results in an R_{TH} of 1.49 kΩ. The Thévenin's equivalent circuit is shown in series with the load in Figure 10-42(c).

$$I_L = \frac{V_{OC}}{R_{TH} + R_L} = \frac{3 \text{ V}}{2.49 \text{ k}\Omega} = 1.2 \text{ mA}$$

$$V_L = I_L R_L = 1.2 \text{ mA} \times 1 \text{ k}\Omega = 1.2 \text{ V}$$

3. E_1 and E_2 are connected series opposing in Figure 10-10(c). Therefore, with the load (R_3) removed as in Figure 10-43(a), 40 V will drop across R_1, R_2, and R_4 in series. Let's work with the loop containing V_{OC}, R_4, and E_2.

$$V_4 = \frac{40 \text{ V} \times R_4}{R_T} = \frac{40 \text{ V} \times 180 \text{ k}\Omega}{347 \text{ k}\Omega} = 20.7 \text{ V}$$

$$V_{OC} = E_2 - V_4 = 60 \text{ V} - 20.7 \text{ V} = 39.3 \text{ V}$$

The source is replaced with a short circuit, and the resulting circuit is shown in Figure 10-43(b).

$$R_{TH} = R_4 \| (R_1 + R_2) = 86.6 \text{ k}\Omega$$

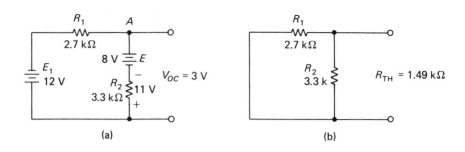

(a)

(b)

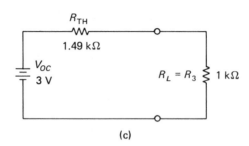

(c)

Figure 10-42

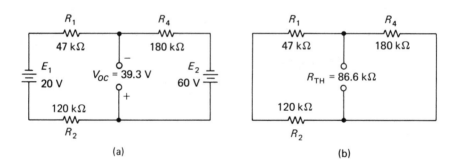

(a)

(b)

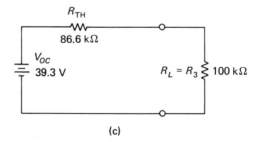

(c)

Figure 10-43

Connecting the load resistor to Thévenin's equivalent circuit in Figure 10-43(c), we get:

$$I_L = \frac{V_{OC}}{R_{TH} + R_L} = \frac{39.3 \text{ V}}{187 \text{ k}\Omega} = 210 \text{ } \mu\text{A}$$

$$V_L = I_L R_L = 210 \text{ } \mu\text{A} \times 100 \text{ k}\Omega = 21 \text{ V}$$

4. In Figure 10-44(a) the load has been removed and V_2 and V_4 are calculated.

$$V_2 = \frac{ER_2}{R_1 + R_2} = \frac{25 \text{ V} \times 27 \text{ k}\Omega}{37 \text{ k}\Omega} = 18.2 \text{ V}$$

$$V_4 = \frac{ER_4}{R_3 + R_4} = \frac{25 \text{ V} \times 12 \text{ k}\Omega}{45 \text{ k}\Omega} = 6.67 \text{ V}$$

Using the circuit consisting of V_2, V_4, and V_{OC} in Figure 10-44(b), we see that $V_{OC} = 11.5$ V and the polarity is as shown. The circuit, with the source shorted, is redrawn in Figure 10-44(c) and R_{TH} is calculated.

$$R_{TH} = R_1 \| R_2 + R_3 \| R_4 = 7.3 \text{ k}\Omega + 8.8 \text{ k}\Omega = 16.1 \text{ k}\Omega$$

Thévenin's equivalent circuit with the load connected is shown in Figure 10-44(d).

$$I_L = \frac{V_{OC}}{R_{TH} + R_L} = \frac{11.5 \text{ V}}{26.1 \text{ k}\Omega} = 441 \text{ } \mu\text{A}$$

$$V_L = I_L R_L = 441 \text{ } \mu\text{A} \times 10 \text{ k}\Omega = 4.41 \text{ V}$$

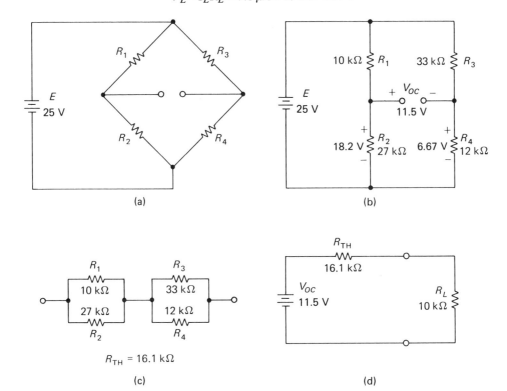

Figure 10-44

5. In Figure 10-45(a), with the load removed, $V_{OC} = V_2$.

$$V_2 = \frac{V_{CC}R_2}{R_1 + R_2} = \frac{20 \text{ V} \times 4.7 \text{ k}\Omega}{56 \text{ k}\Omega + 4.7 \text{ k}\Omega} = 1.55 \text{ V}$$

With the source shorted as in Figure 10-45(b), $R_{TH} = 4.34$ kΩ.

$$R_{TH} = R_1 \| R_2 = 4.34 \text{ k}\Omega$$

Thévenin's equivalent circuit with the load connected is shown in Figure 10-45(c).

$$I_L = I_B = \frac{V_{OC} - V_{BE}}{R_{TH} + R_L} = \frac{1.55 \text{ V} - 0.6 \text{ V}}{77.5 \text{ k}\Omega} = 12.3 \text{ } \mu\text{A}$$

$$I_C = H_{FE}I_B = 60 \times 12.3 \text{ } \mu\text{A} = 735 \text{ } \mu\text{A}$$

$$V_C = V_{CC} - V_3 = 20 \text{ V} - (735 \text{ } \mu\text{A} \times 10 \text{ k}\Omega) = 20 \text{ V} - 7.35 \text{ V} = 12.6 \text{ V}$$

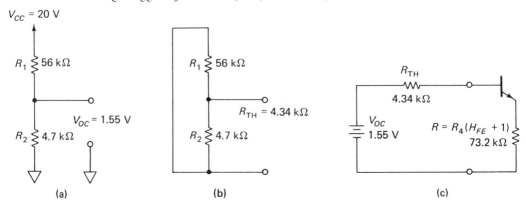

(a) (b) (c)

Figure 10-45

Additional practice problems are at the end of the chapter.

SELF-TEST 10-2

1. In Figure 10-46 assume that R_4 is the load. Thévenize the circuit and find V_L and I_L.   **2.** In Figure 10-47 assume that R_3 is the load. Thévenize the circuit and find V_L and I_L.

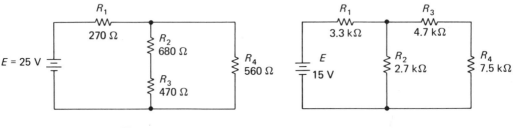

Figure 10-46   **Figure 10-47**

3. In Figure 10-14(a) assume that R_3 is the load. Thévenize the circuit and find V_L and I_L.

4. In Figure 10-14(d) assume that R_3 is the load. Thévenize the circuit and find V_L and I_L.

5. In Figure 10-48 Thévenize the circuit and find V_L and I_L.

6. In Figure 10-49 Thévenize the circuit at the base and find I_B, I_C, and V_C.

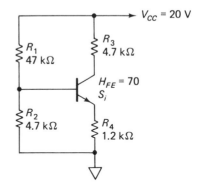

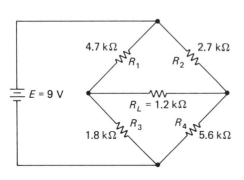

Figure 10-48   **Figure 10-49**

Answers to Self-Test 10-2 are at the end of the chapter.

10.3 NORTON'S THEOREM

Norton's theorem provides the means by which we can reduce a complex circuit to an equivalent current source in parallel with some conductance. The current source is labeled I_{SC} and the parallel conductance is labeled G_N as illustrated in Figure 10-50.

Consider the circuit in Figure 10-51. This circuit was shown in Figure 10-16 and was Thévenized in Figure 10-17. Assuming that R_3 is the load, Norton's equivalent circuit is found by using the following procedure:

1. Replace the load with a short circuit and calculate the current through the short circuit. This is the current source in our equivalent circuit. In Figure 10-52 the load (R_3) has been replaced with a short

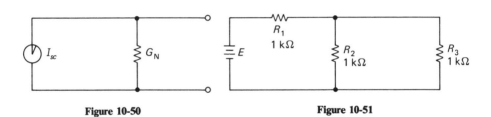

Figure 10-50   **Figure 10-51**

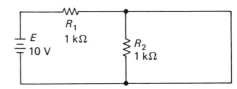

Figure 10-52

circuit. This reduces the circuit resistance to R_1 and the current is 10 mA.

$$I = \frac{V_1}{R_1} = \frac{10\ V}{1\ k\Omega} = 10\ mA$$

2. Open circuit the load. Replace the source with a short circuit and determine the conductance at the open-circuit terminals. This conductance is the reciprocal of R_{TH}. In some circuits it will be easier to find R_{TH} and then G_N. $G_N = 1/R_{TH}$. The conductance, G_N, is 2 mS.
3. Connect the load across the output terminals of the Norton's equivalent circuit. Calculate V_L and I_L. Norton's equivalent circuit of Figure 10-51 with the load connected is shown in Figure 10-53.

Using Ohm's law, we get:

$$V_L = \frac{I_{SC}}{G_N + G_L}$$

$$V_L = \frac{10\ mA}{3\ mS} = 3.33\ V$$

$$I_L = \frac{I_{SC} G_L}{G_N + G_L}$$

$$I_L = \frac{10\ mA \times 1\ mS}{3\ mS} = 3.33\ mA$$

or $\qquad I_L = V_L G_L = 3.33\ V \times 1\ mS = 3.33\ mA$

Norton's equivalent circuit causes the same current through R_3 and the same voltage drop across it, as does the original circuit. Further, the direction of current through the short circuit in Figure 10-52 is the same as in the original circuit.

Knowing V_3 and I_3 (remember we called R_3 the load), we can find the rest of the voltage drops and currents by using Ohm's law and Kirchhoff's laws.

Let's Nortonize some rather simple series-parallel circuits. We will find only V_L and I_L.

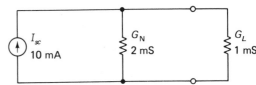

Figure 10-53

PRACTICE PROBLEMS 10-4 For the problems in Figure 10-22:

1. Develop Norton's equivalent circuit.

2. Find V_L and I_L.

Solutions: In Figure 10-22(a) if we short circuit the load, we put a short circuit across R_2 and reduce the circuit resistance to R_1. All the circuit current flows through the short circuit.

$$I_{SC} = \frac{V_1}{R_1} = \frac{20 \text{ V}}{12 \text{ k}\Omega} = 1.67 \text{ mA}$$

$$G_N = G_1 + G_2 = 83.3 \ \mu\text{S} + 50 \ \mu\text{S} = 133 \ \mu\text{S}$$

Norton's equivalent circuit with the load connected is shown in Figure 10-54. Applying Ohm's law, we get:

$$V_L = \frac{I_{SC}}{G_N + G_L} = \frac{1.67 \text{ mA}}{163 \ \mu\text{S}} = 10.2 \text{ V}$$

$$I_L = V_L G_L = 10.2 \text{ V} \times 30.3 \ \mu\text{A} = 309 \ \mu\text{A}$$

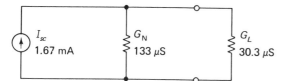

Figure 10-54

In Figure 10-22(b) if we short circuit the load, the short circuit current is the current through R_2.

$$I_{SC} = I_2 = \frac{V_2}{R_2} = \frac{10 \text{ V}}{4.7 \text{ k}\Omega} = 2.13 \text{ mA}$$

If we open circuit the load and short circuit the source,

$$R_{TH} = R_2 = 4.7 \text{ k}\Omega$$

$$G_N = \frac{1}{R_{TH}} = 213 \ \mu\text{S}$$

Norton's equivalent circuit with the load connected is shown in Figure 10-55. Applying Ohm's law, we get:

$$V_L = \frac{I_{SC}}{G_N + G_L} = \frac{2.13 \text{ mA}}{1.21 \text{ mS}} = 1.76 \text{ V}$$

$$I_L = V_L G_L = 1.76 \text{ mA}$$

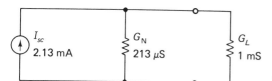

Figure 10-55

In Figure 10-22(c) if we short circuit the load, $R_T = R_1 + R_2 \| R_3$ or

$$R_T = 100 \text{ k}\Omega + 50 \text{ k}\Omega = 150 \text{ k}\Omega$$

$$I_{SC} = I_T = \frac{E}{R_T} = \frac{15 \text{ V}}{150 \text{ k}\Omega} = 100 \text{ }\mu\text{A}$$

$$R_{TH} = R_1 + R_2 \| R_3 = 150 \text{ k}\Omega$$

$$G_N = \frac{1}{R_{TH}} = \frac{1}{150 \text{ k}\Omega} = 6.67 \text{ }\mu\text{S}$$

Norton's equivalent circuit with the load connected is shown in Figure 10-56.

$$V_L = \frac{I_{SC}}{G_N + G_L} = \frac{100 \text{ }\mu\text{A}}{15 \text{ }\mu\text{S}} = 6.67 \text{ V}$$

$$I_L = V_L G_L = 6.67 \text{ V} \times 8.33 \text{ }\mu\text{S} = 55.5 \text{ }\mu\text{A}$$

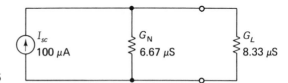

Figure 10-56

If we short circuit the load in Figure 10-22(d), the current through R_1 is the short circuit current.

$$I_{SC} = I_1 = \frac{V_1}{R_1} = \frac{25 \text{ V}}{47 \text{ k}\Omega} = 532 \text{ }\mu\text{A}$$

$$G_N = \frac{1}{R_1} = 21.3 \text{ }\mu\text{S}$$

Norton's equivalent circuit with the load connected is shown in Figure 10-57.

$$V_L = \frac{I_{SC}}{G_N + G_L} = \frac{532 \text{ }\mu\text{A}}{51.6 \text{ }\mu\text{S}} = 10.3 \text{ V}$$

$$I_L = V_L G_L = 10.3 \text{ V} \times 30.3 \text{ }\mu\text{S} = 312 \text{ }\mu\text{A}$$

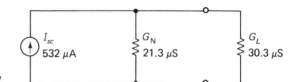

Figure 10-57

If we short circuit the load in Figure 10-22(e), R_3 is also shorted. Then $R_T = R_1 + R_2$. I_T in the resulting circuit is I_{SC}.

$$I_{SC} = \frac{E}{R_T} = \frac{25 \text{ V}}{25 \text{ k}\Omega} = 1 \text{ mA}$$

$$G_N = \frac{1}{R_3} + \frac{1}{R_1 + R_2} = 37 \text{ }\mu\text{S} + 40 \text{ }\mu\text{S} = 77 \text{ }\mu\text{S}$$

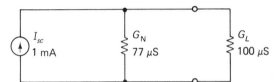

Figure 10-58

Norton's equivalent circuit with the load connected is shown in Figure 10-58.

$$V_L = \frac{I_{SC}}{G_N + G_L} = \frac{1\text{ mA}}{177\ \mu\text{S}} = 5.65 \text{ V}$$

$$I_L = V_L G_L = 5.56\text{ V} \times 100\ \mu\text{S} = 565\ \mu\text{A}$$

Additional problems are at the end of the chapter.

SELF-TEST 10-3

1. Refer to Figure 10-22(a). Let $E = 15$ V, $R_1 = 68$ kΩ, $R_2 = 100$ kΩ, and $R_L = 75$ kΩ. Nortonize the circuit and find V_L and I_L.

2. Refer to Figure 10-22(d). Let $E = 20$ V, $R_1 = 680$ Ω, $R_2 = 1.2$ kΩ, and $R_L = 560$ Ω. Nortonize the circuit and find V_L and I_L.

Answers to Self-Test 10-3 are at the end of the chapter.

END OF CHAPTER PROBLEMS 10-1

Use the Superposition theorem to find the various currents and voltage drops in the following problems:

1. Refer to Figure 10-10(a). Let $E_1 = 25$ V, $E_2 = 50$ V, $R_1 = 2.2$ kΩ, $R_2 = 1.8$ kΩ, and $R_3 = 1.2$ kΩ.

2. Refer to Figure 10-10(b). Let $E_1 = 10$ V, $E_2 = 25$ V, $R_1 = 5.6$ kΩ, $R_2 = 1$ kΩ, and $R_3 = 4.7$ kΩ.

3. Refer to Figure 10-10(c). Let $E_1 = 100$ V, $E_2 = 40$ V, $R_1 = 2.7$ kΩ, $R_2 = 3.3$ kΩ, $R_3 = 1$ kΩ, and $R_4 = 4.7$ kΩ.

4. Refer to Figure 10-10(b). Let $E_1 = 30$ V, $E_2 = 15$ V, $R_1 = 15$ kΩ, $R_2 = 12$ kΩ, and $R_3 = 10$ kΩ.

5. Refer to Figure 10-14(a). Let $E_1 = 9$ V, $E_2 = 18$ V, $R_1 = 3.3$ kΩ, $R_2 = 4.7$ kΩ, and $R_3 = 3.3$ kΩ.

6. Refer to Figure 10-14(b). Let $E_1 = 40$ V, $E_2 = 20$ V, $R_1 = 4.7$ kΩ, $R_2 = 7.5$ kΩ, and $R_3 = 10$ kΩ.

7. Refer to Figure 10-14(c). Let $E_1 = 25$ V, $E_2 = 10$ V, $R_1 = 680$ Ω, $R_2 = 470$ Ω, and $R_3 = 1$ kΩ.

8. Refer to Figure 10-14(c). Let $E_1 = 12$ V, $E_2 = 30$ V, $R_1 = 18$ kΩ, $R_2 = 27$ kΩ, $R_3 = 47$ kΩ.

9. Refer to Figure 10-14(d). Let $E_1 = 40$ V, $E_2 = 15$ V, $R_1 = 2.2$ kΩ, $R_2 = 1.2$ kΩ, $R_3 = 3.3$ kΩ, and $R_4 = 3.3$ kΩ.

10. Refer to Figure 10-14(d). Let $E_1 = 50$ V, $E_2 = 20$ V, $R_1 = 10$ kΩ, $R_2 = 20$ kΩ, $R_3 = 30$ kΩ, and $R_4 = 30$ kΩ.

END OF CHAPTER PROBLEMS 10-2

In problems 1 through 10 below Thévenize each circuit and find V_L and I_L.

1. Refer to Figure 10-22(a) where $E = 30$ V, $R_1 = 22$ kΩ, $R_2 = 47$ kΩ, and $R_L = 10$ kΩ.

2. Refer to Figure 10-22(b) where $E = 20$ V, $R_1 = 20$ kΩ, $R_2 = 56$ kΩ, and $R_L = 33$ kΩ.

3. Refer to Figure 10-22(c) where $E = 12$ V, $R_1 = 680$ Ω, $R_2 = 470$ Ω, $R_3 = 1$ kΩ, and $R_L = 750$ Ω.

4. Refer to Figure 10-22(d) where $E = 40$ V, $R_1 = 5.6$ kΩ, $R_2 = 3.9$ kΩ, and $R_L = 2.7$ kΩ.

5. Refer to Figure 10-22(e) where $E = 50$ V, $R_1 = 22$ kΩ, $R_2 = 10$ kΩ, $R_3 = 33$ kΩ, and $R_L = 15$ kΩ.

6. Refer to Figure 10-10(a) where $E_1 = 25$ V, $E_2 = 50$ V, $R_1 = 2.2$ kΩ, $R_2 = 1.8$ kΩ, and $R_3 = 1.2$ kΩ. Assume that R_2 is the load.

7. Refer to Figure 10-10(b) where $E_1 = 10$ V, $E_2 = 25$ V, $R_1 = 5.6$ kΩ, $R_2 = 1$ kΩ, and $R_3 = 4.7$ kΩ. Assume that R_2 is the load.

8. Refer to Figure 10-10(c) where $E_1 = 100$ V, $E_2 = 40$ V, $R_1 = 2.7$ kΩ, $R_2 = 3.3$ kΩ, $R_3 = 1$ kΩ, and $R_4 = 4.7$ kΩ. Assume that R_3 is the load.

9. Refer to Figure 10-32 where $E = 20$ V, $R_1 = 4.7$ kΩ, $R_2 = 6.8$ kΩ, $R_3 = 1.5$ kΩ, $R_4 = 3$ kΩ, and $R_L = 1$ kΩ.

10. Refer to Figure 10-32 where $E = 25$ V, $R_1 = 680$ Ω, $R_2 = 1$ kΩ, $R_3 = 2$ kΩ, $R_4 = 1.2$ kΩ, and $R_L = 1.2$ kΩ.

11. Refer to Figure 10-36 where $V_{CC} = 20$ V, $R_1 = 47$ kΩ, $R_2 = 6.8$ kΩ, $R_3 = 2$ kΩ, $R_4 = 470$ Ω, and $H_{FE} = 80$. Thévenize and find I_B, I_C, and V_C.

12. Refer to Figure 10-36 where $V_{CC} = 25$ V, $R_1 = 56$ kΩ, $R_2 = 10$ kΩ, $R_3 = 8.2$ kΩ, $R_4 = 2$ kΩ, and $H_{FE} = 60$. Thévenize and find I_B, I_C, and V_C.

END OF CHAPTER PROBLEMS 10-3

1. Refer to Figure 10-22(a). Nortonize the circuit and find V_L and I_L where (a) $E = 30$ V, $R_1 = 22$ kΩ, $R_2 = 47$ kΩ, and $R_L = 10$ kΩ and (b) $E = 50$ V, $R_1 = 18$ kΩ, $R_2 = 33$ kΩ, and $R_L = 39$ kΩ.

2. Refer to Figure 10-22(b). Nortonize the circuit and find V_L and I_L where (a) $E = 20$ V, $R_1 = 20$ kΩ, $R_2 = 56$ kΩ, and $R_L = 33$ kΩ and (b) $E = 25$ V, $R_1 = 330$ Ω, $R_2 = 270$ Ω, and $R_L = 560$ Ω.

3. Refer to Figure 10-22(c). Nortonize the circuit and find V_L and I_L where (a) $E = 12$ V, $R_1 = 680$ Ω, $R_2 = 470$ Ω, $R_3 = 1$ kΩ and $R_L = 750$ Ω and (b) $E = 10$ V, $R_1 = 12$ kΩ, $R_2 = 15$ kΩ, $R_3 = 18$ kΩ, and $R_L = 10$ kΩ.

4. Refer to Figure 10-22(d). Nortonize the circuit and find V_L and I_L where (a) $E = 40$ V, $R_1 = 5.6$ kΩ, $R_2 = 3.9$ kΩ, and $R_L = 2.7$ kΩ and (b) $E = 9$ V, $R_1 = 330$ Ω, $R_2 = 680$ Ω, and $R_L = 390$ Ω.

5. Refer to Figure 10-22(e). Nortonize the circuit and find V_L and I_L where (a) $E = 50$ V, $R_1 = 22$ kΩ, $R_2 = 10$ kΩ, $R_3 = 33$ kΩ, $R_L = 15$ kΩ and (b) $E = 10$ V, $R_1 = 3.3$ kΩ, $R_2 = 3.3$ kΩ, $R_3 = 6.8$ kΩ, and $R_L = 2$ kΩ.

ANSWERS TO SELF-TESTS

Self-Test 10-1

(a) Short E_1:

$$R_T = 4.7 \text{ k}\Omega + 1.65 \text{ k}\Omega = 6.35 \text{ k}\Omega$$

$$I_T = \frac{10 \text{ V}}{6.35 \text{ k}\Omega} = 1.57 \text{ mA}$$

$$I_3 = \frac{1.57 \text{ mA} \times 1.65 \text{ k}\Omega}{3.3 \text{ k}\Omega} = 787 \text{ } \mu\text{A}$$

Short E_2:

$$R_T = 3.3 \text{ k}\Omega + 1.94 \text{ k}\Omega = 5.24 \text{ k}\Omega$$

$$I_T = \frac{30 \text{ V}}{5.24 \text{ k}\Omega} = 5.73 \text{ mA}$$

$$I_3 = \frac{5.73 \text{ mA} \times 1.94 \text{ k}\Omega}{3.3 \text{ k}\Omega} = 3.37 \text{ mA}$$

In the original circuit:

$$I_3 = 787 \text{ } \mu\text{A} + 3.37 \text{ mA} = 4.15 \text{ mA}$$
$$V_3 = 13.7 \text{ V}$$
$$V_2 = 3.7 \text{ V}$$
$$I_2 = 787 \text{ } \mu\text{A}$$
$$V_1 = 16.3 \text{ V}$$
$$I_1 = 4.94 \text{ mA}$$

(b) Short E_1:

$$R_T = 10 \text{ k}\Omega + 5 \text{ k}\Omega = 15 \text{ k}\Omega$$

$$I_T = \frac{60 \text{ V}}{15 \text{ k}\Omega} = 4 \text{ mA}$$

$$I_3 = 2 \text{ mA}$$

Short E_2:

$$R_T = 15 \text{ k}\Omega$$

$$I_T = \frac{100 \text{ V}}{15 \text{ k}\Omega} = 6.67 \text{ mA}$$

$$I_3 = 3.33 \text{ mA}$$

In the original circuit:

$$I_3 = 3.33 \text{ mA} - 2 \text{ mA} = 1.33 \text{ mA}$$
$$V_3 = 13.3 \text{ V}$$
$$V_2 = 73.3 \text{ V}$$
$$I_2 = 7.33 \text{ mA}$$
$$V_1 = 86.7 \text{ V}$$
$$I_1 = 8.67 \text{ mA}$$

(c) Short E_1:

$$R_T = 180\ \Omega + 68.7\ \Omega = 249\ \Omega$$

$$I_T = \frac{12\ \text{V}}{249\ \Omega} = 48.2\ \text{mA}$$

$$I_2 = \frac{48.2\ \text{mA} \times 68.7\ \Omega}{220\ \Omega} = 15.1\ \text{mA}$$

Short E_2:

$$R_T = 100\ \Omega + 99\ \Omega = 199\ \Omega$$

$$I_T = \frac{10\ \text{V}}{199\ \Omega} = 50.3\ \text{mA}$$

$$I_2 = \frac{50.3\ \text{mA} \times 99\ \Omega}{220\ \Omega} = 22.6\ \text{mA}$$

In the original circuit:

$$I_2 = 15.1\ \text{mA} + 22.6\ \text{mA} = 37.7\ \text{mA}$$
$$V_2 = 8.29\ \text{V}$$
$$V_3 = 3.71\ \text{V}$$
$$I_3 = 20.6\ \text{mA}$$
$$V_1 = 1.71\ \text{V}$$
$$I_1 = 17.1\ \text{mA}$$

(d) Short E_2:

$$R_T = R_1 + R_2 + R_3 \| R_4 = 1\ \text{k}\Omega + 2.7\ \text{k}\Omega + 956\ \Omega = 4.66\ \text{k}\Omega$$

$$I_T = \frac{E_1}{R_T} = \frac{30\ \text{V}}{4.66\ \text{k}\Omega} = 6.44\ \text{mA}$$

$$I_3 = \frac{I_T R_X}{R_3} = \frac{6.44\ \text{mA} \times 956\ \Omega}{1.2\ \text{k}\Omega} = 5.13\ \text{mA}$$

Short E_1:

$$R_T = R_4 + R_3 \| (R_1 + R_2) = 4.7\ \text{k}\Omega + 906\ \Omega = 5.61\ \text{k}\Omega$$

$$I_T = \frac{E_2}{R_T} = \frac{10\ \text{V}}{5.61\ \text{k}\Omega} = 1.78\ \text{mA}$$

$$I_3 = \frac{I_T R_X}{R_3} = \frac{1.78\ \text{mA} \times 906\ \Omega}{1.2\ \text{k}\Omega} = 1.35\ \text{mA}$$

In the original circuit:

$$I_3 = 1.35\ \text{mA} + 5.13\ \text{mA} = 6.48\ \text{mA}$$
$$V_3 = 6.48\ \text{mA} \times 1.2\ \text{k}\Omega = 7.77\ \text{V}$$
$$V_4 = E_2 - V_3 = 10\ \text{V} - 7.77\ \text{V} = 2.23\ \text{V}$$
$$I_4 = \frac{2.23\ \text{V}}{4.7\ \text{k}\Omega} = 474\ \mu\text{A}$$
$$I_1 = I_2 = I_3 - I_4 = 6.01\ \text{mA}$$
$$V_1 = 6.01\ \text{mA} \times 1\ \text{k}\Omega = 6.01\ \text{V}$$
$$V_2 = 6.01\ \text{mA} \times 2.7\ \text{k}\Omega = 16.2\ \text{V}$$

Self-Test 10-2

1. The Thévenized circuit with the load connected is shown in Figure 10-59.

$$V_{OC} = V_2 + V_3 = \frac{E(R_2 + R_3)}{R_1 + R_2 + R_3} = 20.2 \text{ V}$$

$$R_{TH} = R_1 \| (R_2 + R_3) = 219 \text{ }\Omega$$

$$V_L = \frac{V_{OC}R_L}{R_{TH} + R_L} = \frac{20.2 \text{ V} \times 560\Omega}{779 \text{ }\Omega} = 14.5 \text{ V}$$

$$I_L = \frac{V_{OC}}{R_{TH} + R_L} = \frac{20.2 \text{ V}}{779 \text{ }\Omega} = 25.9 \text{ mA}$$

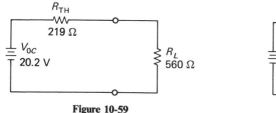

Figure 10-59

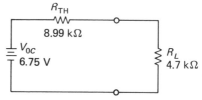

Figure 10-60

2. The Thévenized circuit with the load connected is shown in Figure 10-60.

$$V_{OC} = V_2 = \frac{ER_2}{R_1 + R_2} = \frac{15 \text{ V} \times 2.7 \text{ k}\Omega}{6 \text{ k}\Omega} = 6.75 \text{ V}$$

$$R_{TH} = R_4 + R_1 \| R_2 = 7.5 \text{ k}\Omega + 1.49 \text{ k}\Omega = 8.99 \text{ k}\Omega$$

$$V_L = \frac{V_{OC}R_L}{R_{TH} + R_L} = \frac{6.75 \text{ V} \times 4.7 \text{ k}\Omega}{13.7 \text{ k}\Omega} = 2.32 \text{ V}$$

$$I_L = \frac{V_{OC}}{R_{TH} + R_L} = \frac{6.75 \text{ V}}{13.7 \text{ k}\Omega} = 493 \text{ }\mu\text{A}$$

3. The Thévenized circuit with the load connected is shown in Figure 10-61(b). The circuit used to find V_{OC} is shown in Figure 10-61(a).

$$E_T = E_1 - E_2 = 20 \text{ V}$$

$$V_2 = \frac{20 \text{ V} \times 4.7 \text{ k}\Omega}{8 \text{ k}\Omega} = 11.75 \text{ V}$$

$$V_{OC} = E_2 + V_2 = 21.8 \text{ V}$$

$$R_{TH} = R_1 \| R_2 = 1.94 \text{ k}\Omega$$

$$V_L = \frac{V_{OC}R_L}{R_{TH} + R_L} = \frac{21.8 \text{ V} \times 3.3 \text{ k}\Omega}{5.24 \text{ k}\Omega} = 13.7 \text{ V}$$

$$I_L = \frac{V_{OC}}{R_{TH} + R_L} = \frac{21.8 \text{ V}}{5.24 \text{ k}\Omega} = 4.15 \text{ mA}$$

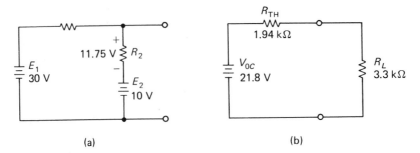

(a)   (b)

Figure 10-61

4. The Thévenized circuit with the load connected is shown in Figure 10-62(b). The circuit used to find V_{OC} is shown in Figure 10-62(a).

$$E_T = E_1 - E_2 = 20 \text{ V}$$

$$V_4 = \frac{E_T R_4}{R_1 + R_2 + R_4} = \frac{20 \text{ V} \times 4.7 \text{ k}\Omega}{8.4 \text{ k}\Omega} = 11.2 \text{ V}$$

$$V_{OC} = V_4 + E_2 = 21.2 \text{ V}$$

$$R_{TH} = R_4 \| (R_1 + R_2) = 4.7 \text{ k}\Omega \| 3.7 \text{ k}\Omega = 2.07 \text{ k}\Omega$$

$$V_L = \frac{V_{OC} R_L}{R_{TH} + R_L} = \frac{21.2 \text{ V} \times 1.2 \text{ k}\Omega}{3.27 \text{ k}\Omega} = 7.78 \text{ V}$$

$$I_L = \frac{V_{OC}}{R_{TH} + R_L} = \frac{21.2 \text{ V}}{3.27 \text{ k}\Omega} = 6.48 \text{ mA}$$

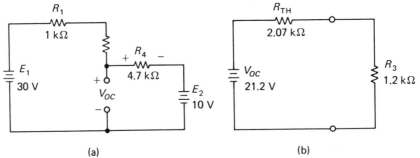

(a)   (b)

Figure 10-62

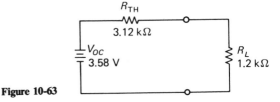

Figure 10-63

5. The Thévenized circuit with the load connected is shown in Figure 10-63. Various solutions are possible. We will use a circuit which includes E, R_3, R_L, and R_2 with the load removed:

$$V_3 = \frac{ER_3}{R_1 + R_3} = \frac{9 \text{ V} \times 1.8 \text{ k}\Omega}{5.5 \text{ k}\Omega} = 2.49 \text{ V}$$

$$V_2 = \frac{ER_2}{R_2 + R_4} = \frac{9 \text{ V} \times 2.7 \text{ k}\Omega}{8.3 \text{ k}\Omega} = 2.93 \text{ V}$$

$$V_{OC} = E - V_2 - V_3 = 9 \text{ V} - 5.42 = 3.58 \text{ V}$$

$$R_{TH} = R_1 \| R_3 + R_2 \| R_4 = 1.3 \text{ k}\Omega + 1.82 \text{ k}\Omega = 3.12 \text{ k}\Omega$$

$$V_L = \frac{V_{OC}R_L}{R_{TH} + R_L} = \frac{3.58 \text{ V} \times 1.2 \text{ k}\Omega}{4.32 \text{ k}\Omega} = 0.994 \text{ V}$$

$$I_L = \frac{V_{OC}}{R_{TH} + R_L} = \frac{3.58 \text{ V}}{4.32 \text{ k}\Omega} = 0.829 \text{ mA}$$

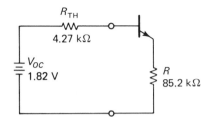

Figure 10-64

6. The Thévenized circuit with the load connected is shown in Figure 10-64.

$$V_{OC} = \frac{V_{CC}R_2}{R_1 + R_2} = \frac{20 \text{ V} \times 4.7 \text{ k}\Omega}{47 \text{ k}\Omega + 4.7 \text{ k}\Omega} = 1.82 \text{ V}$$

$$R_{TH} = R_1 \| R_2 = 4.27 \text{ k}\Omega$$

$$R = R_4(H_{FE} + 1) = 85.2 \text{ k}\Omega$$

$$I_B = \frac{V_{OC} - V_{BE}}{R_{TH} + R} = \frac{1.22 \text{ V}}{89.5 \text{ k}\Omega} = 13.6 \text{ } \mu\text{A}$$

$$I_C = H_{FE}I_B = 955 \text{ } \mu\text{A}$$

$$V_C = V_{CC} - V_3 = 20 \text{ V} - 4.49 \text{ V} = 15.5 \text{ V}$$

Self-Test 10-3

1. $I_{SC} = 221 \text{ } \mu\text{A}$
   $G_N = 24.7 \text{ } \mu\text{S}$
   $V_L = 5.82 \text{ V}$
   $I_L = 77.5 \text{ } \mu\text{A}$

2. $I_{SC} = 29.4 \text{ mA}$
   $G_N = 1.47 \text{ mS}$
   $V_L = 9.03 \text{ V}$
   $I_L = 16.1 \text{ mA}$

CHAPTER **11**

FACTORING ALGEBRAIC EXPRESSIONS

In previous chapters we have multiplied or divided monomials. Many of the algebraic expressions we work with are polynomials—algebraic expressions that contain more than one term. In this chapter we will learn how to find the factors of various polynomials. This skill will be used in later chapters to solve complex equations.

11.1 MULTIPLICATION OF POLYNOMIALS BY MONOMIALS

There are two ways to show how to multiply a polynomial by a monomial. We will examine both ways. Consider the expression $3x(4x+2)$. The sign of grouping (the parentheses) tells us that each term within the group must be multiplied by $3x$. One way to do this is shown in Example 11.1. Notice that each term is multiplied by $3x$. Because there are two terms within the group, there will be two terms in the answer. Both terms will be positive because all signs were positive.

EXAMPLE 11.1

$$3x(4x+2)=3x\cdot4x+3x\cdot2=12x^2+6x$$

EXAMPLE 11.2

$$\begin{array}{r} 4x+2 \\ \times\quad 3x \\ \hline 12x^2+6x \end{array}$$

Example 11.2 is set up the way we set up a multiplication problem in arithmetic. Each term in the multiplicand $(4x+2)$ is multiplied by the multiplier $(3x)$. Most students prefer the method shown in Example 11.1 because it is faster since most problems are given in this form. Let's try a few more.

EXAMPLE 11.3: Multiply $4(2x^2+3x-4)$.

Solution: Multiply each term in the group by 4. This yields:

$$4\cdot2x^2+4\cdot3x-4\cdot4=8x^2+12x-16$$

Notice we put a minus sign in front of $4\cdot4$ because the sign of the term was negative and the sign of the multiplier was positive. Remember $(+)\cdot(-)=(-)$.

EXAMPLE 11.4: Perform the indicated multiplication and combine like terms.

$$4a(a+2b)-3a(a+3b)$$

Solution: We have two different multiplications to perform:

$$4a(a+2b) \quad \text{and} \quad -3a(a+3b)$$

$$4a(a+2b)=4a^2+8ab$$

$$-3a(a+3b)=-3a^2-9ab$$

Putting them together, we get:

$$4a(a+2b)-3a(a+3b)=4a^2+8ab-3a^2-9ab$$

When we combine like terms, we get a^2-ab.

$$4a(a+2b)-3a(a+3b)=a^2-ab$$

EXAMPLE 11.5: Perform the indicated multiplication and combine like terms:

$$3x^2+6x(2x-3y-2)-2x(3x-4y-3)$$

Solution:

$$6x(2x-3y-2)=12x^2-18xy-12x$$

$$-2x(3x-4y-3)=-6x^2+8xy+6x$$

Putting it all together, we get:

$$3x^2+6x(2x-3y-2)-2x(3x-4y-3)$$
$$=3x^2+12x^2-18xy-12x-6x^2$$
$$+8xy+6x$$
$$=9x^2-10xy-6x$$

PRACTICE PROBLEMS 11-1   Perform the indicated operations. Combine like terms.

1. $4(x+2)$ **2.** $3(a-4)$

3. $4a-3(4a-3)$ **4.** $2x(3+4y)$

5. $2x(3x+2y)$ **6.** $4(2x^2+3x-4)$

7. $3a(a+b)-2a(a+4b)$ **8.** $x(xy+4y)-2(x^2y-3xy)$

9. $2a(a^2+9a+20)+3a(a^2-a-12)$ **10.** $4x^2+2x(x+y-3)-2x(2x-y)$

Solutions:

1. $4x+8$ **2.** $3a-12$

3. $-8a+9$ **4.** $6x+8xy$

5. $6x^2+4xy$ **6.** $8x^2+12x-16$

7. a^2-5ab **8.** $-x^2y+10xy$

9. $5a^3+15a^2+4a$ **10.** $2x^2+4xy-6x$

Additional practice problems are at the end of the chapter.

11.2   MULTIPLICATION OF BINOMIALS BY BINOMIALS

As we did in Section 11.1, we can find the product of two binomials in two ways. Consider the expression $(a+4)(a-2)$. Using the method we learned in arithmetic, we will set it up like this:

EXAMPLE 11.6

$$\begin{array}{r} a+4 \\ (\times) \quad a-2 \\ \hline \end{array}$$

Solution: We must multiply each term in the multiplicand by each term in the multiplier.

$$\begin{array}{r} a+4 \\ a-2 \\ \hline a^2+4a \\ -2a-8 \\ \hline a^2+2a-8 \end{array}$$

      result of $a(a+4)$
      result of $-2(a+4)$

Notice that like terms are placed in the same column. This lets us combine like terms when the two rows are added together. No matter how many terms are in the multiplicand and multiplier, this method can be used.

Let's look at it again.

EXAMPLE 11.7

$$(a+4)(a-2)$$

Solution: This time let's do it mentally, the procedure is to multiply each term in the second binomial by each term in the first binomial and combine like terms:

$$a \cdot a = a^2$$

$$a \cdot -2 = -2a$$

$$(a+4)(a-2)$$

$$4 \cdot a = 4a$$

$$4 \cdot -2 = -8$$

The four resulting terms are $a^2 - 2a + 4a - 8$. Combining like terms, we get $a^2 + 2a - 8$.

Let's try two more.

EXAMPLE 11.8

$$(x-3y)(x+7y)$$

Solution: Multiplying each term in the second binomial by each term in the first yields:

$$x \cdot x = x^2$$
$$x \cdot 7y = 7xy$$
$$-3y \cdot x = -3xy$$
$$-3y \cdot 7y = -21y^2$$
$$(x-3y)(x+7y) = x^2 + 7xy - 3xy - 21y^2 = x^2 + 4xy - 21y^2$$

EXAMPLE 11.9

$$(3a-2b)(4a-4b)$$

Solution: Multiplying each term in the second binomial by each term in the first binomial yields:

$$3a \cdot 4a = 12a^2$$
$$3a \cdot -4b = -12ab$$
$$-2b \cdot 4a = -8ab$$
$$-2b \cdot -4b = 8b^2$$
$$(3a-2b)(4a-4b) = 12a^2 - 12ab - 8ab + 8b^2$$
$$= 12a^2 - 20ab + 8b^2$$

PRACTICE PROBLEMS 11-2   Perform the indicated operations:

1. $(a+2)^2$

2. $(a-4)^2$

3. $(2a-3)^2$

4. $(3a-2)^2$

5. $(x+2y)^2$

6. $(2x-3y)^2$

7. $(a+6)(a+2)$

8. $(a+2)(a-5)$

9. $(a-3)(a+4)$

10. $(a-4)(a-5)$

11. $(x+3)(x-3)$

12. $(x-5)(x+5)$

13. $(a+b)(a+2b)$

14. $(a+3b)(a-4b)$

15. $(2a+3b)^2$

Solutions:

1. a^2+4a+4

2. $a^2-8a+16$

3. $4a^2-12a+9$

4. $9a^2-12a+4$

5. $x^2+4xy+4y^2$

6. $4x^2-12xy+9y^2$

7. $a^2+8a+12$

8. $a^2-3a-10$

9. a^2+a-12

10. $a^2-9a+20$

11. x^2-9

12. x^2-25

13. $a^2+3ab+2b^2$

14. $a^2-ab-12b^2$

15. $4a^2+12ab+9b^2$

Additional practice problems are at the end of the chapter.

11.3 DIVISION OF POLYNOMIALS

The process of division of algebraic expressions is much the same as long division in arithmetic. If we are dividing by a monomial, we simply divide each term of the polynomial by the monomial. Such a problem is shown in Example 11.10.

EXAMPLE 11.10: Perform the indicated division:

$$(6x^3+12x^2-3x)\div 3x$$

Solution: Each term in the numerator is divided by $3x$, the denominator. We could set the problem as we would a

long-division problem and solve like this:

$$
\begin{array}{r}
2x^2+4x-1 \\
3x\overline{)6x^3+12x^2-3x} \\
6x^3 \\
\hline
0\ +12x^2 \\
12x^2 \\
\hline
0-3x \\
-3x \\
\hline
0
\end{array}
$$

$3x\cdot2x^2=6x^3$

$3x\cdot4x=12x^2$

$3x\cdot-1=-3x$

no remainder

$3x$ is divided into the first term. The result of this division is $2x^2$. $3x\cdot2x^2=6x^3$. The process continues until all terms have been divided by $3x$. Usually a problem like this can be done by inspection. If there is a remainder, it is written as a fraction. The following example illustrates this.

EXAMPLE 11.11

$$(7x^2+4x)\div4x$$

Solution:

$$
\begin{array}{r}
x+1\ \ +\dfrac{3x^2}{4x} \\
\hline
4x\overline{)7x^2+4x} \\
4x^2 \\
\hline
3x^2+4x \\
4x \\
\hline
3x^2+0
\end{array}
$$

$3x^2$ is left over; therefore, the remainder is $\dfrac{3x^2}{4x}$.

Let's try dividing a polynomial by a binomial.

EXAMPLE 11.12

$$(a^2+3a-54)\div(a+9)$$

Solution: First, we set it up as a long-division problem and then we make sure that the literal numbers are in descending order as shown:

$$a+9\overline{)a^2+3a-54}$$

Then we determine how many times the first term in the

divisor goes into the first term of the dividend. a goes into a^2 a times. Multiplying this a times each term in the divisor results in a^2+9a. Subtracting this result from the dividend leaves a remainder of $-6a$. After we bring the next term down, we have the partial solution shown:

$$
\begin{array}{r}
a \\
a+9\overline{)a^2+3a-54} \\
\underline{a^2+9a} \\
0\ -6a-54
\end{array}
$$

Next we determine how many times a goes into $-6a$. The answer is -6. Now we multiply -6 by $(a+9)$. There is no remainder and the solution is complete.

$$
\begin{array}{r}
a-6 \\
a+9\overline{)a^2+3a-54} \\
\underline{a^2+9a} \\
-6a-54 \\
\underline{-6a-54} \\
0
\end{array}
$$

We can check our answer:

$$(a+9)(a-6)=a^2+3a-54$$

Let's try another one.

EXAMPLE 11.13

$$(6a^2-23ab+20b^2)\div(3a-4b)$$

Solution:

$$
\begin{array}{r}
2a\ -5b \\
3a-4b\overline{)6a^2-23ab+20b^2} \\
\underline{6a^2-\ 8ab} \\
-15ab+20b^2 \\
\underline{-15ab+20b^2} \\
0
\end{array}
$$

Check: $(3a-4b)(2a-5b)=6a^2-23ab+20b^2$

Let's do one more.

EXAMPLE 11.14

$$(x^2 - 49) \div (x + 7)$$

Solution:

$$
\begin{array}{r}
x \quad\; -7 \\
x+7\overline{)x^2 \quad\quad -49} \\
\underline{x^2 + 7x} \\
-7x - 49 \\
\underline{-7x - 49} \\
0
\end{array}
$$

Check: $(x+7)(x-7) = x^2 - 49$

EXAMPLE 11.15: Here's one with a remainder.

$$(x^2 + 4x + 7) \div (x - 3)$$

Solution:

$$
\begin{array}{r}
x \quad\quad +7 \\
x-3\overline{)x^2 + 4x + 7} \\
\underline{x^2 - 3x} \\
7x + 7 \\
\underline{7x - 21} \\
28
\end{array}
$$

the remainder is $\dfrac{28}{x-3}$ so the answer is

$$x + 7 + \frac{28}{x-3}$$

Check the answer:

$$(x-3)\left(x+7+\frac{28}{x-3}\right) = (x-3)\left(\frac{x^2 - 3x + 7x - 21 + 28}{x-3}\right)$$

$$= x^2 - 3x + 7x - 21 + 28 = x^2 + 4x + 7$$

PRACTICE PROBLEMS 11-3   Perform the indicated operations:

1. $(a^3 + 2a^2 - a) \div a$

2. $(4a^4 - 6a^3 + 8a^2) \div 2a$

3. $(x^2 + 13x + 42) \div (x + 6)$

4. $(x^2 - 36) \div (x + 6)$

5. $(9x^2 - 12x + 4) \div (3x - 2)$

6. $(4a^2 + 12ab + 9b^2) \div (2a + 3b)$

7. $(a^2 - ab - 12b^2) \div (a - 4b)$

8. $(x^2 - 9x + 20) \div (x - 5)$

9. $(3ab + a^2 + 2b^2) \div (b + a)$

10. $(9 + 4x^2 + 12x) \div (3 + 2x)$

11. $(2x^2 + 3x + 14) \div (x + 1)$

12. $(6x^2 + 2x - 7) \div (2x + 4)$

Solutions:

1. $a^2 + 2a - 1$

2. $2a^3 - 3a^2 + 4a$

3. $x + 7$

4. $x - 6$

5. $3x - 2$

6. $2a + 3b$

7. $a + 3b$

8. $x - 4$

9. $a + 2b$

10. $2x + 3$

11. $2x + 1 + \dfrac{13}{x + 1}$

12. $3x - 5 + \dfrac{13}{2x + 4}$

Additional practice problems are at the end of the chapter.

SELF-TEST 11-1   Perform the indicated operations:

1. $3a(a - 6)$

2. $2xy(3x + 2y)$

3. $(a - 9)^2$

4. $(4a + 2)^2$

5. $(a + 6)(a - 1)$

6. $(x + 2y)(x - 6y)$

7. $(2x + 3y)(4x + 4y)$

8. $3(a - 2b)(2a - 2b)$

9. $(6a^3 - 12a^2) \div 3a$

10. $(x^2 + 2x - 48) \div (x + 8)$

11. $(x^2 + 13x + 36) \div (x + 9)$

12. $(a^2 + 12ab + 20b^2) \div (a + 2b)$

13. $(13x + 30 + x^2) \div (x + 9)$

14. $(10ab + a^2 + 20b^2) \div (a + 2b)$

Answers to Self-Test 11-1 are at the end of the chapter.

11.4   FACTORING POLYNOMIALS

Earlier in this chapter we multiplied monomials times polynomials. Here we are going to reverse the process. We are going to determine what the common factors are of various algebraic expressions. This is simply a process of determining what factors are common to each term in the expression. Consider the expression $6a^2 + 12a - 18a^3$. If we factor each term, we get:

$$6a^2 = 2 \cdot 3 \cdot a^2$$
$$12a = 2 \cdot 2 \cdot 3a$$
$$18a^3 = 2 \cdot 3 \cdot 3 \cdot a^3$$

The common factors are the factors that appear in each term. These common factors are $2 \cdot 3 \cdot a$ or $6a$. If $6a$ is common to each term, we can divide the expression by $6a$ and get $a + 2 - 3a^2$.

$$\frac{6a^2 + 12a - 18a^3}{6a} = a + 2 - 3a^2$$

The factors of $6a^2 + 12a - 18a^3$ are $6a$ and $a + 2 - 3a^2$.

$$6a^2 + 12a - 18a^3 = 6a(a + 2 - 3a^2)$$

Since we normally write algebraic expressions either in descending order or in ascending order, let's rearrange the answer so that it looks like this:

$$6a(-3a^2 + a + 2)$$

EXAMPLE 11.16: Factor $28a^3 - 35a^2$.

Solution: The prime factors are:

$$28a^3 = 2 \cdot 2 \cdot 7 \cdot a^3$$
$$35a^2 = 5 \cdot 7 \cdot a^2$$

The common factors are $7a^2$; therefore, we divide each term by $7a^2$.

$$\frac{28a^3 - 35a^2}{7a^2} = 4a - 5$$

The factors of $28a^3 - 35a^2$ are $7a^2$ and $4a - 5$.

$$28a^3 - 35a^2 = 7a^2(4a - 5)$$

PRACTICE PROBLEMS 11-4   Factor the following problems:

1. $4a + 2$
2. $6a + 8$
3. $16 - 2y$
4. $21 - 3y$
5. $x^3 + 2x^2$
6. $4x^4 + 2x^2$
7. $18a^2 - 9a$
8. $15a^2 - 3a$
9. $14x^2y - 7x^2z + 20yz$
10. $15a^2b - 21a^3b^2 + 9a^2b^2$

Solutions:

1. $2(2a + 1)$
2. $2(3a + 4)$
3. $2(8 - y)$
4. $3(7 - y)$
5. $x^2(x + 2)$
6. $2x^2(2x^2 + 1)$
7. $9a(2a - 1)$
8. $3a(5a - 1)$
9. $2(7x^2y - 7x^2z + 10yz)$
10. $3a^2b(5 - 7ab + 3b)$

Additional practice problems are at the end of the chapter.

11.5   FACTORS OF TRINOMIALS

A trinomial of the form $ax^2 + bx + c$ can be factored if we can find two numbers whose sum is the coefficient of the second term and whose product is the third term. For example, consider the expression $x^2 + 13x + 42$. From the work we did earlier in the chapter we know that the answer will be in the form $(x + n)(x + m)$. In this expression the coefficient of the second term is 13 and the third term is

42. We must find two numbers whose sum is 13 and whose product is 42. We find these numbers by trial and error. That is, we consider all the combination of numbers whose sum is 13 and select the one combination whose product is 42. In this example, the combinations are:

$$12+1 \qquad 12\times1=12$$
$$11+2 \qquad 11\times2=22$$
$$10+3 \qquad 10\times3=30$$
$$9+4 \qquad 9\times4=36$$
$$8+5 \qquad 8\times5=40$$
$$7+6 \qquad 7\times6=42$$

The two numbers must be 7 and 6:

$$x^2+13x+42=(x+7)(x+6)$$

Of course, it would also be correct to write the answer as $(x+6)(x+7)$. With practice we can usually determine the correct combination by inspection.

EXAMPLE 11.17: Factor $x^2-6x-27$.

Solution: The coefficient of the middle term is -6 and the product is -27. Any time the third term is negative, the factors must be of this form: $(x+n)(x-m)$. The unknown number m must be greater than n because the middle term is negative. There are so many combinations of numbers that equal -6 that it is simpler in this case to consider numbers whose product is -27.

$$1\cdot-27=-27 \qquad 1+(-27)=-26$$
$$-1\cdot27=-27 \qquad -1+27=26$$
$$3\cdot-9=-27 \qquad 3+(-9)=-6$$
$$-3\cdot9=-27 \qquad -3+9=6$$

Of the four possibilities, only 3 and -9 equals -6 when added together:

$$x^2-6x-27=(x+3)(x-9)$$

EXAMPLE 11.18: Factor x^2+2x-8.

Solution: The third term is again negative; therefore, the answer is again of the form $(x+m)(x-n)$. Let's again consider all the possible combinations which yield a product of -8:

$$2 \cdot -4 = -8 \qquad 2 + (-4) = -2$$
$$-2 \cdot 4 = -8 \qquad -2 + 4 = 2$$
$$1 \cdot -8 = -8 \qquad 1 + (-8) = -7$$
$$-1 \cdot 8 = -8 \qquad -1 + 8 = 7$$

Of the four possibilities, only -2 and 4 equals 2 when added together:

$$x^2 + 2x - 8 = (x+4)(x-2)$$

EXAMPLE 11.19: Factor $x^2 + 12x + 36$.

Solution: Inspection shows that the factors are $(x+6)$ and $(x+6)$ or $(x+6)^2$:

$$x^2 + 12x + 36 = (x+6)^2$$

EXAMPLE 11.20: Factor $x^2 - 36$.

Solution: Note that there is no middle term and that the last term is negative. The only way the middle term can drop out is for the two numbers to be equal. The only two equal numbers whose product is -36 is 6 and -6. Then:

$$x^2 - 36 = (x+6)(x-6)$$

EXAMPLE 11.21: Factor $x^2 - 12x + 36$.

Solution: The middle term is negative and the last term is positive. This combination can only result if both numbers are negative. The numbers must be -6 because $-6 + (-6) = -12$ and $-6 \cdot -6 = 36$.

$$x^2 - 12x + 36 = (x-6)^2$$

In the example below we have added another squared term, y^2. When trinomials are of the form $ax^2 + bxy + cy^2$, the factors must be $(px + my)$ $(qx + ny)$. Otherwise, the method of solution is the same.

EXAMPLE 11.22: Factor $x^2 + 2xy + y^2$.

Solution: By inspection we see that the factors must be $(x+y)$ and $(x+y)$ or $(x+y)^2$.

EXAMPLE 11.23: Factor $x^2 - 4xy - 32y^2$.

Solution: The coefficient of $y^2 = -32$ and the coefficient of the middle term is -4. The possible solutions are:

$$-32 \cdot 1 \qquad -32 + 1 = -31$$

$$-16 \cdot 2 \qquad -16 + 2 = -14$$

$$-8 \cdot 4 \qquad -8 + 4 = -4$$

The correct combination is -8 and $+4$. Because the middle term is negative, the larger of the two digits must be the negative digit. Therefore, the three combinations in which the smaller of the two digits is negative was not considered.

$$x^2 - 4xy - 32y^2 = (x + 4y)(x - 8y)$$

EXAMPLE 11.24: Factor $4x^2 + 10xy + 6y^2$.

Solution: This example shows a coefficient of x^2 that is some number other than 1, in this case 4. The coefficient of y^2 is 6. We are looking for two numbers whose product is 4 (because the first term is $4x^2$) and two numbers whose product is 6 (the last term is $6y^2$). The product of the means plus the product of the extremes must equal 10, the coefficient of the middle term. The possible combinations are:

$$(1+2)(4+3) \qquad 8+3=11$$
$$(1+3)(4+2) \qquad 12+2=14$$
$$(4+2)(1+3) \qquad 12+2=14$$
$$(4+3)(1+2) \qquad 8+3=11$$
$$(2+3)(2+2) \qquad 6+4=10$$

We have found the solution on our fifth attempt; therefore, we won't bother to examine the rest of the possibilities, but the student may want to examine the rest of the possible combinations.

$$4x^2 + 10xy + 6y^2 = (2x+3y)(2x+2y)$$

PRACTICE PROBLEMS 11-5   Factor the following:

1. $x^2 + 3x + 2$ **2.** $x^2 + 7x + 12$

3. $a^2 + a - 6$ **4.** $a^2 - a - 12$

5. $a^2 + 15a + 56$ **6.** $a^2 + 2a - 15$

7. $x^2 + 12x + 36$ **8.** $x^2 - 64$

9. $x^2 + 12xy + 35y^2$ **10.** $x^2 + 16xy + 63y^2$

11. $2a^2 - 12ab - 32b^2$ **12.** $2a^2 + 4ab - 48b^2$

Solutions:

1. $(x+2)(x+1)$ **2.** $(x+3)(x+4)$

3. $(a+3)(a-2)$ **4.** $(a-4)(a+3)$

5. $(a+7)(a+8)$ **6.** $(a+5)(a-3)$

7. $(x+6)^2$ **8.** $(x+8)(x-8)$

9. $(x+5y)(x+7y)$ **10.** $(x+7y)(x+9y)$

11. $2(a+2b)(a-8b)$ **12.** $2(a+6b)(a-4b)$

Additional practice problems are at the end of the chapter.

SELF-TEST 11-2   Factor the following:

1. $9x - 36$ **2.** $56y^2 + 8y$

3. $a^2 + 15a + 36$ **4.** $a^2 - 7a - 30$

5. $x^2 + 2xy - 35y^2$ **6.** $x^2 + 12xy + 36y^2$

7. $2x^2 + 16x + 32$ **8.** $3a^2 - 3b^2$

9. $a^2 - 3ab - 54b^2$ **10.** $a^2 - 6ab + 9b^2$

Answers to Self-Test 11-2 are at the end of the chapter.

END OF CHAPTER PROBLEMS 11-1

Perform the indicated operations. Combine like terms.

1. $3(a+3)$ **2.** $5(x+2)$

3. $4(a-1)$ **4.** $6(a-3)$

5. $3a - 2(3a-1)$ **6.** $4a - 3(2a-3)$

7. $3x(2+2y)$

9. $3a(2a+3b)$

11. $3(3x^2+6x-12)$

13. $4a(a+b)-3a(a+5b)$

15. $a(2ab+2b)-3(a^2b-2ab)$

17. $x(x^2+8x+15)+2x(x^2-x-6)$

19. $3a^2+2a(a-b+2)-2a(2a+b)$

8. $6x(3-2y)$

10. $4a(3a+2b)$

12. $4(2x^2+3x-6)$

14. $5a(a+b)-5a(a+2b)$

16. $2a(ab+2b)-4(2a^2b-ab)$

18. $2x(x^2+7x+12)+x(x^2-6x+9)$

20. $2x^2+3x(x-2y+3)-3x(2x+2y)$

END OF CHAPTER PROBLEMS 11-2

Perform the indicated operations:

1. $(a+3)^2$

3. $(a-6)^2$

5. $(2x-4)^2$

7. $(4x+3)^2$

9. $(a+1)(a+3)$

11. $(y+3)(y-6)$

13. $(a-2)(a+6)$

15. $(x-4)(x-1)$

17. $(a+4)(a-4)$

19. $(2x+3)(2x-3)$

21. $(x+y)(x+3y)$

23. $(2x-3y)^2$

25. $(x-5y)(x+4y)$

27. $(2a+3b)(3a-4b)$

2. $(a+5)^2$

4. $(a-2)^2$

6. $(3x-2)^2$

8. $(3x+3)^2$

10. $(y+2)(y+7)$

12. $(a+5)(a-1)$

14. $(a-8)(a+3)$

16. $(x-6)(x-5)$

18. $(a+8)(a-8)$

20. $(3x-2)(3x+2)$

22. $(x+2y)(x+3y)$

24. $(3x-4y)^2$

26. $(x-6y)(x+2y)$

28. $(4a+2b)(3a-3b)$

END OF CHAPTER PROBLEMS 11-3

Perform the indicated operations:

1. $(x^3+3x^2+2x)\div x$

3. $(x^2+5xy+6y^2)\div(x+2y)$

5. $(x^2+4x-21)\div(x+7)$

7. $(x^2-49)\div(x+7)$

9. $(9x^2-12x+4)\div(3x-2)$

2. $(x^4+2x^3+x^2)\div x^2$

4. $(x^2+4xy+3y^2)\div(x+y)$

6. $(x^2+2x-8)\div(x-2)$

8. $(x^2-9)\div(x-3)$

10. $(9x^2+18x+9)\div(3x+3)$

11. $(a^2 - 25) \div (a + 5)$

12. $(b^2 - 11b + 30) \div (b - 6)$

13. $(6x^2 + xy - 12y^2) \div (3x - 4y)$

14. $(12x^2 - 6xy - 6y^2) \div (4x + 2y)$

15. $(9a^2 - 24ab + 16b^2) \div (3a - 4b)$

16. $(24a + 16a^2 + 9) \div (3 + 4a)$

17. $(x^2 - 6xy - 27y^2) \div (x + 3y)$

18. $(8b^2 + 15a^2 - 26ab) \div (5a - 2b)$

19. $(16 - 12x^2 - 16x) \div (2 - 3x)$

20. $(12 - 14x - 10x^2) \div (4 + 2x)$

21. $(12x^2 - 4x - 5) \div (3x + 2)$

22. $(20 + 24a^2 - 50a) \div (3a - 4)$

23. $(8x^2 + 20xy + 15y^2) \div (4x + 5y)$

24. $(10a^2 + 4b^2 - 17ab) \div (5a - 4b)$

END OF CHAPTER PROBLEMS 11-4

Factor the following:

1. $3a + 9$

2. $12b - 18$

3. $7b - 21$

4. $6x + 12$

5. $4x^2y + 6xy$

6. $6x^2y - 3xy$

7. $16a^2 + 2a$

8. $18a^2 + 3a$

9. $x^2 + 5x + 6$

10. $x^2 + 9x + 14$

11. $x^2 + 4x - 32$

12. $a^2 - 3a - 28$

13. $a^2 + 6a + 9$

14. $a^2 + 18a + 81$

15. $a^2 - 16$

16. $x^2 - 64$

17. $x^2 - 8x + 15$

18. $y^2 - 13y + 42$

19. $y^2 + 5y - 36$

20. $y^2 + 3y - 40$

21. $2x^2 + 20x + 42$

22. $3a^2 + 12a + 9$

23. $4a^2 - 4a - 24$

24. $5a^2 - 20a - 25$

25. $3x^2 + 18x - 81$

26. $5x^2 + 15x - 20$

27. $x^2 - 4xy - 12y^2$

28. $x^2 - 16y^2$

29. $6x^2 + xy - y^2$

30. $3x^2 + 4xy + y^2$

ANSWERS TO SELF-TESTS

Self-Test 11-1

1. $3a^2 - 18a$

2. $6x^2y + 4xy^2$

3. $a^2 - 18a + 81$

4. $16a^2 + 16a + 4$

5. $a^2 + 5a - 6$

6. $x^2 - 4xy - 12y^2$

7. $8x^2 + 20xy + 12y^2$

8. $6a^2 - 18ab + 12b^2$

9. $2a^2 - 4a$

10. $x - 6$

11. $x + 4$

12. $a + 10b$

13. $x + 4 - \dfrac{4}{x+9}$

14. $a + 10b - \dfrac{2ab}{a+2b}$

Self-Test 11-2

1. $9(x-4)$

2. $8y(7y+1)$

3. $(a+12)(a+3)$

4. $(a+3)(a-10)$

5. $(x+7y)(x-5y)$

6. $(x+6y)^2$

7. $2(x+4)^2$

8. $3(a+b)(a-b)$

9. $(a+6b)(a-9b)$

10. $(a-3b)^2$

CHAPTER 12

FRACTIONAL EQUATIONS

Many of the equations we work with in electronics are more complex than the ones we worked with in Chapter 7. Sometimes the unknown is part of a polynomial that can be in either the numerator or in the denominator of a term. In this chapter we will solve some equations similar to ones that may be encountered in the study of circuit analysis. Later in the chapter we will solve some typical circuit problems.

12.1 GENERAL EQUATIONS

When we have an equation to solve that involves fractions or mixed numbers, we usually have to get rid of the fraction in order to simplify the solution. Consider the equation in Example 12.1.

> EXAMPLE 12.1: Solve for a.
>
> $$\frac{1}{3} + \frac{a}{4} = 2$$
>
> *Solution:* The first step in the solution is to get rid of the fraction. We get rid of the fraction by first finding the CD (common denominator) and then multiplying each term by it. In this example, the CD is 12. Now we multiply each term by 12.
>
> $$\frac{1}{3}(12) + \frac{a}{4}(12) = 2(12)$$
>
> $$4 + 3a = 24$$

Remember, whatever we do to one term, we must do to *all* terms on both sides of the equation.

The last step is to solve for a.

$$4 + 3a = 24$$
$$3a = 20$$
$$a = \frac{20}{3} = 6\frac{2}{3}$$

(When we have an improper fraction, we change it to a mixed number.)

Check:

$$\frac{1}{3} + \frac{\frac{20}{3}}{4} = 2$$

$$\frac{1}{3} + \frac{5}{3} = 2$$

$$2 = 2$$

Note:

$$\frac{\frac{20}{3}}{4} = \frac{20}{3} \div 4 = \frac{20}{3} \times \frac{1}{4} = \frac{20}{12} = \frac{5}{3}$$

Let's try one in which the unknown is in the denominator.

EXAMPLE 12.2: Solve for x.

$$\frac{5}{x-1} - \frac{1}{2} = 2.$$

Solution: In this example the denominators are 2 and $x-1$. This makes the CD $2(x-1)$. To get rid of the fractions, we multiply both sides by this CD. Remember, we must multiply *each* term by this CD.

$$\frac{5 \cdot 2(x-1)}{x-1} - \frac{2(x-1)}{2} = 2 \cdot 2(x-1)$$

$$10 - x + 1 = 4x - 4$$

$$11 - x = 4x - 4$$

$$11 = 5x - 4 \quad \text{(move } -x \text{ to the right side and it becomes } +x, \text{ that is, } 4x + x = 5x)$$

$$15 = 5x \quad \text{(move } -4 \text{ to the left side and it becomes } +4, \text{ that is, } 11 + 4 = 15)$$

$$3 = x$$

or $\qquad x = 3$

Check:
$$\frac{5}{3-1} - \frac{1}{2} = 2$$

$$\frac{5}{2} - \frac{1}{2} = 2$$

$$2 = 2$$

EXAMPLE 12.3

$$\frac{2}{x} - \frac{3}{y} = 4 \qquad \text{Solve for } x \text{ and } y.$$

Solution: In this example the CD is xy. Multiplying each term by xy, we get:

$$\frac{2xy}{x} - \frac{3xy}{y} = 4xy$$

$$2y - 3x = 4xy$$

This cleared the fractions and we now have a simple equation. Let's first solve for x. To solve for x, we first put all terms containing x on one side of the equation.

$$2y = 4xy + 3x \qquad \text{(move } -3x \text{ to the right side)}$$
or $\quad 4xy + 3x = 2y$

Now we factor out the x.

$$x(4y + 3) = 2y$$

Next we move the $4y + 3$ to the right side.

$$\frac{x(4y+3)}{4y+3} = \frac{2y}{4y+3} \qquad \text{(divide both sides by } 4y+3)$$

$$x = \frac{2y}{4y+3}$$

Now let's solve for y. After multiplying all terms of the original equation by xy, we had:

$$2y - 3x = 4xy$$

Now we put all terms containing y on one side and move all other terms to the other side.

$$2y - 3x - 4xy = 0 \qquad \text{(move } 4xy \text{ to the left side)}$$

$$2y - 4xy = 3x \qquad \text{(move } -3x \text{ to the right side)}$$

Next we factor out y.

$$y(2 - 4x) = 3x$$

Now we move $2-4x$ to the right side.

$$\frac{y(2-4x)}{2-4x} = \frac{3x}{2-4x} \qquad \text{(divide each side by } 2-4x)$$

$$y = \frac{3x}{2-4x}$$

PRACTICE PROBLEMS 12-1 Solve for the unknown in the following problems. Express your answers as fractions or mixed numbers.

1. $\dfrac{8x}{2} - 4 = 10$

2. $\dfrac{R}{2} + \dfrac{R}{4} = 3$

3. $\dfrac{Z+1}{3} - Z = 4 - 2Z$

4. $\dfrac{4}{I+2} + 2 = 5$

5. $\dfrac{3I}{2} - 2 = \dfrac{I}{2} + 4$

6. $\dfrac{4x}{2} - \dfrac{3+x}{5} = \dfrac{-2x+3}{4}$

7. $\dfrac{6-x}{x} - \dfrac{4}{x} = \dfrac{3}{x}$

8. $\dfrac{4}{x+3} - \dfrac{1}{3} = 4$

9. $\dfrac{x}{a} + \dfrac{x}{b} = 2$. Solve for a, b, and x.

10. $3(2x+1) = 2y + x - 3$. Solve for x and y.

Solutions:

1. $x = 3\frac{1}{2}$

2. $R = 4$

3. $Z = 2\frac{3}{4}$

4. $I = -\frac{2}{3}$

5. $I = 6$

6. $x = \dfrac{27}{46}$

7. $x = -1$

8. $x = -2\frac{1}{13}$

9. $a = \dfrac{bx}{2b-x} \qquad b = \dfrac{ax}{2a-x} \qquad x = \dfrac{2ab}{a+b}$

10. $x = \dfrac{2y-6}{5} \qquad y = \dfrac{5x+6}{2}$

12.2 SOME REAL EQUATIONS

In this section we will deal with equations that are used in circuit analysis. We have already used some of the equations in previous chapters. For instance, we have solved this equation for V_1:

$$V_1 = \frac{ER_1}{R_1 + R_2}$$

Suppose that from this basic equation we need to solve for R_1. Example 12.4 shows how to do it.

EXAMPLE 12.4: Given the equation $V_1 = \dfrac{ER_1}{R_1 + R_2}$, solve for R_1.

Solution: Notice that R_1 appears in both the numerator and denominator of the right side. The R_1's can't cancel because there are two terms in the denominator and R_1 appears in only one of them. R_1 would have to be common to *all* terms in the fraction to be canceled. Our first step will be to get rid of the fraction. We can do that by multiplying each term by $R_1 + R_2$.

$$V_1(R_1 + R_2) = \frac{ER_1(R_1 + R_2)}{R_1 + R_2} \qquad \left(\begin{array}{l}\text{multiply both sides}\\ \text{by } R_1 + R_2\end{array}\right)$$

$$V_1(R_1 + R_2) = ER_1$$

We need to remove the parentheses in the left side because R_1 is inside. We could remove the need for the parentheses by dividing each side by V_1, but we would still have R_1 in the right side as part of a fraction. A better step is to perform the indicated multiplication.

$$V_1 R_1 + V_1 R_2 = ER_1$$

Next we put all terms containing R_1 on the left side and move all other terms to the right side.

$$V_1 R_1 + V_1 R_2 - ER_1 = 0 \qquad \text{(move } ER_1 \text{ to the left side)}$$

$$V_1 R_1 - ER_1 = -V_1 R_2 \qquad \left(\begin{array}{l}\text{move } V_1 R_2 \text{ to the}\\ \text{right side}\end{array}\right)$$

Now we factor out the R_1.

$$R_1(V_1 - E) = -V_1 R_2$$

and finally divide by $V_1 - E$.

$$R_1 = \frac{-V_1 R_2}{V_1 - E} \qquad \text{(divide both sides by } V_1 - E)$$

We can get rid of the minus sign in the numerator by changing that sign and the sign of the denominator. Remember, there are three signs associated with a fraction: the sign of the numerator, the sign of the denominator, and the sign of the fraction. When we change the sign of the numerator or the denominator, we must change the sign of each term.

$$R_1 = \frac{V_1 R_2}{E - V_1} \qquad \left(\begin{array}{l}\text{change the signs of both numerator}\\ \text{and denominator}\end{array}\right)$$

We could have avoided the problem with the signs if we had moved $V_1 R_1$ to the right side in our second step. Then:

$$V_1 R_1 + V_1 R_2 = ER_1$$

$$V_1 R_2 = ER_1 - V_1 R_1$$

$$V_1 R_2 = R_1(E - V_1)$$

$$\frac{V_1 R_2}{E - V_1} = R_1$$

or

$$R_1 = \frac{V_1 R_2}{E - V_1}$$

EXAMPLE 12.5: Given the equation $r_i = \beta\left(r_e + \dfrac{r_b}{\beta}\right)$. Solve for r_e.

Solution: In this example, as in many others, we find that there is more than one approach to the solution. The trick is to find the easiest one—the one that requires the fewest steps. In this case, let's divide both sides by β. This step isolates r_e in one term.

$$\frac{r_i}{\beta} = r_e + \frac{r_b}{\beta} \qquad \text{(divide both sides by } \beta)$$

$$r_e = \frac{r_i}{\beta} - \frac{r_b}{\beta} \qquad \left(\text{subtract } \frac{r_b}{\beta} \text{ from both sides}\right)$$

PRACTICE PROBLEMS 12-2

1. $I_2 = \dfrac{I_T G_2}{G_1 + G_2}$. Solve for I_T, G_1, and G_2.

2. $V_1 = \dfrac{ER_1}{R_1 + R_2}$. Solve for E, R_1, and R_2.

3. $F = \dfrac{9}{5}C + 32$. Solve for C.

4. $A_i = \dfrac{h_{fe}}{h_{oe} R_L + 1}$. Solve for R_L.

5. $S = \dfrac{R_E + R_B}{R_E + R_B(1 - \alpha)}$. Solve for R_B.

6. $R_S = \dfrac{R_i R_o}{R_i + R_o}$. Solve for R_o.

7. $r_i = \beta\left(r_e + \dfrac{r_b}{\beta}\right)$. Solve for r_b.

8. $A_V = \dfrac{r_e}{r_e + r_e'}$. Solve for r_e.

Solutions:

1. $I_T = \dfrac{I_2(G_1 + G_2)}{G_2}$

$G_1 = \dfrac{I_T G_2 - I_2 G_2}{I_2}$

$G_2 = \dfrac{I_2 G_1}{I_T - I_2}$

2. $E = \dfrac{V_1(R_1 + R_2)}{R_1}$

$R_1 = \dfrac{V_1 R_2}{E - V_1}$

$R_2 = \dfrac{ER_1 - V_1 R_1}{V_1}$

3. $C = \dfrac{5F - 160}{9}$ **4.** $R_L = \dfrac{h_{fe} - A_i}{A_i h_{oe}}$

5. $R_B = \dfrac{R_E - SR_E}{S - S\alpha - 1}$ **6.** $R_o = \dfrac{R_S R_i}{R_i - R_S}$

7. $r_b = r_i - \beta r_e$ **8.** $r_e = \dfrac{A_V r'_e}{1 - A_V}$

12.3 APPLICATIONS

Now let's carry the work in the previous section one step further. Let's determine actual values of unknowns just as we would do in circuit analysis.

EXAMPLE 12-6: Given the equation $I_1 = \dfrac{I_T G_1}{G_1 + G_2}$, where $I_1 = 2.73$ mA, $I_T = 6.75$ mA, and $G_1 = 370$ μS, find G_2.

Solution: Let's first multiply both sides by $G_1 + G_2$. This results in

$$I_1(G_1 + G_2) = I_T G_1 \qquad (12\text{-}1)$$

Now we have two possible solutions. Let's examine both.

First solution:

$$G_1 + G_2 = \frac{I_T G_1}{I_1} \qquad \text{(divide both sides by } I_1)$$

$$G_2 = \frac{I_T G_1}{I_1} - G_1 \qquad \text{(move } G_1 \text{ to the right side)}$$

Then we plug in known values.

$$G_2 = \frac{6.75 \text{ mA} \times 370 \text{ } \mu\text{S}}{2.73 \text{ mA}} - 370 \text{ } \mu\text{S}$$

$$G_2 = 545 \text{ } \mu\text{S}$$

Second solution:

$$I_1(G_1 + G_2) = I_T G_1 \qquad (12\text{-}1)$$

$$I_1 G_1 + I_1 G_2 = I_T G_1 \qquad \text{(remove parentheses in left side)}$$

$$I_1 G_2 = I_T G_1 - I_1 G_1 \qquad \text{(move } I_1 G_1 \text{ term to right side)}$$

$$G_2 = \frac{I_T G_1 - I_1 G_1}{I_1} \qquad \text{(divide by } I_1)$$

Then we plug in known values.

$$G_2 = \frac{6.75 \text{ mA} \times 370 \text{ } \mu S - 2.73 \text{ mA} \times 370 \text{ } \mu S}{2.73 \text{ mA}} = 545 \text{ } \mu S$$

The first solution appears to be easier, but of course both resulted in the same answer.

EXAMPLE 12.7: Given the equation $R_{TH} = \dfrac{V_{OC} - V_L}{I_L}$, solve for

V_{OC} when $R_{TH} = 1.5 \text{ k}\Omega$, $V_L = 3.75 \text{ V}$, and $I_L = 5 \text{ mA}$.

Solution: We need to rearrange the equation and solve for V_{OC}; therefore, let's do it this way:

$$R_{TH} I_L = V_{OC} - V_L \qquad \text{(multiply by } I_L)$$

$$R_{TH} I_L + V_L = V_{OC} \qquad \text{(move } V_L \text{ to the left side)}$$

We then plug in known values.

$$V_{OC} = 1.5 \text{ k}\Omega \times 5 \text{ mA} + 3.75 \text{ V}$$

$$V_{OC} = 7.5 \text{ V} + 3.75 \text{ V} = 11.25 \text{ V}$$

PRACTICE PROBLEMS 12-3

1. $V_1 = \dfrac{E R_1}{R_1 + R_2}$

(a) Solve for E when $V_1 = 34$ V, $R_1 = 270$ Ω, and $R_2 = 330$ Ω and, (b) solve for R_1 when $E = 25$ V, $V_1 = 14.7$ V, and $R_2 = 3.3$ kΩ.

2. $I_L = \dfrac{I_{SC} G_L}{G_L + G_N}$

(a) Solve for G_L when $I_{SC} = 10$ mA, $I_L = 3$ mA, and $G_N = 300$ μS and (b) solve for G_N when $I_{SC} = 600$ μA, $I_L = 200$ μA, and $G_L = 2.7$ mS.

3. $A_i = \dfrac{h_{fe}}{1 + h_{oe} R_L}$

Solve for h_{fe} when $A_i = 65$, $h_{oe} = 10$ μS, and $R_L = 5$ kΩ.

4. $\beta = \dfrac{\alpha}{1 - \alpha}$

Solve for α when $\beta = 90$.

5. $R_{TH} = \dfrac{V_{OC} - V_L}{I_L}$

Solve for V_L when $R_{TH} = 1.7$ kΩ, $V_{OC} = 12$ V, and $I_L = 750$ μA.

Solutions:

1. (a) $E = 75.6$ V
   (b) $R_1 = 4.71$ kΩ

2. (a) $G_L = 129$ μS
   (b) $G_N = 5.4$ mS

3. $h_{fe} = 68.3$

4. 0.989

5. 10.7 V

SELF-TEST 12-1

1. $\dfrac{R+2}{3} - R = 2 - R$

Solve for R.

2. $\dfrac{2}{I-2} + 3 = 5$

Solve for I.

3. $\dfrac{a}{2} + \dfrac{3}{b} = \dfrac{2}{x}$

Solve for a, b, and x.

4. $I_L = \dfrac{I_{SC} G_L}{G_L + G_N}$

Solve for G_L when $I_{SC} = 750$ μA, $I_L = 200$ μA, and $G_N = 300$ μS.

5. $R_T = \dfrac{R_1 R_2}{R_1 + R_2}$

Solve for R_2 when $R_T = 38.7$ kΩ and $R_1 = 68$ kΩ.

Answers to Self-Test 12-1 are at the end of the chapter.

12.4 QUADRATIC EQUATIONS

Quadratic equations of interest to us are second-degree equations. In other words, the unknown is raised to the second power. Some examples of these quadratic equations are: $x^2 = 25$, $2x^2 + x + 2 = 0$, and $x^2 + x = 4$. A quadratic equation in standard form looks like this:

$$ax^2 + bx + c = 0 \qquad (12\text{-}2)$$

where a and b are the coefficients of x^2 and x and where c is a constant. If the equation is solved for x, the result is:

$$x = \frac{-b \pm \sqrt{b^2 - 4ac}}{2a} \qquad (12\text{-}3)$$

EXAMPLE 12.8: Solve the equation $x^2 - 25 = 0$.

Solution: Quadratic equations which have no x term are called *pure quadratics*. Such equations are readily solved by moving the constant to the right member and finding the square root of both members.

$$x^2 = 25$$

$$x = \sqrt{25} = 5$$

The equation is satisfied when $x = 5$ or ± 5.

12.4.1 Solving by Factoring. Whenever the quadratic is factorable, the solution is found in the following manner.

EXAMPLE 12.9: Find x in the equation $x^2 + 7x + 12 = 0$.

Solution:

$$x^2 + 7x + 12 = 0$$
$$(x+3)(x+4) = 0 \quad \text{(factor the left side)}$$

Let each factor equal zero and solve for x.

$$x + 3 = 0$$
$$x = -3$$
$$x + 4 = 0$$
$$x = -4$$

The equation is satisfied when $x = -3$ or $x = -4$.

EXAMPLE 12.10: $x^2 - 3x - 6 = 4$. Solve for x.

Solution: Put the equation in standard form and then factor the left member.

$$x^2 - 3x - 10 = 0$$
$$(x+2)(x-5) = 0$$

Equating each factor to zero, we get:

$$x + 2 = 0$$
$$x = -2$$
$$x - 5 = 0$$
$$x = 5$$

The equation is satisfied when $x = -2$ or $x = 5$.

PRACTICE PROBLEMS 12-4   Solve the following equations for x.

1. $x^2 = 49$
2. $x^2 - x - 30 = 0$
3. $x^2 + 5x + 6 = 0$
4. $x^2 - 4x - 10 = 11$
5. $x^2 + 10x + 12 = -12$

Solutions:

1. $x = 7, x = -7$
2. $x = -5, x = 6$
3. $x = -2, x = -3$
4. $x = 7, x = -3$
5. $x = -4, x = -6$

Additional practice problems are at the end of the chapter.

12.4.2   *Solving by the Quadratic Equation.*   When we have a second-degree equation to solve, we first put it into the form $ax^2 + bx + c = 0$. If the equation is factorable, we merely factor as before. If the equation is not factorable, then we resort to using the quadratic equation.

EXAMPLE 12.11: Given the equation $x^2 + 4x + 5 = 13$, solve for x.

Solution: First we put the equation in standard form:

$$x^2 + 4x - 8 = 0$$

Next we determine if the equation can be factored. Since this one does not factor, we will solve it by using the quadratic equation:

$$x = \frac{-b \pm \sqrt{b^2 - 4ac}}{2a} \qquad (12\text{-}2)$$

An examination of the equation shows that a (the coefficient of x^2) $= 1$; b (the coefficient of x) $= 4$; and c (the constant) $= -8$. Substituting these values in our equation results in the following:

$$x = \frac{-4 \pm \sqrt{4^2 - (4)(1)(-8)}}{2(1)} = \frac{-4 \pm \sqrt{16 + 32}}{2}$$

$$= \frac{-4 \pm \sqrt{48}}{2} = \frac{-4 \pm 6.93}{2}$$

At this point we see that there are two solutions:

1. $x = \dfrac{-4 + 6.93}{2} = \dfrac{2.93}{2} = 1.47$

2. $x = \dfrac{-4 - 6.93}{2} = \dfrac{-10.93}{2} = -5.47$

PRACTICE PROBLEMS 12.5   Solve the following equations for x:

1. $2x^2 + 3x - 9 = 0$ **2.** $x^2 - 5x + 2 = 10$

3. $3x^2 - 6x = 7$ **4.** $x^2 - 3x - 12 = -5$

Solutions:

1. $x = 1.5, x = -3$ **2.** $x = 6.28, x = -1.28$

3. $x = 2.83, x = -0.83$ **4.** $x = -1.54, x = 4.54$

Additional practice problems are at the end of the chapter.

SELF-TEST 12-2   Solve for x in the following problems:

1. $x^2 - 121 = 0$ **2.** $x^2 - 6x - 27 = 0$

3. $x^2 - 10x + 30 = 3x - 12$ **4.** $2x^2 + 4x - 5 = 0$

5. $2x^2 - 5x - 3 = 0$

Answers to Self-Test 12-2 are at the end of the chapter.

Solve for the unknown in the following problems. Express your answers as fractions or mixed numbers.

1. $\dfrac{5x}{2} + 3 = 6$

2. $\dfrac{6x}{4} - 2 = 5$

3. $\dfrac{2I}{4} + \dfrac{I}{6} = 2$

4. $\dfrac{3R}{2} + \dfrac{2R}{3} = 6$

5. $\dfrac{R-2}{5} + 3 = \dfrac{2R+2}{3} + 4$

6. $\dfrac{R+2}{3} - \dfrac{R-1}{2} = 3 - 2R$

7. $\dfrac{3}{I-3} + 3 = 4$

8. $\dfrac{4}{I+4} - 2 = 5$

9. $\dfrac{4R}{2} = 3 + \dfrac{R}{3}$

10. $\dfrac{3R}{4} - 2 = \dfrac{2R}{3}$

11. $\dfrac{5x}{2} + \dfrac{x-2}{6} = \dfrac{3x-6}{4}$

12. $\dfrac{2x}{4} - \dfrac{3-x}{2} = \dfrac{2x-1}{6}$

13. $\dfrac{4-x}{2x} - \dfrac{3}{x} = \dfrac{1}{x}$

14. $\dfrac{3x-1}{2x} + \dfrac{1}{x} = \dfrac{2}{x}$

15. $\dfrac{3}{G-3} - \dfrac{1}{4} = 5$

16. $\dfrac{2}{R+3} - \dfrac{1}{3} = 1$

17. $\dfrac{1}{a} + \dfrac{1}{b} = 2$. Solve for a and b.

18. $\dfrac{2}{a} - 3 = \dfrac{1}{b}$. Solve for a and b.

19. $\dfrac{x}{2a} + \dfrac{x}{3b} = 2$. Solve for x, a, and b.

20. $\dfrac{2x}{3a} - \dfrac{x}{b} = 3$. Solve for x, a, and b.

21. $\dfrac{3}{a-2} + \dfrac{2}{a+2} = \dfrac{b}{a+2}$. Solve for a and b.

22. $\dfrac{1}{x+3} - \dfrac{3}{x-4} = \dfrac{2y}{x-4}$. Solve for x and y.

23. $2(3R_1 - 1) = 3R_2 + R_1 + 2$. Solve for R_1 and R_2.

24. $4(R_1 + 3) - R_2 = 2(R_1 + R_2)$. Solve for R_1 and R_2.

1. $V_L = \dfrac{V_{OC} R_L}{R_{TH} + R_L}$
Solve for V_{OC}, R_L, and R_{TH}.

2. $V_2 = \dfrac{E R_2}{R_1 + R_2}$
Solve for E, R_1, and R_2.

3. $I_L = \dfrac{I_{SC} G_L}{G_L + G_N}$
Solve for I_{SC}, G_L, and G_N.

4. $I_1 = \dfrac{I_T G_1}{G_1 + G_2}$
Solve for I_T, G_1, and G_2.

5. $R_S = \dfrac{R_i R_o}{R_i + R_o}$
Solve for R_i.

6. $S = \dfrac{R_E + \dfrac{R}{\beta}}{R_E + I(1 - \alpha)}$
Solve for R_E.

7. $r_i = \beta(r_e + \dfrac{r_b}{\beta})$

Solve for r_e.

8. $A_V = \dfrac{r_e}{r_e + r_i'}$

Solve for r_e.

9. $R_T = \dfrac{R_1 R_2}{R_1 + R_2}$

Solve for R_1 and R_2.

10. $Z_T = \dfrac{Z_1 Z_2}{Z_1 + Z_2}$

Solve for Z_1 and Z_2.

11. $A = \dfrac{B}{1 + \dfrac{1}{A}}$

Solve for A.

12. $B = \dfrac{A}{2 + \dfrac{1}{B}}$

Solve for B.

13. $x + 2 = \dfrac{3 + \dfrac{2}{y}}{4}$

Solve for y.

14. $V_2 = \dfrac{ER_1}{R_1 + R_2}$

Solve for E, R_1, and R_2.

END OF CHAPTER PROBLEMS 12-3

1. $V_2 = \dfrac{ER_2}{R_1 + R_2}$

(a) Solve for R_2 when $E = 35$ V, $V_2 = 10$ V, and $R_1 = 12$ kΩ and (b) solve for R_1 when $E = 15$ V, $V_2 = 6.3$ V, and $R_2 = 560$ Ω.

2. $I_1 = \dfrac{I_T G_1}{G_1 + G_2}$

(a) Solve for I_T when $I_1 = 6.35$ mA, $G_1 = 3$ mS, and $G_2 = 5.83$ mS and (b) solve for G_2 when $I_T = 275$ μA, and $I_1 = 120$ μA, and $G_1 = 50$ μS.

3. $A_i = \dfrac{h_{fe}}{1 + h_{oe} R_L}$

(a) Solve for h_{oe} when $A_i = 37$, $h_{fe} = 70$, and $R_L = 12$ kΩ (h_{oe} is measured in Siemens) and (b) solve for R_L when $A_i = 80$, $h_{fe} = 100$, and $h_{oe} = 12$ μS.

4. $\alpha = \dfrac{\beta}{1 + \beta}$

Solve for β when $\alpha = 0.993$.

$\beta = \dfrac{\alpha}{1 - \alpha}$

5. $R_T = \dfrac{R_1 R_2}{R_1 + R_2}$

Solve for R_1 when $R_T = 1.83$ kΩ and $R_2 = 4.7$ kΩ.

6. $Z_T = \dfrac{Z_1 Z_2}{Z_1 + Z_2}$

Solve for Z_2 when $Z_T = 27.6$ kΩ and $Z_1 = 56$ kΩ.

7. $S = \dfrac{R_E + R_B}{R_E + R_B(1 - \alpha)}$

Solve for R_E when $S = 3.41$, $R_B = 3$ kΩ, and $\alpha = 0.99$.

END OF CHAPTER PROBLEMS 12-4

Solve the following quadratics by factoring:

1. $x^2 - 16 = 0$

2. $x^2 = 81$

3. $x^2 + 6x + 8 = 0$

4. $x^2 + 12x + 35 = 0$

5. $x^2 + 3x - 18 = 0$ **6.** $x^2 - x - 20 = 0$

7. $x^2 - 6x + 8 = 0$ **8.** $x^2 - 14x + 48 = 0$

9. $x^2 - x - 56 = 0$ **10.** $x^2 - 4x - 32 = 0$

11. $x^2 - 15 = 3x + 25$ **12.** $x^2 - 2 = 10 - 4x$

13. $x^2 - 3x + 16 = 10x - 20$ **14.** $x^2 - 3x = 4x - 12$

END OF CHAPTER PROBLEMS 12-5

Solve for x in the following problems:

1. $x^2 + 8x + 12 = 0$ **2.** $x^2 + 7x + 3 = 0$

3. $3x^2 + 4x = -3x - 3$ **4.** $3x^2 - 8x + 4 = 0$

5. $3x^2 + 2x - 1 = 0$ **6.** $x^2 - 6x = 2x + 5$

7. $2x^2 - 5x - 5 = 0$ **8.** $2x^2 - 7x + 5 = 0$

ANSWERS TO SELF-TESTS

Self-Test 12-1

1. $R = 4$ **2.** $I = 3$

3. $x = \dfrac{4b}{ab+6}$, $a = \dfrac{4b-6x}{bx}$, $b = \dfrac{6x}{4-ax}$ **4.** $G_L = 109 \ \mu S$

5. $R_2 = 89.8 \ k\Omega$

Self-Test 12-2

1. $x = \pm 11$ **2.** $x = 9, \ x = -3$

3. $x = 6, \ x = 7$ **4.** $x = 0.87, \ x = -2.87$

5. $x = 3, x = -\frac{1}{2}$

CHAPTER 13

GRAPHING

13.1 GRAPHING LINEAR EQUATIONS

Recall from previous chapters that linear equations are equations in which a term contains only *one* variable. Further, the variable exponent is one. For example, $2x + 1 = 9$; $y - 6 = 0$; $5 + 6a = 23$ are all linear equations. $6x + 7y = 14$ is also a linear equation. Even though the variables $x + y$ are different, they appear in different terms. $x + 6 + 7y + 14Z = 10$ would be a linear equation because no term contains more than one variable. $6xy + 7 = 0$ and $4x^2 + 2y = 14$ are *not* linear equations. In the first equation one term contains two variables (x and y). In the second equation one of the variables is squared ($4x^2$).

The graph of any linear equation is a straight line. Let's see why this is true. Consider the graph in Figure 13-1. This is called a system of *rectangular coordinates*. This kind of graph will be used many times throughout the text. The point where the x-axis (the horizontal axis) and the y-axis (the vertical axis) intersect (cross) is called the *origin*. Notice that the origin is "0" for both vertical and horizontal axes. Along the horizontal or x-axis, positive values are plotted to the right of the origin and negative values are plotted to the left of the origin. Along the vertical, or y-axis, positive values are plotted from the origin upward and negative values are plotted from the origin downward.

Let's identify some points using the graph in Figure 13-2. We identify points by giving the x and y values or coordinates. Suppose we were to identify a point whose coordinates are $(5, -2)$. The first number identifies the horizontal or x value and the second number identifies the vertical or y value. The point is located by moving 5 units to the right along the x-axis because the number is

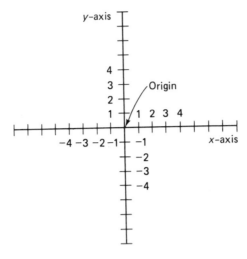

Figure 13-1. System of rectangular coordinates.

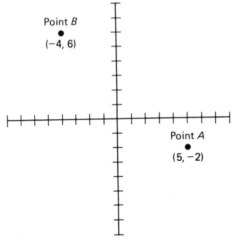

Figure 13-2.

positive. We move 2 units down along the y-axis because the number is negative. The point is shown as point A. Let's find the point whose coordinates are $(-4,6)$. This point is identified as point B in Figure 13-2. The point was found by moving 4 units *left* from the origin along the x-axis. We moved left because $x = -4$. Then we moved upward 6 units. We moved upward because $y = 6$.

PRACTICE PROBLEMS 13-1   Identify the following points on the graph of Figure 13-1:

1. $(2,6)$

2. $(-1,7)$

3. $(0,-8)$

4. $(-6,0)$

5. $(-1,2)$

6. $(6,-3)$

7. $(-3,-4)$

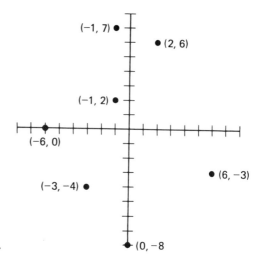

Figure 13-3.

Solutions:

See Figure 13-3 for the identification of points.

Consider the equation $2x - y = 6$. Usually, the graph of a linear equation intersects (crosses) both the horizontal and vertical axes. Therefore, we may assume that one point on the graph will be where $y = 0$ and x equals a value to be determined by solving the resulting equation. That is:

$$2x - y = 6$$
If $y = 0$
then $2x = 6$
and $x = 3$

One point on the graph then is where $y = 0$ and $x = 3$. On the graph the point would be identified as $(3,0)$. (The x value is always given first.) Another point would be where $x = 0$.

$$2x - y = 6$$
If $x = 0$
then $-y = 6$
$$y = -6$$

A second point on the graph is where $x = 0$ and $y = -6$. On the graph the point would be identified as $(0, -6)$. The graph is shown in Figure 13-4. Point A is called the *x-intercept* because that is where the line crosses the x-axis. Its coordinates are $(3,0)$. The x-coordinate is called the *abscissa*. The y-coordinate is called the *ordinate*.

Point B is called the y-intercept. Its coordinates are $(0, -6)$. Of course, there are many points along the graph which may be plotted or identified. These

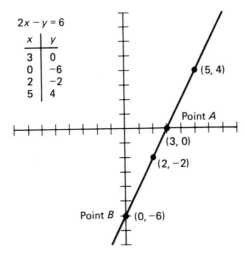

Figure 13-4.

points may be found by letting x equal some value and then solving for y. Or we could let y equal some value and solve for x. For example, using the same equation, we get:

$$2x - y = 6$$
$$\text{If } x = 2$$
$$\text{then } 2(2) - y = 6$$
$$4 - y = 6$$
$$-y = 2$$
$$y = -2$$

The coordinates are $(2, -2)$.

$$\text{If } y = 4$$
$$\text{then } 2x - (4) = 6$$
$$2x = 10$$
$$x = 5$$

The coordinates are $(5, 4)$.

These two points are also plotted in Figure 13-4.

A table showing values of x and corresponding values of y is also included in Figure 13-4. Such a table should be drawn whenever points are to be plotted because it helps identify coordinates.

EXAMPLE 13.1: Given the equation $y = 2x + 2$. Draw the graph. Find the x- and y-intercepts. Identify 4 points on the graph.

Solution: First let's rearrange the terms so that both variables are in the left-hand member. This will make it easier for us to solve for unknowns as we go along.

$$y = 2x + 2$$
$$y - 2x = 2$$

Next, let's find the x-intercept. This is the point where the line passes through the x-axis.

$$y - 2x = 2$$
$$\text{Let } y = 0$$
$$\text{then } -2x = 2$$
$$x = -1$$

The x-intercept is at coordinates $(-1, 0)$.
Now find the y-intercept.

$$y - 2x = 2$$
$$\text{Let } x = 0$$
$$\text{then } y = 2$$

The y-intercept is at coordinates $(0, 2)$. Use a straight edge and draw a line extending through the two points as shown in Figure 13-5. A third point is found by letting $x =$ some number. Upon examination of the straight line, we see that the point $(2, 6)$ should fall on the line. Let's see if it does.

$$y - 2x = 2$$

Let $x = 2$. Then:

$$y - 2(2) = 2$$
$$y - 4 = 2$$
$$y = 6$$

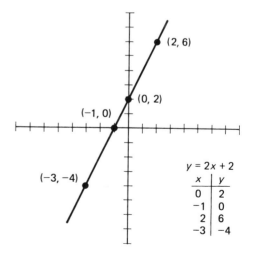

Figure 13-5.

The fact that the point $(2,6)$ falls on the line tells us that our line is in the correct location. Let $x = -3$. Then:

$$y - 2(-3) = 2$$
$$y + 6 = 2$$
$$y = -4$$

When we plot $(-3, -4)$ we see that this point falls on our line. The point $(2,6)$ or the point $(-3, -4)$ would be proof enough that our line is in the correct location. *Three* points are needed to verify our line. If the line passes through all three points, we can assume it to be correct.

PRACTICE PROBLEMS 13-2   Plot the following equations on linear graph paper. Identify the x- and y-intercepts. Plot at least 3 points on the graph:

1. $x - 4y = 8$ **2.** $3x + 6y = 12$

3. $y = 4x + 4$ **4.** $x = 3y - 6$

Solutions:

1. The graph is shown in Figure 13-6(a). The x-intercept $= (8,0)$. The y-intercept $= (0, -2)$.

2. The graph is shown in Figure 13-6(b). The x-intercept $= (4,0)$. The y-intercept $= (0,2)$.

3. The graph is shown in Figure 13-6(c). The x-intercept $= (-1,0)$. The y-intercept $= (0,4)$.

4. The graph is shown in Figure 13-6(d). The x-intercept $= (-6,0)$. The y-intercept $= (0,2)$.

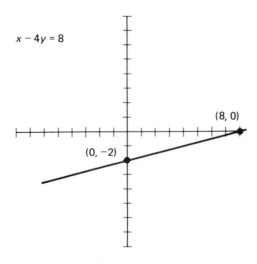

Figure 13-6(a).

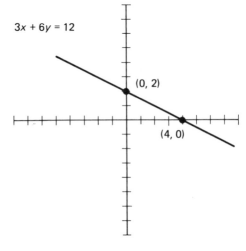

Figure 13-6(b).

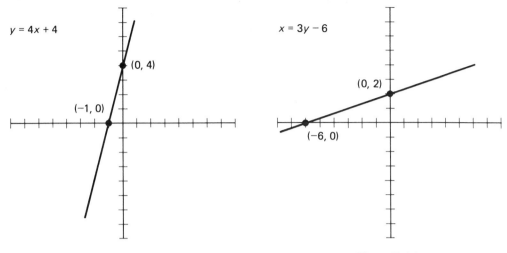

Figure 13-6(c).   Figure 13-6(d).

Additional practice problems are at the end of the chapter.

13.2 SLOPE OF A LINE

The linear equation in Figure 13-4 is redrawn in Figure 13-7. Consider the point $(3,0)$. If x is changed from 3 to 5, y changes from 0 to 4. The change in x is 2. This would be written $\Delta x = 2$ (delta $x = 2$). The change in y is 4. This would be written $\Delta y = 4$ (*delta y* = 4). Upon inspection of the curve, we find that for any

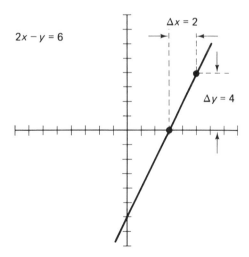

Figure 13-7.

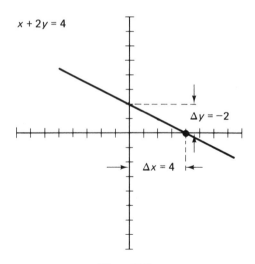

$x + 2y = 4$

$\Delta y = -2$

$\Delta x = 4$

Figure 13-8.

increase of 2 units horizontally ($\Delta x = 2$), y increases 4 units ($\Delta y = 4$). For example, if x increases from -2 units to 0, y increases from -10 units to -6 units. If x increases from 1 unit to 3 units, y increases from -4 units to 0.

Consider the straight line in Figure 13-8 which results from the equation $x + 2y = 4$. If x increases from 0 to 4 ($\Delta x = 4$), y *decreases* from 2 to 0 ($\Delta y = -2$). Any increase in the value of x causes a corresponding decrease in the value of y. Any *4*-unit increase in x causes a *2*-unit decrease in y. Any *8*-unit increase in x causes a *4*-unit decrease in y.

The *slope* of a line is defined as the change in y which results from an *increase* in x.

$$\text{slope} = \frac{\Delta y}{\Delta x} = \frac{\text{change in } y}{\text{increase in } x}$$

When we read a graph, we always make Δx positive. The resulting Δy will be either positive or negative depending on whether y increased or decreased as x increased. The slope is positive if Δy is positive and it is negative if Δy is negative. In Figure 13-8 the slope equals $\frac{\Delta y}{\Delta x} = \frac{-2}{4} = -\frac{1}{2}$. When the slope is negative, the line falls from left to right. When the slope is positive the line rises from left to right.

EXAMPLE 13.2: Find the slopes of the lines in practice problems 13-2.

Solution: If we use the x- and y-intercepts to determine the slope, then in problem 1 the slope is:

$$\text{slope} = \frac{\Delta y}{\Delta x} = \frac{1}{4}$$

The slope is positive because when x is increased 4 units, y

increases 1 unit. In problem 2:

$$\text{slope} = \frac{\Delta y}{\Delta x} = -\frac{1}{2}$$

The slope is negative because when x is increased 2 units, y decreases 1 unit. In problem 3:

$$\text{slope} = \frac{\Delta y}{\Delta x} = \frac{4}{1} = 4$$

The slope is positive because when x is increased 1 unit, y increases 4 units. In problem 4:

$$\text{slope} = \frac{\Delta y}{\Delta x} = \frac{1}{3}$$

The slope is positive because when x is increased 3 units, y increases 1 unit.

PRACTICE PROBLEMS 13-3   Plot the following equations on linear graph paper. Identify the x- and y-intercepts. Determine the slope.

1. $x + y = 4$

2. $x - 3y = -6$

3. $4x + y = -8$

4. $8x - 4y = 16$

Solutions:

1. Figure 13-9(a) shows the graph of $x + y = 4$. The x-intercept is at coordinates $(4,0)$. The y-intercept is at coordinates $(0,4)$.

$$\text{slope} = \frac{\Delta y}{\Delta x} = \frac{-4}{4} = -1$$

2. Figure 13-9(b) shows the graph of $x - 3y = -6$. The x-intercept is at coordinates $(-6,0)$. The y-intercept is at coordinates $(0,2)$.

$$\text{slope} = \frac{\Delta y}{\Delta x} = \frac{2}{6} = \frac{1}{3}$$

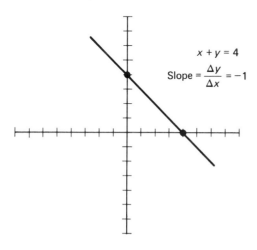

Figure 13-9(a).

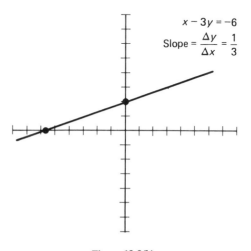

Figure 13-9(b).

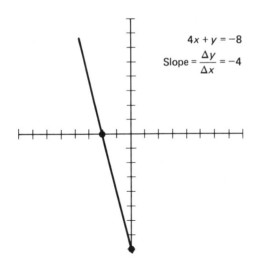

$$4x + y = -8$$
$$\text{Slope} = \frac{\Delta y}{\Delta x} = -4$$

Figure 13-9(c).

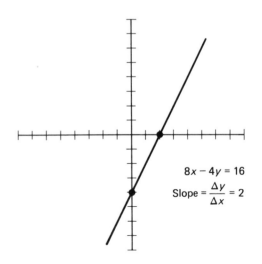

$$8x - 4y = 16$$
$$\text{Slope} = \frac{\Delta y}{\Delta x} = 2$$

Figure 13-9(d).

3. Figure 13-9(c) shows the graph of $4x + y = -8$. The x-intercept is at coordinates $(-2,0)$. The y-intercept is $(0, -8)$.

$$\text{slope} = \frac{\Delta y}{\Delta x} = \frac{-8}{2} = -4$$

4. Figure 13-9(d) shows the graph of $8x - 4y = 16$. The x-intercept is $(2,0)$. The y-intercept is $(0, -4)$.

$$\text{slope} = \frac{\Delta y}{\Delta x} = \frac{4}{2} = 2$$

Additional practice problems are at the end of the chapter.

13.3 SLOPE-INTERCEPT FORM

Any linear equation containing two variables can be written like this:

$$y = mx + b$$

y and x are the variables. m and b are constants. When the equation is written in this form, it is said to be written in the *slope-intercept* form because "m" is the slope and "b" is the vertical or y-intercept.

EXAMPLE 13.3: Given the equation $3y - 2x = 9$, find the slope and the y-intercept.

Solution: Using the rules of algebra, we can change to slope-intercept form:

$$3y - 2x = 9$$
$$3y = 2x + 9$$
$$y = \frac{2}{3}x + 3$$

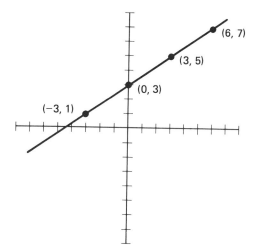

Figure 13-10.

The slope is $\frac{2}{3}$ and the y-intercept is 3. A slope of $\frac{2}{3}$ means that if x is increased 3 units, y increases 2 units. The y-intercept coordinates are $(0,3)$. With this information we can graph the equation. In Figure 13-10 we first plot the point $(0,3)$, the y-intercept. Next we find a second point. The slope is $\frac{2}{3}$ which means that if we change x from 0 to 3, y will increase from 3 to 5. This point is $(3,5)$ and is plotted. A third point can be plotted by changing x from 3 to 6 and changing y from 5 to 7. This point is $(6,7)$. Starting at the y-intercept we could have plotted the point $(-3, -1)$. This point results from changing x from 0 to -3 which causes a change in y from 3 to 1.

PRACTICE PROBLEMS 13-4 Change the following equations into the slope-intercept form. Determine the slope and y-intercept. Graph the equation.

1. $4x - 3y = -15$

2. $3y - 4x = 18$

3. $3x + 3y = 5$

4. $6y - 5x = 24$

5. $3y - 7x = -12$

6. $4x - y = 5$

Solutions:

1. $y = \dfrac{4}{3}x + 5$

Slope $= \frac{4}{3}$, y-intercept $= (0,5)$.

The graph of the equation is shown in Figure 13-11(a).

2. $y = \dfrac{4}{3}x + 6$

Slope $= \frac{4}{3}$, y-intercept $= (0,6)$.

The graph of the equation is shown in Figure 13-11(b).

$y = \frac{4}{3}x + 5$

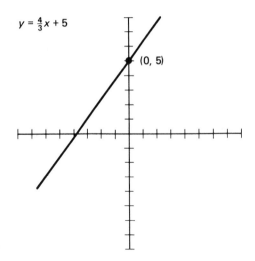

(0, 5)

Figure 13-11(a).

$y = \frac{4}{3}x + 6$

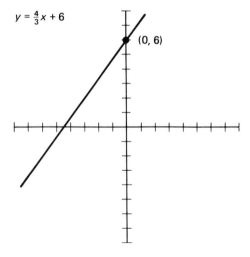

(0, 6)

Figure 13-11(b).

3. $y = -x + \dfrac{5}{3}$

Slope $= -1$, y-intercept $= (0, \frac{5}{3})$.

The graph of the equation is shown in Figure 13.11(c).

4. $y = \dfrac{5}{6}x + 4$

Slope $= \frac{5}{6}$, y-intercept $= (0, 4)$.

The graph of the equation is shown in Figure 13.11(d).

$y = -x + \frac{5}{3}$

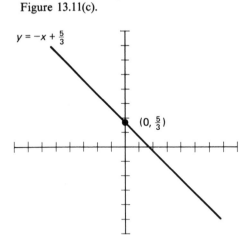

(0, $\frac{5}{3}$)

Figure 13-11(c).

$y = \frac{5}{6}x + 4$

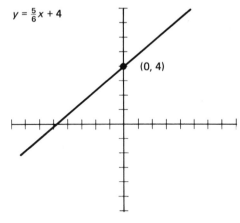

(0, 4)

Figure 13-11(d).

5. $y = \dfrac{7}{3}x - 4$

Slope $= \frac{7}{3}$, y-intercept $= (0, -4)$.

The graph of the equation is shown in Figure 13.11(e).

6. $y = 4x - 5$

Slope $= 4$, y-intercept $= (0, -5)$.

The graph of the equation is shown in Figure 13.11(f).

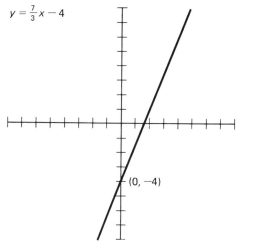

Figure 13-11(e).

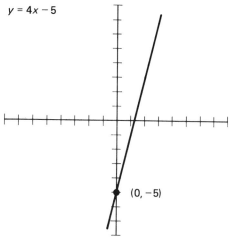

Figure 13-11(f).

Additional practice problems are at the end of the chapter.

SELF-TEST 13-1

1. Plot the equation $5x + 3y = 15$ on linear graph paper. Identify the x- and y-intercepts.

2. Plot the equation $3x - y = 6$ on linear graph paper. Identify the x- and y-intercepts. Determine the slope.

3. Change the equation $2x - 3y = 6$ into slope-intercept form. Determine the slope. Identify the y-intercept. Graph the equation.

Answers to Self-Test 13-1 are at the end of the chapter.

13.4 INTERPRETING GRAPHS

In Figure 13-12 we have plotted current versus voltage of a resistor for values of voltage from 0 V to 20 V. The resulting curve is a straight line. We can determine the resistance by using the techniques learned previously. Notice that I is plotted along the y-axis and V is plotted along the x-axis. Relating this curve to a system of rectangular coordinates, we get:

$$\text{slope} = \frac{\Delta y}{\Delta x} = \frac{\Delta I}{\Delta V} = \frac{4\ \text{mA}}{4\ \text{V}} = 1\ \text{mS}$$

The slope is equal to the conductance. That is, the slope defines the conductance

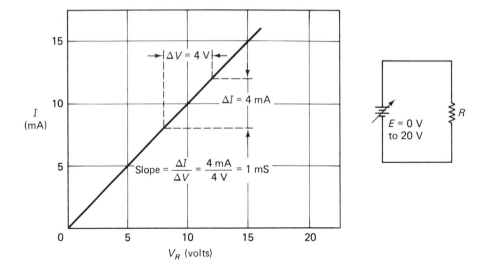

Figure 13-12.

characteristic of the curve. To find R:

$$R = \frac{1}{G} = \frac{1}{1 \text{ mS}} = 1 \text{ k}\Omega$$

The curve, then, is a curve of current versus voltage for a 1-kΩ resistor. With such a curve we can easily find the resulting current for any value of voltage plotted. Because the curve is linear, the slope is the same anywhere along the curve. Therefore, any convenient place along the curve can be chosen to determine the slope.

Figure 13-13 is the curve of a forward-biased silicon diode. The resulting curve of current versus voltage is non-linear. When the curve is non-linear, the slope of the line depends on where we choose to measure. The most accurate method of determining the slope of such a curve is to first select a point on the curve and then construct a tangent to that point. From geometry a *tangent* is defined as a line that touches a curve at only one point. In Figure 13-14 we have constructed a tangent to the curve where $V_f = 0.5$ V. The slope of this line is the same as the slope of the curve at that point.

$$\text{slope} = \frac{\Delta I}{\Delta V} = \frac{8 \text{ mA}}{0.3 \text{ V}} = 26.7 \text{ mS}$$

The resistance determined from the slope is symbolized r_d (lower-case "r" signifies a dynamic or AC resistance).

$$r_d = \frac{1}{\text{slope}} = \frac{1}{26.7 \text{ mS}} = 37.5 \text{ }\Omega$$

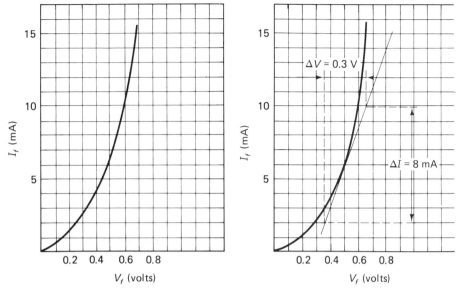

Figure 13-13.   **Figure 13-14.**

In addition to the method discussed previously, r_d can be found by the $\dfrac{\Delta I}{\Delta V}$ method if we consider that small increments of change along the curve will be nearly a straight line. Notice that r_d changes drastically as V_{DC} changes. This tells us that when we are required to find r_d, we must specify where on the curve the measurement is to be made.

At the point in question, there are really two resistances. If we simply plot the point, we have a voltage and some resulting current. Since we are examining one point, we consider this to be a DC resistance. This DC resistance is symbolized R_D. The general equation is:

$$R_D = \frac{V_{DC}}{I_{DC}}$$

Solving our problem for R_D, we get:

$$R_D = \frac{V_{DC}}{I_{DC}} = \frac{0.5\ \text{V}}{6\ \text{mA}} = 83.3\ \Omega$$

Figure 13-15 is a universal time constant curve. Notice that for about the first 20% of the curve, the curve is relatively linear. After that the non-linearity becomes obvious. This curve is called an *exponential curve*. Exponential curves will be analyzed at length in Chapters 22 and 23. In this curve we are primarily interested in some point on the curve.

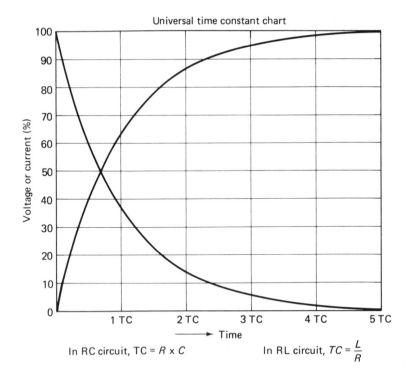

Universal time constant chart

In RC circuit, TC = $R \times C$ In RL circuit, $TC = \dfrac{L}{R}$

Figure 13-15.

EXAMPLE 13.4: If the applied voltage is 50 V, find V_C and V_R after 1.5 TC.

Solution: From the curve we see that the capacitor has charged to 78% of maximum in 1.5 TC. V_R is 22% of maximum. Then:

$$V_C = 78\% \times 50 \text{ V} = 39 \text{ V}$$
$$V_R = 22\% \times 50 \text{ V} = 11 \text{ V}$$

EXAMPLE 13.5: In Figure 13-15 (a) how many time constants are required for the capacitor to charge to 60% of maximum and (b) if the maximum voltage is 60 V, how many time constants are required to charge the capacitor to 48 V?

Solution: (a) From the curve we see that it takes approximately 0.9 TC for the capacitor to charge to 60% of maximum.
(b) We must first find what percent 48 V is of 60 V.

$$\frac{48 \text{ V}}{60 \text{ V}} \times 100 = 80\%$$

Approximately 1.6 TC are required to charge the capacitor to 48 V (80% of maximum).

PRACTICE PROBLEMS 13-5

1. Refer to Figure 13-16. Find the slope. Find R.   **2.** Refer to Figure 13-13. Find R_D and r_d if $V_f = 0.4$ V.

3. Refer to Figure 13-15. Let $E = 10$ V. Find V_C and V_R at 0.7 TC and at 2 TC.

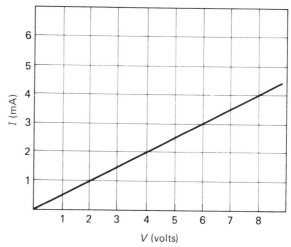

Figure 13-16.

Solutions:

1. Slope $= 500$ μS, $R = 2$ kΩ **2.** $R_D \simeq 90$ Ω, $r_d \simeq 55$ Ω

3. At 0.7 TC, $V_C = 5$ V and $V_R = 5$ V
    At 2 TC, $V_C = 8.6$ V and $V_R = 1.4$ V

Additional practice problems are at the end of the chapter.

13.5 PLOTTING CURVES

Refer again to Figure 13-12. A voltage source was connected in series with the resistor. As we vary the applied voltage in the circuit, the current changes. In this circuit the variable V is called the *independent* variable. I is the *dependent* variable. Its value depends on the value of V selected. In plotting curves it is standard practice to plot the independent variable along the x-axis. This leaves the dependent variable to be plotted along the y-axis. Whenever we are required to plot a curve, we must determine the position of the variables in this manner.

EXAMPLE 13.6: In a resistive circuit, plot a curve of V versus I where $R = 10$ kΩ. Vary V from 0 V to 10 V.

V (volts)	I (μA)
2	200
5	500
8	800

(a)

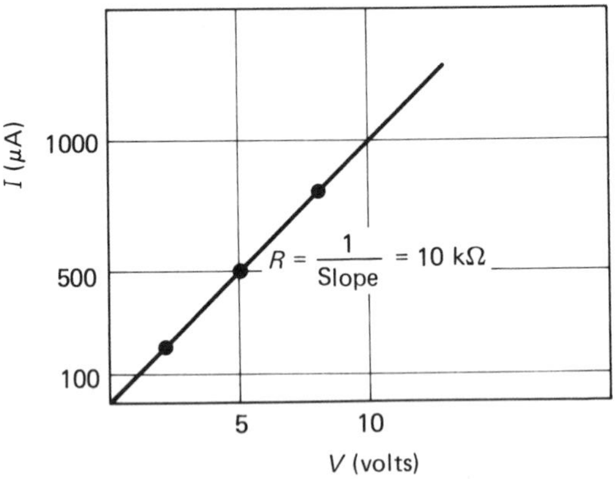

(b)

Figure 13-17.

Solution: Select an appropriate scale. It would be convenient to use a scale of 1 V/div for voltage and 100 μA/div for current. Of course, this is just one of many scales that could be selected. Since this is a resistive circuit, the curve will be linear. We could draw the curve by using two points but we will use three. The third point is used to check for errors. Using Ohm's law, we can construct a data table of variable values. Figure 13-17(a) is such a table. The resulting curve is shown in Figure 13-17(b).

Let's check our work. The reciprocal of the slope should equal 10 kΩ.

$$\text{Slope} = \frac{\Delta I}{\Delta V}$$

Let $\Delta V = 2$ V

Then $\Delta I = 200$ μA

$$\frac{\Delta I}{\Delta V} = \frac{200 \ \mu A}{2 \ V} = 100 \ \mu S$$

$$R = \frac{1}{G} = \frac{1}{100 \ \mu S} = 10 \ k\Omega$$

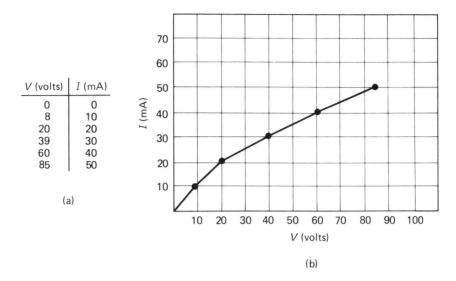

V (volts)	I (mA)
0	0
8	10
20	20
39	30
60	40
85	50

(a)

(b)

Figure 13-18.

EXAMPLE 13.7: Plot a curve of current versus voltage from the given data in Figure 13-18(a). *V* is the independent variable.

Solution: Select an appropriate scale. 10 V/div for voltage and 10 mA/div for current were chosen for our solution. The resulting curve is shown in Figure 13-18(b).

PRACTICE PROBLEMS 13-6

1. Plot a curve of current versus voltage for $R = 20$ kΩ for values of voltage from 0 V to 50 V.

2. Plot a curve of current versus voltage from the data in Figure 13-19.

V (volts)	I (mA)
0	0
0.2	0.2
0.3	0.4
0.4	0.65
0.5	1.0
0.6	1.5
0.7	2.4

Figure 13-19.

Solutions:

1. See Figure 13-20.

2. See Figure 13-21.

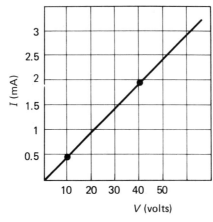

V (volts)	I (mA)
0	0
10	0.5
30	1.5
50	2.5

Figure 13-20.

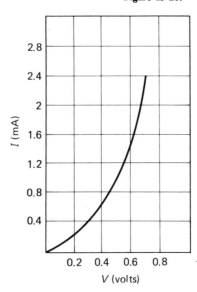

Figure 13-21.

Additional practice problems are at the end of the chapter.

SELF-TEST 13-2

1. Refer to Figure 13-22. Find R.

2. Refer to Figure 13-13. Find R_D and r_d if $V_f = 0.5$ V.

3. Refer to Figure 13-15. Let $E = 25$ V. Find V_C and V_R 2.3 TC after the capacitor starts to charge.

4. Plot a curve of current versus voltage for $R = 100$ kΩ. Plot values of voltage from 0 V to 10 V.

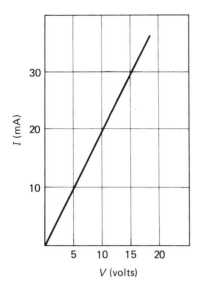

V (volts)	I (mA)
0	0
1	1.8
2	3
3	3.8
4	4.3
6	5.3
8	6.2
10	7

Figure 13-23.

Figure 13-22.

5. Plot a curve of current versus voltage from the data in Figure 13-23.

Answers to Self-Test 13-2 are at the end of the chapter.

END OF CHAPTER PROBLEMS 13-1

Construct systems of rectangular coordinates on linear graph paper and plot the following points and equations:

1. Plot the following points (a) $(0,6)$, (b) $(-5,0)$, and (c) $(3,-5)$.

2. Plot the following points: (a) $(-5,-5)$, (b) $(-4,4)$, and (c) $(6,3)$.

3. Graph the following equations: (a) $x+y=4$, (b) $2x-y=6$, and (c) $x+3y=3$.

4. Graph the following equations: (a) $2y=4x+8$, (b) $x-2y=-4$, and (c) $x+2y=-6$.

END OF CHAPTER PROBLEMS 13-2

Plot the following equations on linear graph paper. Identify the x- and y-intercepts. Determine the slope.

1. $2x+y=6$

2. $3x-6y=12$

3. $2y=5x+5$

4. $2x=2y+7$

5. $3x+5y=15$

6. $-4x-6y=24$

7. $2x-7y=-14$

8. $6x+3y=-18$

9. $-3x+4y=-24$

10. $3y=7x-21$

END OF CHAPTER PROBLEMS 13-3

Change the following equations into the slope-intercept form. Determine the slope and the y-intercept. Graph the equation.

1. $x + 3y = 3$
2. $x - 5y = 7$
3. $3x + 4y = -5$
4. $y - x = -4$
5. $x - 4y = 12$
6. $x + 4y = -5$
7. $5y - 2x = 10$
8. $x - y = 6$
9. $3x - 7y = 21$
10. $8x - 6y = 24$

END OF CHAPTER PROBLEMS 13-4

1. Refer to Figure 13-24. Find R_{DC} and r_{AC} where $V = 3$ V.
2. Refer to Figure 13-24. Find R_{DC} and r_{AC} where $V = 6.5$ V.

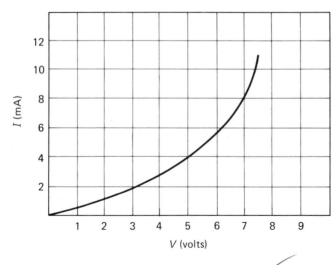

Figure 13-24.

3. Refer to Figure 13-15. Let $E = 100$ V. Find V_C and V_R at (a) 0.5 TC and (b) 3 TC.
4. Refer to Figure 13-15. Let $E = 60$ V. Find V_C and V_R at (a) 1 TC and (b) 3.5 TC.

END OF CHAPTER PROBLEMS 13-5

1. Plot a curve of I versus V_R for $R = 100$ Ω for values of V_R from 0 V to 50 V.
2. Plot a curve of I versus V_R for $R = 2$ kΩ for values of V_R from 0 V to 20 V.
3. Plot a curve of I versus V from the data in Figure 13-25.
4. Plot a curve of I versus V from the data in Figure 13-26.

V (volts)	I (mA)
0	0
1	15
2	23
3	30
4	34
5	38
6	42

Figure 13-25.

V (mV)	I (mA)
0	0
40	1
80	2
120	3.2
200	6
240	8
300	11

Figure 13-26.

ANSWERS TO SELF-TESTS

Self-Test 13-1

1. See Figure 13-27.

2. See Figure 13-28.

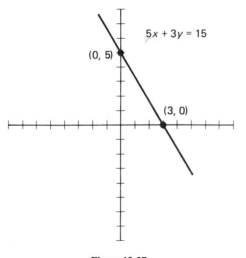

$5x + 3y = 15$

$(0, 5)$

$(3, 0)$

Figure 13-27.

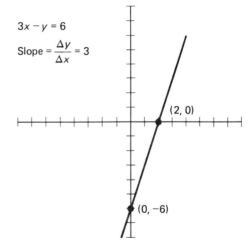

$3x - y = 6$

$\text{Slope} = \dfrac{\Delta y}{\Delta x} = 3$

$(2, 0)$

$(0, -6)$

Figure 13-28.

3. See Figure 13-29.

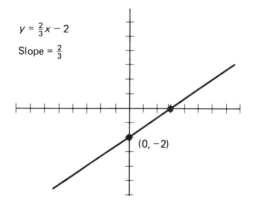

$y = \frac{2}{3}x - 2$

$\text{Slope} = \frac{2}{3}$

$(0, -2)$

Figure 13-29.

Self-Test 13-2

1. $R = 500 \ \Omega$

2. $R_D \simeq 83 \ \Omega$, $r_d \simeq 38 \ \Omega$

3. $V_C \simeq 22.5$ V, $V_R \simeq 2.5$ V

4. See Figure 13-30.

5. See Figure 13-31.

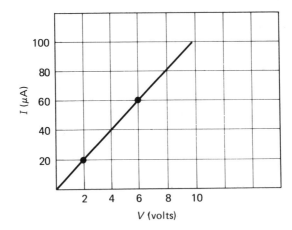

Figure 13-30.

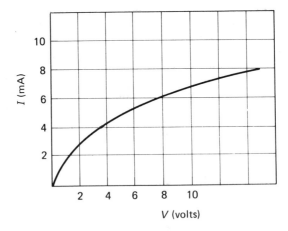

Figure 13-31.

CHAPTER 14

SIMULTANEOUS LINEAR EQUATIONS

In the chapters on circuit analysis we used various laws and theorems in solving circuit problems. In this chapter we will discuss methods of solutions using simultaneous equations and determinants.

14.1 LINEAR EQUATIONS IN TWO UNKNOWNS—
GRAPHICAL SOLUTION

As we saw in Chapter 13, every linear equation in two unknowns has an unlimited number of solutions. If we have two equations, each equation still has an unlimited number of solutions. However, if the two equations are plotted on the same system of rectangular coordinates, they will intersect at some common point provided they (1) are not parallel lines and (2) are not the same line. The coordinates of this common point will satisfy both equations.

> EXAMPLE 14.1: Plot the equations $x - 2y = 4$ and $3x + 2y = 4$ on the same system of rectangular coordinates. Find the point of intersection.

> *Solution:* The plot of the equations is shown in Figure 14-1. The point of intersection is at $(2, -1)$. This point is common to both equations. Because the lines intersect, there is a common solution which in this case is $(2, -1)$. Further, this is the only solution. $x = 2$ and $y = -1$ are the only values common to both equations.

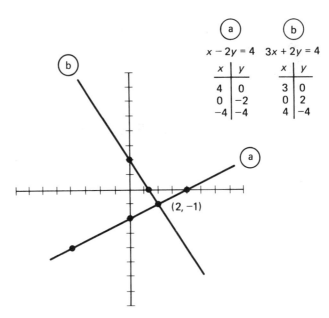

Figure 14-1.

PRACTICE PROBLEMS 14-1   Plot the following equations on the same system of rectangular coordinates. Find the coordinates of the point of intersection.

1. $x+y=5$
   $x-y=3$

2. $x+y=3$
   $2x+y=7$

3. $x+2y=7$
   $2x+2y=10$

4. $\dfrac{x}{4}+\dfrac{y}{3}=\dfrac{7}{12}$

   $\dfrac{x}{2}-\dfrac{y}{4}=\dfrac{1}{4}$

5. $x-y=1.5$
   $x+y=3.75$

Solutions:

1. (4, 1). The graph is shown in Figure 14-2.

2. (4, −1). The graph is shown in Figure 14-3.

3. (3, 2). The graph is shown in Figure 14-4.

4. (1, 1). The graph is shown in Figure 14-5.

5. (2.6, 1.1). The graph is shown in Figure 14-6.

In graphical analysis all answers are approximations because we cannot read exact values from graphs. We were able to determine accurately the answers to the first four problems because the point of intersection was a whole number and the graphs are scaled in whole numbers. In problem 5 we had to approximate the answer.

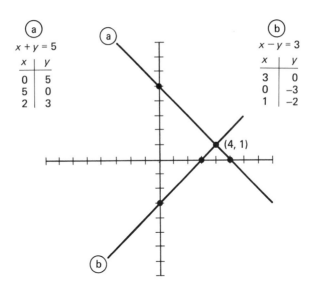

Figure 14-2.

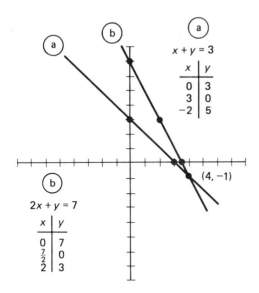

Figure 14-3.

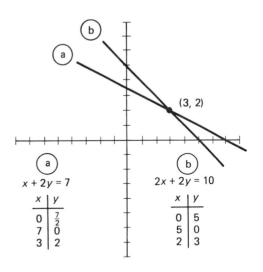

Figure 14-4.

311

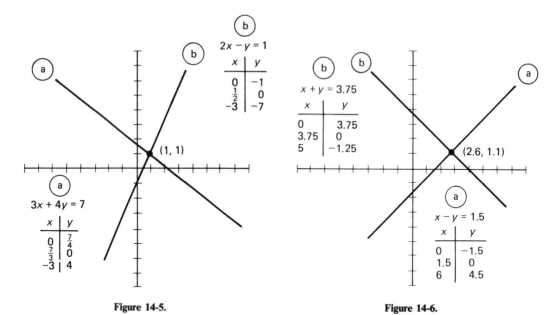

Figure 14-5.   Figure 14-6.

Additional practice problems are at the end of the chapter.

14.2 LINEAR EQUATIONS IN TWO UNKNOWNS— ALGEBRAIC SOLUTION

14.2.1 Solution by Addition or Subtraction.   Solving the two equations in Example 14.1 graphically resulted in $x=2$ and $y=-1$. Let's solve the same problem by using the addition or subtraction method. This method uses rules of algebra we have already developed. We will change one or the other of the equations so that both x values, or both y values, have the same coefficient. If the signs of the coefficients are the same, we will subtract one equation from the other. If the signs are different, we will add the equations. The subtraction (or addition) results in a new equation which has only one unknown. We can now find the value of this unknown. This value is common to both original equations. The value of this unknown is then substituted back in one of the original equations and the other unknown is then found. Let's see how it works.

EXAMPLE 14.2: Solve the following set of equations for x and y by the addition or subtraction method.

$$x-2y=4 \qquad (14\text{-}1)$$

$$3x+2y=4 \qquad (14\text{-}2)$$

Solution: We can eliminate the y term by addition:

$$x - 2y = 4 \qquad (14\text{-}1)$$

$$\underline{3x + 2y = 4} \qquad (14\text{-}2)$$

$$(\text{add}) \quad 4x \quad\quad = 8 \qquad (14\text{-}3)$$

Solve Equation 14-3 for x:

$$4x = 8$$

$$x = 2$$

Substitute $x = 2$ in Equation 14-1 and solve for y.

$$x - 2y = 4 \qquad (14\text{-}1)$$

$$2 - 2y = 4$$

$$-2y = 2$$

$$y = -1$$

The common values of x and y are $x = 2$ and $y = -1$.
Check Equation 14-1.

$$x - 2y = 4$$

$$2 - 2(-1) = 4$$

$$2 + 2 = 4$$

Check Equation 14-2

$$3x + 2y = 4$$

$$3(2) + 2(-1) = 4$$

$$6 - 2 = 4$$

The solution is verified.

EXAMPLE 14.3: Solve the following set of equations for x and y by the addition or subtraction method:

$$2x - y = 4 \qquad (14\text{-}4)$$

$$x - 2y = -1 \qquad (14\text{-}5)$$

Solution:

Rewrite 14.4.	$2x - y = 4$	
Multiply 14.5 by 2.	$\underline{2x - 4y = -2}$	(14-6)
Subtract.	$0 + 3y = 6$	(14-7)

Solve Equation 14-7 for y:

$$3y = 6$$

$$y = 2$$

Substitute $y = 2$ in Equation 14-4 and solve for x.

$$2x - y = 4 \tag{14-4}$$
$$2x - 2 = 4$$
$$2x = 6$$
$$x = 3$$

The common values of x and y are $x = 3$ and $y = 2$.
Check Equation 14-4.

$$2x - y = 4 \tag{14-4}$$
$$2(3) - 2 = 4$$
$$6 - 2 = 4$$

Check Equation 14-5.

$$x - 2y = -1 \tag{14-5}$$
$$3 - 2(2) = -1$$
$$3 - 4 = -1$$

PRACTICE PROBLEMS 14-2   Solve the following sets of equations for x and y by the addition or subtraction method.

1. $x + y = 5$
$\quad x - y = 3$

2. $x + y = 3$
$\quad 2x + y = 7$

3. $x + 2y = 7$
$\quad 2x + 2y = 10$

4. $\dfrac{x}{4} + \dfrac{y}{3} = \dfrac{7}{12}$
$\quad \dfrac{x}{2} - \dfrac{y}{4} = \dfrac{1}{4}$

5. $x - y = 1.5$
$\quad x + y = 3.75$

Solutions:

1. $x = 4, y = 1$

2. $x = 4, y = -1$

3. $x = 3, y = 2$

4. $x = 1, y = 1$

5. $x = 2.625, y = 1.125$

Additional practice problems are at the end of the chapter.

14.2.2   Solution by Substitution.   Let's work our example problems again, but this time we will solve the equations by substitution. This involves finding the value of x (or y) in one equation and then substituting that value back into the other equation. This eliminates one of the unknowns and we can then find the value of the other. Example 14.4 shows how it works.

EXAMPLE 14.4: Solve the following set of equations for x and y by using the substitution method:

$$x - 2y = 4 \qquad (14\text{-}1)$$

$$3x + 2y = 4 \qquad (14\text{-}2)$$

Solution: Solve Equation 14-1 for x:

$$x - 2y = 4 \qquad (14\text{-}1)$$

$$x = 4 + 2y \qquad (14\text{-}8)$$

Substitute the value of x in Equation 14-8 back into Equation 14-2 and then solve the equation for y.

$$3x + 2y = 4 \qquad (14\text{-}2)$$

$$3(4 + 2y) + 2y = 4$$

$$12 + 6y + 2y = 4$$

$$8y = -8$$

$$y = -1$$

Now substitute -1 for y in either equation and solve for x. (We will choose Equation 14-1.)

$$x - 2y = 4 \qquad (14\text{-}1)$$

$$x - 2(-1) = 4$$

$$x + 2 = 4$$

$$x = 2$$

The solution is $x = 2$ and $y = -1$. We could check the solutions as we did in the previous section.

EXAMPLE 14.5: Solve the following set of equations for x and y by using the substitution method:

$$2x - y = 4 \qquad (14\text{-}4)$$

$$x - 2y = -1 \qquad (14\text{-}5)$$

Solution: Solve for x in Equation 14-4:

$$2x - y = 4 \qquad (14\text{-}4)$$

$$2x = y + 4$$

$$x = \frac{y + 4}{2} \qquad (14\text{-}9)$$

Substitute the value of x in Equation 14-9 for the value of x in Equation 14-5:

$$x - 2y = -1 \qquad \text{(14-5)}$$

$$\frac{y+4}{2} - 2y = -1$$

Multiply both sides by 2:

$$y + 4 - 4y = -2$$

$$-3y = -6$$

$$y = 2$$

Now substitute this value for y in Equation 14-4:

$$2x - y = 4 \qquad \text{(14-4)}$$

$$2x - 2 = 4$$

$$2x = 6$$

$$x = 3$$

The solution is $x = 3$ and $y = 2$. We could check the solutions as we did in the previous section.

PRACTICE PROBLEMS 14-3   Solve the sets of equations in practice problems 14-2 by using the substitution method.

Solutions:

The solutions are the same as for practice problems 14-2.

Additional practice problems are at the end of the chapter.

14.3   DETERMINANTS

Another way to solve simultaneous equations is by using *determinants*. Determinants are arrays of numbers. These numbers correspond to the coefficients and constants of the different equations.

Consider the following general equations with two unknowns:

$$a_1 x + b_1 y = k_1 \qquad \text{(14-10)}$$

$$a_2 x + b_2 y = k_2 \qquad \text{(14-11)}$$

a_1, b_1, a_2, and b_2 are the coefficients of the unknowns. k_1 and k_2 are the constants. Notice that the equations are set up the same way we set them up when we were using the addition or subtraction method. The determinant for the denominator looks like this:

$$\begin{vmatrix} a_1 & b_1 \\ a_2 & b_2 \end{vmatrix}$$

This determinant has two rows and two columns and is called a *second-order determinant*. The numbers a_1, a_2, b_1, and b_2 are called the *elements* of the determinants. $a_1 b_2 - a_2 b_1$ is called the *expansion* of the determinants. The numbers a_1 and b_2 are the *principal diagonal*. The numbers a_2 and b_1 are the secondary diagonal. The expansion of the determinant results from the product of the elements in the principal diagonal $(a_1 b_2)$ minus the product of the elements in the secondary diagonal $(a_2 b_1)$. This process is illustrated below:

$$\begin{vmatrix} a_1 & b_1 \\ a_2 & b_2 \end{vmatrix} = a_1 b_2 - a_2 b_1$$

This is called the *determinant of the denominator*. This is the denominator for both unknowns. The determinant for the numerator of the unknown, x, is written:

$$\begin{vmatrix} k_1 & b_1 \\ k_2 & b_2 \end{vmatrix} = k_1 b_2 - k_2 b_1$$

The determinant for the numerator of the unknown, y, is written:

$$\begin{vmatrix} a_1 & k_1 \\ a_2 & k_2 \end{vmatrix} = a_1 k_2 - a_2 k_1$$

When we put it all together it looks like this:

$$x = \frac{\begin{vmatrix} k_1 & b_1 \\ k_2 & b_2 \end{vmatrix}}{\begin{vmatrix} a_1 & b_1 \\ a_2 & b_2 \end{vmatrix}} = \frac{k_1 b_2 - k_2 b_1}{a_1 b_2 - a_2 b_1}$$

$$y = \frac{\begin{vmatrix} a_1 & k_1 \\ a_2 & k_2 \end{vmatrix}}{\begin{vmatrix} a_1 & b_1 \\ a_2 & b_2 \end{vmatrix}} = \frac{a_1 k_2 - a_2 k_1}{a_1 b_2 - a_2 b_1}$$

EXAMPLE 14.6: Solve for x and y in the set of equations below by using determinants. (This is the same set of equations used in Example 14.2.)

$$x - 2y = 4 \qquad (14\text{-}12)$$
$$3x + 2y = 4 \qquad (14\text{-}13)$$

Solution: Determine the numbers to be used in the following arrays:

$$a_1 = 1, \; b_1 = -2, \; k_1 = 4$$
$$a_2 = 3, \; b_2 = 2, \; k_2 = 4$$

Now substitute these known values in our general equations for x and y.

$$x = \frac{\begin{vmatrix} 4 & -2 \\ 4 & 2 \end{vmatrix}}{\begin{vmatrix} 1 & -2 \\ 3 & 2 \end{vmatrix}} = \frac{8-(-8)}{2-(-6)} = \frac{16}{8} = 2$$

$$y = \frac{\begin{vmatrix} 1 & 4 \\ 3 & 4 \end{vmatrix}}{\begin{vmatrix} 1 & -2 \\ 3 & 2 \end{vmatrix}} = \frac{4-12}{2-(-6)} = \frac{-8}{8} = -1$$

The solution is $x = 2$, $y = -1$.

Notice that the denominators are identical. This means that we need calculate the denominator only once. This method may be less laborious than either the addition or subtraction method or the substitution method in problem-solving situations.

EXAMPLE 14.7: Solve for x and y in the set of equations below. (This is the same set of equations used in Example 14.5.)

$$2x - y = 4$$
$$x - 2y = -1$$

Solution: Determine the numbers to be used in the following arrays:

$$a_1 = 2, \ b_1 = -1, \ k_1 = 4$$
$$a_2 = 1, \ b_2 = -2, \ k_2 = -1$$

$$x = \frac{\begin{vmatrix} 4 & -1 \\ -1 & -2 \end{vmatrix}}{\begin{vmatrix} 2 & -1 \\ 1 & -2 \end{vmatrix}} = \frac{-8-1}{-4-(-1)} = \frac{-9}{-3} = 3$$

$$y = \frac{\begin{vmatrix} 2 & 4 \\ 1 & -1 \end{vmatrix}}{\begin{vmatrix} 2 & -1 \\ 1 & -2 \end{vmatrix}} = \frac{-2-4}{-3} = \frac{-6}{-3} = 2$$

The solution is $x = 3$, $y = 2$.

PRACTICE PROBLEMS 14-4   Solve the sets of equations in practice problems 14-2 by using determinants.

Solutions:

The solutions are the same as for practice problems 14.2.

Additional practice problems are at the end of the chapter.

SELF-TEST 14-1 Solve the following sets of equations (a) graphically, (b) by the addition or subtraction method, (c) by the substitution method, and (d) by using determinants.

1. $4x + y = 9$
   $x + y = 6$

2. $2x + 4y = 4$
   $x - y = 5$

3. $3x - 5y = 11$
   $x + 8y = -6$

Answers to Self-Test 14-1 are at the end of the chapter.

Let's now consider three equations and three unknowns. Our general equations look like this:

$$a_1 x + b_1 y + c_1 z = k_1$$
$$a_2 x + b_2 y + c_2 z = k_2$$
$$a_3 x + b_3 y + c_3 z = k_3$$

The determinants for the three unknowns look like this:

$$x = \frac{\begin{vmatrix} k_1 & b_1 & c_1 \\ k_2 & b_2 & c_2 \\ k_3 & b_3 & c_3 \end{vmatrix}}{\begin{vmatrix} a_1 & b_1 & c_1 \\ a_2 & b_2 & c_2 \\ a_3 & b_3 & c_3 \end{vmatrix}} = \frac{k_1 b_2 c_3 + b_1 c_2 k_3 + c_1 k_2 b_3 - c_1 b_2 k_3 - k_1 c_2 b_3 - b_1 k_2 c_3}{a_1 b_2 c_3 + b_1 c_2 a_3 + c_1 a_2 b_3 - c_1 b_2 a_3 - a_1 c_2 b_3 - b_1 a_2 c_3} \quad (14\text{-}14)$$

$$y = \frac{\begin{vmatrix} a_1 & k_1 & c_1 \\ a_2 & k_2 & c_2 \\ a_3 & k_3 & c_3 \end{vmatrix}}{\begin{vmatrix} a_1 & b_1 & c_1 \\ a_2 & b_2 & c_2 \\ a_3 & b_3 & c_3 \end{vmatrix}} = \frac{a_1 k_2 c_3 + k_1 c_2 a_3 + c_1 a_2 k_3 - c_1 k_2 a_3 - a_1 c_2 k_3 - k_1 a_2 c_3}{a_1 b_2 c_3 + b_1 c_2 a_3 + c_1 a_2 b_3 - c_1 b_2 a_3 - a_1 c_2 b_3 - b_1 a_2 c_3} \quad (14\text{-}15)$$

$$z = \frac{\begin{vmatrix} a_1 & b_1 & k_1 \\ a_2 & b_2 & k_2 \\ a_3 & b_3 & k_3 \end{vmatrix}}{\begin{vmatrix} a_1 & b_1 & c_1 \\ a_2 & b_2 & c_2 \\ a_3 & b_3 & c_3 \end{vmatrix}} = \frac{a_1 b_2 k_3 + b_1 k_2 a_3 + k_1 a_2 b_3 - k_1 b_2 a_3 - a_1 k_2 b_3 - b_1 a_2 k_3}{a_1 b_2 c_3 + b_1 c_2 a_3 + c_1 a_2 b_3 - c_1 b_2 a_3 - a_1 c_2 b_3 - b_1 a_2 c_3} \quad (14\text{-}16)$$

There are various methods used to find values of x, y, and z. Here is one of them:

Rewrite columns one and two to the right of the determinant. For the unknown, x, the numerator would look like this

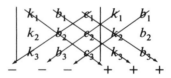

Diagonal lines have been drawn through the elements wherever there are three elements in the diagonal group. Where there are fewer than three elements in the diagonal group, that diagonal group is not used. The result is three diagonal groups running down from left to right and three diagonal groups running down from right to left. The products of each diagonal group running down from left to right are added. The products of the groups which run down from right to left are subtracted. This procedure is used for each numerator and each denominator. Since all the denominators are alike, however, the denominators need to be determined only once.

EXAMPLE 14.8: Solve for the unknowns in the following set of equations by using determinants:

$$x + 2y + 3z = 14$$
$$2x + y + 2z = 10$$
$$3x + 4y - 3z = 2$$

Solution: Let's first write down the coefficients:

$$a_1 = 1, b_1 = 2, c_1 = 3, k_1 = 14$$
$$a_2 = 2, b_2 = 1, c_2 = 2, k_2 = 10$$
$$a_3 = 3, b_3 = 4, c_3 = -3, k_3 = 2$$

Next let's find the denominator since it will be common to all solutions. Referring to Equation 14-14 and replacing the literal numbers with real numbers, we get:

$$= (1)(1)(-3) + (2)(2)(3) + (3)(2)(4) - (3)(1)(3) - (1)(2)(4)$$
$$- (2)(2)(-3)$$
$$= -3 + 12 + 24 - 9 - 8 + 12 = 28$$

Using the same method, let's find x.

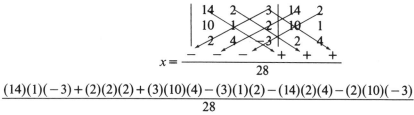

$$x = \frac{-\ -\ -\ +\ +\ +}{28}$$

$$\frac{(14)(1)(-3)+(2)(2)(2)+(3)(10)(4)-(3)(1)(2)-(14)(2)(4)-(2)(10)(-3)}{28}$$

$$= \frac{-42+8+120-6-112+60}{28} = \frac{28}{28} = 1$$

Now let's find y the same way.

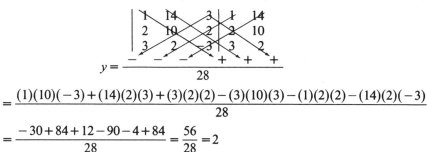

$$y = \frac{-\ -\ -\ +\ +\ +}{28}$$

$$= \frac{(1)(10)(-3)+(14)(2)(3)+(3)(2)(2)-(3)(10)(3)-(1)(2)(2)-(14)(2)(-3)}{28}$$

$$= \frac{-30+84+12-90-4+84}{28} = \frac{56}{28} = 2$$

Now we solve for z and are finished.

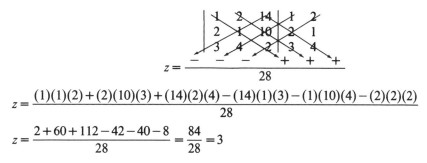

$$z = \frac{-\ -\ -\ +\ +\ +}{28}$$

$$z = \frac{(1)(1)(2)+(2)(10)(3)+(14)(2)(4)-(14)(1)(3)-(1)(10)(4)-(2)(2)(2)}{28}$$

$$z = \frac{2+60+112-42-40-8}{28} = \frac{84}{28} = 3$$

PRACTICE PROBLEMS 14-5 Use determinants to solve for the unknowns in the following sets of equations:

1. $x+2y+z=6$
$2x-y+3z=-13$
$3x-2y+3z=-16$

2. $3x+2y=12$
$4x+2z=16$
$4y+3z=24$

(Even though only two of the three unknowns appear in each equation, we must use all three in our solution. When the third unknown is missing, we assign a value of zero to that coefficient. For example, $c_1 = 0$ in the first equation of problem 2.

Solutions:

1. $x = 1.25, y = 4.25, z = -3.75$ **2.** $x = 2, y = 3, z = 4$

14.4 APPLICATIONS

In Chapter 10 we learned how to use various circuit theorems to solve complex circuit problems. An alternate method is solution by simultaneous linear equations.

As a first example of solution by simultaneous equations, let's solve for the currents in Figure 14-7. In this example we will solve the problem using all three methods: by addition and subtraction, by substitution, and by determinants.

> EXAMPLE 14.9: Find the currents and the voltage drops in the circuit of Figure 14-7.

> *Solution:* This circuit can easily be solved by using Ohm's law and Kirchhoff's laws. This solution would yield:

$$R_T = R_1 + R_2 \| R_3 = 1 \text{ k}\Omega + 1.2 \text{ k}\Omega = 2.2 \text{ k}\Omega$$

$$I_T = \frac{E}{R_T} = \frac{20 \text{ V}}{2.2 \text{ k}\Omega} = 9.09 \text{ mA}$$

$$V_1 = IR_1 = 9.09 \text{ mA} \times 1 \text{ k}\Omega = 9.09 \text{ V}$$

$$V_2 = V_3 = IR_x = 9.09 \text{ mA} \times 1.2 \text{ k}\Omega = 10.91 \text{ V}$$

$$(R_x = R_2 \| R_3)$$

$$I_2 = \frac{V_2}{R_2} = \frac{10.91 \text{ V}}{2 \text{ k}\Omega} = 5.45 \text{ mA}$$

$$I_3 = \frac{V_3}{R_3} = \frac{10.91 \text{ V}}{3 \text{ k}\Omega} = 3.64 \text{ mA}$$

$$I_1 = I_T = 9.09 \text{ mA}$$

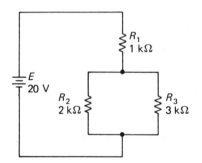

Figure 14-7.

Now let's work the same problem using simultaneous linear equations. Let's first establish polarities of the voltage drops across each resistor as in Figure 14-8. Using Kirchhoff's voltage law, let's start at point A and move counterclockwise around the loop containing E, R_1, and R_2.

$$E - V_2 - V_1 = 0$$

or
$$E = V_1 + V_2 \tag{14-17}$$

Starting at point A and moving counterclockwise around the loop containing E, R_1, and R_3, we get:

$$E - V_3 - V_1 = 0$$

or
$$E = V_1 + V_3 \tag{14-18}$$

If we express these equations in terms of IR drops, then Equation 14-17 would look like this:

$$E = I_1 R_1 + I_2 R_2$$

Substituting known values gives:

$$20 = 1kI_1 + 2kI_2 \tag{14-19}$$

Equation 14-18 would look like this:

$$E = I_1 R_1 + I_3 R_3$$

Substituting known values gives:

$$20 = 1kI_1 + 3kI_3 \tag{14-20}$$

Now we have two equations that we can solve simultaneously.

$$20 = 1kI_1 + 2kI_2 \tag{14-19}$$

$$20 = 1kI_1 + 3kI_3 \tag{14-20}$$

In Equations 14-19 and 14-20 we have three unknowns. We can simplify the solution by reducing the number of unknowns to two. Since $I_1 = I_2 + I_3$, then $I_3 = I_1 - I_2$. Let's

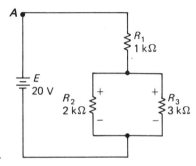

Figure 14-8.

substitute $(I_1 - I_2)$ for I_3 in Equation 14-20.

$$20 \text{ V} = 1kI_1 + 3kI_3$$

$$20 \text{ V} = 1kI_1 + 3k(I_1 - I_2)$$

$$20 \text{ V} = 1kI_1 + 3kI_1 - 3kI_2$$

$$20 \text{ V} = 4kI_1 - 3kI_2 \qquad (14\text{-}21)$$

Now let's look at Equations 14-19 and 14-21. We now have two equations and two unknowns and can now solve the problem.

$$20 \text{ V} = 1kI_1 + 2kI_2 \qquad (14\text{-}19)$$

$$20 \text{ V} = 4kI_1 - 3kI_2 \qquad (14\text{-}21)$$

Solution by addition and subtraction:

To eliminate one unknown, we will multiply Equation 14-19 by 4.

$$80 \text{ V} = 4kI_1 + 8kI_2 \qquad (14\text{-}22)$$

Subtracting Equation 14-22 from Equation 14-21 gives:

$$20 \text{ V} = 4kI_1 - 3kI_2 \qquad (14\text{-}21)$$

$$\underline{80 \text{ V} = 4kI_1 + 8kI_2} \qquad (14\text{-}22)$$

$$-60 \text{ V} = 0 \quad -11kI_2$$

Solving for I_2, we get:

$$I_2 = \frac{-60 \text{ V}}{-11k} = 5.45 \text{ mA}$$

Now let's plug the known value of I_2 back into Equation 14-21 and solve for I_1.

$$20 \text{ V} = 4kI_1 - 3kI_2$$

$$20 \text{ V} = 4kI_1 - 3k(5.45 \text{ mA})$$

$$20 \text{ V} = 4kI_1 - 16.35 \text{ V}$$

$$36.35 \text{ V} = 4kI_1$$

$$\frac{36.35 \text{ V}}{4k} = 9.09 \text{ mA} = I_1$$

Since $\quad I_1 = I_2 + I_3$

then $\quad\ I_3 = I_1 - I_2$

$\qquad\quad I_3 = 9.09 \text{ mA} - 5.45 \text{ mA} = 3.64 \text{ mA}$

Knowing I_1, I_2, and I_3, we can solve for V_1, V_2, and V_3:

$$V_1 = I_1 R_1 = 9.09 \text{ mA} \times 1k\Omega = 9.09 \text{ V}$$
$$V_2 = I_2 R_2 = 5.45 \text{ mA} \times 2k\Omega = 10.9 \text{ V}$$
$$V_3 = I_3 R_3 = 3.64 \text{ mA} \times 3k\Omega = 10.9 \text{ V}$$

Verifying, we get:

$$E = V_1 + V_2 = 9.09 \text{ V} + 10.9 \text{ V} = 20 \text{ V}$$

Solution by substitution:

$$20 \text{ V} = 1kI_1 + 2kI_2 \qquad\qquad (14\text{-}19)$$
$$20 \text{ V} = 4kI_1 - 3kI_2 \qquad\qquad (14\text{-}21)$$

In Equation 14-19 solve for one of the variables. Let's use I_1.

$$1kI_1 = 20 \text{ V} - 2kI_2$$
$$I_1 = \frac{20 \text{ V} - 2kI_2}{1k} = 20 \text{ mA} - 2I_2$$

Substituting this value of I_1 in Equation 14-21 we get:

$$20 \text{ V} = 4k(20 \text{ mA} - 2I_2) - 3kI_2 = 80 \text{ V} - 8kI_2 - 3kI_2$$
$$20 \text{ V} = 80 \text{ V} - 11kI_2$$
$$11kI_2 = 60 \text{ V}$$
$$I_2 = 5.45 \text{ mA}$$

Substituting this value for I_2 in Equation 14-19 and solving for I_1:

$$20 \text{ V} = 1kI_1 + 2k(5.45 \text{ mA}) = 1kI_1 + 10.9 \text{ V}$$
$$20 \text{ V} - 10.9 \text{ V} = 1kI_1$$
$$I_1 = \frac{9.1}{1k} = 9.10 \text{ mA}$$

I_3 and the voltage drops can be found as before.

Solution by determinants:

$$20 \text{ V} = 1kI_1 + 2kI_2 \qquad\qquad (14\text{-}19)$$
$$20 \text{ V} = 4kI_1 - 3kI_2 \qquad\qquad (14\text{-}21)$$

In the array:

$$a_1 = 1k, \ b_1 = 2k, \text{ and } k_1 = 20$$

$$a_2 = 4k, \ b_2 = -3k, \text{ and } k_2 = 20$$

$$I_1 = \frac{k_1 b_2 - k_2 b_1}{a_1 b_2 - a_2 b_1} = \frac{-60.0 \times 10^3 - 40.0 \times 10^3}{-11.0 \times 10^6} = 9.09 \text{ mA}$$

$$I_2 = \frac{a_1 k_2 - a_2 k_1}{a_1 b_2 - a_2 b_1} = \frac{20.0 \times 10^3 - 80.0 \times 10^3}{-11.0 \times 10^6} = 5.45 \text{ mA}$$

I_3 and the voltage drops can be found as before.

Graphical solution:

The problem may also be solved graphically. Although this method is seldom used in solving electrical problems such as this, it is useful in some applications and is presented here. Figure 14-9 is the graphical solution.

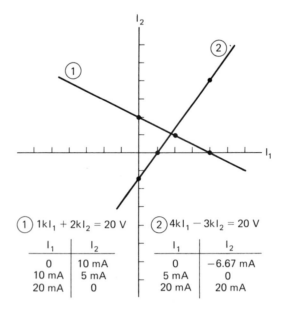

① 1kI₁ + 2kI₂ = 20 V	
I_1	I_2
0	10 mA
10 mA	5 mA
20 mA	0

② 4kI₁ − 3kI₂ = 20 V	
I_1	I_2
0	−6.67 mA
5 mA	0
20 mA	20 mA

Figure 14-9.   Graphical solution. Notice that while other solutions yield exact values for I_1 and I_2, the graphical solution provides only approximate values. That is, the intersect occurs where $I_1 \approx 5$ mA and $I_2 \approx 9$ mA.

PRACTICE PROBLEMS 14-6   Solve for the currents and voltage drops in Figure 14-10 by using simultaneous linear equations.

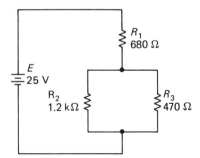

Figure 14-10.

Solution:

The two basic equations should look like this:

$$E = V_1 + V_2 \quad \text{and} \quad E = V_1 + V_3$$

Substituting known values, we get:

$$25 \text{ V} = 680I_1 + 1.2kI_2 \tag{14-23}$$

and

$$25 \text{ V} = 680I_1 + 470I_3 \tag{14-24}$$

Letting $I_3 = I_1 - I_2$, Equation 14-24 becomes:

$$25 \text{ V} = 680I_1 + 470I_3$$
$$25 \text{ V} = 680I_1 + 470(I_1 - I_2)$$
$$25 \text{ V} = 680I_1 + 470I_1 - 470I_2$$
$$25 \text{ V} = 1150I_1 - 470I_2 \tag{14-25}$$

Now we need to multiply (or divide) Equations 14-23 and 14-25 by some factor or factors to make I_1 or I_2 equal in both equations. As we know, there are an infinite number of factors we could use. We choose to multiply Equation 14-25 by 2.55 so that the coefficient of I_2 in both equations equals 1.2

$$25 \text{ V} = 1150I_1 - 470I_2 \tag{14-25}$$

$$25 \text{ V}(2.55) = 1150(2.55)I_1 - 470(2.55)I_2$$

$$63.8 \text{ V} = 2930I_1 - 1200I_2 \tag{14-26}$$

Putting Equations 14-23 and 14-26 together and adding, we get:

$$25 \text{ V} = 680I_1 + 1200I_2 \tag{14-23}$$

$$\underline{63.8 \text{ V} = 2930I_1 - 1200I_2} \tag{14-26}$$

$$88.8 \text{ V} = 3610I_1$$

$$\frac{88.8 \text{ V}}{3610} = I_1 = 24.6 \text{ mA}$$

Substituting 24.6 mA for I_1 in Equation 14.23, we get:

$$25\ V = 680(24.6\ mA) + 1200I_2$$

$$25\ V = 16.7\ V + 1200I_2$$

$$8.3\ V = 1200I_2$$

$$\frac{8.3\ V}{1200} = I_2 = 6.92\ mA$$

In this problem:

$$I_1 = I_2 + I_3$$

$$I_3 = I_1 - I_2$$

Then $\qquad\qquad I_3 = 24.6\ mA - 6.92\ mA = 17.7\ mA$

Using these currents, we get:

$$V_1 = I_1 R_1 = 24.6\ mA \times 680\ \Omega = 16.7\ V$$

$$V_2 = I_2 R_2 = 6.92\ mA \times 1.2\ k\Omega = 8.3\ V$$

$$V_3 = I_3 R_3 = 17.7\ mA \times 470\ \Omega = 8.32\ V$$

Let's solve the same problem using determinates:

$$25\ V = 680 I_1 + 1200 I_2 \tag{14-23}$$

$$25\ V = 1150 I_1 - 470 I_2 \tag{14-25}$$

In the array:

$$a_1 = 680,\ b_1 = 1200 \text{ and } k_1 = 25$$
$$a_2 = 1150,\ b_2 = -470 \text{ and } k_2 = 25$$

$$I_1 = \frac{\begin{vmatrix} 25 & 1200 \\ 25 & -470 \end{vmatrix}}{\begin{vmatrix} 680 & 1200 \\ 1150 & -470 \end{vmatrix}} = \frac{-11.8 \times 10^3 - 30 \times 10^3}{-320 \times 10^3 - 1.38 \times 10^6} = \frac{-41.8 \times 10^3}{-1.70 \times 10^6} = 24.6\ mA$$

$$I_2 = \frac{\begin{vmatrix} 680 & 25 \\ 1150 & 25 \end{vmatrix}}{\begin{vmatrix} 680 & 1200 \\ 1150 & -470 \end{vmatrix}} = \frac{17.0 \times 10^3 - 28.8 \times 10^3}{-320 \times 10^3 - 1.38 \times 10^6} = \frac{-11.8 \times 10^3}{-1.7 \times 10^3} = 6.91\ mA$$

These values for I_1 and I_2 agree with the values found using the substitution method. The calculations to find I_3 and the voltage drops are the same as when using the substitution method.

Differences between voltage drops or currents using the various methods are due to rounding. You should understand these methods of solution before proceeding to the more complex problems.

Additional practice problems are at the end of the chapter.

In Figure 14-11 other methods than just Ohm's law and Kirchhoff's laws must be used. Let's find the currents and voltage drops by using simultaneous linear equations.

We first have to determine the direction of current in the different branches. If the direction of current is not obvious, we simply assume a direction. If our assumption is wrong, when we solve for current, its magnitude will be correct but it will be negative.

Let's assume that the direction of current results from E_2. The resulting polarities would be as shown in Figure 14-12. There are three loops. One loop includes E_1, R_1, E_2, and R_3. A second loop includes E_1, R_1, and R_2, and a third loop includes E_2, R_2, and R_3. Even though there are three loops, we need use only two. Let's use the loops that contain E_2.

$$E_2 - V_2 - V_3 = 0$$

$$E_2 = V_2 + V_3 \tag{14-27}$$

$$E_2 - V_1 + E_1 - V_3 = 0$$

$$E_2 + E_1 = V_1 + V_3 \tag{14-28}$$

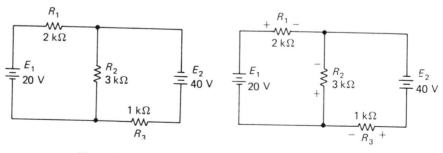

Figure 14-11.   Figure 14-12.

E_1 and E_2 are both included in the left-hand member of Equation 14.28 only because they are both known. Substituting known values, we get:

$$E_2 = V_2 + V_3 \tag{14-27}$$

$$40 \text{ V} = V_2 + V_3 \tag{14-29}$$

$$E_1 + E_2 = V_1 + V_3 \tag{14-28}$$

$$60 \text{ V} = V_1 + V_3 \tag{14-30}$$

Expressing Equations 14-29 and 14-30 in terms of IR drops, we get:

$$40 \text{ V} = V_2 + V_3 \tag{14-29}$$

$$40 \text{ V} = I_2 R_2 + I_3 R_3$$

$$40 \text{ V} = 3kI_2 + 1kI_3 \tag{14-31}$$

$$60 \text{ V} = V_1 + V_3 \tag{14-30}$$

$$60 \text{ V} = I_1 R_1 + I_3 R_3$$

$$60 \text{ V} = 2kI_1 + 1kI_3 \tag{14-32}$$

Let's look at Equations 14-31 and 14-32 together.

$$40 \text{ V} = 3000 I_2 + 1000 I_3 \tag{14-31}$$

$$60 \text{ V} = 2000 I_1 + 1000 I_3 \tag{14-32}$$

If we assume that the current through R_3 is the total current, then $I_3 = I_1 + I_2$ which makes $I_1 = I_3 - I_2$. Substituting $I_3 - I_2$ for I_1 in Equation 14-32 to eliminate one unknown yields:

$$60 \text{ V} = 2000 I_1 + 1000 I_3 \tag{14-32}$$

$$60 \text{ V} = 2000(I_3 - I_2) + 1000 I_3$$

$$60 \text{ V} = 2000 I_3 - 2000 I_2 + 1000 I_3$$

$$60 \text{ V} = -2000 I_2 + 3000 I_3 \tag{14-33}$$

Now let's put Equations 14-31 and 14-33 together and solve.

$$40 \text{ V} = 3000 I_2 + 1000 I_3 \tag{14-31}$$

$$60 \text{ V} = -2000 I_2 + 3000 I_3 \tag{14-33}$$

Solution by addition and subtraction:

Multiplying Equation 14-31 by 3 causes I_3 in both equations to be equal.

$$40 \text{ V}(3) = 3000 I_2(3) + 1000 I_3(3)$$
$$120 \text{ V} = 9000 I_2 + 3000 I_3 \tag{14-34}$$

Now we can put Equations 14-33 and 14-34 together and solve by subtracting.

$$120 \text{ V} = 9000 I_2 + 3000 I_3 \tag{14-34}$$
$$\underline{60 \text{ V} = -2000 I_2 + 3000 I_3} \tag{14-33}$$
$$60 \text{ V} = 11,000 I_2$$

$$\frac{60 \text{ V}}{11,000 \ \Omega} = 5.45 \text{ mA} = I_2$$

Substituting 5.45 mA for I_2 in Equation 14-34 yields:

$$120 \text{ V} = 9000(5.45 \text{ mA}) + 3000 I_3$$
$$120 \text{ V} = 49.1 \text{ V} + 3000 I_3$$
$$70.9 \text{ V} = 3000 I_3$$
$$\frac{70.9 \text{ V}}{3000 \ \Omega} = I_3 = 23.6 \text{ mA}$$

If $\qquad I_3 = 23.6 \text{ mA} \quad$ and $\quad I_2 = 5.45 \text{ mA}$

then $\qquad I_1 = I_3 - I_2 = 23.6 \text{ mA} - 5.45 \text{ mA} = 18.2 \text{ mA}$

Referring to Figure 14-11 and solving for the various voltage drops, we get:

$$V_1 = I_1 R_1 = 18.2 \text{ mA} \times 2 \text{ k}\Omega = 36.4 \text{ V}$$
$$V_2 = I_2 R_2 = 5.45 \text{ mA} \times 3 \text{ k}\Omega = 16.4 \text{ V}$$
$$V_3 = I_3 R_3 = 23.6 \text{ mA} \times 1 \text{ k}\Omega = 23.6 \text{ V}$$

To verify, we apply Kirchhoff's voltage law to the loops.

Solution by substitution:

$$40 \text{ V} = 3000 I_2 + 1000 I_3 \tag{14-31}$$
$$60 \text{ V} = -2000 I_2 + 3000 I_3 \tag{14-33}$$

Solving Equation 14-31 for I_2 we get

$$I_2 = \frac{40 \text{ V} - 1000 I_3}{3000} = 13.3 \times 10^{-3} \text{ mA} - 3.33 \times 10^{-1} I_3$$

Substituting this value for I_2 in Equation 14-33 yields

$$60 \text{ V} = -2000\left(13.3 \times 10^{-3} - 3.33 \times 10^{-1} I_3\right)$$

$$= -26.7 \text{ V} = 667 I_3 + 3000 I_3 = -26.7 \text{ V} = 3670 I_3$$

$$86.7 \text{ V} = 3670 I_3$$

$$L_3 = 23.6 \text{ mA}$$

Substituting this value back into Equation 14-33 and solving for I_2

$$60 \text{ V} = -2000 I_2 + 3000(23.6 \times 10^{-3}) = -2000 I_2 + 70.8 \text{ V}$$

$$-10.8 \text{ V} = -2000 I_2$$

$$I_2 = 5.4 \text{ mA}$$

I_1 and the voltage drops would be the same as those calculated when the solution was by addition and subtraction (some differences may exist due to rounding).

Solution by determinants

$$40 \text{ V} = 3000 I_2 + 1000 I_3 \tag{14-31}$$

$$60 \text{ V} = -2000 I_2 + 3000 I_3 \tag{14-33}$$

In the array:

$$a_1 = 3000, \ b_1 = 1000, \text{ and } k_1 = 40$$

$$a_2 = -2000, \ b_2 = 3000, \text{ and } k_2 = 60$$

$$I_2 = \frac{k_1 b_2 - k_2 b_1}{a_1 b_2 - a_2 b_1} = \frac{120 \times 10^3 - 60 \times 10^3}{9.00 \times 10^6 - (-2 \times 10^6)} = 5.45 \text{ mA}$$

$$I_3 = \frac{a_1 k_2 - a_2 k_1}{a_1 b_2 - a_2 b_1} = \frac{180 \times 10^3 - (-80 \times 10^3)}{9.00 \times 10^6 - (-2 \times 10^6)} = 23.6 \text{ mA}$$

Values for I_1 and the voltage drops may be calculated as before.

PRACTICE PROBLEMS 14-7   Use simultaneous equations to find the currents and voltage drops of the circuits in Figure 14-13(a), (b), and (c).

Solutions:

Remember, your method of solution may be different from the author's because you may not assume the same direction of currents or you may not use the same two loops.

Refer to Figure 14-13(a). This solution uses the loops that contain E_1, R_1, and R_2 and E_1, E_2, R_1, and R_3. Assume that the direction of current results from E_1. Starting at the positive terminal of E_1 and moving in a clockwise direction, gives:

$$E_1 - V_1 - V_2 = 0$$

$$E_1 = V_1 + V_2$$

$$40 \text{ V} = V_1 + V_2$$

$$40 \text{ V} = 100 I_1 + 330 I_2 \tag{14-35}$$

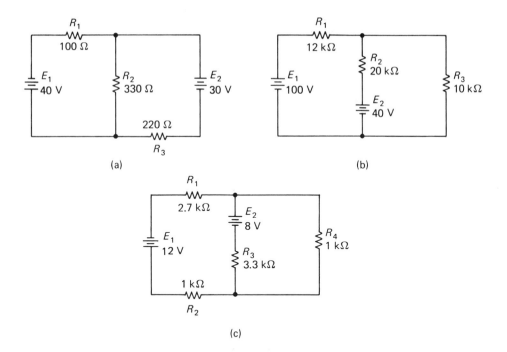

Figure 14-13.

$$E_1 - V_1 + E_2 - V_3 = 0$$
$$E_1 + E_2 = V_1 + V_3$$
$$40 \text{ V} + 30 \text{ V} = V_1 + V_3$$
$$70 \text{ V} = V_1 + V_3$$
$$70 \text{ V} = 100 I_1 + 220 I_3 \qquad (14\text{-}36)$$

Since we assumed that I_1 is the total current, then $I_2 = I_1 - I_3$ and we can substitute $I_1 - I_3$ for I_2 in Equation 14-35:

$$40 \text{ V} = 100 I_1 + 330 I_2 \qquad (14\text{-}35)$$
$$40 \text{ V} = 100 I_1 + 330 (I_1 - I_3)$$
$$40 \text{ V} = 430 I_1 - 330 I_3 \qquad (14\text{-}37)$$

Multiplying Equation 14-36 by 4.3 yields:

$$301 \text{ V} = 430 I_1 + 946 I_3 \qquad (14\text{-}38)$$

Subtracting Equation 14-37 from Equation 14-38 yields

$$40 \text{ V} = 430 I_1 - 330 I_3 \qquad (14\text{-}37)$$
$$301 \text{ V} = 430 I_1 + 946 I_3 \qquad (14\text{-}38)$$
$$-261 \text{ V} = 0 - 1276 I_3$$
$$261 \text{ V} = 1267 I_3$$

Then
$$I_3 = \frac{261 \text{ V}}{1276 \, \Omega} = 205 \text{ mA}$$

Substituting in Equation 14-36 gives:
$$70 \text{ V} = 100I_1 + 220(205 \text{ mA})$$
$$70 \text{ V} = 100I_1 + 45.1 \text{ V}$$
$$24.9 \text{ V} = 100I_1$$
$$I_1 = 249 \text{ mA}$$

Substituting in Equation 14-35 gives:
$$40 \text{ V} = 100I_1 + 330I_2 \qquad (14\text{-}37)$$
$$40 \text{ V} = 100(249 \text{ mA}) + 330I_2$$
$$40 \text{ V} = 24.9 \text{ V} + 330I_2$$
$$15.1 \text{ V} = 330I_2$$
$$I_2 = 45.8 \text{ mA}$$
$$V_1 = I_1R_1 = 249 \text{ mA} \times 100 \, \Omega = 24.9 \text{ V}$$
$$V_2 = I_2R_2 = 45.8 \text{ mA} \times 330 \, \Omega = 15.1 \text{ V}$$
$$V_3 = I_3R_3 = 205 \text{ mA} \times 220 \, \Omega = 45.1 \text{ V}$$

Apply Kirchhoff's laws to verify the solution.

Refer to Figure 14-13(b). This solution uses the loop that includes E_1, E_2, R_1, and R_2, and the loop that includes E_2, R_2, and R_3. Assume that the current in each loop results from E_1. Starting at the negative terminal of E_1 and moving clockwise, for the first loop we get
$$-E_1 + V_1 + V_2 + E_2 = 0$$
$$-100 + V_1 + V_2 + 40 \text{ V} = 0$$
$$V_1 + V_2 = 60 \text{ V}$$
$$12kI_1 + 20kI_2 = 60 \qquad (14\text{-}38)$$

For the second loop, start at the negative terminal of E_2 and move clockwise:
$$-E_2 - V_2 - V_3 = 0$$
$$-40 \text{ V} - V_2 + V_3 = 0$$
$$-V_2 + V_3 = 40 \text{ V}$$
$$-20kI_2 + 10kI_3 = 40 \text{ V} \qquad (14\text{-}39)$$

If we assume that I_1 is the total current, then $I_1 = I_2 + I_3$ and we can substitute $I_2 + I_3$ for I_1 in Equation 14-38.
$$12k(I_2 + I_3) + 20kI_2 = 60 \text{ V}$$
$$12kI_2 + 12kI_3 + 20kI_2 = 60 \text{ V}$$
$$32kI_2 + 12kI_3 = 60 \text{ V} \qquad (14\text{-}40)$$

Using Equations 14-39 and 14-40 we will solve for the currents using the substitution method.
$$-20kI_2 + 10kI_3 = 40 \text{ V} \qquad (14\text{-}39)$$
$$32kI_2 + 12kI_3 = 60 \text{ V} \qquad (14\text{-}49)$$

Solving for I_2 in Equation 14-39 we get:

$$I_2 = \frac{40 \text{ V} - 10kI_3}{-20k} = -2 \text{ mA} + 0.5I_3$$

Substituting this value into Equation 14-40

$$32(-2 \text{ mA} + 0.5I_3) + 12kI_3 = 60 \text{ V}$$
$$-64 \text{ V} + 16kI_3 + 12kI_3 = 60 \text{ V}$$
$$28kI_3 = 124 \text{ V}$$
$$I_3 = 4.43 \text{ mA}$$

Substituting this value of I_3 back into Equation 14-39:

$$-20kI_2 + 10k(4.43 \text{ mA}) = 40 \text{ V}$$
$$-20kI_2 + 44.3 \text{ V} = 40 \text{ V}$$
$$-20kI_2 = -4.3 \text{ V}$$
$$I_2 = 215 \text{ } \mu\text{A}$$

Knowing I_2 and I_3, we can solve for the rest of the unknowns.

$$I_1 = I_2 + I_3 = 215 \text{ } \mu\text{A} + 4.43 \text{ mA} = 4.65 \text{ mA}$$
$$V_1 = I_1 R_1 = 4.65 \text{ mA} \times 12 \text{ k}\Omega = 55.7 \text{ V}$$
$$V_2 = I_2 R_2 = 215 \text{ } \mu\text{A} \times 20 \text{ k}\Omega = 4.3 \text{ V}$$
$$V_3 = I_3 R_3 = 4.43 \text{ mA} \times 10 \text{ k}\Omega = 44.3 \text{ V}$$

Apply Kirchhoff's laws to verify solutions.

Refer to Figure 14−13(c). This solution uses the loop that includes E_1, E_2, R_1, R_2, and R_3, and the loop that includes E_2, R_3, and R_4. Assume that the current in each loop results from E_2. Starting at the positive terminal of E_2 and moving counterclockwise, for the first loop we get:

$$E_2 - I_3 R_3 - I_1 R_2 + E_1 - I_1 R_1 = 0$$
$$8 \text{ V} - 3.3kI_3 - 1kI_1 + 12 \text{ V} - 2.7kI_1 = 0$$
$$20 \text{ V} = 3.7kI_1 + 3.3kI_3 \qquad (14\text{-}41)$$

(Because the current is the same through R_1 and R_2 we call that current I_1.)

For the second loop, starting again at the positive terminal of E_2 and moving counterclockwise:

$$E_2 - I_3 R_3 - I_4 R_4 = 0$$
$$8 \text{ V} - 3.3kI_3 - 1kI_4 = 0$$
$$8 \text{ V} = 3.3kI_3 + 1kI_4 \qquad (14\text{-}42)$$

If we assume that I_3 is the total current, then $I_3 = I_1 + I_4$, or $I_4 = I_3 - I_1$, and we can substitute $I_3 - I_1$ for I_4 in Equation 14-42.

$$8 \text{ V} = 3.3kI_3 + 1k(I_3 - I_1) = 3.3kI_3 + 1kI_3 - 1kI_1$$
$$8 \text{ V} = 4.3kI_3 - 1kI_1 \qquad (14\text{-}43)$$

Using Equations 14-41 and 14-43, we will solve for the currents using determinants.

$$20 \text{ V} = 3.7kI_1 + 3.3kI_3 \tag{14-41}$$

$$8 \text{ V} = -1kI_1 + 4.3kI_3 \tag{14-43}$$

In the array:

$$a_1 = 3.7k, \ b_1 = 3.3k \text{ and } k_1 = 20 \ V$$

$$a_2 = -1k, \ b_2 = 4.3k \text{ and } k_2 = 8 \text{ V}$$

$$I_1 = \frac{k_1 b_2 - k_2 b_1}{a_1 b_2 - a_2 b_1} = \frac{20 \text{ V} \times 4.3k - 8 \times 3.3k}{3.7k \times 4.3k - (-1k \times 3.3k)}$$

$$= \frac{86.0 \times 10^3 - 26.4 \times 10^3}{15.9 \times 10^6 + 3.3 \times 10^3} = \frac{59.6 \times 10^3}{19.2 \times 10^6} = 3.1 \text{ mA}$$

$$I_3 = \frac{a_1 k_2 - a_2 k_1}{a_1 b_2 - a_2 b_1} = \frac{3.7k \times 8 - (-1k \times 20)}{3.7k \times 4.3k - (-1k \times 3.3k)}$$

$$= \frac{29.6 \times 10^3 + 20 \times 10^3}{15.9 \times 10^6 + 3.3 \times 10^6} = \frac{49.6 \times 10^3}{19.2 \times 10^6} = 2.58 \text{ mA}$$

We assumed that I_3 was total current. Then:

$$I_4 = I_3 - I_1 = 2.58 \text{ mA} - 3.1 \text{ mA} = -520 \ \mu\text{A}$$

The negative value for I_4 tells us that we assumed the wrong direction for current through R_4. I_1 is the total current, not I_3. The values calculated are correct. We just have to reverse the direction of current through R_4.

$$V_1 = I_1 R_1 = 3.1 \text{ mA} \times 2.7 \text{ k}\Omega = 8.37 \text{ V}$$

$$V_2 = I_2 R_2 = 3.1 \text{ mA} \times 1 \text{ k}\Omega = 3.1 \text{ V}$$

$$V_3 = I_3 R_3 = 2.58 \text{ mA} \times 3.3 \text{ k}\Omega = 8.51 \text{ V}$$

$$V_4 = I_4 R_4 = 520 \ \mu\text{A} \times 1 \text{ k}\Omega = 0.52 \text{ V}$$

Apply Kirchhoff's law to verify the solutions. Additional problems are at the end of the chapter.

SELF-TEST 14-2

1. Use the addition or subtraction method to solve for the currents and voltage drops of the circuit in Figure 14-14.

2. Use the substitution method to solve for the currents and voltage drops of the circuit in Figure 14-15.

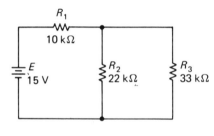

Figure 14-14.

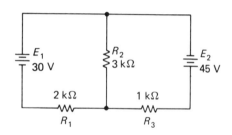

Figure 14-15.

Answers to Self-Test 14-2 are at the end of the chapter.

END OF CHAPTER PROBLEMS 14-1

Solve the following sets of equations graphically:

1. $2x - y = 4$
   $2y = x + 1$

2. $3x + 3y = 5$
   $x - y = -5$

3. $5x + 2y = 2$
   $3x - y = 10$

4. $x + 2y = 6$
   $4x + 3y = 4$

5. $8x - 4y = 12$
   $x + 2y = 4$

6. $9x - y = 6$
   $x + y = 4$

END OF CHAPTER PROBLEMS 14-2

Solve the following equations for x and y by using the addition or subtraction method:

1. $2x - y = 4$
   $2y = x + 1$

2. $3x + 3y = 5$
   $x - y = -5$

3. $5x + 2y = 2$
   $3x - y = 10$

4. $x + 2y = 6$
   $4x + 3y = 4$

5. $8x - 4y = 12$
   $x + 2y = 4$

6. $9x - y = 6$
   $x + y = 4$

7. $2x - 5y = 1$
   $3x - 8y = 2$

8. $3x - 2y = 3$
   $2x + y = 2$

9. $\dfrac{x}{2} + \dfrac{y}{3} = 3$
   $\dfrac{x}{2} - \dfrac{y}{3} = 1$

10. $x + 3y = 6$
    $2x - 3y = 9$

END OF CHAPTER PROBLEMS 14-3

Solve for the currents and voltage drops in the following problems by using simultaneous equations:

1. Figure 14-16

2. Figure 14-17

Figure 14-16.

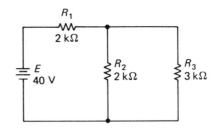

Figure 14-17.

3. Figure 14-18

4. Figure 14-19

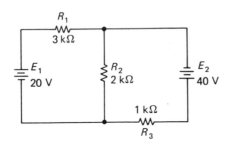

Figure 14-18.

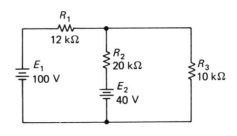

Figure 14-19.

5. Figure 14-20

6. Figure 14-21

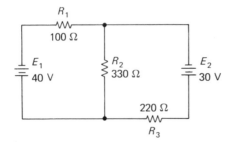

Figure 14-20.

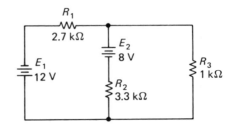

Figure 14-21.

338

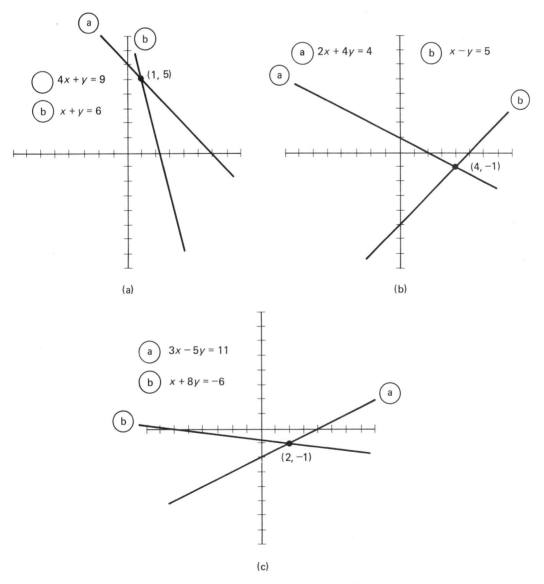

Figure 14-22. Answers to Self-Test 14-1: (a) problem 1; (b) problem 2; and (c) problem 3.

ANSWERS TO SELF-TESTS

Self-Test 14-1

1. (a) See Figure 14-22 for graphical solution.
   (b) Addition-subtraction method:

   $$4x + y = 9 \quad (1)$$
   $$x + y = 6 \quad (2)$$

   Multiply (2) by 4: $\quad 4x + 4y = 24$
   Rewrite (1): $\qquad\quad 4x + y = 9$
   Subtract: $\qquad\qquad\quad \overline{3y = 15}$
   $$y = 5$$

   Substitute in (1)
   $$4x + 5 = 9$$
   $$4x = 4$$
   $$x = 1$$
   The solution is $x = 1$, $y = 5$.

   (c) Substitution method:
   Solve for x in (2)
   $$x + y = 6$$
   $$x = 6 - y$$
   Substitute for x in (1)
   $$4(6 - y) + y = 9$$
   $$24 - 4y + y = 9$$
   $$-3y = -15$$
   $$y = 5$$
   Substitute for y in (1)
   $$4x + 5 = 9$$
   $$4x = 4$$
   $$x = 1$$
   The solution is $x = 1$, $y = 5$.

2. (a) See Figure 14-22 for graphical solution.
   (b) Addition-subtraction method:

   $$2x + 4y = 4 \quad (1)$$
   $$x - y = 5 \quad (2)$$

   Multiply (2) by 2: $\quad 2x - 2y = 10$
   Rewrite (1): $\qquad\quad 2x + 4y = 4$
   Subtract: $\qquad\qquad\quad \overline{-6y = 6}$
   $$y = -1$$

   Substitute for y in (1)
   $$2x + 4(-1) = 4$$
   $$2x - 4 = 4$$
   $$2x = 8$$
   $$x = 4$$
   The solution is $x = 4$, $y = -1$.

   (c) Substitution method:
   Solve for x in (2)
   $$x - y = 5$$
   $$x = y + 5$$
   Substitute for x in (1)
   $$2(y + 5) + 4y = 4$$
   $$2y + 10 + 4y = 4$$
   $$6y = -6$$
   $$y = -1$$
   Substitute for y in (2)
   $$x - (-1) = 5$$
   $$x + 1 = 5$$
   $$x = 4$$
   The solution is $x = 4$, $y = -1$.

3. (a) See Figure 14-22 for graphical solution.
   (b) Addition-subtraction method:

   $$3x - 5y = 11 \quad (1)$$
   $$x + 8y = -6 \quad (2)$$

   Multiply (2) by 3: $\quad 3x + 24y = -18$
   Rewrite (1): $\qquad\quad 3x - 5y = 11$
   Subtract: $\qquad\qquad\quad \overline{29y = -29}$
   $$y = -1$$

   Substitute for y in (2)
   $$x + 8(-1) = -6$$
   $$x = 2$$
   The solution is $x = 2$, $y = -1$.

   (c) Substitution method:
   Solve for x in (2)
   $$x + 8y = -6$$
   $$x = -6 - 8y$$
   Substitute for x in (1)
   $$3(-6 - 8y) - 5y = 11$$
   $$-18 - 24y - 5y = 11$$
   $$-29y = 29$$
   $$y = -1$$
   Substitute for y in (2)
   $$x + 8(-1) = -6$$
   $$x = 2$$
   The solution is $x = 2$, $y = -1$.

Self-Test 14-2

1. $I_1 = 647$ μA, $I_2 = 388$ μA, $I_3 = 259$ μA, $V_1 = 6.47$ V, $V_2 = V_3 = 8.53$ V

2. $I_1 = 23.2$ mA, $I_2 = 5.47$ mA, $I_3 = 28.6$ mA, $V_1 = 46.4$ V, $V_2 = 16.4$ V, $V_3 = 28.6$ V

CHAPTER 15

COMPLEX NUMBERS

When we worked with squares and square roots we found that whenever numbers are squared the answer is always positive. The number to be squared can be either positive or negative: $3^2 = 9$ or $-3^2 = 9$. Up to now, when we took the square root of a number we always took the square root of a positive number.

15.1 IMAGINARY NUMBERS

What is the result if we find the square root of a negative number? What is $\sqrt{-36}$? The answer cannot be 6 or -6 because the square of either of these numbers results in 36. We cannot take the square root of a negative member and come up with a real number answer. We can find a solution to the problem in the following manner:

$$\sqrt{-36} = \sqrt{(36)(-1)} = \sqrt{36} \cdot \sqrt{-1} = 6\sqrt{-1}$$

A number in this form is called an *imaginary* number. In mathematics the letter i is used to represent $\sqrt{-1}$.

$$\sqrt{-36} = \sqrt{36}\ i = 6i$$

The letter i in electronics denotes current. Therefore, we use the letter j to represent $\sqrt{-1}$. We place the "j" in front of the number to make it easier to recognize the presence of an imaginary number.

$$\sqrt{-36} = \sqrt{(36)(-1)} = \sqrt{36} \cdot \sqrt{-1} = j6$$

Whenever we see a number preceded by the letter "j," we know that the number is multiplied by $\sqrt{-1}$.

The word "imaginary" is used simply as a means of separating these numbers from "real" numbers. Imaginary numbers exist, and they are very useful in analyzing AC circuits, as we shall see in later chapters.

15.2 COMPLEX NUMBERS

It is not uncommon for numbers to consist of real numbers *and* imaginary numbers. Such numbers are called *complex* numbers. $4 + j3$ is a complex number. 4 is the real number part and $j3$ is the imaginary number part. These two parts cannot be added together.

Consider the imaginary part:

$$j = \sqrt{-1}$$
$$j^2 = \sqrt{-1} \cdot \sqrt{-1} = -1$$
$$j^3 = \sqrt{-1} \cdot \sqrt{-1} \cdot \sqrt{-1} = -1\sqrt{-1} = -j$$
$$j^4 = \sqrt{-1} \cdot \sqrt{-1} \cdot \sqrt{-1} \cdot \sqrt{-1} = (-1)(-1) = 1$$

EXAMPLE 15.1: Find $\sqrt{-144}$.

Solution:

$$\sqrt{-144} = j12$$

EXAMPLE 15.2: $\sqrt{36} + \sqrt{-36}$.

Solution:

$$\sqrt{36} = 6$$
$$\sqrt{-36} = j6$$
$$\sqrt{36} + \sqrt{-36} = 6 + j6$$

Imaginary numbers can be added or subtracted together, just as real numbers can be added or subtracted.

EXAMPLE 15.3: $\sqrt{-36} + \sqrt{-36}$.

Solution:

$$j6 + j6 = j12$$

Multiplication of real and complex numbers is performed as a simple multiplication of two numbers. The operator j is treated as though it were a literal number.

EXAMPLE 15.4: $\sqrt{-36} \cdot \sqrt{36}$.

Solution:

$$j6\cdot6 = j36$$

EXAMPLE 15.5: $\sqrt{-36} \cdot \sqrt{-36}$.

Solution:

$$j6\cdot j6 = j^2 36 = (-1)\cdot36 = -36$$

EXAMPLE 15.6: $j^2 6 \cdot j5$.

Solution:

$$j^2 6 = (-1)6 = -6$$
$$j^2 6 \cdot j5 = -6\cdot j5 = -j30$$

EXAMPLE 15.7: Find $\dfrac{\sqrt{-36}}{3}$.

Solution:

$$\frac{\sqrt{-36}}{3} = \frac{j6}{3} = j2$$

EXAMPLE 15.8: Find $\dfrac{j^2 36}{j9}$.

Solution:

$$\frac{j^2 36}{j9} = j4$$

EXAMPLE 15.9: Find $\dfrac{j24}{j^2 6}$.

Solution:

$$-j4$$

When we divide j by j^2, we get $-j$. That is, $\dfrac{1}{j} = j^{-1} = -j$.

EXAMPLE 15.10: $j6+j^212+j^32$.

Solution:

$$j^212=-12$$
and
$$j^32=-j2$$
Then
$$j6+j^212+j^32=j6-12-j2=-12+j4$$

PRACTICE PROBLEMS 15-1   Perform the indicated operations:

1. $\sqrt{-81}+\sqrt{-81}$

2. $\sqrt{-36}+\sqrt{-49}$

3. $\sqrt{-49}-\sqrt{-16}$

4. j^3+j^23

5. $j^26\cdot j4$

6. $\dfrac{\sqrt{81}}{\sqrt{-9}}$

7. $\dfrac{j^372}{j8}$

8. $j^24+j^35-j^26+j3$

Solutions:

1. $j9+j9=j18$

2. $j6+j7=j13$

3. $j7-j4=j3$

4. $-3-j$

5. $j^324=-j24$

6. $\dfrac{9}{j3}=-j3$

7. $j^29=-9$

8. $-4-j5+6+j3=2-j2$

Additional practice problems are at the end of the chapter.

15.2.1 Multiplication and Division of Complex Numbers.   We multiply complex numbers in the same way we multiply algebraic terms.

EXAMPLE 15.11: Find the product of $(2+j3)(2+j4)$.

Solution:

$$(2+j3)(2+j4)=4+j6+j8+j^212$$

$$=4+j14+(-1)12=-8+j14$$

We divide by multiplying the expression by the *conjugate* of the denominator. The conjugate of a complex number is the same number but the sign of the imaginary part is changed. The conjugate of $2+j3$ is $2-j3$. The conjugate of $4-j5$ is $4+j5$, and so on.

EXAMPLE 15.12: Perform the following division:

$$\frac{2+j3}{4-j6}$$

Solution: The conjugate of the denominator is $4+j6$. If we multiply by $\frac{4+j6}{4+j6}$, we are really multiplying by 1 since any number divided by itself is 1. We perform this multiplication in order to simplify the denominator.

$$\frac{2+j3}{4-j6} \times \frac{4+j6}{4+j6} = \frac{(2+j3)(4+j6)}{(4-j6)(4+j6)}$$

$$= \frac{8+j12+j12+j^2 18}{16-j24+j24-j^2 36} = \frac{-10+j24}{16-j^2 36}$$

$$= \frac{-10+j24}{52} = -\frac{10}{52} + j\frac{24}{52}$$

$$= -0.192 + j0.462$$

PRACTICE PROBLEMS 15-2   Perform the following multiplications:

1. $(3+j4)(3+j6)$ **2.** $(3+j2)(3-j3)$

3. $(4+j6)(4-j6)$ **4.** $(-2+j5)(3-j6)$

Perform the following divisions:

5. $\dfrac{2+j3}{3-j2}$ **6.** $\dfrac{6-j2}{2+j4}$

Solutions:

1. $-15+j30$ **2.** $15-j3$

3. 52 **4.** $24+j27$

5. j **6.** $\dfrac{1}{5} - j\dfrac{7}{5}$

Additional practice problems are at the end of the chapter.

15.2.2 Graphing Complex Numbers.   In Chapter 13 we developed a system of rectangular coordinates and graphed linear equations. We can develop a similar set of coordinates for plotting complex numbers. Such a set is shown in Figure 15-1. Notice that the operator j is plotted along the positive y-axis. j^2 (or -1) is plotted along the negative x-axis; j^3 or $-j$ is plotted along the negative

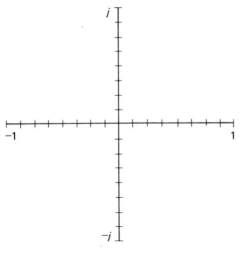

Figure 15-1.

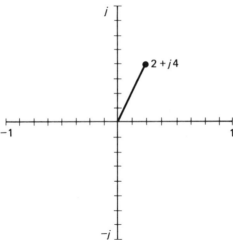

Figure 15-2.

y-axis. j^4 or 1 is plotted along the *x*-axis. Real numbers, then, are plotted along the *x*-axis and imaginary numbers are plotted along the *y*-axis.

Consider the complex number $2+j4$. A point is plotted on the set of coordinates in Figure 15-2 which represents this number. The line drawn from the origin to this point is called a *vector*. Thus, 2 and $j4$ are the rectangular coordinates of the number. The vector expresses the magnitude and direction. Such a graph has wide use in the solution of AC circuit problems and is discussed at length in Chapter 19.

PRACTICE PROBLEMS 15-3.   Graph the following complex numbers:

1. $6+j3$

2. $3-j6$

3. $-4+j3$

4. $-3-j5$

Solutions:

See Figure 15-3.

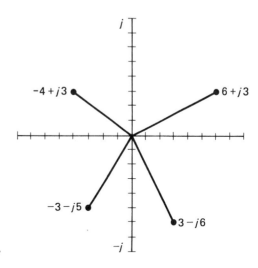

$-4+j3$

$6+j3$

$-3-j5$

$3-j6$

Figure 15-3.

SELF-TEST 15-1   Perform the indicated operations:

1. $j^2 + j^2 4$

2. $j^3 4 - j6$

3. $\dfrac{\sqrt{-81}}{\sqrt{9}}$

4. $\dfrac{\sqrt{121}}{\sqrt{-25}}$

5. $\dfrac{j^2 56}{j^3 8}$

6. $(6 - j3)(3 + j7)$

7. $(3 - j6)(4 - j2)$

8. $(4 + j3)(4 - j3)$

9. $\dfrac{3 - j2}{1 + j2}$

10. $\dfrac{2 + j5}{2 - j2}$

Answers to Self-Test 15-1 are at the end of the chapter.

END OF CHAPTER PROBLEMS 15-1

1. $\sqrt{-49} + \sqrt{-36}$

2. $\sqrt{-16} + \sqrt{-64}$

3. $\sqrt{-64} - \sqrt{-49}$

4. $\sqrt{-81} - \sqrt{-49}$

5. $j^3 4 + j^2 4$

6. $j^3 6 + j^2 5$

7. $j^3 8 - j4$

8. $j^3 6 - j5$

9. $j^2 6 \cdot j3$

10. $j^2 4 \cdot j5$

11. $j^3 4 \cdot j3$

12. $j^3 5 \cdot j5$

13. $\dfrac{\sqrt{-144}}{\sqrt{16}}$

14. $\dfrac{\sqrt{-64}}{\sqrt{4}}$

15. $\dfrac{\sqrt{100}}{\sqrt{-25}}$

16. $\dfrac{\sqrt{144}}{\sqrt{-16}}$

17. $\dfrac{j^3 42}{j6}$

18. $\dfrac{j^3 72}{j9}$

19. $\dfrac{j^2 100}{j^3 5}$

20. $\dfrac{j^2 64}{j^3 4}$

END OF CHAPTER PROBLEMS 15-2

Perform the following multiplications:

1. $(5+j2)(3-j2)$

2. $(6-j2)(2-j4)$

3. $(2+j3)(2-j3)$

4. $(4-j3)(4+j3)$

5. $(4-j3)(4-j3)$

6. $(6+j2)(6+j3)$

Perform the following divisions:

7. $\dfrac{5+j3}{1+j4}$

8. $\dfrac{1+j}{1-j}$

9. $\dfrac{4+j4}{3-j}$

10. $\dfrac{4-j5}{3-j5}$

11. $\dfrac{1+j2}{5+j2}$

12. $\dfrac{2-j3}{3+j2}$

13. $\dfrac{2-j5}{-3+j4}$

14. $\dfrac{-8-j2}{4-j5}$

ANSWERS TO SELF-TEST

Self-Test 15-1

1. -5

2. $-j10$

3. $j3$

4. $-j2.2$

5. $-j7$

6. $39+j33$

7. $0-j30$

8. 25

9. $-0.2-j1.6$

10. $-0.75+j1.75$

CHAPTER 16

THE RIGHT TRIANGLE

There are three kinds of triangles: those that contain an obtuse angle and two acute angles; those that contain three acute angles; and those that contain a right angle and two acute angles. Examples of the three triangles are shown in Figure 16-1. In any triangle the sum of the three angles equals 180°.

We will limit our study of triangles to the right triangle because the relationship between resistance and reactance or between conductance and susceptance always results in a right triangle. Also, AC waveforms can be evaluated mathematically by using the right triangle.

16.1 SIDES AND ANGLES

Figure 16-2 is a right triangle shown in two common positions. For purposes of discussion, the triangles are labeled so that A denotes angle A, B denotes angle B, and C denotes the 90°- or right angle. Side a lies opposite angle A, side b lies opposite angle B, and side c lies opposite angle C. In a right triangle the side opposite the right angle is called the *hypotenuse*. In Figure 16.2 side c is the hypotenuse.

Let's develop some basic relations between the sides and angles in Figure 16-2. First, *the sum of angles A and B equals 90°*. Second, *the hypotenuse is the side of greatest length*. Third, when comparing sides a and b, *the greater side lies opposite the greater angle*. Further, when discussing angle A, side b is the *adjacent* side (the side next to angle A) and side a is the *opposite* side. Of course, side c is also adjacent to angle A, but side c has already been defined as the hypotenuse.

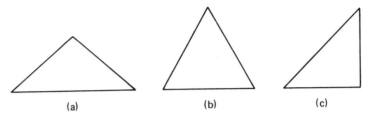

Figure 16-1. (a) Triangle with one obtuse and two acute angles; (b) triangle with three acute angles; and (c) triangle with one right angle and two acute angles.

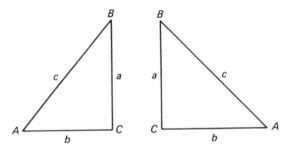

Figure 16-2. Right triangles.

Let's look at the right triangles in Figure 16-3. In Figure 16-3(a), angle A is 30°; therefore, angle B must equal 60° since the sum of angles A and B must equal 90°. The greater side lies opposite the greater angle. Therefore, since angle B is greater than angle A, side b is greater than side a. Since the hypotenuse is always the side of greatest length, our comparison of length is limited to the relationship between sides a and b. In Figure 16-3(b), angle A is the greater angle; therefore, side a is greater than side b. Finally, in Figure 16-3(c), B is 40°; therefore, angle A must be 50° making side a the greater side.

Of course, not all triangles are labeled as in Figure 16-3 and not all triangles are shown in this position. Notice in Figure 16-4(a) that the hypotenuse is labeled "Z" and the angles are θ (the Greek letter *theta*) and ϕ (the Greek letter

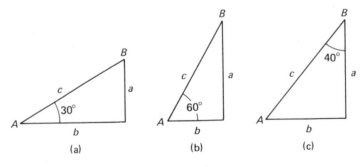

Figure 16-3.

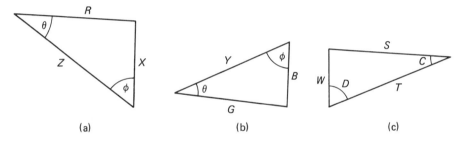

Figure 16-4.

phi). Side R lies opposite angle ϕ and is adjacent to angle θ. Side X lies opposite angle θ and is adjacent to angle ϕ. In Figure 16-4(b), side Y is the hypotenuse. Side G lies opposite angle ϕ and is adjacent to angle θ. Side B lies opposite angle θ and is adjacent to angle ϕ.

In Figure 16-4(c), side T is the hypotenuse. Side W lies opposite angle C and is adjacent to angle D. Side S lies opposite angle D and is adjacent to angle C. One should be able to identify the relationship between sides and angles regardless of the position of the triangle.

PRACTICE PROBLEMS 16-1

1. In the triangles in Figure 16-5 identify which of sides a and b is greater. Determine the size of the unknown angles.

2. Refer to Figure 16-6(a).
   (a) Which side is adjacent to angle θ?
   (b) Which side lies opposite angle ϕ?
   (c) Identify the hypotenuse.

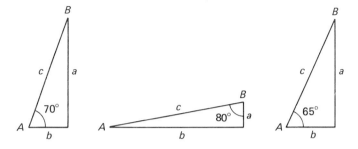

Figure 16-5.

3. Refer to Figure 16-6(b).
   (a) Which side lies opposite angle B?
   (b) Side m is adjacent to which angle?
   (c) Identify the hypotenuse.

4. Refer to Figure 16-6(c).
   (a) Side G is opposite which angle?
   (b) Which side is adjacent to angle θ?
   (c) Identify the hypotenuse.

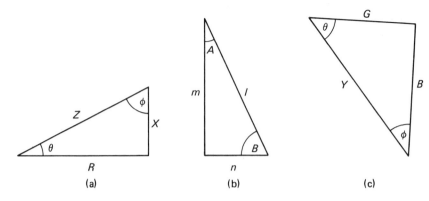

Figure 16-6.

Solutions:

1. In Figure 16-5(a) angle B is 20°. Side a is greater than side b. In (b) angle A is 10°; therefore, side b is greater. In (c) angle B is 25°; therefore, side a is greater.

2. (a) Side R is adjacent to angle θ.
(b) Side R is opposite angle ϕ.
(c) Side Z is the hypotenuse.

3. (a) Side m lies opposite angle B.
(b) Side m is adjacent to angle A.
(c) Side l is the hypotenuse.

4. (a) Side G is opposite angle ϕ.
(b) Side G is adjacent to angle θ.
(c) Side Y is the hypotenuse.

Additional practice problems are at the end of the chapter.

16.2 PYTHAGOREAN THEOREM

If two sides of a right triangle are known, the third side can be found by using the Pythagorean theorem which states that "in a right triangle, the square of the hypotenuse is equal to the sum of the squares of the other two sides." Let's express that statement mathematically for the triangle in Figure 16-7.

$$c^2 = a^2 + b^2 \tag{16-1}$$

Taking the square root of both sides, we get:

$$c = \sqrt{a^2 + b^2} \tag{16-2}$$

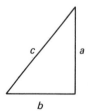

Figure 16-7.

EXAMPLE 16.1: Suppose sides a and b in Figure 16-7 are of length 6 and 8, respectively. What is the length of the hypotenuse?

Solution:

$$c = \sqrt{a^2 + b^2} = \sqrt{6^2 + 8^2} = \sqrt{36 + 64} = \sqrt{100} = 10$$

EXAMPLE 16.2: If side $a = 5$ and side $c = 8$ in Figure 16-7, what is the value of side b?

Solution:

$$c^2 = a^2 + b^2 \tag{16-1}$$
$$b^2 = c^2 - a^2$$
$$b = \sqrt{c^2 - a^2} \tag{16-3}$$
$$b = \sqrt{8^2 - 5^2} = \sqrt{64 - 25} = \sqrt{39} = 6.24$$

EXAMPLE 16.3: If side $b = 75$ and side $c = 150$ in Figure 16-7, what is the value of side a?

Solution:

$$c^2 = a^2 + b^2 \tag{16-1}$$
$$a^2 = c^2 - b^2$$
$$a = \sqrt{c^2 - b^2} \tag{16-4}$$
$$a = \sqrt{150^2 - 75^2} = \sqrt{16{,}875} = 130$$

In an AC circuit the sides of the triangle could represent circuit resistance (R), reactance (X), and impedance (Z), whose unit of measure is the ohm. Z is the hypotenuse; therefore, the equation is:

$$Z^2 = R^2 + X^2 \tag{16-5}$$

EXAMPLE 16.4: Refer to Figure 16-8. Find X if $R = 68$ kΩ and $Z = 80$ kΩ.

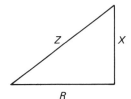

Figure 16-8.

Solution:

$$Z^2 = R^2 + X^2$$
$$X^2 = Z^2 - R^2$$
$$X = \sqrt{Z^2 - R^2} = \sqrt{(80 \text{ k}\Omega)^2 - (68 \text{ k}\Omega)^2} = 42.1 \text{ k}\Omega$$

The triangle could also represent circuit conductance (G), susceptance (B), and admittance (Y) whose unit of measure is the Siemen. Y is the hypotenuse; therefore, the equation is:

$$Y^2 = G^2 + B^2 \qquad (16\text{-}6)$$

EXAMPLE 16.5: Refer to Figure 16-9. Find Y where $G = 400\,\mu\text{S}$ and $B = 600\ \mu\text{S}$.

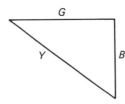

Figure 16-9.

Solution:

$$Y^2 = G^2 + B^2$$
$$Y = \sqrt{G^2 + B^2} = \sqrt{(400\ \mu\text{S})^2 + (600\ \mu\text{S})^2} = 721\ \mu\text{S}$$

PRACTICE PROBLEMS 16-2

1. Find the length of the unknown side in the triangles in Figure 16-10(a), (b), and (c). Determine whether angle A or angle B is greater.

2. In Figure 16-11 find the following:
   (a) Side X when $Z = 17.3$ kΩ and $R = 10$ kΩ.
   (b) Side R when $Z = 800$ Ω and $X = 500$ Ω.
   (c) Side Z when $R = 6.8$ kΩ and $X = 3$ kΩ.

3. In Figure 16-12 find the following:
   (a) Side Y when $B = 3$ mS and $G = 2$ mS.
   (b) Side B when $Y = 400$ μS and $G = 150$ μS.
   (c) Side G when $Y = 1.3$ mS and $B = 600$ μS.

Solutions:

1. For triangle (a), use Equation 16-4.
$$a = \sqrt{c^2 - b^2} = \sqrt{40^2 - 10^2} = 38.7$$
Angle A is greater than angle B because side a is greater than side b. Remember, the greater angle lies opposite the greater side. For triangle (b), use Equation 16-2.

$$c = \sqrt{a^2 + b^2} = \sqrt{60^2 + 60^2} = 84.9$$

Angles A and B are equal because the sides are equal. For triangle (c), use Equation 16-3.
$$b = \sqrt{c^2 - a^2} = \sqrt{14^2 - 8^2} = 11.5$$

Side b is greater than side a; therefore, angle B is greater than angle A.

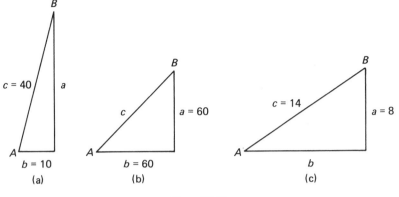

Figure 16-10.

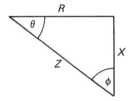

Figure 16-11.

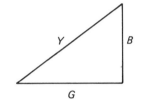

Figure 16-12.

2. (a) $x = 14.1$ kΩ
   (b) $R = 624$ Ω
   (c) $Z = 7.43$ kΩ

3. (a) $Y = 3.61$ mS
   (b) $B = 371$ μS
   (c) $G = 1.15$ mS

Additional practice problems are at the end of this chapter.

16.3 TRIGONOMETRIC FUNCTIONS

The Pythagorean theorem has its limitations. We must know two sides in order to find the third side, and we can't determine the actual size of the angles.

We could find the length of the sides *and* the angles by using graph paper, a protractor, and a rule. Let's draw a line 3 units long, as in Figure 16-13. Let this

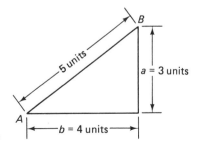

Figure 16-13.

be side *a*. Another line is drawn at right angles to side *a* and is 4 units in length (side *b*). If we complete the triangle and measure the length of side *c* (the hypotenuse), we will find that side *c* is 5 units as measured with the rule. $\angle A$ would equal approximately 37° and $\angle B$ would equal approximately 53° as determined by the protractor.

If we made side *a* 6 units long and side *b* 8 units long, how long would the hypotenuse be? What would $\angle A$ and $\angle B$ equal? If the length of *a* and *b* is doubled, the length of the hypotenuse would also double. This could be demonstrated either graphically (as in Figure 16-14) or by using the Pythagorean theorem. Again using the protractor, we could measure $\angle A$ and $\angle B$. They would still equal approximately 37° and 53°. They did not change because the relative lengths of the sides did not change. In our examples the ratio of side *a* to side *b* was 4 to 3 and 8 to 6 which in both cases is 1.33 to 1. 4 to 3 = 1.33 and 8 to 6 = 1.33. As long as the ratio of side *a* to side *b* is 1.33 to 1, that is, side *a* is 1.33 times longer than side *b*, $\angle A$ is 37°. We could expand on this by selecting other lengths of sides *a* and *b* and measuring the angles and the hypotenuse. In this manner, we could determine the approximate value for angles *A* and *B* for any ratio of side *a* to side *b*. We must consider the angles measured with the protractor as approximate since most protractors are not precision instruments.

It is rarely necessary to solve problems graphically though, because mathematicians have prepared tables of these and other ratios for us to use. These tables are called *trigonometric tables*.

Since a triangle has three sides, there are three possible trigonometric ratios. In Figure 16-15 these ratios would be: side *a* to side *b* (as previously discussed); side *a* to side *c*; and side *b* to side *c*. These trigonometric ratios are called *functions* and are identified as *sine, cosine,* and *tangent*. (Actually, there are three additional functions—secant, cosecant, and cotangent which are reciprocals of the sine, cosine, and tangent functions. But we will confine our discussion to the sine, cosine, and tangent functions since they are the only ones we need to solve for angles and sides of a right triangle.)

The sine of an angle $= \dfrac{\text{opposite side}}{\text{hypotenuse}}$ (16-7)

$\sin\theta = \dfrac{a}{c}$

The cosine of an angle $= \dfrac{\text{adjacent side}}{\text{hypotenuse}}$ (16-8)

$\cos\theta = \dfrac{b}{c}$

The tangent of an angle $= \dfrac{\text{opposite side}}{\text{adjacent side}}$ (16-9)

$\tan\theta = \dfrac{a}{b}$

The sine function expresses the ratio of the length of the opposite side to the length of the hypotenuse. The equation would read: $\sin\theta = \dfrac{a}{c}$. The cosine

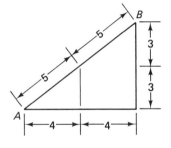

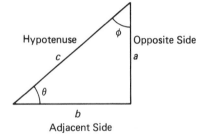

Figure 16-14.   Figure 16-15.

function expresses the ratio of the length of the adjacent side to the length of the hypotenuse. The equation would read: $\cos\theta = \dfrac{b}{c}$. The tangent function expresses the ratio of the length of the opposite side to the length of the adjacent side. The equation would read: $\tan\theta = \dfrac{a}{b}$.

PRACTICE PROBLEMS 16-3

1. Refer to Figure 16-16(a). Write the equation for the three functions with reference to the given angle.

2. Refer to Figure 16-16(b). Write the equation for the three functions with reference to the given angle.

3. Refer to Figure 16-17(a). Write the equations for the three functions with reference to the given angle.

4. Refer to Figure 16-17(b). Write the equations for the three functions with reference to the given angle.

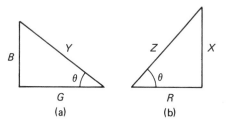

(a)   (b)

Figure 16-16.

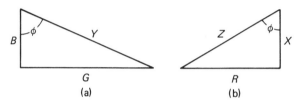

(a)   (b)

Figure 16-17.

Solutions:

In Figure 16-16(a) side G is the adjacent side, side B is the opposite side, and side Y is the hypotenuse when considering the angle θ. The sine function expresses the ratio of the opposite side to the hypotenuse; therefore, $\sin\theta = \dfrac{B}{Y}$. The cosine function expresses the ratio of the adjacent side to the hypotenuse; therefore $\cos\theta = \dfrac{G}{Y}$. Finally, $\tan\theta = \dfrac{B}{G}$ since the tangent expresses the ratio of the opposite side to the adjacent side.

In Figure 16-16(b), side R is the adjacent side, side X is the opposite side, and side Z is the hypotenuse when considering the angle θ. Therefore, $\sin\theta = \dfrac{X}{Z}$, $\cos\theta = \dfrac{R}{Z}$, and $\tan\theta = \dfrac{X}{R}$.

If we determined that in Figure 16-17(a) the adjacent side is side B and the opposite side is side G, we are right. Then, $\sin\phi = \dfrac{G}{Y}$, $\cos\phi = \dfrac{B}{Y}$, and $\tan\phi = \dfrac{G}{B}$.

You should have determined that in Figure 16-17(b) the adjacent side is side X and the opposite side is side R. Then, $\sin\phi = \dfrac{R}{Z}$, $\cos\phi = \dfrac{X}{Z}$, and $\tan\phi = \dfrac{R}{X}$.

Additional practice problems are at the end of the chapter.

SELF-TEST 16-1   Refer to Figure 16-18.

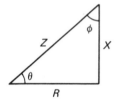

Figure 16-18.

1. $\angle\phi = 47.3°$. Find θ.

2. If $\angle\theta = 37.5°$, which side is longer?

3. If $Z = 73.6$ kΩ and $R = 47$ kΩ, find X.

4. In problem 3, which angle is greater?

5. Write the equations for the three functions with reference to $\angle\theta$.

Answers to Self-Test 16-1 are at the end of the chapter.

16.4 TRIGONOMETRIC TABLES

Now that we understand what a trig function is, let's see how we can use trig tables to find unknown angles or ratios.

Table 16-1 is a partial table of the trig functions. Table A in Appendix A is a complete table of trig functions.

Let's take a look at Table 16-1. It shows angles from 30° to 60° and the sine, cosine, and tangent functions. What is the tangent of a 40°-angle? First, we find 40° and then we find the *tan* column. Tan 40° = 0.8391. What is tan 50°? Tan 50° = 1.1918. What is the sine of 35°? This time we find 35° and locate the sine

TABLE 16-1. TRIGONOMETRIC FUNCTIONS
(PARTIAL TABLE)

Angle°	sin	tan	cot	cos	Angle°
30.0	.50000	.57735	1.7321	.86603	**60.0**
.1	.50151	.57968	1.7251	.86515	.9
.2	.50302	.58201	1.7182	.86427	.8
.3	.50453	.58435	1.7113	.86340	.7
.4	.50603	.58670	1.7045	.86251	.6
.5	.50754	.58905	1.6977	.86163	.5
.6	.50904	.59140	1.6909	.86074	.4
.7	.51054	.59376	1.6842	.85985	.3
.8	.51204	.59612	1.6775	.85895	.2
.9	.51354	.59849	1.6709	.85806	.1
31.0	.51504	.60086	1.6643	.85717	**59.0**
.1	.51653	.60324	1.6577	.85627	.9
.2	.51803	.60562	1.6512	.85536	.8
.3	.51952	.60801	1.6447	.85446	.7
.4	.52101	.61040	1.6383	.85355	.6
.5	.52250	.61280	1.6319	.85264	.5
.6	.52399	.61520	1.6255	.85173	.4
.7	.52547	.61761	1.6191	.85081	.3
.8	.52697	.62003	1.6128	.84989	.2
.9	.52844	.62245	1.6066	.84897	.1
32.0	.52992	.62487	1.6003	.84805	**58.0**
.1	.53140	.62730	1.5941	.84712	.9
.2	.53288	.62973	1.5880	.84619	.8
.3	.53435	.63217	1.5818	.84526	.7
.4	.53583	.63462	1.5757	.84433	.6
.5	.53730	.63707	1.5697	.84339	.5
.6	.53877	.63953	1.5637	.84245	.4
.7	.54024	.64199	1.5577	.84151	.3
.8	.54171	.64446	1.5517	.84057	.2
.9	.54317	.64693	1.5458	.83962	.1
33.0	.54464	.64941	1.5399	.83867	**57.0**
.1	.54610	.65189	1.5340	.83772	.9
.2	.54756	.65438	1.5282	.83676	.8
.3	.54902	.65688	1.5224	.83581	.7
.4	.55048	.65938	1.5166	.83485	.6
.5	.55194	.66189	1.5108	.83389	.5
.6	.55339	.66440	1.5051	.83292	.4
.7	.55484	.66692	1.4994	.83195	.3
.8	.55630	.66944	1.4938	.83098	.2
.9	.55775	.67197	1.4882	.83001	.1
34.0	.55919	.67451	1.4826	.82904	**56.0**
.1	.56064	.67705	1.4770	.82806	.9
.2	.56208	.67960	1.4715	.82708	.8
.3	.56353	.68215	1.4659	.82610	.7
.4	.56497	.68471	1.4605	.82511	.6
.5	.56641	.68728	1.4550	.82413	.5
.6	.56784	.68985	1.4496	.82314	.4
.7	.56928	.69243	1.4442	.82214	.3
.8	.57071	.69502	1.4388	.82115	.2
.9	.57215	.69761	1.4335	.82015	.1
	cos	cot	tan	sin	

TABLE 16-1. (Cont'd.)

Angle°	sin	tan	cot	cos	Angle°
35.0	.57358	.70021	1.4281	.81915	**55.0**
.1	.57501	.70281	1.4229	.81815	.9
.2	.57643	.70542	1.4176	.81714	.8
.3	.57786	.70804	1.4124	.81614	.7
.4	.57928	.71066	1.4071	.81513	.6
.5	.58070	.71329	1.4019	.81412	.5
.6	.58212	.71593	1.3968	.81310	.4
.7	.58354	.71857	1.3916	.81208	.3
.8	.58496	.72122	1.3865	.81106	.2
.9	.58637	.72388	1.3814	.81004	.1
36.0	.58779	.72654	1.3764	.80902	**54.0**
.1	.58920	.72921	1.3713	.80799	.9
.2	.59061	.73189	1.3663	.80696	.8
.3	.59201	.73457	1.3613	.80593	.7
.4	.59342	.73726	1.3564	.80489	.6
.5	.59482	.73996	1.3514	.80386	.5
.6	.59622	.74267	1.3465	.80282	.4
.7	.59763	.74538	1.3416	.80178	.3
.8	.59902	.74810	1.3367	.80073	.2
.9	.60042	.75082	1.3319	.79968	.1
37.0	.60182	.75355	1.3270	.79864	**53.0**
.1	.60321	.75629	1.3222	.79758	.9
.2	.60460	.75904	1.3175	.79653	.8
.3	.60599	.76180	1.3127	.79547	.7
.4	.60738	.76456	1.3079	.79441	.6
.5	.60876	.76733	1.3032	.79335	.5
.6	.61015	.77010	1.2985	.79229	.4
.7	.61153	.77289	1.2938	.79122	.3
.8	.61291	.77568	1.2892	.79016	.2
.9	.61429	.77848	1.2846	.78908	.1
38.0	.61566	.78129	1.2799	.78801	**52.0**
.1	.61704	.78410	1.2753	.78694	.9
.2	.61841	.78692	1.2708	.78586	.8
.3	.61978	.78975	1.2662	.78478	.7
.4	.62115	.79259	1.2617	.78369	.6
.5	.62251	.79544	1.2572	.78261	.5
.6	.62388	.79829	1.2527	.78152	.4
.7	.62524	.80115	1.2482	.78043	.3
.8	.62660	.80402	1.2437	.77934	.2
.9	.62796	.80690	1.2393	.77824	.1
39.0	.62932	.80978	1.2349	.77715	**51.0**
.1	.63068	.81268	1.2305	.77605	.9
.2	.63203	.81558	1.2261	.77494	.8
.3	.63338	.81849	1.2218	.77384	.7
.4	.63473	.82141	1.2174	.77273	.6
.5	.63608	.82434	1.2131	.77162	.5
.6	.63742	.82727	1.2088	.77051	.4
.7	.63877	.83022	1.2045	.76940	.3
.8	.64011	.83317	1.2002	.76828	.2
.9	.64145	.83613	1.1960	.76717	.1
	cos	cot	tan	sin	

TABLE 16-1. (Cont'd.)

Angle°	sin	tan	cot	cos	Angle°
40.0	.64279	.83910	1.1918	.76604	**50.0**
.1	.64412	.84208	1.1875	.76492	.9
.2	.64546	.84507	1.1833	.76380	.8
.3	.64679	.84806	1.1792	.76267	.7
.4	.64812	.85107	1.1750	.76154	.6
.5	.64945	.85408	1.1708	.76041	.5
.6	.65077	.85710	1.1667	.75927	.4
.7	.65210	.86014	1.1626	.75813	.3
.8	.65342	.86318	1.1585	.75700	.2
.9	.65474	.86623	1.1544	.75585	.1
41.0	.65606	.86929	1.1504	.75471	**49.0**
.1	.65738	.87236	1.1463	.75356	.9
.2	.65869	.87543	1.1423	.75241	.8
.3	.66000	.87852	1.1383	.75126	.7
.4	.66131	.88162	1.1343	.75011	.6
.5	.66262	.88473	1.1303	.74896	.5
.6	.66393	.88784	1.1263	.74780	.4
.7	.66523	.89097	1.1224	.74664	.3
.8	.66653	.89410	1.1184	.74548	.2
.9	.66783	.89725	1.1145	.74431	.1
42.0	.66913	.90040	1.1106	.74314	**48.0**
.1	.67043	.90357	1.1067	.74198	.9
.2	.67172	.90674	1.1028	.74080	.8
.3	.67301	.90993	1.0990	.73963	.7
.4	.67430	.91313	1.0951	.73846	.6
.5	.67559	.91633	1.0913	.73728	.5
.6	.67688	.91955	1.0875	.73610	.4
.7	.67816	.92277	1.0837	.73491	.3
.8	.67944	.92601	1.0799	.73373	.2
.9	.68072	.92926	1.0761	.73254	.1
43.0	.68200	.93252	1.0724	.73135	**47.0**
.1	.68327	.93578	1.0686	.73016	.9
.2	.68455	.93906	1.0649	.72897	.8
.3	.68582	.94235	1.0612	.72777	.7
.4	.68709	.94565	1.0575	.72657	.6
.5	.68835	.94896	1.0538	.72537	.5
.6	.68962	.95229	1.0501	.72417	.4
.7	.69088	.95562	1.0464	.72294	.3
.8	.69214	.95897	1.0428	.72176	.2
.9	.69340	.96232	1.0392	.72055	.1
44.0	.69466	.96569	1.0355	.71934	**46.0**
.1	.69591	.96907	1.0319	.71813	.9
.2	.69717	.97246	1.0283	.71691	.8
.3	.69842	.97586	1.0247	.71569	.7
.4	.69966	.97927	1.0212	.71447	.6
.5	.70091	.98270	1.0176	.71325	.5
.6	.70215	.98613	1.0141	.71203	.4
.7	.70339	.98958	1.0105	.71080	.3
.8	.70463	.99304	1.0070	.70957	.2
.9	.70587	.99652	1.0035	.70834	.1
45.0	.70711	1.00000	1.0000	.70711	**45.0**
	cos	cot	tan	sin	

361

under the column marked *sin*. Sin 35° = 0.57358. What is the cosine of 54°? The cosine column is marked cos. Cos 54° = 0.58779. What is the sine of 33.7°? We note that each degree is divided into tenths of a degree. First we find 33° and then move down to .7. This is 7 tenths or 33.7. 33.7° = 0.55484.

PRACTICE PROBLEMS 16-4   Use the trig table to find the following. Verify your answers by using a calculator. The algorithm for finding trig functions by using a calculator is in Appendix A.

1. sin 42°	**2.** cos 58°
3. tan 28°	**4.** tan 45°
5. sin 45°	**6.** cos 45°
7. sin 21.6°	**8.** cos 48.4°
9. tan 50.1°	**10.** cos 55.6°
11. sin 55.6°	**12.** tan 31.9°

Solutions:

1. 0.66913	**2.** 0.52992
3. 0.53171	**4.** 1.0000
5. 0.70711	**6.** 0.70711
7. 0.36812	**8.** 0.66393
9. 1.1960	**10.** 0.56497
11. 0.82511	**12.** 0.62244

Additional practice problems are at the end of the chapter.

16.4.1 Inverse Trig Functions.

Suppose we know the length of the sides and need to find the angles. One of the basic relationships was:

$$\sin \theta = \frac{\text{opposite side}}{\text{hypotenuse}}$$

If $\theta = 30°$, then $\sin \theta = \sin 30° = 0.5$. This tells us that if the angle is 30°, the ratio of the opposite side to the hypotenuse is 0.5.

Now let's turn it around and say, "Given some ratio of sides, what is the angle?" For example, suppose the ratio of the opposite side to the hypotenuse is 0.74314. What is the angle? The solution requires the use of inverse trig functions. *Inverse trig functions* are written *arc sin*, *arc cos*, and *arc tan*. Or they are written $\sin^{-1}$, $\cos^{-1}$, and $\tan^{-1}$. Most of the time we will use $\sin^{-1}$, $\cos^{-1}$, and $\tan^{-1}$.

In our example the ratio was opposite side to hypotenuse, which indicates a sine function. Therefore, our inverse trig function is $\sin^{-1}$. $\sin^{-1} 0.74314 = \angle \theta$.

The expression $\sin^{-1} 0.74314$ asks the question, "What is the angle whose sin is 0.74314?" To find the answer, we follow along the sin column in Table 16-1 until we find 0.74314 and then we find the angle that corresponds to that value. In this case, the angle is 48°.

$$\sin^{-1} 0.74314 = 48°$$

EXAMPLE 16.6: Find the angle whose cosine is 0.90183.

Solution: $\cos^{-1} 0.90183 = 25.6°$

EXAMPLE 16.7: Find the angle whose tangent is 1.4994.

Solution: $\tan^{-1} 1.4994 = 56.3°$

PRACTICE PROBLEMS 16-5 Use the trig table in Appendix A to find the following angles. Verify your answers by using a calculator. The algorithm for finding inverse trig functions with a calculator is also in Appendix A.

1. $\sin^{-1} 0.86602$
2. $\sin^{-1} 0.70711$
3. $\sin^{-1} 0.35184$
4. $\cos^{-1} 0.83580$
5. $\tan^{-1} 1.7320$
6. $\cos^{-1} 0.25207$
7. $\cos^{-1} 0.5$
8. $\cos^{-1} 0.86602$
9. $\tan^{-1} 1.0$
10. $\tan^{-1} 0.57735$
11. $\tan^{-1} 1.2437$
12. $\sin^{-1} 0.22325$

Solutions:

1. $\sin^{-1} 0.86602 = 60°$
2. $\sin^{-1} 0.70711 = 45°$
3. $\sin^{-1} 0.35184 = 20.6°$
4. $\cos^{-1} 0.83580 = 33.3°$
5. $\tan^{-1} 1.7320 = 60°$
6. $\cos^{-1} 0.25207 = 75.4°$
7. $\cos^{-1} 0.5 = 60°$
8. $\cos^{-1} 0.86602 = 30°$
9. $\tan^{-1} 1.0 = 45°$
10. $\tan^{-1} 0.57735 = 30°$
11. $\tan^{-1} 1.2437 = 51.2°$
12. $\sin^{-1} 0.22325 = 12.9°$

Additional practice problems are at the end of the chapter.

16.5 TRIGONOMETRIC EQUATIONS

Now let's see how we can use both trig functions and inverse trig functions to find unknown sides and angles of a right triangle. Consider the triangle in Figure 16-15. Let's assume that side b equals 25 and angle $\theta = 40°$. How long are sides a and c and what is the angle ϕ? First, let's write the equations of the three functions.

$$\sin\theta = \frac{\text{opposite side}}{\text{hypotenuse}} = \frac{a}{c} \qquad (16\text{-}7)$$

$$\cos\theta = \frac{\text{adjacent side}}{\text{hypotenuse}} = \frac{b}{c} \qquad (16\text{-}8)$$

$$\tan\theta = \frac{\text{opposite side}}{\text{adjacent side}} = \frac{a}{b} \qquad (16\text{-}9)$$

As with any linear equation, if we know two of the quantities, we can manipulate the equation to solve for the unknown. In our problem we know the length of side b and the angle θ. An examination of the equations shows that to find side c, the hypotenuse, we would use the cosine function. To find the opposite side, we would use the tangent function. To find the angle ϕ, we would simply subtract the angle θ from $90°$. $\phi = 90° - 40°$. Let's find the hypotenuse first.

$$\cos\theta = \frac{b}{c} \qquad (16\text{-}8)$$

$$c = \frac{b}{\cos\theta} \qquad (16\text{-}10)$$

$Cos\,\theta$ is one quantity and cannot be separated. That is, we could not move "cos" and leave θ as one member of the equation. And we could not move "θ" and leave cos as one member. $Cos\,\theta$ would move from one member to the other. In Equation 16.8, since c is unknown, we solve for c by multiplying both members by c and dividing both members by $\cos\theta$. Plugging in known values, we get:

$$c = \frac{25}{\cos 40°} = \frac{25}{0.766} = 32.6$$

Even though we learned how to use the trig tables, most of our problem solving is done with the calculator. Since the calculator has a trig table stored in its memory, we need the table in the book only as a reference. Since three-place accuracy in answers is almost always close enough, we will continue that practice here. The algorithm for solving this problem by using a calculator is in Appendix A.

Now let's find side a. Since we now know θ, side b, and the hypotenuse, we have a choice. We can use the tangent function or we can use the sine function. We will do it both ways.

$$\tan\theta = \frac{a}{b} \qquad (16\text{-}9)$$

Solve for a:

$$a = b\tan\theta \qquad (16\text{-}11)$$
$$a = 25\tan 40° = 25 \times 0.839 = 21$$

or
$$\sin\theta = \frac{a}{c} \tag{16-7}$$

Solve for a:

$$a = c\sin\theta$$
$$a = 32.6\sin 40° = 32.6 \times 0.643 = 21 \tag{16-12}$$
$$\angle\phi = 90° - \angle\theta = 90° - 40° = 50°$$

EXAMPLE 16.8: Side $a = 150$ and side $b = 300$ in Figure 16-15. Find the hypotenuse and $\angle\phi$.

Solution: Only the tangent function contains one unknown.

$$\tan\phi = \frac{b}{a}$$

$$\tan\phi = \frac{300}{150} = 2$$

Now that we know that $\tan\phi = 2$, our next step is to find the angle whose tangent is 2.

$$\tan^{-1}2 = \phi = 63.4°$$

We were given sides a and b and we found $\angle\phi$. We can find the hypotenuse by using the Pythagorean theorem:

$$c = \sqrt{a^2 + b^2} = \sqrt{1.125 \times 10^5} = 335$$

Another solution would use the sine function:

$$\sin\phi = \frac{b}{c}$$

$$c = \frac{b}{\sin\phi} = \frac{300}{\sin 63.4°} = 335$$

A third solution would use the cosine function:

$$\cos\phi = \frac{a}{c}$$

$$c = \frac{a}{\cos\phi} = \frac{150}{\cos 63.4°} = 335$$

Since all solutions yield the same result, they are all valid.

EXAMPLE 16.9: Refer to Figure 16-18. Find R, Z, and $\angle \phi$ if $X = 2.83$ kΩ and $\theta = 33.4°$.

Solution:

$$\tan \theta = \frac{X}{R}$$

$$R = \frac{X}{\tan \theta} = \frac{2.83 \text{ k}\Omega}{\tan 33.4°} = \frac{2.83 \text{ k}\Omega}{0.659} = 4.29 \text{ k}\Omega$$

$$\sin \theta = \frac{X}{Z}$$

$$Z = \frac{X}{\sin \theta} = \frac{2.83 \text{ k}\Omega}{\sin 33.4°} = 5.14 \text{ k}\Omega$$

$$\phi = 90° - \theta = 90° - 33.4° = 56.6°$$

EXAMPLE 16.10: Refer to Figure 16-17(a). Find B, θ, and ϕ if $Y = 670$ μS and $G = 430$ μS.
Find θ:

Solution:

$$\cos \theta = \frac{G}{Y}$$

$$\cos \theta = \frac{430 \text{ }\mu\text{S}}{670 \text{ }\mu\text{S}} = 0.642$$

$$\cos^{-1} 0.642 = \theta = 50.1°$$
$$\phi = 90° - \theta = 90° - 50.1° = 39.9°$$

We will use the sine function to find B.

$$\sin \theta = \frac{B}{Y}$$

$$Y \sin \theta = B$$
$$670 \text{ }\mu\text{S} \sin 50.1° = 670 \text{ }\mu\text{S} \times 0.767 = 514 \text{ }\mu\text{S}$$

PRACTICE PROBLEMS 16-6

1. Refer to Figure 16-15.
   (a) $a = 25.5$, $\theta = 20°$. Find b, c, and ϕ.
   (b) $b = 1400$, $\theta = 65.4°$. Find a, c, and ϕ.
   (c) $a = 350$, $b = 500$. Find θ and c.
   (d) $a = 7.6$, $c = 11.2$. Find θ and b.
   (e) $a = 120$, $\theta = 45°$. Find b and c.
   (f) $\phi = 40°$, $c = 850$. Find a and b.

2. Refer to Figure 16-18.
   (a) $X = 700$ Ω, $R = 900$ Ω. Find Z and θ.
   (b) $\theta = 68.3°$, $X = 30$ kΩ. Find R and Z.
   (c) $Z = 73.6$ kΩ, $R = 56$ kΩ. Find X and θ.
   (d) $\theta = 27°$, $R = 300$ Ω. Find X and Z.
   (e) $Z = 12.3$ kΩ, $X = 8.77$ kΩ. Find R and θ.
   (f) $\theta = 30°$, $Z = 3$ kΩ. Find X and R.

3. Refer to Figure 16-17(a).
(a) $G = 2.8$ mS, $B = 3.65$ mS. Find Y and θ.
(b) $\theta = 30°$, $B = 250$ μS. Find G and Y.
(c) $G = 10.3$ mS, $Y = 15.8$ mS. Find B and θ.
(d) $\theta = 52°$, $Y = 650$ μS. Find G and B.
(e) $B = 140$ μS, $Y = 270$ μS. Find G and θ.
(f) $\theta = 45°$, $G = 100$ μS. Find B and Y.

Solutions:

1. (a) $b = 70$, $c = 74.5$, $\phi = 70°$
(b) $a = 3060$, $c = 3360$, $\phi = 24.6°$
(c) $c = 610$, $\theta = 35°$
(d) $b = 8.23$, $\theta = 42.7°$
(e) $b = 120$, $c = 170$
(f) $a = 651$, $b = 546$

2. (a) $Z = 1.14$ kΩ, $\theta = 37.9°$
(b) $R = 11.9$ kΩ, $Z = 32.3$ kΩ
(c) $X = 47.8$ kΩ, $\theta = 40.5°$
(d) $X = 153$ Ω, $Z = 337$ Ω
(e) $R = 8.62$ kΩ, $\theta = 45.5°$
(f) $R = 2.6$ kΩ, $X = 1.5$ kΩ

3. (a) $Y = 4.6$ mS, $\theta = 52.5°$
(b) $G = 433$ μS, $Y = 500$ μS
(c) $B = 12$ mS, $\theta = 49.3°$
(d) $B = 512$ μS, $G = 400$ μS
(e) $G = 231$ μS, $\theta = 31.2°$
(f) $B = 100$ μS, $Y = 141$ μS

SELF-TEST 16-2

1. Find the sine, cosine, and tangent of $62.4°$.

2. Find $\sin^{-1} 0.74548$.

3. Find $\cos^{-1} 0.75927$.

4. Find $\tan^{-1} 1.42815$.

5. Refer to Figure 16-15. Find b and θ if $a = 27$ and $c = 42.3$.

6. Refer to Figure 16-18. Find Z and θ if $R = 2.7$ kΩ and $X = 3.85$ kΩ.

7. Refer to Figure 16-17(a). Find B and Y if $G = 220$ μS and $\theta = 38.2°$.

Answers to Self-Test 16-2 are at the end of the chapter.

END OF CHAPTER PROBLEMS 16-1

1. Refer to Figure 16-18.
(a) If $\angle\theta = 35°$, find $\angle\phi$.
(b) Which side (R or X) is greater?
(c) Which side lies opposite $\angle\phi$?
(d) Which side is adjacent to $\angle\theta$?

2. Refer to Figure 16-17(a).
(a) If $\angle\phi = 70°$, find $\angle\theta$.
(b) Which side (G or B) is greater?
(c) Which side lies opposite $\angle\theta$? Which side is adjacent to $\angle\phi$?

3. Refer to Figure 16-18.
(a) Find Z when $R = 2.7$ kΩ and $X = 4$ kΩ.
(b) Find R when $Z = 600$ Ω and $X = 350$ Ω.
(c) Find X when $Z = 120$ kΩ and $R = 68$ kΩ.

4. Refer to Figure 16-18.
(a) Find Z when $R = 20$ kΩ and $X = 25$ kΩ.
(b) Find R when $Z = 600$ kΩ and $X = 280$ kΩ.
(c) Find X when $Z = 11.7$ kΩ and $R = 7.8$ kΩ.

5. Refer to Figure 16-17(a).
   (a) Find Y when $G=1.83$ mS and B $=2.37$ mS.
   (b) Find G when $Y=1$ mS and $B=600$ μS.
   (c) Find B when $Y=460$ μS and $G=175$ μS.

6. Refer to Figure 16-17(a).
   (a) Find Y when $G=2.8$ mS and $B=4.35$ mS.
   (b) Find G when $Y=170$ μS and $B=95$ μS.
   (c) Find B when $Y=10.7$ mS and G $=6.73$ mS.

END OF CHAPTER PROBLEMS 16-2

1. Refer to Figure 16-18. Write the equation for the three functions with references to (a) $\angle\theta$ and (b) $\angle\phi$.

2. Refer to Figure 16-17(a). Write the equation for the three functions with reference to (a) $\angle\theta$ and (b) $\angle\phi$.

END OF CHAPTER PROBLEMS 16-3

Use the trig table in Appendix A to find the following. Verify your answers by using the calculator.

1. $\sin 27.3°$

2. $\sin 53.2°$

3. $\cos 78.3°$

4. $\cos 41.7°$

5. $\tan 67.3°$

6. $\tan 12.7°$

7. $\sin^{-1} 0.39715$

8. $\sin^{-1} 0.95106$

9. $\cos^{-1} 0.55630$

10. $\cos^{-1} 0.45088$

11. $\tan^{-1} 1.47146$

12. $\tan^{-1} 0.71329$

END OF CHAPTER PROBLEMS 16-4

Refer to Figure 16-15.

1. $a=25$, $b=40$. Find θ, ϕ, and c.

2. $a=67.5$, $b=33$. Find θ, ϕ, and c.

3. $a=200$, $\phi=35°$. Find θ, b, and c.

4. $a=6.35$, $\phi=61°$. Find θ, b, and c.

5. $c=17.3$, $\theta=40°$. Find a and b.

6. $c=680$, $\theta=78°$. Find a and b.

Refer to Figure 16-18.

7. $X=17.6$ kΩ, $\theta=22.3°$. Find R and Z.

8. $X=27.8$ kΩ, $\theta=53°$. Find R and Z.

9. $R=5.6$ kΩ, $X=10$ kΩ. Find Z and θ.

10. $R=100$ kΩ, $X=200$ kΩ. Find Z and θ.

11. $R=20$ kΩ, $\theta=30°$. Find X and Z.

12. $R=3.3$ kΩ, $\theta=60°$. Find X and Z.

13. $Z=40$ kΩ, $\theta=67.5°$. Find R and X.

14. $Z=6.2$ kΩ, $\theta=17.8°$. Find R and X.

15. $R=12$ kΩ, $Z=22$ kΩ. Find X and θ.

16. $R=39$ kΩ, $Z=65$ kΩ. Find X and θ.

17. $X=17.5$ kΩ, $Z=38$ kΩ. Find R and θ.

18. $X=450$ Ω, $Z=750$ Ω. Find R and θ.

Refer to Figure 16-17(a)

19. $G=60$ μS, $\theta=25°$. Find B and Y.

20. $G=6.7$ mS, $\theta=73°$. Find B and Y.

21. $G=12.5$ mS, $B=8.73$ mS. Find Y and θ.

22. $G=440$ μS, $B=560$ μS. Find Y and θ.

23. $B = 35$ mS, $\theta = 10.7°$. Find G and Y.

24. $B = 800$ μS, $\theta = 75°$. Find G and Y.

25. $G = 6.35$ mS, $Y = 9.35$ mS. Find B and θ.

26. $G = 300$ μS, $Y = 500$ μS. Find B and θ.

27. $Y = 70$ μS, $\theta = 32°$. Find G and B.

28. $Y = 600$ μS, $\theta = 58°$. Find G and B.

29. $B = 7.5$ mS, $Y = 11.5$ mS. Find G and θ.

30. $B = 6.35$ mS, $Y = 9.5$ mS. Find G and θ.

ANSWERS TO SELF-TESTS

Self-Test 16-1

1. $\theta = 42.7°$

2. Side R is longer

3. $X = 56.6$ kΩ

4. $\angle \theta$

5. $\sin\theta = \dfrac{X}{Z}$, $\cos\theta = \dfrac{R}{Z}$, $\tan\theta = \dfrac{X}{R}$

Self-Test 16-2

1. $\sin 62.4° = 0.88620$, $\cos 62.4° = 0.46330$, $\tan 62.4° = 1.91282$

2. $48.2°$

3. $40.6°$

4. $55°$

5. $b = 32.6$, $\theta = 39.7°$

6. $Z = 4.7$ kΩ, $\theta = 55°$

7. $B = 173$ μS, $Y = 280$ μS

CHAPTER 17

AC FUNDAMENTALS

17.1 GENERATING ANGLES

In order to develop an understanding of AC concepts we need to discuss some basic mathematical and physical ideas. In Figure 17-1 two lines are drawn at right angles to each other. The lines are xx^1 and yy^1. Such a figure has been defined as a system of rectangular coordinates. The origin is the point where the lines cross. A line extending outward from the origin is called a *vector*.

Let's consider a line or vector which extends along the x-axis as in Figure 17-2. If the vector were rotated either clockwise or counterclockwise until it were again along the x-axis, the vector would have gone through 360° or a complete circle. It is standard practice to consider the x-axis as 0° of the circle and to rotate the vector counterclockwise. The y-axis is displaced 90° from the x-axis, the x^1-axis is displaced 180° from the x-axis, and the y^1-axis is displaced 270°. Thus, counterclockwise rotation is considered positive rotation. Clockwise rotation is considered negative rotation. From this we can say that y^1 is displaced 270° or −90° from the x-axis or 0°.

An angle is in its standard position when its initial side is on the positive x-axis and its vertex is at the origin. The x-y coordinates divide the system into fourths or quadrants as labeled in Figure 17-1. Note that in the first quadrant both x and y are positive. In the second quadrant x is negative and y is positive. In the third quadrant both x and y are negative and in the fourth quadrant x is positive and y is negative.

In Figure 17-3 a vector has been rotated through 30°. Since the resulting angle has its vertex at the origin, the vector has a length equal to the radius of the circle. The angle was formed by extending a line from point P to point A. In

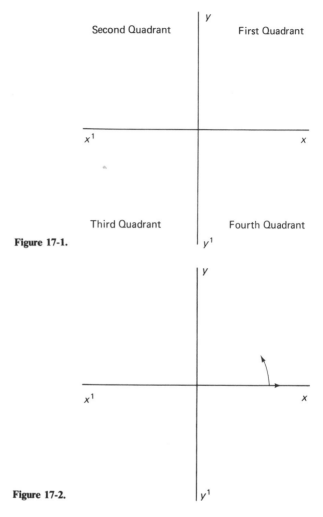

Figure 17-1.

Figure 17-2.

this position the line is in its initial position. If we rotate the line to point *B*, the line is in its terminal position. Notice that we rotated the line counterclockwise. The angle θ results. The angle can be measured in degrees or radians. If the line were rotated through 360°, a complete circle would be drawn. The line from point *P* to any point on the circumference is the radius. When the length of arc *AB* equals the radius, the angle is equal to 1 radian, as shown in Figure 17-4. A circle or 360° contains 6.28 rad. It follows then that there are 3.14 rad in 180°. The Greek letter π (pi) is used to denote 3.14 rad. Since there are 3.14 rad in 180°, then

$$\frac{180°}{3.14 \text{ rad}} = 57.3°$$

That is, 1 rad = 57.3°.

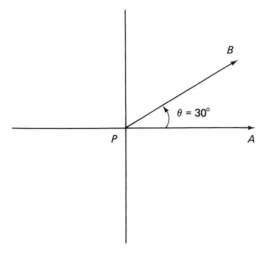

Figure 17-3.

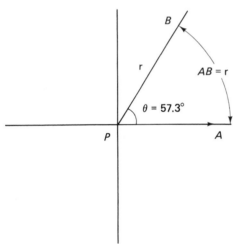

Figure 17-4.

EXAMPLE 17.1: Convert an angle of 85° into radians.

$$1 \text{ rad} = 57.3°$$

Solution: To convert from degrees to radians, divide by 57.3°.

$$\frac{85°}{57.3°} = 1.48 \text{ rad}$$

EXAMPLE 17.2: Convert an angle of 3.42 rad to degrees.

$$1 \text{ rad} = 57.3°$$

Solution: To convert from radians to degrees, multiply 57.3° by the number of radians.

$$3.42 \text{ rad} = 3.42 \times 57.3° = 196°$$

Let's consider the system of rectangular coordinates shown in Figure 17-5(a). If we start at 0° (the positive x-axis) and rotate in a counterclockwise direction, we will have drawn a complete circle. This rotation could represent one complete cycle of a generated sine wave of voltage or current. The radius of the circle could equal the maximum or peak value of the sine wave. Suppose this represents a voltage waveform whose maximum voltage is 100 V. $V = 100$ V$_{pk}$. As we rotate our vector from 0°, the radius, V_{pk}, is the hypotenuse of a right triangle. The vertical component, y, is the instantaneous value of voltage.

The instantaneous voltage is increasing from 0 V at 0° to 100 V at 90°. This is shown graphically in Figure 17-5(b). t_0 on the graph corresponds to 0° on our system of rectangular coordinates. As we increase to 90°, we can calculate the voltage generated at any instant by using the equation $v = V_{pk} \sin \theta$. (Lower-case v and i are used to denote instantaneous values of voltage and current.)

Continuing on from 90° to 180°, we could calculate instantaneous values and plot our graph. At this point we have completed one half-cycle or one-half revolution and are back to 0 V. As we continue on past 180°, v is negative. We can show mathematically why this is so. Look at the angles generated in Figure 17-6. We have generated an angle in each of the four quadrants. Remember that

$$\sin \theta = \frac{\text{opposite side}}{\text{hypotenuse}}$$

The side opposite the angle θ is the value in the y direction which we have labeled v. The hypotenuse is the radius vector which has been labeled V_{pk}. The

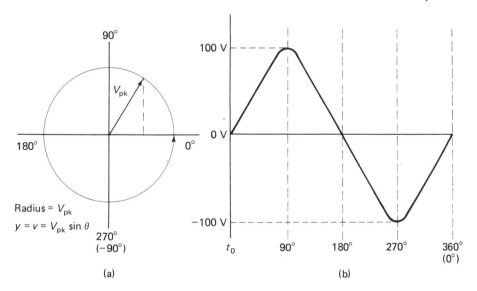

(a)   (b)

Figure 17-5.

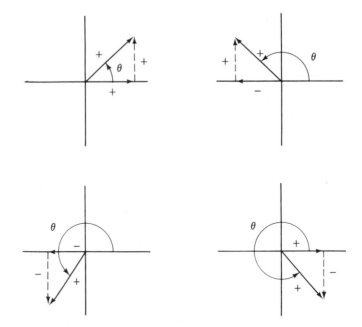

Figure 17-6.

radius vector is always considered to be positive. The opposite side, v, is positive in the first and second quadrants because the y-axis is positive. When we are in the third or fourth quadrant, the y-axis is negative; therefore, v is negative.

A look at our graph in Figure 17-5(b) shows that as we complete the last half-cycle, v goes from 0 V to -100 V as we rotate from 180° to 270°. v changes from a value equal to -100 V to 0 V as we rotate from 270° back to 0°. When AC voltages and currents are generated, this is the waveform produced. This waveform results from the constant rotation of a generator or from the natural oscillation of an electronics circuit. This sine curve could also represent the back and forth movement of a pendulum or the ripple created when a rock is dropped into still water.

In Figure 17-5(b) at time zero (t_0) the amplitude of the waveform is zero. Notice that time is plotted along the x-axis and amplitude is plotted along the y-axis. This is standard notation. t_0 on our graph corresponds to 0° on our system of rectangular coordinates. The maximum amplitude in the y direction corresponds to the radius of our circle.

17.3 FREQUENCY AND TIME

The number of times per second that we generate the sine wave is called *frequency*. The time it takes to generate one complete sine wave is called the *period*. We can express the relationship between frequency and time as an equation:

$$f = \frac{1}{T} \tag{17-1}$$

where f is the frequency in cycles per second and T is the time in seconds. Frequency and time are reciprocals of each other. The unit of frequency is the *hertz*. One hertz (Hz) equals one cycle per second. The unit of time is the second.

EXAMPLE 17.3: If the frequency of rotation is 250 Hz, what is the period?

Solution: Solve for T:

$$f = \frac{1}{T} \tag{17-1}$$

$$T = \frac{1}{f} \tag{17-2}$$

$$T = \frac{1}{250 \text{ Hz}} = 4 \text{ ms}$$

If there are 250 cycles per second (250 Hz), the time for 1 cycle (the period) is 4 ms.

EXAMPLE 17.4: If it takes 100 μs to complete one cycle, what is the frequency?

Solution:

$$f = \frac{1}{T}$$

$$f = \frac{1}{100 \ \mu s} = 10 \text{ kHz}$$

If the period is 100 μs, we will generate 10,000 cycles in one second or 10 kHz.

PRACTICE PROBLEMS 17-1

1. $f = 15$ kHz. Find T.

2. $f = 500$ Hz. Find T.

3. $t = 400 \ \mu$s. Find f.

4. $T = 16.7$ ms. Find f.

5. Convert 145° to radians.

6. Convert 300° to radians.

7. Convert 2 rad to degrees.

8. Convert 5.73 rad to degrees.

Solutions:

1. 66.7 μs

2. 2 ms

3. 2.5 kHz

4. 59.9 Hz

5. 2.53 rad   **6.** 5.24 rad

7. 115°   **8.** 328°

Additional practice problems are at the end of the chapter.

17.4 ANGULAR VELOCITY

Let's generate an angle by rotating our radius vector in Figure 17-3 at a constant speed. In angular motion this speed is called *angular velocity*. Angular velocity is measured in radians per second (rad/sec) or in degrees per second. The symbol for angular velocity is ω (the Greek letter *omega*). In radian measure, ω equals 2π times the frequency

$$\omega = 2\pi f$$

If the angle is measured in degrees, ω equals 360° times the frequency.

$$\omega = 360° f$$

ω is more commonly measured in radians per second.

When the frequency is 1 Hz (one cycle per second), the angular velocity is 2π rad/sec or 360°/sec. If the frequency is 10 Hz, the angular velocity is

$$\omega = 2\pi f$$
$$\omega = 2\pi \times 10 \text{ Hz} = 20\pi \text{ rad/sec}$$
$$= 62.8 \text{ rad/sec}$$

or $$\omega = 360° f = 360° \times 10 \text{ Hz} = 3600°/\text{sec}$$

The total angle swept by a vector in a given time (t) is ωt.

$$\theta = \omega t \tag{17-3}$$

In radian measure the equation would be:

$$\theta = 2\pi f t \tag{17-4}$$

In measuring in degrees the equation would be:

$$\theta = 360° f t \tag{17-5}$$

EXAMPLE 17.5: Find the angle generated after 5 ms when the frequency is 500 Hz. Express your answer in degrees and radians.

Solution:
$$\theta = \omega t = 2\pi f t \tag{17-4}$$
$$\theta = 2\pi \times 500 \times 5 \text{ ms} = 15.7 \text{ rad}$$
$$\theta = 57.3 \times 15.7 \text{ rad} = 900°$$

EXAMPLE 17.6: Find the frequency if an angle of 2.35 rad is generated in 150 μs.

Solution:

$$\theta = \omega t = 2\pi f t \qquad (17\text{-}4)$$

$$f = \frac{\theta}{2\pi t} = \frac{2.35 \text{ rad}}{2\pi \times 150 \text{ } \mu S} = 2.49 \text{ kHz}$$

EXAMPLE 17.7: How much time is required to generate an angle of 75° when the frequency is 1 kHz?

Solution:

$$\theta = \omega t = 360° f t \qquad (17\text{-}5)$$

$$t = \frac{\theta}{360° f} = \frac{75°}{360 \times 1 \text{ kHz}} = 208 \text{ } \mu s$$

PRACTICE PROBLEMS 17-2

1. If the frequency is 2.5 kHz, find the angle in both degrees and radians after (a) 50μs, (b) 100 μs, (c) 200 μs, and (d) 1 ms.

2. Find the frequency if the angular velocity is (a) 670 rad/sec, (b) 2000 rad/sec, (c) 7200°/sec, and (d) 12,000°/sec.

3. What is the frequency if an angle of 0.56 rad is generated in 55 μs?

4. What is the frequency if an angle of 285° is generated in 200 μs?

5. How much time is required to generate an angle of 1.7 rad when the frequency is 25 kHz?

6. How much time is required to generate an angle of 400° when the frequency is 1.5 kHz?

Solutions:

1. (a) 0.785 rad or 45°
   (b) 1.57 rad or 90°
   (c) 3.14 rad or 180°
   (d) 15.7 rad or 900°

2. (a) 107 Hz
   (b) 318 Hz
   (c) 20 Hz
   (d) 33.3 Hz

3. 1.62 kHz

4. 3.96 kHz

5. 10.8 μs

6. 741 μs

Additional practice problems are at the end of the chapter.

SELF-TEST 17-1

1. $f = 50$ kHz. Find T.

2. Convert 160° to radians.

3. Convert 0.628 rad to degrees.

4. If the frequency is 400 Hz, find the angle generated after 750 μs. Express your answer in both degrees and radians.

5. What is the frequency if an angle of 120° is generated in 200 μs?

6. How much time is required to generate an angle of 1.3 rad when the frequency is 5 kHz?

Answers to Self-Test 17-1 are at the end of the chapter.

In Figure 17-5, at t_0 the voltage is 0 V and at 90° the voltage is at its maximum value which, in this example, is 100 V. If the frequency is 1 kHz, how long does it take to reach this 100 V? To solve the problem, the first thing we have to do is find the period—the time it takes to complete one cycle

$$t = \frac{1}{f} \qquad (17\text{-}2)$$

$$t = \frac{1}{1\ \text{kHz}} = 1\ \text{ms}$$

Since 90° is $\frac{1}{4}$ of a complete cycle, then it must take $\frac{1}{4}$ the time for a complete cycle to reach 90°:

$$\text{time to complete } 90° = \frac{\text{period}}{4} = \frac{1\ \text{ms}}{4} = 250\ \mu s$$

During the time from t_1 to $t = 250\ \mu s$, V is increasing from 0 V to 100 V. This is not a linear change but a sinusoidal change. That is, the amplitude at any instant is a function of the sine of the generated angle. We have already determined that V_{pk} is the hypotenuse of a right triangle and the opposite side is the magnitude of the voltage at some instant in time. If we display the sine wave on a system of rectangular coordinates, the resulting equation for finding v at any instant is:

$$v = V_{pk} \sin \theta \qquad (17\text{-}6)$$

When working with an EMF, our equation would be:

$$e = E_{pk} \sin \theta \qquad (17\text{-}7)$$

The equation for instantaneous circuit current would be:

$$i = I_{pk} \sin \theta$$

EXAMPLE 17.8: $e = 100\ V_{pk} \sin \theta$. If $f = 1$ kHz, find e after (a) 50 μs, (b) 250 μs, and (c) 600 μs.

Solution:

$$e = E_{pk} \sin \theta \qquad (17\text{-}7)$$

$$e = 100\ V \sin 2\pi ft$$

$$(a) \quad e = 100\ V \sin 2\pi \times 1\ \text{kHz} \times 50\ \mu s$$

$$= 100\ V \sin 0.314\ \text{rad}$$

$$= 100\ V \times 0.309 = 30.9\ V$$

Since the angle θ is expressed in radians, the calculator must be in the radians mode to find $\sin \theta$.

(b) $e = 100 \text{ V} \sin 2\pi \times 1 \text{ kHz} \times 250 \text{ } \mu\text{s}$

$= 100 \text{ V} \sin 1.57 \text{ rad}$

$= 100 \text{ V} \times 1 = 100 \text{ V}$

We had determined previously that e would equal E_{pk} 250 μs after t_0.

(c) $e = 100 \text{ V} \sin 2\pi \times 1 \text{ kHz} \times 600 \text{ } \mu\text{s}$

$= 100 \text{ V} \sin 3.77$

$= 100 \text{ V} \times (-0.588) = -58.8 \text{ V}$

e is negative because 600 μs put us in the third quadrant.

PRACTICE PROBLEMS 17-3

1. $V = 60 \text{ } V_{\text{pk}}$, $f = 500$ Hz. Find v after (a) 100 μs, (b) 250 μs, (c) 500 μs, (d) 800 μs, (e) 1.2 ms, and (f) 1.6 ms.

2. $I = 30 \text{ mA}_{\text{pk}}$, $f = 250$ Hz. Find i after (a) 400 μs, (b) 1 ms, (c) 1.7 ms, (d) 2.1 ms, (e) 3 ms, and (f) 11.3 ms.

Solutions:

1. (a) 18.5 V (b) 42.4 V (c) 60 V (d) 35.3 V (e) −35.3 V (f) −57.1 V

2. (a) 17.6 mA (b) 30 mA (c) 13.6 mA (d) −4.69 mA (e) −30 mA (f) −26.7 mA

Additional practice problems are at the end of the chapter.

17.6 RMS VALUES OF VOLTAGE AND CURRENT

In Figure 17-5, at t_0 the voltage is 0 V. After some time has lapsed, we reach 90° of rotation and v is at its maximum value. As we continue through the complete cycle, v decreases to 0 at 180°, increases to its maximum negative value at 270°, and decreases to zero at 360°. If we took an average of all the instantaneous values of voltage for the entire cycle, we would get 0 V. This would be true because one-half the time the voltages are positive and one-half the time they are negative. For every positive instantaneous voltage there is a negative instantaneous voltage. If we consider *half* a cycle, then we could compute an average voltage. This average voltage would be 0.319 of the peak value.

$$V_{\text{AVE}} = 0.319 \text{ V}_{\text{pk}} = \frac{V_{\text{pk}}}{\pi} \qquad (17\text{-}8)$$

If both half-cycles were of the same polarity, then:

$$V_{\text{AVE}} = 0.637 \text{ V}_{\text{pk}} = \frac{2V_{\text{pk}}}{\pi} \qquad (17\text{-}9)$$

Average values of voltage and current are used primarily in determining component values in rectifier-type power supplies.

In determining currents and voltage drops in AC circuits, RMS values of current and voltage are used. The RMS value of a current or voltage is that value which converts the same energy as does a DC value. For example, in a DC circuit, 100 mA flows through a resistance of 10 Ω. The power dissipated is 100 mW. For an AC current to dissipate 100 mW in that same resistor, the RMS current would also have to be 100 mA.

The RMS value of an AC voltage or current is found by taking the square root of the average (mean) of the squares of the instantaneous values. Because of the method of calculation, this value is also called the *effective value*. This RMS or effective value is the same for all sine waves and equals 0.707 of the maximum or peak value.

$$I_{RMS} = 0.707\, I_{pk} = \frac{I_{pk}}{\sqrt{2}} \qquad (17\text{-}10)$$

$$V_{RMS} = 0.707\, V_{pk} \qquad (17\text{-}11)$$

If we solve for peak values, we get:

$$I_{pk} = 1.414\, I_{RMS} = \sqrt{2}\, I_{RMS} \qquad (17\text{-}12)$$

$$V_{pk} = 1.414\, V_{RMS} \qquad (17\text{-}13)$$

Since peak-to-peak values of voltage and current equal twice the peak values ($V_{p\text{-}p} = 2\, V_{pk}$), then:

$$V_{p\text{-}p} = 2.828\, V_{RMS} = 2\sqrt{2}\, V_{RMS} \qquad (17\text{-}14)$$

$$I_{p\text{-}p} = 2.828\, I_{RMS} \qquad (17\text{-}15)$$

Solving for effective values, we get:

$$V_{RMS} = 0.3535\, V_{p\text{-}p} = \frac{V_{p\text{-}p}}{2\sqrt{2}}$$

$$I_{RMS} = 0.3535\, I_{p\text{-}p}$$

When AC voltages or currents are given, values are always RMS values. Thus, if we see an AC voltage written as 120 V, we know that it is 120 V_{RMS}. If the voltage or current is either a peak or a peak-to-peak value, the proper subscript must be used. Therefore, ten volts RMS is written 10 V. Ten volts peak is written 10 V_{pk}. Ten volts peak-to-peak is written 10 $V_{p\text{-}p}$.

EXAMPLE 17.9: Convert 12 V_{pk} to RMS and p-p.

Solution:

$$V_{RMS} = 0.707\, V_{pk}$$

$$V_{RMS} = 0.707 \times 12 = 8.48 \text{ V}$$

$$V_{p\text{-}p} = 2\, V_{pk}$$

$$V_{p\text{-}p} = 2 \times 12 = 24 \text{ V}_{p\text{-}p}$$

Example 17.10: Convert 20 mA to pk and p-p.

Solution:

$$I_{pk} = 1.414\ I_{RMS}$$
$$I_{pk} = 1.414 \times 20\ mA = 28.3\ mA_{pk}$$
$$I_{p\text{-}p} = 2.828\ I_{RMS}$$
$$I_{p\text{-}p} = 2.828 \times 20\ mA = 56.6\ mA_{p\text{-}p}$$

Example 17.11: Convert 20 $V_{p\text{-}p}$ to RMS and pk.

Solution:

$$V_{RMS} = 0.3535\ V_{p\text{-}p}$$
$$V_{RMS} = 0.3535 \times 20 = 7.07\ V$$
$$V_{pk} = \frac{V_{p\text{-}p}}{2}$$
$$V_{pk} = \frac{20\ V}{2} = 10\ V$$

PRACTICE PROBLEMS 17-4

1. An AC voltage is 30 V. Convert to pk and p-p. **2.** An AC current is 150 mA. Convert to pk and p-p.

3. An AC voltage is 50 V_{pk}. Convert to p-p and RMS. **4.** An AC current is 70 mA_{pk}. Convert to p-p and RMS.

5. An AC voltage is 200 $V_{p\text{-}p}$. Convert to pk and RMS. **6.** An AC current is 1.5 $A_{p\text{-}p}$. Convert to pk and RMS.

Solutions:

1. 30 V = 42.4 V_{pk} = 84.8 $V_{p\text{-}p}$ **2.** 150 mA = 212 mA_{pk} = 424 $mA_{p\text{-}p}$
3. 50 V_{pk} = 100 $V_{p\text{-}p}$ = 35.4 V **4.** 70 mA_{pk} = 140 $mA_{p\text{-}p}$ = 49.5 mA
5. 200 $V_{p\text{-}p}$ = 100 V_{pk} = 70.7 V **6.** 1.5 $A_{p\text{-}p}$ = 750 mA_{pk} = 530 mA

Additional practice problems are at the end of the chapter.

17.7 ANGLE OF LEAD OR LAG

When we have a purely resistive AC circuit, the applied voltage and circuit current equal zero at the same time. They also reach their peak values at the same time. In such a circuit we say that the voltage and current are *in phase*. The

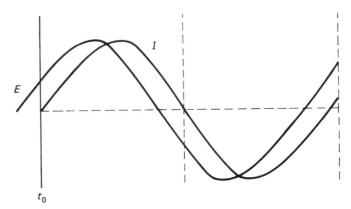

Figure 17-7.

voltage drops across the various resistances are also in phase with the current and applied voltage.

Whenever we have capacitance or inductance in an AC circuit, the current and the applied voltage are *out of phase*. The applied voltage either leads or lags the current. If the applied voltage *leads* the current, this means that E reaches its values of 0 V and E_{pk} at an earlier time than does the current. The phase difference can be anywhere from 0° to 90°. Suppose E leads I by 30° as illustrated in Figure 17-7. Using the current wave form as our reference, we could write the equations for E and I at any instant.

$$i = I_{pk} \sin \omega t$$

$$e = E_{pk} \sin(\omega t + 30°)$$

In a series circuit, current would be used as the reference because current is the same throughout the circuit. In a parallel circuit we would use voltage as the reference.

EXAMPLE 17.12: $I = 50$ mA$_{pk}$, $E = 100$ V$_{pk}$, and the frequency is 2 kHz. Using current as the reference, compute the instantaneous values of i and e 40 μs after t_0 when I leads E by 60°.

Solution:

$$\theta = \omega t = 360° f t = 28.8°$$

Since the angle of lead was given in degrees, we found θ in degrees.

$$i = I_{pk} \sin 28.8° = 24.1 \text{ mA}$$

E lags I by 60°; therefore,

$$e = E_{pk} \sin(\theta - 60°) = E_{pk} \sin(28.8° - 60°)$$

$$= E_{pk} \sin - 31.2° = -51.8 \text{ V}$$

EXAMPLE 17.13: $I = 20$ mA$_{pk}$, $E = 20$ V$_{pk}$, $f = 12$ kHz, and E leads I by 20°. Compute i and e 5 μs after t_0. Use current as the reference.

Solution:

$$\theta = \omega t = 360° \times 12 \text{ kHz} \times 5 \text{ μs} = 21.6°$$

$$i = I_{pk} \sin \theta = 20 \text{ mA} \sin 21.6° = 7.36 \text{ mA}$$

$$e = E_{pk} \sin(\theta + 20°) = 20 \text{ V} \sin 41.6° = 13.3 \text{ V}$$

PRACTICE PROBLEMS 17-5

1. $I = 100$ mA$_{pk}$, $E = 30$ V$_{pk}$, and $f = 500$ Hz. Use current as the reference. I leads E by 40°. Find i and e after (a) 50 μs, (b) 225 μs, (c) 800 μs, and (d) 1.2 ms.

2. $I = 12$ mA$_{pk}$, $E = 15$ V$_{pk}$, and $f = 7.5$ kHz. Use voltage as the reference. I leads E by 60°. Find i and e after (a) 10 μs, (b) 40 μs, (c) 100 μs, and (d) 150 μs.

Solutions:

1. (a) $i = 15.6$ mA, $e = -15.5$ V
(b) $i = 64.9$ mA, $e = 0.262$ V
(c) $i = 58.8$ mA, $e = 29.1$ V
(d) $i = -58.8$ mA, $e = 2.09$ V

2. (a) $i = 12$ mA, $e = 6.81$ V
(b) $i = 2.49$ mA, $e = 14.3$ V
(c) $i = -6$ mA, $e = -15$ V
(d) $i = 11.6$ mA, $e = 10.6$ V

Additional practice problems are at the end of the chapter.

SELF-TEST 17-2

1. If $V = 25$ V$_{pk}$ and $f = 8$ kHz, find v after 25 μs.

2. What is the instantaneous value of current after 2 ms if the frequency is 400 Hz and I_{pk} is 20 mA?

3. Convert 25 V to peak values and peak-to-peak values.

4. Convert 60 mA$_{p-p}$ to peak values and RMS values.

5. I leads E by 25° in an AC circuit. If $E = 50$ V and the frequency is 25 kHz, find e after 7 μs. Use current as the reference.

Answers to Self-Test 17-2 are at the end of the chapter.

END OF CHAPTER PROBLEMS 17-1

1. $f = 1$ kHz. Find T.

2. $f = 27.5$ kHz. Find T.

3. $f = 1.76$ MHz. Find T.

4. $f = 120$ Hz. Find T.

5. $T = 1.33$ ms. Find f.

6. $T = 83.3$ μs. Find f.

7. $T = 8 \ \mu s$. Find f.

8. $T = 10$ ms. Find f.

9. Convert 25° to radians.

10. Convert 90° to radians.

11. Convert 180° to radians.

12. Convert 120° to radians.

13. Convert 75° to radians.

14. Convert 315° to radians.

15. Convert 600° to radians.

16. Convert 830° to radians.

17. Convert 0.146 rad to degrees.

18. Convert 2.5 rad to degrees.

19. Convert 0.5 rad to degrees.

20. Convert 4.3 rad to degrees.

21. Convert 1.75 rad to degrees.

22. Convert 3 rad to degrees.

23. Convert 10 rad to degrees.

24. Convert 12 rad to degrees.

END OF CHAPTER PROBLEMS 17-2

1. Find the angular velocity in both degrees per second and radians per second if the frequency is (a) 60 Hz and (b) 150 Hz.

2. Find the angular velocity in both degrees per second and radians per second if the frequency is (a) 30 Hz and (b) 100 Hz.

3. Find the frequency if the angular velocity is (a) 900 rad/sec, (b) 5000 rad/sec, (c) 10,000°/sec, and (d) 120,000°/sec.

4. Find the frequency if the angular velocity is (a) 1500 rad/sec, (b) 7500 rad/sec, (c) 56,000°/sec, and (d) 100,000°/sec.

5. If the frequency is 500 Hz, find the angle generated in both degrees and radians after (a) 100 μs, (b) 250 μs, (c) 1 ms, and (d) 2.8 ms.

6. If the frequency is 35 kHz, find the angle generated in both degrees and radians after (a) 5 μs, (b) 12 μs, (c) 40 μs, and (d) 75 μs.

7. What is the frequency when an angle of (a) 1.8 rad is generated in 20 μs, (b) 8.73 rad is generated in 150 μs, (c) 145° is generated in 2.3 ms, and (d) 310° is generated in 400 μs?

8. What is the frequency when an angle of (a) 0.25 rad is generated in 5 μs, (b) 4.6 rad is generated in 75 μs, (c) 75° is generated in 20 μs, and (d) 500° is generated in 150 μs?

9. If the frequency is 5 kHz, how much time is required to generate an angle of (a) 1 rad, (b) 0.125 rad, (c) 4.73 rad, (d) 20°, (e) 90°, and (f) 300°?

10. If the frequency is 125 Hz, how much time is required to generate an angle of (a) 0.25 rad, (b) 2.35 rad, (c) 5.6 rad, (d) 300°, (e) 135°, and (f) 270°?

END OF CHAPTER PROBLEMS 17-3

1. $E = 25 \ V_{pk}$, $f = 2.5$ kHz. Find e after (a) 10 μs, (b) 25 μs, (c) 150 μs, (d) 225 μs, (e) 300 μs, and (f) 375 μs.

2. $E = 15 \ V_{pk}$, $f = 12$ kHz. Find e after (a) 2 μs, (b) 10 μs, (c) 20 μs, (d) 45 μs, (e) 70 μs, and (f) 100 μs.

3. $I = 700 \ \mu A_{pk}$, $f = 50$ kHz. Find i after (a) 5 μs, (b) 10 μs, (c) 15 μs, (d) 20 μs, (e) 25 μs, and (f) 30 μs.

4. $I = 50 \ mA_{pk}$, $f = 2$ kHz. Find i after (a) 50 μs, (b) 120 μs, (c) 200 μs, (d) 300 μs, (e) 400 μs, and (f) 510 μs.

END OF CHAPTER PROBLEMS 17-4

1. An AC voltage is 65 V. Convert to pk and p-p.

2. An AC voltage is 400 mV. Convert to pk and p-p.

3. An AC current is 3 A. Convert to pk and p-p.

4. An AC current is 350 mA. Convert to pk and p-p.

5. An AC voltage is 80 V_{pk}. Convert to p-p and RMS.

6. An AC voltage is 9 V_{pk}. Convert to p-p and RMS.

7. An AC current is 500 mA_{pk}. Convert to p-p and RMS.

8. An AC current is 1.3 A_{pk}. Convert to p-p and RMS.

9. An AC voltage is 17 V_{p-p}. Convert to pk and RMS.

10. An AC voltage is 600 V_{p-p}. Convert to pk and RMS.

11. An AC current is 700 μA_{p-p}. Convert to pk and RMS.

12. An AC current is 7.8 mA_{p-p}. Convert to pk and RMS.

END OF CHAPTER PROBLEMS 17-5

1. $I=4$ mA_{pk}, $E=10$ V_{pk}, and $f=20$ kHz. Use current as the reference. E leads I by 25°. Find i and e after (a) 5 μs, (b) 12 μs, (c) 25 μs, and (d) 40 μs.

2. $I=150$ mA_{pk}, $E=50$ V, and $f=3$ kHz. Use voltage as the reference. E leads I by 45°. Find i and e after (a) 25 μs, (b) 60 μs, (c) 160 μs, and (d) 300 μs.

3. $I=25$ mA_{pk}, $E=20$ V, and $f=100$ Hz. Use voltage as the reference. I leads E by 35°. Find i and e after (a) 1 ms, (b) 2.5 ms, (c) 3.2 ms, and (d) 8.3 ms.

4. $I=600$ μA_{pk}, $E=9$ V_{pk}, and $f=5$ kHz. Use current as the reference. I leads E by 30°. Find i and e after (a) 15 μs, (b) 40 μs, (c) 140 μs, and (d) 250 μs.

ANSWERS TO SELF-TESTS

Self-Test 17-1

1. $T=20$ μs

2. 2.79

3. 36°

4. 1.88 rad $=108°$

5. 1.67 kHz

6. 41.4 μs

Self-Test 17-2

1. 23.8 V

2. -19 mA

3. 35.4 $V_{pk}=70.7$ V_{p-p}

4. 30 $mA_{pk}=21.2$ mA

5. 43.5 V

CHAPTER 18

AC CIRCUIT ANALYSIS—
SERIES CIRCUITS

18.1 SERIES RC CIRCUITS

Capacitance in AC circuits causes the current to lead the applied voltage. In a purely capacitive circuit ($R = 0\ \Omega$), current would lead the voltage by $90°$. At the other extreme, if the circuit is purely resistive, current and applied voltage are in phase. Let's see what happens in a circuit that contains both resistance and capacitance. Consider the circuit in Figure 18-1.

Because current is constant in a series circuit, we use I as our reference and plot it along the reference or x-axis, as shown in Figure 18-1(b). Since there is no phase difference between the current through and the voltage across a resistor,

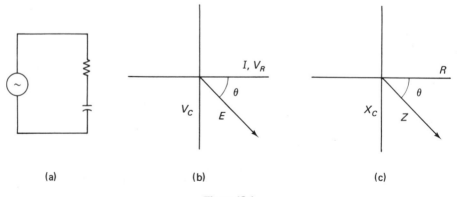

(a)   (b)   (c)

Figure 18-1

V_R is also plotted along the x-axis. The circuit current leads the voltage drop across the capacitor by 90°. Therefore, V_C is plotted along the negative y-axis. **E** is the resultant vector of these coordinates. **E** then is the hypotenuse of a right triangle and θ is the phase angle. V_R is the side adjacent to the angle θ and V_C is the opposite side.

EXAMPLE 18.1: Refer to Figure 18-1(a). Let $V_R = 10$ V and $V_C = -12$ V. Find **E** and θ.

Solution: Refer to Figure 18-2.

$$\tan \theta = \frac{\text{opposite side}}{\text{adjacent side}} = \frac{V_C}{V_R} = \frac{-12 \text{ V}}{10 \text{ V}} = -1.2$$

$$\tan^{-1} -1.2 = \theta = -50.2°$$

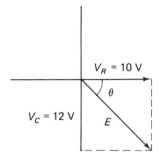

Figure 18-2

(Remember, V_C is negative because we plotted V_C along the negative y-axis.) We can now find **E** by using either the sine or cosine function. We will use the sine function.

$$\sin \theta = \frac{\text{opposite side}}{\text{hypotenuse}} = \frac{V_C}{E}$$

$$E = \frac{V_C}{\sin \theta} = \frac{-12 \text{ V}}{\sin -50.2°} = \frac{-12 \text{ V}}{-0.768} = 15.6 \text{ V}$$

EXAMPLE 18.2: Refer to Figure 18-1(a). Let **E** = 25 V and $V_R = 18$ V. Find V_C and θ.

Solution:

$$\cos \theta = \frac{V_R}{E} = \frac{18 \text{ V}}{25 \text{ V}} = 0.72$$

$$\cos^{-1} 0.72 = \theta = -43.9°$$

From the given information we know that θ is negative

only because we know that we are operating in the fourth quadrant.

$$\sin\theta = \frac{V_C}{E}$$

$$V_C = E\sin\theta = 25 \text{ V} \sin -43.9° = -17.3 \text{ V}$$

PRACTICE PROBLEMS 18-1   Refer to Figure 18-1 for the following problems:

1. $V_R = 6.8$ V, $V_C = 10$ V. Find E and θ.

2. $V_R = 30$ V, $E = 50$ V. Find V_c and θ.

3. $V_R = 10$ V, $\theta = -60°$. Find V_C and E.

4. $E = 20$ V, $\theta = -35°$. Find V_R and V_C.

5. $E = 50$ V, $V_C = 40$ V. Find V_R and θ.

6. $V_C = 11$ V, $\theta = -65°$. Find E and V_R.

Solutions:

1. $E = 12.1$ V
   $\theta = -55.8°$

2. $V_C = -40$ V
   $\theta = -53.1°$

3. $V_C = -17.3$ V
   $E = 20$ V

4. $V_R = 16.4$ V
   $V_C = 11.5$ V

5. $V_R = 30$ V
   $\theta = -53.1°$

6. $E = 12.1$ V
   $V_R = 5.11$ V

In Figure 18-1(c), since V_R and I are in phase, R is plotted along the x-axis. X_C, the capacitive reactance, must be plotted along the negative y-axis to show the 90°-phase difference between capacitive values and resistive values. The capacitive reactance is determined by the following equation:

$$X_C = \frac{1}{2\pi fC} \tag{18-1}$$

EXAMPLE 18.3: Refer to Figure 18-1(a). Let $R = 2$ kΩ, $C = 20$ nF, and $f = 3$ kHz. Find X_C, Z, and θ.

Solution:

$$X_C = \frac{1}{2\pi fC} = \frac{1}{2\pi \times 3 \text{ k}\Omega \times 20 \text{ nF}} = 2.65 \text{ k}\Omega$$

$$\tan\theta = \frac{\text{opposite side}}{\text{adjacent side}} = \frac{X_C}{R} = \frac{-2.65 \text{ k}\Omega}{2 \text{ k}\Omega} = -1.33$$

$$\tan^{-1} -1.33 = \theta = -53°$$

θ is negative because we are in the fourth quadrant.

$$\cos\theta = \frac{R}{Z}$$

$$Z = \frac{R}{\cos\theta} = \frac{2 \text{ k}\Omega}{\cos -53°} = 3.32 \text{ k}\Omega$$

EXAMPLE 18.4: Refer to Figure 18-1(a). Let $R = 10$ kΩ, $C = 50$ nF, $f = 500$ Hz, and $E = 10$ V. Find X_C, Z, θ, V_R, V_C, and I.

Solution:

$$X_C = \frac{1}{2\pi f C} = \frac{1}{2\pi \times 500 \times 50 \text{ nF}} = 6.37 \text{ k}\Omega$$

$$\tan\theta = \frac{\text{opposite side}}{\text{adjacent side}} = \frac{X_C}{R} = \frac{-6.37 \text{ k}\Omega}{10 \text{ k}\Omega} = -0.637$$

$$\tan^{-1} -0.637 = -32.5°$$

$$\sin\theta = \frac{X_C}{Z}$$

$$Z = \frac{X_C}{\sin\theta} = \frac{-6.37 \text{ k}\Omega}{\sin -32.5°} = 11.9 \text{ k}\Omega$$

Now that we know Z, we can find I.

$$I = \frac{E}{Z}$$

$$I = \frac{10 \text{ V}}{11.9 \text{ k}\Omega} = 840 \text{ }\mu\text{A}$$

To find V_C and V_R, we can use Ohm's law:

$$V_C = IX_C = 840 \text{ }\mu\text{A} \times 6.37 \text{ k}\Omega = 5.35 \text{ V}$$

$$V_R = IR = 840 \text{ }\mu\text{A} \times 10 \text{ k}\Omega = 8.4 \text{ V}$$

We could also find V_C and V_R by using trig functions since we know E and θ.

$$V_R = E\cos\theta = 10 \text{ V} \cos -32.5° = 8.43 \text{ V}$$

$$V_C = E\sin\theta = 10 \text{ V} \sin -32.5° = -5.37 \text{ V}$$

(Differences in the values of V_R and V_C in the two methods are a result of rounding the value of Z to three places.)

EXAMPLE 18.5: $R = 5$ kΩ, $C = 10$ nF, $Z = 8.3$ kΩ, and $E = 10$ V. Find I, θ, V_C, V_R, f, and X_C.

Solution:

$$I = \frac{E}{Z} = \frac{10 \text{ V}}{8.3 \text{ k}\Omega} = 1.2 \text{ mA}$$

Since we know R and Z, we can find X_C and θ. We will use the cosine function to find θ and the sine function to find X_C.

$$\cos\theta = \frac{R}{Z} = \frac{5 \text{ k}\Omega}{8.3 \text{ k}\Omega} = 0.602 \text{ k}\Omega$$

$$\cos^{-1} 0.602 = \theta = -53.1°$$

$$\sin\theta = \frac{X_C}{Z}$$

$$X_C = Z\sin\theta = 8.3 \text{ k}\Omega \sin -53.1° = -6.64 \text{ k}\Omega$$

(Remember, mathematically the value of X_C is negative because we are operating in the fourth quadrant. Electrically, X_C is treated as a positive value.)

Now we can find the frequency because we know C and X_C. If we solve for f in the equation

$$X_C = \frac{1}{2\pi f C}$$

we get:

$$f = \frac{1}{2\pi C X_C} = \frac{1}{2\pi \times 10 \text{ nF} \times 6.64 \text{ k}\Omega} = 2.4 \text{ kHz}$$

We can find V_R and V_C by using Ohm's law or by using trig functions. We should solve both ways to check our work. We will use Ohm's law here.

$$V_R = IR = 1.2 \text{ mA} \times 5 \text{ k}\Omega = 6 \text{ V}$$

$$V_C = IX_C = 1.2 \text{ mA} \times 6.64 \text{ k}\Omega = 7.97 \text{ V}$$

PRACTICE PROBLEMS 18-2   Refer to Figure 18-1 for the following problems and find X_C, Z, θ, V_R, V_C, and I.

1. $R = 470 \text{ }\Omega$, $C = 0.2 \text{ }\mu\text{F}$, $f = 1 \text{ kHz}$, and $E = 9 \text{ V}$

2. $R = 2 \text{ k}\Omega$, $C = 30 \text{ nF}$, $f = 3 \text{ kHz}$, and $E = 20 \text{ V}$

3. $R = 27 \text{ k}\Omega$, $C = 100 \text{ pF}$, $f = 35 \text{ kHz}$, and $E = 15 \text{ V}$

Refer to Figure 18-1 for the following problems and find I, θ, f, X_C, V_R, and V_C.

4. $R = 12 \text{ k}\Omega$, $C = 200 \text{ pF}$, $Z = 20 \text{ k}\Omega$, and $E = 15 \text{ V}$

5. $R = 680 \text{ }\Omega$, $C = 1 \text{ }\mu\text{F}$, $Z = 800 \text{ }\Omega$, and $E = 25 \text{ V}$

6. $R = 3 \text{ k}\Omega$, $C = 50 \text{ nF}$, $Z = 4.5 \text{ k}\Omega$, and $E = 30 \text{ V}$

Solutions:

1. $X_C = 796 \text{ }\Omega$, $Z = 924 \text{ }\Omega$, $\theta = -59.4°$, $V_R = 4.58 \text{ V}$, $V_C = 7.75 \text{ V}$, $I = 9.74 \text{ mA}$

2. $X_C = 1.77 \text{ k}\Omega$, $Z = 2.67 \text{ k}\Omega$, $\theta = -41.5°$, $V_R = 15 \text{ V}$, $V_C = 13.2 \text{ V}$, $I = 7.49 \text{ mA}$

3. $X_C = 45.5 \text{ k}\Omega$, $Z = 52.9 \text{ k}\Omega$, $\theta = -59.3°$, $V_R = 7.66$ V, $V_C = 12.9$ V, $I = 284$ μA

4. $I = 750$ μA, $\theta = -53.1°$, $f = 49.7$ kHz, $X_C = 16 \text{ k}\Omega$, $V_R = 9$ V, $V_C = 12$ V

5. $I = 31.3$ mA, $\theta = -31.8°$, $f = 144$ Hz, $X_C = 421 \, \Omega$, $V_R = 21.2$ V, $V_C = 13.2$ V

6. $I = 6.67$ mA, $\theta = -48.2°$, $f = 501$ Hz, $X_C = 3.35 \text{ k}\Omega$, $V_R = 20$ V, $V_C = 22.4$ V

18.2   SERIES RL CIRCUITS

Inductance in AC circuits causes the applied voltage to lead the circuit current. In a purely inductive circuit the voltage drop across the inductance, V_L, leads the current by 90°. Consider the circuit in Figure 18-3(a). Current is plotted along the reference axis just as in the RC circuit problem. V_L leads I by 90° and is plotted along the y-axis. **E** is the resultant vector of these coordinates. **E** is the hypotenuse of a right triangle and θ is the phase angle. V_R is the adjacent side and V_L is the opposite side.

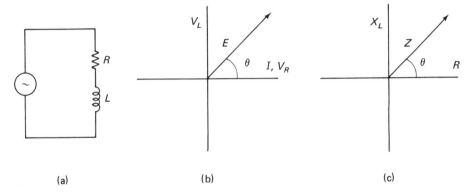

(a)   (b)   (c)

Figure 18-3

EXAMPLE 18.6: Let $V_R = 15$ V and $V_L = 10$ V in Figure 18-3(a). Find **E** and θ.

Solution: Refer to Figure 18-4.

$$\tan \theta = \frac{V_L}{V_R} = \frac{10 \text{ V}}{15 \text{ V}} = 0.667$$

$$\tan^{-1} 0.667 = \theta = 33.7°$$

E can be found by using either the sine function or the cosine function. Let's use the cosine function.

$$\cos \theta = \frac{V_R}{\mathbf{E}}$$

$$\mathbf{E} = \frac{V_R}{\cos \theta} = \frac{15 \text{ V}}{\cos 33.7°} = 18 \text{ V}$$

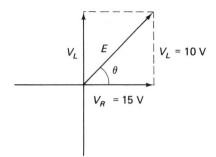

Figure 18-4

EXAMPLE 18.7: Let $E = 10$ V and $V_L = 5$ V in Figure 18-3(a). Find V_R and θ.

Solution:

$$\sin \theta = \frac{V_L}{E} = \frac{5 \text{ V}}{10 \text{ V}} = 0.5$$

$$\sin^{-1} 0.5 = \theta = 30°$$

Now that we know θ, we can use either the cosine function or the tangent function to find V_R. Let's use the tangent function.

$$\tan \theta = \frac{V_L}{V_R}$$

$$V_R = \frac{V_L}{\tan \theta} = \frac{5 \text{ V}}{\tan 30°} = 8.66 \text{ V}$$

PRACTICE PROBLEMS 18-3   Refer to Figure 18-3 for the following problems.

1. $V_R = 12$ V, $V_L = 18$ V. Find **E** and θ.
2. $V_R = 27$ V, $\theta = 65°$. Find **E** and V_L.
3. $V_R = 35$ V, $E = 50$ V. Find V_L and θ.
4. $V_L = 6$ V, $E = 10$ V. Find V_R and θ.
5. $V_L = 15$ V, $\theta = 27°$. Find V_R and **E**.
6. $E = 25$ V, $\theta = 33°$. Find V_R and V_L.

Solutions:

1. $E = 21.6$ V
   $\theta = 56.3°$
2. $E = 63.9$ V
   $V_L = 57.9$ V
3. $V_L = 35.7$ V
   $\theta = 45.6°$
4. $V_R = 8$ V
   $\theta = 36.9°$
5. $V_R = 29.4$ V
   $E = 33$ V
6. $V_R = 21$ V
   $V_C = 13.6$ V

In Figure 18-3(c), since V_R and I are in phase, R is plotted along the x-axis. X_L, the inductive reactance, must be plotted along the positive y-axis to show the 90°-phase difference between inductive and resistive values. The inductive reactance is determined by the equation

$$X_L = 2\pi fL \qquad (18-2)$$

EXAMPLE 18.8: Refer to Figure 18-3(a). Let $R = 2.7$ kΩ, $L = 200$ mH, and $f = 3$ kHz. Find X_L, θ, and Z.

Solution:

$$X_L = 2\pi fL = 2\pi \times 3 \text{ kΩ} \times 200 \text{ mH} = 3.77 \text{ kΩ}$$

Knowing R and X_L, we can use the tangent function to find θ.

$$\tan\theta = \frac{X_L}{R} = \frac{3.77 \text{ kΩ}}{2.7 \text{ kΩ}} = 1.4$$

$$\tan^{-1} 1.4 = \theta = 54.4°$$

We can use either the sine function or the cosine function to find Z. Let's use the sine function.

$$\sin\theta = \frac{X_L}{Z}$$

$$Z = \frac{X_L}{\sin\theta} = \frac{3.77 \text{ kHz}}{\sin 54.4°} = 4.64 \text{ kΩ}$$

EXAMPLE 18.9: Refer to Figure 18-3. Let $R = 10$ kΩ, $L = 100$ mH, $f = 20$ kHz, and $\mathbf{E} = 15$ V. Find X_L, θ, Z, V_R, V_L, and I.

Solution:

$$X_L = 2\pi fL = 2\pi \times 20 \text{ kHz} \times 100 \text{ mH} = 12.6 \text{ kΩ}$$

Knowing R and X_L, we can find Z and θ.

$$\tan\theta = \frac{X_L}{R} = \frac{12.6 \text{ kΩ}}{10 \text{ kΩ}} = 1.26$$

$$\tan^{-1} 1.26 = \theta = 51.5°$$

$$\cos\theta = \frac{R}{Z}$$

$$Z = \frac{R}{\cos\theta} = \frac{10 \text{ kΩ}}{\cos 51.5°} = 16.1 \text{ kΩ}$$

We can find V_R and V_L by using Ohm's law.

$$I = \frac{E}{Z} = \frac{15 \text{ V}}{16.1 \text{ k}\Omega} = 932 \text{ } \mu\text{A}$$

$$V_R = IR = 932 \text{ } \mu\text{A} \times 10 \text{ k}\Omega = 9.32 \text{ V}$$

$$V_L = IX_L = 932 \text{ } \mu\text{A} \times 12.6 \text{ k}\Omega = 11.7 \text{ V}$$

We can also find V_R and V_L by using trig functions.

$$V_R = E\cos\theta = 15 \text{ V}\cos 51.5° = 9.34 \text{ V}$$

$$V_L = E\sin\theta = 15 \text{ V}\sin 51.5° = 11.7 \text{ V}$$

(Differences in the value of V_R are due to rounding.)

EXAMPLE 18.10: Refer to Figure 18-3. Let $L = 30$ mH, $Z = 12$ kΩ, $X_L = 9$ kΩ, and $V_R = 3.7$ V. Find R, θ, E, I, f, and V_L.

Solution: Let's find θ since we know Z and X_L.

$$\sin\theta = \frac{X_L}{Z} = \frac{9 \text{ k}\Omega}{12 \text{ k}\Omega} = 0.75$$

$$\sin^{-1} 0.75 = \theta = 48.6°$$

Having found θ, we can now find R, E, and V_L.

$$\cos\theta = \frac{R}{Z}$$

$$R = Z\cos\theta = 12 \text{ k}\Omega\cos 48.6° = 7.94 \text{ k}\Omega$$

$$\cos\theta = \frac{V_R}{E}$$

$$E = \frac{V_R}{\cos\theta} = \frac{3.7 \text{ V}}{\cos 48.6°} = 5.59 \text{ V}$$

$$\sin\theta = \frac{V_C}{E}$$

$$V_C = E\sin\theta = 5.59\sin 48.6° = 4.19 \text{ V}$$

We can use Ohm's law to find I.

$$I = \frac{E}{Z} = \frac{5.59 \text{ V}}{12 \text{ k}\Omega} = 466 \text{ } \mu\text{A}$$

Solving for f in the equation $X_L = 2\pi f L$, we get:

$$f = \frac{X_L}{2\pi L} = \frac{9 \text{ k}\Omega}{2\pi \times 30 \text{ mH}} = 47.7 \text{ kHz}$$

18.2.1 Q. In circuits containing inductance, an important variable is Q. The Q of a coil is simply the ratio of its inductive reactance to its resistance.

$$Q_{\text{coil}} = \frac{X_L}{R_L} \tag{18-3}$$

This is also the ratio of energy stored to the energy used. (Inductance is an energy storer. Resistance is an energy user.)

The Q of a circuit is the ratio of circuit inductive reactance to *total* circuit resistance:

$$Q_{\text{ckt}} = \frac{X_L}{R_T} \tag{18-4}$$

when R_L is the only resistance in the circuit, Q_{coil} and Q_{ckt} are equal. In the section on resonance we will see how Q is used to calculate the bandwidth of a circuit.

PRACTICE PROBLEMS 18-4

Refer to Figure 18-3 for the following problems and find X_L, Z, and θ.

1. $R = 2.7$ kΩ, $L = 1$ H, and $f = 600$ Hz **2.** $R = 25$ kΩ, $L = 200$ mH, and $f = 15$ kHz

Refer to Figure 18-3 for the following problems and find X_L, Z, θ, V_R, V_L, Q, and I.

3. $R = 600$ Ω, $L = 400$ mH, $f = 200$ Hz, and $E = 20$ V **4.** $R = 3$ kΩ, $L = 50$ mH, $f = 15$ kHz, and $E = 9$ V

Refer to Figure 18-3 for the following problems and find X_L, θ, f, E, V_R, and I.

5. $L = 100$ mH, $Z = 7.3$ kΩ, $R = 5$ kΩ, and $V_L = 7.2$ V **6.** $L = 20$ mH, $Z = 1$ kΩ, $R = 300$ Ω, and $V_L = 3.2$ V

Solutions:

1. $X_L = 3.77$ kΩ
$Z = 4.64$ kΩ
$\theta = 54.4°$

2. $X_L = 18.8$ kΩ
$Z = 31.3$ kΩ
$\theta = 37°$

3. $X_L = 503$ Ω
$\theta = 40°$
$Z = 783$ Ω
$V_R = 15.3$ V
$V_L = 12.8$ V
$I = 25.5$ mA
$Q = 0.838$

4. $X_L = 4.71$ kΩ
$\theta = 57.5°$
$Z = 5.59$ kΩ
$V_R = 4.83$ V
$V_L = 7.59$ V
$I = 1.61$ mA
$Q = 1.57$

5. $X_L = 5.32$ kΩ
$\theta = 46.8°$
$f = 8.47$ kHz
$E = 9.88$ V
$V_R = 6.75$ V
$I = 1.35$ mA

6. $X_L = 954$ Ω
$\theta = 72.5°$
$f = 7.59$ kHz
$E = 3.36$ V
$V_R = 1.01$ V
$I = 3.36$ mA

SELF-TEST 18-1   Refer to Figure 18-1.

1. $V_R = 17$ V, $\theta = -36°$. Find V_C and E. **2.** $R = 2$ kΩ, $C = 50$ nF, $f = 1$ kHz, and E = 10 V. Find X_C, Z, θ, V_R, V_C, and I.

3. $R = 330\ \Omega$, $C = 1\ \mu F$, $Z = 600\ \Omega$, and $E = 9$ V.
Find X_C, f, θ, V_R, V_C, and I.

Refer to Figure 18-3.

4. $V_L = 3.3$ V, $\theta = 65°$. Find V_R and E.

5. $R = 4.7$ kΩ, $L = 175$ mH, $f = 5$ kHz, and $E = 25$ V. Find X_L, θ, Z, V_R, V_L, and I.

6. $L = 50$ mH, $Z = 1.2$ kΩ, $R = 750\ \Omega$, and $V_L = 7.8$ V. Find X_L, θ, f, E, V_R, and I.

Answers to Self-Test 18-1 are at the end of the chapter.

18.3 POLAR TO RECTANGULAR CONVERSION

Any plane vector may be identified by expressing the vector in terms of its magnitude and direction or in terms of its horizontal and vertical components. For example, if we are given a vector of magnitude 50 at an angle of $-30°$, we could represent the vector graphically as in Figure 18-5. In equation form we could write $r = 50\ \underline{/-30°}$. When the magnitude and direction are given, we say that the quantity is in *polar form*. Polar form notation gives us the magnitude of the hypotenuse of a right triangle and the phase angle. Knowing the hypotenuse and phase angle, we can quickly determine the x and y values necessary to cause such a vector.

$$x = r\cos\theta = 50\cos -30° = 43.3$$

$$y = r\sin\theta = 50\sin -30° = -25$$

In Chapter 15 we saw how the j operator could be used to indicate 90° rotation from the x-axis. $+j$ indicates 90° rotation in a positive direction and $-j$ indicates 90° rotation in a negative direction. $+j$, then, puts us at 90° and $-j$ puts us at 270° or $-90°$. We can use the j operator anytime we want to show 90° rotation. This means that we could write the x and y values above in this manner:

$$r = 43.3 - j25$$

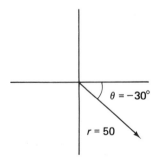

$\theta = -30°$

$r = 50$

Figure 18-5

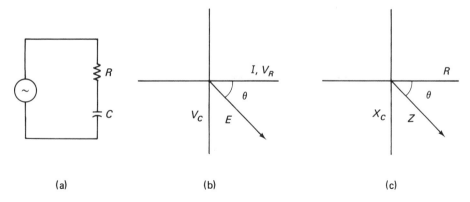

Figure 18-6

We have expressed the magnitude of the hypotenuse in terms of its rectangular coordinates. A vector expressed in these terms is said to be in *rectangular form*.

$$r_{\text{polar}} = 50 \;\underline{/-30°}$$
$$r_{\text{rect}} = 43.3 - j25$$

Either form describes the vector.

Polar and rectangular notation is used in AC circuit analysis to define the impedance (Z) and the applied voltage (E) in a series circuit. It is also used to define admittance (Y) and total current (I) in a parallel circuit.

18.3.1 Series RC and RL Circuits. Consider the circuit in Figure 18-6. We could express E in both polar and rectangular forms.

$$E \;\underline{/\theta} = V_R - jV_C$$

EXAMPLE 18.11: Let $V_R = 20$ V and $V_C = 30$ V in Figure 18-6(a). Find E and θ. Express E in both polar and rectangular forms.

Solution:

$$E_{\text{rect}} = 20 \text{ V} - j30 \text{ V}$$

$$\tan\theta = \frac{V_C}{V_R} = \frac{-30 \text{ V}}{20 \text{ V}} = -1.5$$

$$\tan^{-1} -1.5 = \theta = -56.3°$$

$$\sin\theta = \frac{V_C}{E}$$

$$E = \frac{V_C}{\sin\theta} = \frac{-30 \text{ V}}{\sin -56.3°} = 36.1 \text{ V}$$

$$E_{\text{polar}} = 36.1 \;\underline{/-56.3°} \text{ V}$$

In Figure 18-6(c) we show the relationship between X_C, R, and Z. Z in polar and rectangular forms is:

$$Z \ \underline{/\theta} = R - jX_C$$

EXAMPLE 18-12: Let $R = 7.5$ kΩ and $\theta = -27.5°$ in Figure 18-6(c). Express Z in both polar and rectangular forms.

Solution:

$$\cos\theta = \frac{R}{Z}$$

$$Z = \frac{R}{\cos\theta} = \frac{7.5\ \text{k}\Omega}{\cos -27.5°} = 8.46\ \text{k}\Omega$$

$$Z_{\text{polar}} = 8.46 \ \underline{/-27.5°}\ \text{k}\Omega$$

$$X_C = Z\sin\theta = 8.46\ \text{k}\Omega\sin -27.5° = -3.9\ \text{k}\Omega$$

$$Z_{\text{rect}} = 7.5\ \text{k}\Omega - j3.9\ \text{k}\Omega$$

In the RL circuit of Figure 18-7 we could express E in both polar and rectangular forms.

$$E \ \underline{/\theta} = V_R + jV_L$$

EXAMPLE 18-13: Let $V_R = 15$ V and $V_L = 10$ V in Figure 18-7(a). Find E and θ. Express E in both polar and rectangular forms.

Solution:

$$E_{\text{rect}} = 15\ \text{V} + j10\ \text{V}$$

$$\tan\theta = \frac{V_L}{V_R} = \frac{10\ \text{V}}{15\ \text{V}} = 0.667$$

$$\tan^{-1}0.667 = \theta = 33.7°$$

$$\cos\theta = \frac{V_R}{E}$$

$$E = \frac{V_R}{\cos\theta} = \frac{15\ \text{V}}{\cos 33.7°} = 18\ \text{V}$$

$$E_{\text{polar}} = 18 \ \underline{/33.7°}\ \text{V}$$

In Figure 18-7(c) we show the relationship between X_L, R, and Z. Z in polar and rectangular forms is:

$$Z \ \underline{/\theta} = R + jX_L$$

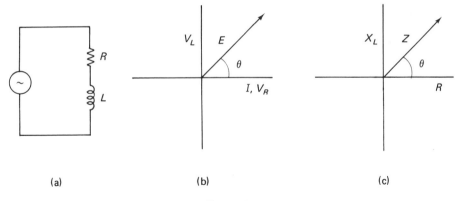

Figure 18-7

EXAMPLE 18.14: Let $X_L = 27$ kΩ and $\theta = 55°$ in Figure 18-7. Express Z in both polar and rectangular forms.

Solution:

$$\sin\theta = \frac{X_L}{Z}$$

$$Z = \frac{X_L}{\sin\theta} = \frac{27 \text{ k}\Omega}{\sin 55°} = 33 \text{ k}\Omega$$

$$Z_{\text{polar}} = 33 \underline{/55°} \text{ k}\Omega$$

$$R = Z\cos\theta = 33 \text{ k}\Omega\cos 55° = 18.9 \text{ k}\Omega$$

$$Z_{\text{rect}} = 18.9 \text{ k}\Omega + j27 \text{ k}\Omega$$

PRACTICE PROBLEMS 18-5   Express your answers to the following problems in (a) polar form and (b) rectangular form:

1. $R = 3.3$ kΩ
$\theta = 30°$

2. $X_C = 12.8$ kΩ
$\theta = -60°$

3. $Z = 17$ kΩ
$X_L = 10$ kΩ

4. $R = 270$ Ω
$X_L = 500$ Ω

5. $V_C = 8.6$ V
$\theta = -35°$

6. $V_R = 7.3$ V
$\theta = 47°$

7. $E = 15$ V
$\theta = -25°$

8. $V_R = 10.7$ V
$V_L = 6.3$ V

Solutions:

1. $Z_{\text{polar}} = 3.81 \underline{/30°}$ kΩ
$Z_{\text{rect}} = 3.3 \text{ k} + j1.91$ kΩ

2. $Z_{\text{polar}} = 14.8 \underline{/-60°}$ kΩ
$Z_{\text{rect}} = 7.39 \text{ k} - j12.8$ kΩ

3. $Z_{polar} = 17 \;/36°\; k\Omega$
   $Z_{rect} = 13.7\; k + j10\; k\Omega$

4. $Z_p = 568 \;/61.6°\; \Omega$
   $Z_{rect} = 270\; \Omega + j500\; \Omega$

5. $E_{polar} = 15 \;/-35°\; V$
   $E_{rect} = 12.3\; V - j8.6\; V$

6. $E_{polar} = 10.7 \;/47°\; V$
   $E_{rect} = 7.3\; V + j7.83\; V$

7. $E_{polar} = 15 \;/-25°\; V$
   $E_{rect} = 13.6\; V - j6.34\; V$

8. $E_{polar} = 12.4 \;/30.5°\; V$
   $E_{rect} = 10.7\; V + j6.3\; V$

Additional practice problems are at the end of the chapter.

18.3.2 Addition and Subtraction.. Only vectors which lie along the same plane can be added or subtracted directly. For example, in Figure 18-8 we have plotted a number of vectors on a system of rectangular coordinates. These vectors are labeled A through G. We could add vectors **A** and **B** algebraically. The resultant vector would be 2 units long and would fall along the positive *x*-axis. Vectors **C** and **D** could be added together. The resultant vector would be 4 units long and would be plotted along the negative *y*-axis. Vectors **E** and **G** could be added together. The resultant vector would be 2 units long and would be in the direction of vector **E**. Since vector **F** does not fall on the same plane as any other vector, it cannot be added to any other vector.

Consider the vectors labeled Z_1 and Z_2 in Figure 18-9. In polar form the impedances are:

$$Z_1 = 1 \;/30°\; k\Omega$$

$$Z_2 = 1.5 \;/50°\; k\Omega$$

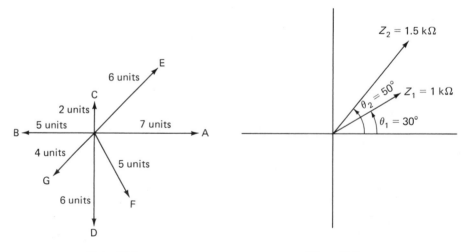

Figure 18-8   Figure 18-9

We cannot find Z_T by adding Z_1 and Z_2 because they don't lie along the same plane. However, their rectangular components do fall along the same plane. R_1, the side adjacent to θ, and R_2, the side adjacent to θ_2, fall on the same plane. XL_1, the side opposite θ, and XL_2, the side opposite θ_2 fall on the same plane. Therefore, if we change Z_1 and Z_2 to rectangular form, we can then add the rectangular components. The result of this addition is Z_T in rectangular form. This is easily converted to polar form so that we know the magnitude of Z_T and the phase angle.

$$Z_1 = 1 \underline{/30°} \text{ k}\Omega = 866 \ \Omega \ + j500 \ \Omega$$
$$Z_2 = 1.5 \underline{/50°} \text{ k}\Omega = 964 \ \Omega \ + j1150 \ \Omega$$
$$Z_T = \overline{1830 \ \Omega + j1650 \ \Omega}$$
$$Z_T = 2.46 \underline{/42°} \text{ k}\Omega$$

EXAMPLE 18.15: $Z_1 = 3 \underline{/30°}$ kΩ and $Z_2 = 2 \underline{/-60°}$ kΩ. Find Z_T. Express your answer in polar form.

Solution:

$$Z_1 = 2 \underline{/30°} \text{ k}\Omega = 1.73 \text{ k}\Omega \ + j1 \text{ k}\Omega$$

$$Z_2 = 2 \underline{/-60°} \text{ k}\Omega = 1 \text{ k}\Omega \qquad - j1.73 \text{ k}\Omega$$

$$Z_T = \overline{2.73 \text{ k}\Omega - j0.73 \text{ k}\Omega}$$

$$Z_t = 2.83 \underline{/-15.0°} \text{ k}\Omega$$

PRACTICE PROBLEMS 18.6   Find Z_T in the following problems. Express your answers in polar form.

1. $Z_1 = 17 \underline{/60°}$ kΩ, $Z_2 = 30 \underline{/20°}$ kΩ

2. $Z_1 = 270 \underline{/-35°}$ Ω, $Z_2 = 1 \underline{/65°}$ kΩ

3. $Z_1 = 100 \underline{/-70°}$ kΩ, $Z_2 = 100 \underline{/-40°}$ kΩ

4. $Z_1 = 48 \underline{/20°}$ kΩ, $Z_2 = 70 \underline{/-65°}$ kΩ

5. $Z_1 = 6.8 \underline{/70°}$ kΩ, $Z_2 = 3 \underline{/25°}$ kΩ

Solutions:

1. $44.4 \underline{/34.3°}$ kΩ

2. $990 \underline{/49.4°}$ Ω

3. $193 \underline{/-55°}$ kΩ

4. $88.3 \underline{/-32.2°}$ kΩ

5. $9.17 \underline{/56.6°}$ kΩ

Additional practice problems are at the end of the chapter.

18.3.3 Equivalent Series Circuit.
Any passive AC network may be reduced to a resistance in series with some reactance. The reactance may be either capacitive or inductive depending on the component values. In some circuits the

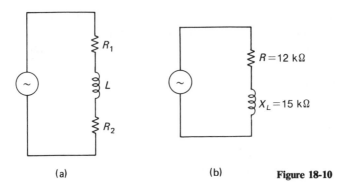

(a)   (b)   **Figure 18-10**

equivalent circuit is purely resistive. Circuits such as these will be discussed a little later in the chapter.

In a circuit such as the one in Figure 18-10(a) the total resistance is the sum of the individual resistances. $R_T = R_1 + R_2 = 12$ kΩ. Once we have found R_T, we can find Z.

$$Z_{\text{rect}} = 12 \text{ k}\Omega + j15 \text{ k}\Omega$$

$$Z_{\text{polar}} = 19.2 \underline{/51.3°} \text{ k}\Omega$$

The equivalent series circuit (ESC) is shown in Figure 18-10(b). In a circuit such as the one in Figure 18-11 we could express Z as the sum of all the resistances and reactances. All the resistances are added together to get R_T. The reactances are added together to find the resultant reactance. Since X_C and X_L are 180° out of phase, the total reactance will be the difference between X_L and X_C.

$$Z_{\text{rect}} = R_1 + R_2 + jX_L - jX_C$$
$$= 7.4 \text{ k}\Omega + j12 \text{ k}\Omega - j3 \text{ k}\Omega$$
$$= 7.4 \text{ k}\Omega + j9 \text{ k}\Omega$$

$$Z_{\text{polar}} = 11.7 \underline{/50.6°} \text{ k}\Omega$$

The ESC is shown in Figure 18-11(b).

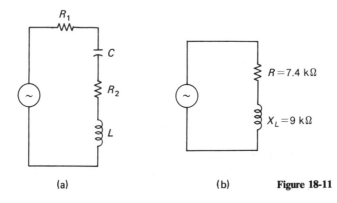

(a)   (b)   **Figure 18-11**

EXAMPLE 18.16: In Figure 18-11(a) let $f = 2.5$ kHz, $C = 10$ nF, $L = 600$ mH, $R_1 = 3.3$ kΩ, and $R_2 = 2$ kΩ. Find the ESC. Include component values.

Solution:

$$X_L = 2\pi f L = 9.42 \text{ k}\Omega$$

$$X_C = \frac{1}{2\pi f C} = 6.37 \text{ k}\Omega$$

$$Z_{\text{rect}} = 3.3 \text{ k}\Omega + 2 \text{ k}\Omega + j9.42 \text{ k}\Omega - j6.37 \text{ k}\Omega$$
$$= 5.3 \text{ k}\Omega + j3.05 \text{ k}\Omega$$

The ESC includes a resistance of 5.3 kΩ and an inductive reactance of 3.05 kΩ. The equivalent inductance is:

$$L = \frac{X_L}{2\pi f} = \frac{3.05 \text{ k}\Omega}{2\pi \times 2.5 \text{ kHz}} = 194 \text{ mH}$$

The ESC is shown in Figure 18-12.

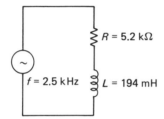

$R = 5.2$ kΩ

$f = 2.5$ kHz $L = 194$ mH

Figure 18-12

PRACTICE PROBLEMS 18-7

1. Refer to Figure 18-10(a). Find the ESC if $R_1 = 27$ kΩ, $R_2 = 12$ kΩ, and $X_L = 30$ kΩ. Express Z in both polar and rectangular forms.

2. Refer to Figure 18-11(a). Let $R_1 = 1.5$ kΩ, $R_2 = 1.2$ kΩ, $X_L = 1.5$ kΩ, and $X_C = 2.7$ kΩ. Find the ESC. Express Z in both polar and rectangular forms.

3. Refer to Figure 18-11(a). Let $R_1 = 1.5$ kΩ, $R_2 = 2.7$ kΩ, $C = 50$ nF, $L = 800$ mH, $f = 400$ Hz. Draw the ESC. Include component values.

4. Refer to Figure 18-13. Find Z_{rect}, Z_{polar}, V_R, V_C, V_L, and I. Draw the ESC.

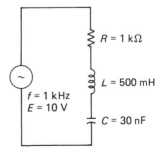

$R = 1$ kΩ

$L = 500$ mH

$f = 1$ kHz
$E = 10$ V

$C = 30$ nF

Figure 18-13

Solutions:

1. $Z_{polar} = 49.2 \; \underline{/37.6°} \; k\Omega$
$Z_{rect} = 39 \; k\Omega + j30 \; k\Omega$. See Figure 18-14.

2. $Z_{rect} = 2.7 \; k\Omega - j1.2 \; k\Omega$
$Z_{polar} = 2.95 \; \underline{/-24°} \; k\Omega$. See Figure 18-15.

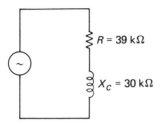

Figure 18-14

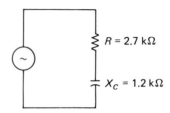

Figure 18-15

3. $Z_{rect} = 4.2 \; k\Omega - j5.95 \; k\Omega$
$Z_{polar} = 7.28 \; \underline{/-54.8°} \; k\Omega$. See Figure 18-16.

4. $Z_{rect} = 1 \; k\Omega - j2.17 \; k\Omega$
$Z_{polar} = 2.39 \; \underline{/-65.3°} \; k\Omega$. See Figure 18-17.
$V_R = 4.19 \; V$
$V_C = 9.08 \; V$
$I = 4.18 \; mA$

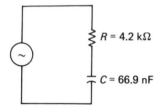

Figure 18-16

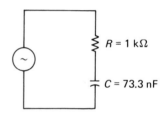

Figure 18-17

Additional practice problems are at the end of the chapter.

___

18.4 SERIES RESONANCE

In circuits that include capacitance and inductance there is a frequency where $X_L = X_C$. This frequency is called the *resonant frequency*. At this frequency (labeled f_r) the phase angle is 0°. The circuit is purely resistive because the resultant reactance is 0 Ω. The impedance is equal to R. If Z equals R, then Z must be at its minimum value. This makes I equal to its maximum value. So far we have listed the following condition at resonance:

$$X_C = X_L$$
$$\theta = 0°$$
$$Z = R$$
I is maximum.

The voltage drop across R must equal the applied voltage. The voltage drop across C would equal the voltage drop across L.

$$V_R = E$$
$$V_C = V_L$$

The resonant frequency can be determined in the following manner: Since resonance is defined as that frequency where $X_C = X_L$, then by substitution we can write:

$$\frac{1}{2\pi fC} = 2\pi fL$$

Solving for f, we get:

$$f = \frac{1}{2\pi\sqrt{LC}}$$

EXAMPLE 18.17: In Figure 18-18(a) find f_r; at this frequency find V_R, V_L, V_C, X_L, X_C, Z, and I. Draw the ESC.

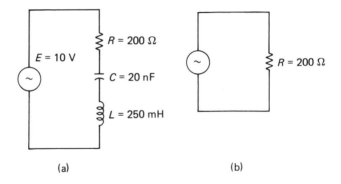

(a)   (b)

Figure 18-18

Solution:

$$f_r = \frac{1}{2\pi\sqrt{LC}} = 2.25 \text{ kHz}$$

The algorithm for this problem using a calculator is in Appendix A.

Knowing f_r, we can solve the rest of the problem by using Ohm's law or trig functions.

$$X_L = 2\pi fL = 3.53 \text{ k}\Omega$$

$$X_C = \frac{1}{2\pi fC} = 3.54 \text{ k}\Omega$$

Our values of X_L and X_C are not exactly equal because we rounded f_r to 2.25 kHz. To five places, $f_r = 2.2508$ kHz which makes $X_C = X_L = 3.5355$ kΩ. Let's stick to our three-place accuracy and remember that some slight differences may be the result. If

$$X_C = X_L \cong 3.54 \text{ k}\Omega$$

then Z is:

$$Z_{\text{rect}} = 200 \ \Omega + j3.54 \text{ k}\Omega - j3.54 \text{ k}\Omega$$

$$= 200 \ \Omega + j0$$

$$Z_{\text{polar}} = 200 \ \underline{/0°} \ \Omega$$

$$I = \frac{E}{Z} = \frac{10 \text{ V}}{200 \ \Omega} = 50 \text{ mA}$$

$$V_R = IR = 10 \text{ V}$$

$$V_C = IX_C = 177 \text{ V}$$

$$V_L = IX_L = 177 \text{ V}$$

The ESC is shown in Figure 18-18(b).

The fact that V_C and V_L are both greater than E does not mean we got something for nothing and it does not mean that we have violated Kirchhoff's laws. Remember, V_C and V_L are 180° out of phase. They are not 177 V at the same time. Added vectorially the resultant voltage drop is 0 V.

The energy stored in both L and C in the previous example is dependent on the circuit Q. When Q is known, V_C and V_L can be determined in this manner:

$$Q = \frac{X_L}{R} = \frac{3.54 \text{ k}\Omega}{200 \ \Omega} = 17.7$$

$$V_L = QE = 17.7 \times 10 \text{ V} = 177 \text{ V}$$

$$V_C = QE = 17.7 \times 10 \text{ V} = 177 \text{ V}$$

Let's see what conditions exist above and below resonance.

EXAMPLE 18.18: Assume that the operating frequency in Figure 18-18 is 2.5 kHz. Find X_L, X_C, V_R, V_L, V_C, Z, θ, and I. Draw the ESC.

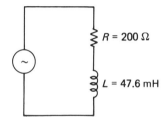

Figure 18-19

Solution:

$$X_C = \frac{1}{2\pi fC} = 3.18 \text{ k}\Omega$$

$$X_L = 2\pi fL = 3.93 \text{ k}\Omega$$

$$Z_{\text{rect}} = 200 \ \Omega + j3.93 \text{ k}\Omega - j3.18 \text{ k}\Omega$$

$$= 200 \ \Omega + j750 \ \Omega$$

$$Z_{\text{polar}} = 776 \ \underline{/75.1^\circ} \ \Omega$$

$$I = \frac{E}{Z} = \frac{10 \text{ V}}{773 \ \Omega} = 12.9 \text{ mA}$$

$$V_L = 50.8 \text{ V}$$

$$V_C = 41.1 \text{ V}$$

$$V_R = 2.59 \text{ V}$$

The ESC is shown in Figure 18-19. The resultant reactance is 750 Ω and is inductive. This inductance is:

$$L = \frac{X_L}{2\pi f} = 47.8 \text{ mH}$$

EXAMPLE 18.19: Change the operating frequency to 2 kHz in Figure 18-18. Find X_L, X_C, V_R, V_L, V_C, Z, θ, and I. Draw the ESC.

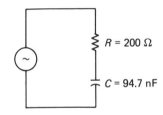

Figure 18-20

Solution:

$$X_L = 2\pi f L = 3.14 \text{ k}\Omega$$

$$X_C = \frac{1}{2\pi f C} = 3.98 \text{ k}\Omega$$

$$Z_{rect} = 200 \ \Omega + j3.14 \text{ k}\Omega - j3.98 \text{ k}\Omega$$

$$= 200 \ \Omega - j840 \ \Omega$$

$$Z_{polar} = 863 \ \underline{/-76.6°} \ \Omega$$

$$I = \frac{E}{Z} = \frac{10 \text{ V}}{863 \ \Omega} = 11.6 \text{ mA}$$

$$V_R = IR = 2.32 \text{ V}$$

$$V_C = IX_C = 46.1 \text{ V}$$

$$V_L = IX_L = 36.4 \text{ V}$$

The ESC is shown in Figure 18-20. The resultant reactance is 840 Ω and is capacitive. The capacitance is:

$$C = \frac{1}{2\pi f X_C} = 94.7 \text{ nF}$$

The examples above show that at frequencies above resonance the equivalent circuit is RL. Below resonance the equivalent circuit is RC. The impedance increased as we moved away from resonance and the current decreased.

PRACTICE PROBLEMS 18-8   Refer to Figure 18-18.

1. $R = 500 \ \Omega$, $C = 30 \text{ nF}$, $L = 100 \text{ mH}$, and $E = 25 \text{ V}$. Find f_r. Determine X_L, X_C, V_R, V_C, V_L, Z, θ, I, and the ESC at (a) $f = f_r$, (b) $f = 3.5 \text{ kHz}$, and (c) $f = 2.7 \text{ kHz}$.

2. $R = 1 \text{ k}\Omega$, $C = 500 \text{ pF}$, $L = 50 \text{ mH}$, and $E = 15 \text{ V}$. Find f_r. Determine X_L, X_C, V_R, V_C, V_L, Z, θ, I, and the ESC at (a) $f = f_r$, (b) $f = 33 \text{ kHz}$, and (c) $f = 29 \text{ kHz}$.

Solutions:

1. $f_r = 2.91 \text{ kHz}$

(a) $X_L = 1.83 \text{ k}\Omega$, $X_C = 1.83 \text{ k}\Omega$, $Z_{rect} = 500 \ \Omega + j0$, $Z_{polar} = 500 \ \underline{/0°} \ \Omega$, $I = 50 \text{ mA}$, $V_R = 25 \text{ V}$, $V_C = 91.5 \text{ V}$, and $V_L = 91.5 \text{ V}$. The ESC consists of 500 Ω of resistance.

(b) $X_L = 2.2 \text{ k}\Omega$, $X_C = 1.52 \text{ k}\Omega$, $Z_{rect} = 500 \ \Omega + j684 \ \Omega$, $Z_{polar} = 847 \ \underline{/53.8°} \ \Omega$, $I = 29.5 \text{ mA}$, $V_R = 14.8 \text{ V}$, $V_L = 64.9 \text{ V}$, and $V_C = 44.9 \text{ V}$.

The ESC consists of 500 Ω of resistance and 31.1 mH of inductance.

(c) $X_L = 1.7 \text{ k}\Omega$, $X_C = 1.96 \text{ k}\Omega$, $Z_{rect} = 500 \ \Omega - j265 \ \Omega$, $Z_{polar} = 566 \ \underline{/-27.9°} \ \Omega$, $I = 44.2 \text{ mA}$, $V_R = 22.1 \text{ V}$, $V_L = 75.1 \text{ V}$, and $V_C = 86.6 \text{ V}$. The ESC consists of 500 Ω of resistance and 222 nF of capacitance.

2. $f_r = 31.8$ kHz

(a) $X_L = 10$ kΩ, $X_C = 10$ kΩ, $Z_{rect} = 1$ kΩ $+ j0$, $Z_{polar} = 100$ $\underline{/0°}$ Ω, $I = 15$ mA, $V_R = 15$ V, $V_L = 150$ V, and $V_C = 150$ V. The ESC consists of 1 kΩ of resistance.

(b) $X_L = 10.4$ kΩ, $X_C = 9.65$ kΩ, $Z_{rect} = 1$ kΩ $+ j750$ Ω, $Z_{polar} = 1.25$ $\underline{/36.9°}$ kΩ, $I = 12$ mA, $V_R = 12$ V, $V_L = 125$ V, and $V_C = 116$ V. The

ESC consists of 1 kΩ of resistance and 3.62 mH of inductance.

(c) $X_L = 9.11$ kΩ, $X_C = 11$ kΩ, $Z_{rect} = 1$ kΩ $- j1.87$ kΩ, $Z_{polar} = 2.12$ $\underline{/-61.8°}$ kΩ, $I = 7.08$ mA, $V_R = 7.08$ V, $V_L = 64.5$ V, and $V_C = 77.8$ V. The ESC consists of 1 kΩ of resistance and 2.93 nF of capacitance.

Additional practice problems are at the end of the chapter.

SELF-TEST 18-2

1. $X_L = 11.3$ kΩ, $\theta = 27.5°$. Find Z in both polar and rectangular forms.

2. $Z_1 = 2.7$ $\underline{/17.5°}$ kΩ, $Z_2 = 4$ $\underline{/-58°}$ kΩ. Find Z_T.

3. Refer to Figure 18-11(a). Let $R_1 = 7.5$ kΩ, $R_2 = 4.7$ kΩ, $X_L = 20$ kΩ, and $X_C = 3.8$ kΩ. Find the ESC. Express Z in both polar and rectangular forms.

4. Refer to Figure 18-13. Let $R = 1.2$ kΩ, $C = 10$ nF, $L = 500$ mH, $E = 20$ V, and $f = 2.1$ kHz. Determine X_L, X_C, V_R, V_L, V_C, Z, θ, I, and the ESC.

Answers to Self-Test 18-2 are at the end of the chapter.

END OF CHAPTER PROBLEMS 18-1

Refer to Figure 18-1 for the following problems:

1. $V_R = 20$ V, $V_C = 20$ V. Find **E** and θ.

2. $V_R = 2.3$ V, $V_C = 1$ V. Find **E** and θ.

3. $V_R = 21$ V, **E** $= 40$ V. Find V_C and θ.

4. $V_R = 6$ V, **E** $= 10$ V. Find V_C and θ.

5. $V_R = 15$ V, $\theta = -40°$. Find **E** and V_C.

6. $V_R = 28$ V, $\theta = -70°$. Find **E** and V_C.

7. **E** $= 70$ V, $\theta = -27°$. Find V_R and V_C.

8. **E** $= 18$ V, $\theta = -50°$. Find V_R and V_C.

9. **E** $= 30$ V, $V_C = 20$ V. Find V_R and θ.

10. **E** $= 9$ V, $V_C = 3$ V. Find V_R and θ.

11. $V_C = 5.5$ V, $\theta = -32°$. Find **E** and V_R.

12. $V_C = 14$ V, $\theta = -55°$. Find **E** and V_R.

Refer to Figure 18-1 for the following problems and find X_C, Z, and θ:

13. $R = 10$ kΩ, $C = 50$ nF, $f = 500$ Hz

14. $R = 4.7$ kΩ, $C = 10$ nF, $f = 3.39$ kHz

15. $R = 700$ Ω, $C = 1$ μF, $f = 200$ Hz

16. $R = 18$ kΩ, $C = 200$ pF, $f = 50$ kHz

17. $R = 2$ kΩ, $C = 20$ nF, $f = 2.5$ kHz

18. $R = 1$ kΩ, $C = 1$ nF, $f = 200$ kHz

END OF CHAPTER PROBLEMS 18-2

Refer to Figure 18-1 for the following problems and find X_C, Z, θ, I, V_R, and V_C:

1. $R = 2$ kΩ, $C = 40$ nF, $f = 3$ kHz, **E** $= 30$ V

2. $R = 25$ kΩ, $C = 470$ pF, $f = 10$ kHz, **E** $= 5$ V

3. $R = 10$ kΩ, $C = 100$ pF, $f = 100$ kHz, **E** $= 10$ V

4. $R = 200$ Ω, $C = 2$ μF, $f = 200$ Hz, **E** $= 20$ V

5. $R = 3.3$ kΩ, $C = 20$ nF, $f = 1.5$ kHz, **E** $= 30$ V

6. $R = 560$ Ω, $C = 0.5$ μF, $f = 1$ kHz, **E** $= 50$ V

Refer to Figure 18-1 for the following problems and find I, θ, f, X_C, V_R, and V_C:

7. $R = 8.2$ kΩ, $C = 200$ pF, $Z = 12$ kΩ, $E = 40$ V   **8.** $R = 1$ kΩ, $C = 50$ nF, $Z = 2$ kΩ, $E = 9$ V

9. $R = 750$ Ω, $C = 10$ nF, $Z = 1.2$ kΩ, $E = 15$ V   **10.** $R = 30$ kΩ, $C = 250$ pF, $Z = 47$ kΩ, $E = 10$ V

11. $R = 2.7$ kΩ, $C = 30$ pF, $Z = 3.5$ kΩ, $E = 30$ V   **12.** $R = 18$ kΩ, $C = 1$ nF, $Z = 25$ kΩ, $E = 5$ V

END OF CHAPTER PROBLEMS 18–3

Refer to Figure 18-3 for the following problems:

1. $V_R = 8$ V, $V_L = 10$ V. Find **E** and θ.   **2.** $V_R = 22$ V, $V_L = 30$ V. Find **E** and θ.

3. $V_R = 6.7$ V, $\theta = 27°$. Find **E** and V_L.   **4.** $V_R = 13.6$ V, $\theta = 50°$. Find **E** and V_L.

5. $V_R = 8.35$ V, **E** $= 12$ V. Find V_L and θ.   **6.** $V_R = 15$ V, **E** $= 30$ V. Find V_L and θ.

7. $V_L = 15$ V, **E** $= 30$ V. Find V_R and θ.   **8.** $V_L = 30$ V, **E** $= 35$ V. Find V_R and θ.

9. $V_L = 3$ V, $\theta = 62°$. Find **E** and V_R.   **10.** $V_L = 10$ V, $\theta = 33°$. Find **E** and V_R.

11. **E** $= 9$ V, $\theta = 58°$. Find V_R and V_L.   **12.** **E** $= 22$ V, $\theta = 38°$. Find V_R and V_L.

Find X_L, Z, and θ in the following problems:

13. $R = 3.9$ kΩ, $L = 600$ mH, $f = 1$ kHz   **14.** $R = 12$ kΩ, $L = 150$ mH, $f = 20$ kHz

15. $R = 500$ Ω, $L = 40$ mH, $f = 1.5$ kHz   **16.** $R = 33$ kΩ, $L = 2$ H, $f = 5$ kHz

Find X_L, Z, θ, V_R, V_L, and I in the following problems:

17. $R = 1$ kΩ, $L = 500$ mH, $f = 250$ Hz,   **18.** $R = 22$ kΩ, $L = 200$ mH, $f = 10$ kHz, **E** $=$
E $= 15$ V   5 V

19. $R = 300$ Ω, $L = 150$ mH, $f = 500$ Hz,   **20.** $R = 15$ kΩ, $L = 40$ mH, $f = 10$ kHz, **E** $=$
E $= 50$ V   25 V

Find X_L, θ, f, **E**, V_R, and I in the following problems:

21. $R = 4.7$ kΩ, $Z = 6$ kΩ, $L = 650$ mH,   **22.** $R = 56$ kΩ, $Z = 100$ kΩ, $L = 2$ H, $V_L = 12$ V
$V_L = 7.5$ V

23. $R = 18$ kΩ, $Z = 32$ kΩ, $L = 75$ mH,   **24.** $R = 750$ Ω, $Z = 1.6$ kΩ, $L = 800$ mH,
$V_L = 3.7$ V $V_L = 8.7$ V

Find X_L, L, θ, V_R, **E**, and I in the following problems:

25. $R = 6.8$ kΩ, $Z = 11$ kΩ, $f = 4.3$ kHz,   **26.** $R = 820$ Ω, $Z = 1.25$ kΩ, $f = 2$ kHz,
$V_L = 6$ V $V_L = 13.6$ V

27. $R = 2.2$ kΩ, $Z = 4.8$ kΩ, $f = 7.5$ kHz,   **28.** $R = 560$ Ω, $Z = 900$ Ω, $f = 25$ kHz,
$V_L = 4.5$ V $V_L = 14$ V

END OF CHAPTER PROBLEMS 18-4

In problems 1 through 12, find Z. Express your answers in both *polar* and *rectangular* forms.

1. $X_L = 23$ kΩ   **2.** $X_C = 17.5$ kΩ
$\theta = 48°$ $\theta = -28°$

3. $R = 820\ \Omega$
   $\theta = -17°$

4. $R = 18\ k\Omega$
   $\theta = 72°$

5. $Z = 6.35\ k\Omega$
   $\theta = -45°$

6. $Z = 38.6\ k\Omega$
   $\theta = 24°$

7. $X_C = 30\ k\Omega$
   $R = 40\ k\Omega$

8. $X_L = 17.5\ k\Omega$
   $R = 11.3\ k\Omega$

9. $X_C = 3\ k\Omega$
   $Z = 4.6\ k\Omega$

10. $X_L = 600\ \Omega$
    $Z = 1\ k\Omega$

11. $X_L = 18\ k\Omega$
    $R = 10\ k\Omega$

12. $X_C = 18\ k\Omega$
    $R = 33\ k\Omega$

In problems 13 through 24, find E. Express your answers in both *polar* and *rectangular* forms.

13. $V_L = 12.7\ V$
    $V_R = 10\ V$

14. $V_C = 17\ V$
    $V_R = 25\ V$

15. $V_C = 5.7\ V$
    $E = 12\ V$

16. $V_L = 24\ V$
    $E = 30\ V$

17. $E = 15\ V$
    $\theta = -40°$

18. $E = 25\ V$
    $\theta = 72°$

19. $V_L = 7.93\ V$
    $\theta = 34.6°$

20. $V_C = 28\ V$
    $\theta = -66°$

21. $V_R = 2.6\ V$
    $E = 5\ V$

22. $V_C = 6\ V$
    $E = 10\ V$

23. $V_L = 16\ V$
    $E = 30\ V$

24. $V_C = 20\ V$
    $E = 25\ V$

END OF CHAPTER PROBLEMS 18-5

Find Z_T in the following problems. Express your answers in polar form.

1. $Z_1 = 2\ \underline{/20°}\ k\Omega,\ Z_2 = 3\ \underline{/65°}\ k\Omega$

2. $Z_1 = 3\ \underline{/60°}\ k\Omega,\ Z_2 = 4.5\ \underline{/25°}\ k\Omega$

3. $Z_1 = 11\ \underline{/-20°}\ k\Omega,\ Z_2 = 15\ \underline{/-30°}\ k\Omega$

4. $Z_1 = 600\ \underline{/-35°}\ \Omega,\ Z_2 = 1\ \underline{/-50°}\ k\Omega$

5. $Z_1 = 7.3\ \underline{/-25°}\ k\Omega,\ Z_2 = 3\ \underline{/40°}\ k\Omega$

6. $Z_1 = 1.73\ \underline{/-60°}\ k\Omega,\ Z_2 = 3.5\ \underline{/-27°}\ k\Omega$

7. $Z_1 = 78\ \underline{/38°}\ k\Omega,\ Z_2 = 50\ \underline{/-65°}\ k\Omega$

8. $Z_1 = 100\ \underline{/70°}\ \Omega,\ Z_2 = 400\ \underline{/-15°}\ \Omega$

9. $Z_1 = 270\ \underline{/30°}\ k\Omega,\ Z_2 = 150\ \underline{/30°}\ k\Omega$

10. $Z_1 = 5\ \underline{/90°}\ k\Omega,\ Z_2 = 3\ \underline{/0°}\ k\Omega$

11. $Z_1 = 7.5\ \underline{/-90°}\ k\Omega,\ Z_2 = 13.5\ \underline{/22°}\ k\Omega$

12. $Z_1 = 10\ \underline{/0°}\ k\Omega,\ Z_2 = 15\ \underline{/-75°}\ k\Omega$

END OF CHAPTER PROBLEMS 18-6

1. Refer to Figure 18-10(a). $R_1 = 560\ \Omega$, $R_2 = 1.2\ k\Omega$, and $X_L = 3\ k\Omega$. Draw the ESC. Express Z in both polar and rectangular forms.

2. Refer to Figure 18-10(a). $R_1 = 1.8\ k\Omega$, $R_2 = 1.2\ k\Omega$, and $X_L = 2\ k\Omega$. Draw the ESC. Express Z in both polar and rectangular forms.

3. Refer to Figure 18-11(a). $R_1 = 470$ Ω, $R_2 = 680$ Ω, $X_C = 2$ kΩ, and $X_L = 3.3$ kΩ. Draw the ESC. Express Z in both polar and rectangular forms.

4. Refer to Figure 18-11(a). $R_1 = 2.2$ kΩ, $R_2 = 4.7$ kΩ, $X_C = 6.1$ kΩ, and $X_L = 1.3$ kΩ. Draw the ESC. Express Z in both polar and rectangular forms.

5. Refer to Figure 18-11(a). $R_1 = 1.2$ kΩ, $R_2 = 2.2$ kΩ, $C = 100$ nF, $L = 800$ mH, and $f = 1.2$ kHz. Draw the ESC. Include component values.

6. Refer to Figure 18-11(a). $R = 100$ Ω, $R_2 = 330$ Ω, $C = 0.2$ μF, $L = 350$ mH, and $f = 700$ Hz. Draw the ESC. Include component values.

7. Refer to Figure 18-13. $R = 200$ Ω, $L = 200$ mH, $C = 350$ nF, $f = 500$ Hz, and $E = 25$ V. Find Z_{rect}, Z_{polar}, V_R, V_C, V_L, and I. Draw the ESC.

8. Refer to Figure 18-13. $R = 500$ Ω, $L = 500$ mH, $C = 5$ nF, $f = 3$ kHz, and $E = 20$ V. Find Z_{rect}, Z_{polar}, V_R, V_C, V_L, and I. Draw the ESC.

END OF CHAPTER PROBLEMS 18-7

Refer to Figure 18-18.

1. $R = 470$ Ω, $C = 5$ nF, $L = 750$ mH, and $E = 10$ V. Find f_r. Determine X_L, X_C, V_R, V_C, V_L, Z, θ, I, and the ESC at (a) $f = f_r$, (b) $f = 2.3$ kHz, and (c) $f = 2.7$ kHz.

2. $R = 1.2$ kΩ, $C = 1$ nF, $L = 50$ mH, and $E = 20$ V. Find f_r. Determine X_L, X_C, V_R, V_C, V_L, Z, θ, I, and the ESC at (a) $f = f_r$, (b) $f = 20$ kHz, and (c) $f = 25$ kHz.

3. $R = 300$ Ω, $C = 500$ nF, $L = 1$ H, and $E = 25$ V. Find f_r. Determine X_L, X_C, V_R, V_C, V_L, Z, θ, I, and the ESC at (a) $f = f_r$, (b) $f = 200$ Hz, and (c) $f = 300$ Hz.

4. $R = 680$ Ω, $C = 50$ nF, $L = 200$ mH, and $E = 10$ V. Find f_r. Determine X_L, X_C, V_R, V_C, V_L, Z, θ, I, and the ESC at (a) $f = f_r$, (b) $f = 1$ kHz, and (c) $f = 2$ kHz.

ANSWERS TO SELF-TESTS

Self-Test 18-1

1. $V_C = 12.4$ V
   $E = 21$ V

2. $X_C = 3.18$ kΩ
   $Z = 3.76$ kΩ
   $\theta = 57.9°$
   $V_R = 5.32$ V
   $V_C = 8.46$ V
   $I = 2.66$ mA

3. $X_C = 501$ Ω
   $f = 318$ Hz
   $\theta = -56.6°$
   $V_R = 4.95$ V
   $V_C = 7.51$ V
   $I = 15$ mA

4. $E = 3.64$ V
   $V_R = 1.54$ V

5. $X_L = 5.5$ kΩ
   $\theta = 49.5°$
   $Z = 7.23$ kΩ
   $V_R = 16.2$ V
   $V_L = 19$ V
   $I = 3.46$ mA

6. $X_L = 937$ Ω
   $\theta = 51.3°$
   $f = 2.98$ kHz
   $E = 10.3$ V
   $V_R = 6.69$ V
   $I = 8.58$ mA

Self-Test 18-2

1. $Z_{rect} = 21.7 \text{ k}\Omega + j11.3 \text{ k}\Omega$
$Z_{polar} = 24.5 \underline{/27.5°} \text{ k}\Omega$

2. $Z_{rect} = 4.7 \text{ k}\Omega - j2.58 \text{ k}\Omega$
$Z_{polar} = 5.36 \underline{/-28.8°} \text{ k}\Omega$

3. $Z_{rect} = 12.2 \text{ k}\Omega + j16.2 \text{ k}\Omega$
$Z_{polar} = 20.3 \underline{/53°} \text{ k}\Omega$
The ESC consists of 12.2 kΩ of resistance and 16.2 kΩ of inductive reactance.

4. $X_C = 7.58$ kΩ, $X_L = 6.6$ kΩ, $Z_{polar} = 1.55 \underline{/-39.3°}$ kΩ, $I = 12.9$ mA, $V_R = 15.5$ V, $V_C = 97.7$ V, $V_L = 85.1$ V. The ESC consists of 1.2 kΩ of resistance and 77.1 nF of capacitance.

AC CIRCUIT ANALYSIS—PARALLEL CIRCUITS

19.1 RC CIRCUIT ANALYSIS

In parallel circuits the voltage drops across the various branches are equal. In Figure 19-1(a) we have a resistive branch in parallel with a capacitive branch. Because the voltage drops are equal, we plot V along the 0° or x-axis as shown in Figure 19-1(b). The current through the capacitor leads the voltage drop across it by 90°; therefore, the capacitive current, I_C, is plotted along the positive y-axis. The current through the resistor, I_R, is plotted along the x-axis since no phase shift occurs due to resistance. This puts the total current, I_T, in the first quadrant. I_T then, is the hypotenuse of the right triangle and θ is the phase angle. I_R is the adjacent side and I_C is the opposite side.

EXAMPLE 19.1: In the circuit of Figure 19-1(a) let $I_R = 7$ mA and $I_C = 10$ mA. Find I_T and θ. Express I_T in both rectangular and polar forms.

Solution: Refer to Figure 19-2.

$$I_{\text{rect}} = 7 \text{ mA} + j10 \text{ mA}$$

$$\tan\theta = \frac{\text{opposite side}}{\text{adjacent side}}$$

$$\tan\theta = \frac{I_C}{I_R} = \frac{10 \text{ mA}}{7 \text{ mA}} = 1.43$$

$$\tan^{-1} 1.43 = 55°$$

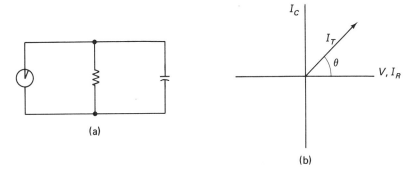

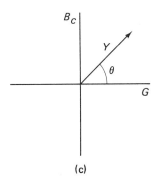

Figure 19-1

We can now find I_T by using either the sine or cosine function. We will use the cosine function.

$$\cos\theta = \frac{\text{adjacent side}}{\text{hypotenuse}}$$

$$\cos\theta = \frac{I_R}{I_T}$$

$$I_T = \frac{I_R}{\cos\theta} = \frac{7\text{ mA}}{\cos 55°} = 12.2\text{ mA}$$

$$I_{\text{polar}} = 12.2\ \underline{/55°}\text{ mA}$$

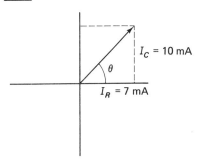

Figure 19-2

EXAMPLE 19.2: In Figure 19-1(a) let $I_T = 300$ μA and $I_C = 100$ μA. Find I_R and θ. Express I_T in both rectangular and polar forms.

Solution:

$$\sin \theta = \frac{I_C}{I_T} = \frac{100 \ \mu A}{300 \ \mu A} = 0.333$$

$$\sin^{-1} 0.333 = 19.5°$$

$$\cos \theta = \frac{I_R}{I_T}$$

$$I_R = I_T \cos \theta = 300 \ \mu A \cos 19.5° = 283 \ \mu A$$

$$I_{\text{rect}} = 283 \ \mu A + j100 \ \mu A$$

$$I_{\text{polar}} = 300 \ \underline{/19.5°} \ \mu A$$

PRACTICE PROBLEMS 19-1   In the problems below refer to Figure 19-1(a). From the information given, express I_T in rectangular and polar forms.

1. $I_T = 30$ mA and $\theta = 40°$

2. $I_R = 3.5$ mA and $I_T = 6$ mA

3. $I_C = 40$ μA and $I_T = 80$ μA

4. $I_T = 500$ mA and $\theta = 63.2°$

5. $I_R = 200$ μA and $I_C = 300$ μA

Solutions:

1. $I_{\text{rect}} = 23$ mA $+ j19.3$ mA
   $I_{\text{polar}} = 30 \ \underline{/40°}$ mA

2. $I_{\text{rect}} = 3.5$ mA $+ j4.87$ mA
   $I_{\text{polar}} = 6 \ \underline{/54.3°}$ mA

3. $I_{\text{rect}} = 69.3$ μA $+ j40$ μA
   $I_{\text{polar}} = 80 \ \underline{/30°}$ μA

4. $I_{\text{rect}} = 225$ mA $+ j446$ mA
   $I_{\text{polar}} = 500 \ \underline{/63.2°}$ mA

5. $I_{\text{rect}} = 200$ μA $+ j300$ μA
   $I_{\text{polar}} = 361 \ \underline{/56.3°}$ μA

We learned in the chapter on Ohm's law that conductance and resistance are reciprocals. Susceptance and reactance are reciprocals as are admittance and impedance. Conductance, susceptance, and admittance are parallel circuit parameters just as resistance, reactance, and impedance are series circuit parameters.

The phase relationship between conductance, G, capacitive susceptance, B_C, and admittance, Y, is shown in Figure 19-1(c). The capacitive susceptance is determined by the equation

$$B_C = 2\pi f C \qquad \qquad (19\text{-}1)$$

EXAMPLE 19.3: Refer to Figure 19-1(a). Let $R = 3.3$ kΩ, $C = 75$ nF, and $f = 500$ Hz. Find G, B_C, Y, and θ. Express Y in rectangular and polar forms.

Solution:

$$B_C = 2\pi f c = 2\pi \times 500 \text{ Hz} \times 75 \text{ nF} = 236 \text{ }\mu\text{S}$$

$$G = \frac{1}{R} = \frac{1}{3.3 \text{ k}\Omega} = 303 \text{ }\mu\text{S}$$

$$Y_{\text{rect}} = 303 \text{ }\mu\text{S} + j236 \text{ }\mu\text{S}$$

$$\tan\theta = \frac{\text{opposite side}}{\text{adjacent side}} = \frac{B_C}{G} = \frac{236 \text{ }\mu\text{S}}{303 \text{ }\mu\text{S}} = 0.779$$

$$\tan^{-1} 0.779 = 37.9°$$

$$\sin\theta = \frac{B_C}{Y}$$

$$Y = \frac{B_C}{\sin\theta} = \frac{236 \text{ }\mu\text{S}}{\sin 37.9°} = 384 \text{ }\mu\text{S}$$

$$Y_{\text{polar}} = 384 \text{ } \underline{/37.9°} \text{ }\mu\text{S}$$

EXAMPLE 19.4: Refer to Figure 19-1(a). Let $R = 680$ Ω, $C = 100$ nF, $f = 5.5$ kHz, and $I_T = 25$ mA. Find G, B_C, Y, θ, I_R, I_C, and V. Express Y in rectangular and polar forms.

Solution:

$$B_C = 2\pi f C = 2\pi \times 5.5 \text{ kHz} \times 100 \text{ nF} = 3.46 \text{ mS}$$

$$G = \frac{1}{R} = \frac{1}{680 \text{ }\Omega} = 1.47 \text{ mS}$$

$$Y_{\text{rect}} = 1.47 \text{ mS} + j3.46 \text{ mS}$$

$$\tan\theta = \frac{B_C}{G} = \frac{3.46 \text{ mS}}{1.47 \text{ mS}} = 2.35$$

$$\tan^{-1} 2.35 = 67°$$

$$\sin\theta = \frac{B_C}{Y}$$

$$Y = \frac{B_C}{\sin\theta} = \frac{3.46 \text{ mS}}{\sin 67°} = 3.76 \text{ mS}$$

$$Y_{\text{polar}} = 3.76 \text{ } \underline{/67°} \text{ mS}$$

$$I_{\text{polar}} = 25 \text{ } \underline{/67°} \text{ mA}$$

$$I_R = I_T \cos\theta = 25 \text{ mA} \cos 67° = 9.77 \text{ mA}$$

$$I_C = I_T \sin\theta = 25 \text{ mA} \sin 67° = 23 \text{ mA}$$

$$V = \frac{I_T}{Y} = 6.65 \text{ V}$$

PRACTICE PROBLEMS 19-2   Refer to Figure 19-1(a) for problems 1 through 4. Find G, B_C, Y, θ, I_R, I_C, and V. Express Y in rectangular and polar forms.

1. $R = 15$ kΩ, $C = 10$ nF, $f = 1.5$ kHz, $I_T = 500$ μA   **2.** $R = 1.8$ kΩ, $C = 200$ nF, $f = 350$ Hz, $I_T = 2$ mA

3. $R = 6.8$ kΩ, $C = 50$ nF, $f = 300$ Hz, $I_T = 10$ mA   **4.** $R = 47$ kΩ, $C = 200$ pF, $f = 15$ kHz, $I_T = 50$ mA

5. Refer to Figure 19-1(a). Let $R = 2$ kΩ, $f = 2.5$ kHz, $\theta = 35°$, and $I_C = 240$ μA. Find G, B_C, C, I_T, I_R, and V.

Solutions:

1. $Y_{rect} = 66.7$ μS $+ j94.2$ μS
$Y_{polar} = 115$ $\underline{/54.7°}$ μS
$I_{rect} = 289$ μA $+ j408$ μA
$V = 4.35$ V

2. $Y_{rect} = 556$ μS $+ j440$ μS
$Y_{polar} = 709$ $\underline{/38.4°}$ μS
$I_{rect} = 1.57$ mA $+ j1.24$ mS
$V = 2.82$ V

3. $Y_{rect} = 147$ μS $+ j94.2$ μS
$Y_{polar} = 175$ $\underline{/32.7°}$ μS
$I_{rect} = 8.42$ mA $+ j5.4$ mA
$V = 57.1$ V

4. $Y_{rect} = 21.3$ μS $+ j18.8$ μS
$Y_{polar} = 28.4$ $\underline{/41.5°}$ μS
$I_{rect} = 37.4$ μA $+ j33.2$ μA
$V = 1.76$ V

5. $G = 500$ μS
$B_C = 350$ μS $(B_C = G\tan\theta)$
$C = 22.3$ nF
$I_T = 418$ μA
$I_R = 343$ μA
$V = 686$ mV

Additional practice problems are at the end of the chapter.

19.2   RL CIRCUIT ANALYSIS

Inductance causes the applied voltage to lead the circuit current. The voltage drop across an inductor leads the current through it by 90°. Consider the circuit in Figure 19-3(a). Voltage is plotted along the reference axis just as in the RC circuit. I_L lags V by 90° and is plotted along the negative y-axis as shown in Figure 19-3(b). I_T is the resultant vector of these coordinates. I_T is the hypotenuse of a right triangle and θ is the phase angle. I_R is the adjacent side and I_L is the opposite side.

EXAMPLE 19.5: In the circuit of Figure 19-3(a) let $I_R = 30$ mA and $I_L = 20$ mA. Find I and θ. Express I_T in rectangular and polar forms.

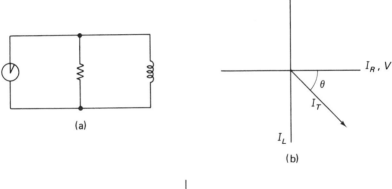

(a)

(b)

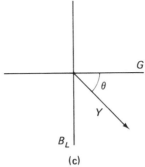

(c)

Figure 19-3

Solution: Refer to Figure 19-4.

$$\tan \theta = \frac{\text{opposite side}}{\text{adjacent side}}$$

$$\tan \theta = \frac{I_L}{I_R} = \frac{-20 \text{ mA}}{30 \text{ mA}} = -0.667$$

$$\tan^{-1} -0.667 = -33.7°$$

$$\sin \theta = \frac{\text{opposite side}}{\text{hypotenuse}} = \frac{I_L}{I_T}$$

$$I_T = \frac{I_L}{\sin \theta} = \frac{-20 \text{ mA}}{\sin -33.7°} = 36 \text{ mA}$$

$$I_{\text{rect}} = 30 \text{ mA} - j20 \text{ mA}$$

$$I_{\text{polar}} = 36 \underline{/-33.7°} \text{ mA}$$

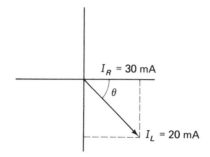

Figure 19-4

EXAMPLE 19.6: In Figure 19-3(a) let $I_T = 7.3$ mA and $I_R = 3.6$ mA. Find I_L and θ. Express I in both rectangular and polar forms.

Solution:

$$\cos \theta = \frac{I_R}{I_T} = \frac{3.6 \text{ mA}}{7.3 \text{ mA}} = 0.493$$

$$\cos^{-1} 0.493 = -60.5°$$

$$I_{\text{polar}} = 7.3 \; \underline{/-60.5°} \text{ mA}$$

$$I_L = I_T \sin \theta = 7.3 \text{ mA} \sin -60.5° = -6.35 \text{ mA}$$

$$I_{\text{rect}} = 3.6 \text{ mA} - j6.35 \text{ mA}$$

PRACTICE PROBLEMS 19-3   Refer to Figure 19-3(a). From the information given, find I_T in both rectangular and polar forms.

1. $I_R = 400 \ \mu\text{A}$
   $I_L = 300 \ \mu\text{A}$

2. $I_R = 17$ mA
   $I_T = 25$ mA

3. $I_R = 4.5$ mA
   $\theta = -65°$

4. $I_T = 50$ mA
   $\theta = -36°$

5. $I_L = 50 \ \mu\text{A}$
   $\theta = -17°$

Solutions:

1. $I_{\text{rect}} = 400 \ \mu\text{A} - j300 \ \mu\text{A}$
   $I_{\text{polar}} = 500 \; \underline{/-36.9°} \ \mu\text{A}$

2. $I_{\text{rect}} = 17$ mA $- j18.3$ mA
   $I_{\text{polar}} = 25 \; \underline{/-47.2°}$ mA

3. $I_{\text{rect}} = 4.5$ mA $- j9.65$ mA
   $I_{\text{polar}} = 10.6 \; \underline{/-65°}$ mA

4. $I_{\text{rect}} = 40.5$ mA $- j29.4$ mA
   $I_{\text{polar}} = 50 \; \underline{/-36°}$ mA

5. $I_{\text{rect}} = 164 \ \mu\text{A} - j50 \ \mu\text{A}$
   $I_{\text{polar}} = 171 \; \underline{/-17°} \ \mu\text{A}$

The phase relationship between conductance, inductive susceptance, B_L, and admittance is shown in Figure 19-3(c). The inductive susceptance is determined by the equation

$$B_L = \frac{1}{2\pi f L} \qquad (19\text{-}2)$$

EXAMPLE 19.7: Refer to Figure 19-3(a). Let $R = 6.8$ kΩ, $L = 200$ mH, $f = 4$ kHz, and $I_T = 10$ mA. Find Y and I_T in both rectangular and polar forms.

Solution:

$$G = \frac{1}{R} = \frac{1}{6.8 \text{ k}\Omega} = 147 \ \mu\text{S}$$

$$B_L = \frac{1}{2\pi f L} = \frac{1}{2\pi \times 4 \text{ kHz} \times 200 \text{ mH}} = 199 \ \mu\text{S}$$

$$Y_{\text{rect}} = 147 \ \mu\text{S} - j199 \ \mu\text{S}$$

$$\tan \theta = \frac{B_L}{G} = \frac{-199 \ \mu\text{S}}{147 \ \mu\text{S}} = -1.35$$

$$\tan^{-1} -1.35 = -53.5°$$

$$\sin \theta = \frac{B_L}{Y}$$

$$Y = \frac{B_L}{\sin \theta} = \frac{-199 \ \mu\text{S}}{\sin -53.5°} = 248 \ \mu\text{S}$$

$$Y = 248 \ \underline{/-53.5°} \ \mu\text{S}$$

$$I_{\text{polar}} = 10 \ \underline{/-53.5°} \ \text{mA}$$

$$I_R = I_T \cos \theta = 10 \text{ mA} \cos -53.5° = 5.95 \text{ mA}$$

$$I_L = I_T \sin \theta = 10 \text{ mA} \sin -53.5° = -8.04 \text{ mA}$$

$$I_{\text{rect}} = 5.95 \text{ mA} - j8.04 \text{ mA}$$

Of course, we could have found I_R and I_L by using Ohm's law:

$$I_R = \frac{I_T G}{Y} = \frac{10 \text{ mA} \times 147 \ \mu\text{S}}{248 \ \mu\text{S}} = 5.93 \text{ mA}$$

$$I_L = \frac{I_T B_L}{Y} = \frac{10 \text{ mA} \times 199 \ \mu\text{S}}{248 \ \mu\text{S}} = 8.02 \text{ mA}$$

Differences in the answers are due to rounding.

EXAMPLE 19.8: Refer to Figure 19-3(a). Let $R=12$ kΩ, $L=100$ mH, $\theta=-40°$, and $I_R=50$ μA. Find f. Find Y and I_T in both rectangular and polar forms.

Solution:

$$G=\frac{1}{R}=\frac{1}{12\text{ k}\Omega}=83.3\ \mu S$$

$$B_L=G\tan\theta=83.3\ \mu S\tan-40°=-69.9\ \mu S$$

$$Y_{rect}=83.3\ \mu S-j69.9\ \mu S$$

$$\cos\theta=\frac{G}{Y}$$

$$Y=\frac{G}{\cos\theta}=\frac{83.3\ \mu S}{\cos-40°}=109\ \mu S$$

$$Y_{polar}=109\ \underline{/-40°}\ \mu S$$

$$f=\frac{1}{2\pi B_L L}=\frac{1}{2\pi\times69.9\ \mu S\times100\text{ mH}}=22.8\text{ kHz}$$

$$I_L=I_R\tan\theta=50\ \mu A\tan-40°=-42\ \mu A$$

$$I_{rect}=50\ \mu A-j42\ \mu A$$

$$I=\frac{I_R}{\cos\theta}=\frac{50\ \mu A}{\cos-40°}=65.3\ \mu A$$

$$I_{polar}=65.3\ \underline{/-40°}\ \mu A$$

PRACTICE PROBLEMS 19-4   Refer to Figure 19-3(a). Find Y and I_T in both rectangular and polar forms in problems 1 through 4.

1. $R=1.2$ kΩ, $L=50$ mH, $f=3$ kHz, $I_T=400$ μA
2. $R=15$ kΩ, $L=20$ mH, $f=100$ kHz, $I_T=1$ mA
3. $R=470$ Ω, $L=400$ mH, $f=200$ Hz, $I_T=20$ mA
4. $R=5.6$ kΩ, $L=100$ mH, $f=9$ kHz, $I_T=50$ mA

5. Refer to Figure 19-3(a). Let $R=7.5$ kΩ, $L=50$ mH, $\theta=-55°$, and $I_R=1.3$ mA. Find f. Find Y and I_T in both rectangular and polar forms.

6. Refer to Figure 19-3(a). $L=10$ mH, $f=50$ kHz, $\theta=-35°$, and $I_L=60$ μA. Find R. Find Y and I_T in both rectangular and polar forms.

Solutions:

1. $Y_{rect}=833\ \mu S-j1.06$ mS
   $Y_{polar}=1.35\ \underline{/-51.9°}$ mS
   $I_{rect}=247\ \mu A-j315\ \mu A$
   $I_{polar}=400\ \underline{/-51.9°}\ \mu A$

2. $Y_{rect}=66.7\ \mu S-j79.6\ \mu S$
   $Y_{polar}=104\ \underline{/-50°}\ \mu S$
   $I_{rect}=642\ \mu A-j766\ \mu A$
   $I_{polar}=1\ \underline{/-50°}$ mA

3. $Y_{rect} = 2.13 \text{ mS} - j1.99 \text{ mS}$
$Y_{polar} = 2.91 \ \underline{/-43.1^6} \text{ mS}$
$I_{rect} = 14.6 \text{ mA} - j13.7 \text{ mA}$
$I_{polar} = 20 \ \underline{/-43.1°} \text{ mA}$

4. $Y_{rect} = 179 \ \mu\text{S} - j177 \ \mu\text{S}$
$Y_{polar} = 252 \ \underline{/-44.7°} \ \mu\text{S}$
$I_{rect} = 35.5 \text{ mA} - j35.2 \text{ mA}$
$I_{polar} = 50 \ \underline{/-44.7°} \text{ mA}$

5. $f = 16.8 \text{ kHz}$
$Y_{rect} = 133 \ \mu\text{S} - j190 \ \mu\text{S}$
$Y_{polar} = 232 \ \underline{/-55°} \ \mu\text{S}$
$I_{rect} = 1.3 \text{ mA} - j1.86 \text{ mA}$
$I_{polar} = 2.27 \ \underline{/-55°} \text{ mA}$

6. $R = 22.4 \text{ k}\Omega$
$Y_{rect} = 455 \ \mu\text{S} - j318 \ \mu\text{S}$
$Y_{polar} = 555 \ \underline{/-35°} \ \mu\text{S}$
$I_{rect} = 85.7 \ \mu\text{A} - j60 \ \mu\text{A}$
$I_{polar} = 105 \ \underline{/-35°} \ \mu\text{A}$

SELF-TEST 19-1

1. Refer to Figure 19-1(a). Let $R = 10 \text{ k}\Omega$, $f = 4.3 \text{ kHz}$, $C = 2.5 \text{ nF}$, and $I_T = 5 \text{ mA}$. Find Y_{rect}, Y_{polar}, I_{rect}, I_{polar}, and V.

2. Refer to Figure 19-1(a). Let $R = 27 \text{ k}\Omega$, $f = 800 \text{ Hz}$, $C = 10 \text{ nF}$, and $I_T = 500 \ \mu\text{A}$. Find Y_{rect}, Y_{polar}, I_{rect}, I_{polar}, and V.

3. Refer to Figure 19-3(a). Let $R = 120 \text{ k}\Omega$, $f = 50 \text{ kHz}$, $L = 0.5 \text{ H}$, and $I_T = 70 \ \mu\text{A}$. Find Y_{rect}, Y_{polar}, I_{rect}, I_{polar}, and V.

4. Refer to Figure 19-3(a). Let $R = 68 \text{ k}\Omega$, $f = 1.5 \text{ kHz}$, $L = 5.3 \text{ H}$, and $I_T = 2 \text{ mA}$. Find Y_{rect}, Y_{polar}, I_{rect}, I_{polar}, and V.

5. Refer to Figure 19-3(a). Let $R = 27 \text{ k}\Omega$, $L = 100 \text{ mH}$, $f = 30 \text{ kHz}$, $\theta = 55°$, and $I_L = 2 \text{ mA}$. Find R. Find Y and I_T in both rectangular and polar forms.

Answers to Self-Test 19-1 are at the end of the chapter.

19.3 EQUIVALENT CIRCUITS

19.3.1 Multiplication and Division of Vectors. We determined in Chapter 15 that addition or subtraction of vectors must be done in rectangular form. When we multiply or divide vectors, the multiplications and divisions must be done in *polar* form. Multiplication of vectors usually occurs when simplifying complex circuits in which impedances are in parallel. Such circuits will be analyzed in Chapter 20. Division of vectors occurs in the simplification of complex circuits and when making impedance–admittance conversions.

When multiplying impedances, we first put the impedances in polar form. The magnitudes are multiplied together. The angles are *added algebraically*. When dividing impedances, the magnitudes are divided. The angles are *subtracted*.

EXAMPLE 19.9: Find the circuit impedance when the circuit admittance is $300 \ \underline{/-30°} \ \mu\text{S}$.

Solution: Impedance and admittance are reciprocals.

$$Z = \frac{1}{Y}$$

$$Z = \frac{1}{300 \ \underline{/-30°} \ \mu S}$$

The rule is that we must divide in polar form; therefore, nothing further need be done to solve for Z.

$$Z = 3.33 \ \underline{/30°} \ k\Omega$$

The phase angle is positive because $(-) \ -30° = 30°$. Fo some students the process of subtraction is simplified by first moving the phase angle to the numerator, changing its sign, and then adding.

$$Z = \frac{1 \ \underline{/30°}}{300 \ \mu S} = 3.33 \ \underline{/30°} \ k\Omega$$

EXAMPLE 19.10: Two impedances Z_1 and Z_2 are in parallel. $Z = 15 \ \underline{/60°} \ k\Omega$ and $Z_2 = 10 \ \underline{/-60°} \ k\Omega$. Find Z_T.

Solution:

$$Z_T = \frac{Z_1 Z_2}{Z_1 + Z_2}$$

We will simplify the denominator first. Remembering that impedances must be added in rectangular form, our first step is to convert Z_1 and Z_2 to rectangular form.

$$Z_{1(polar)} = 15 \ \underline{/60°} \ k\Omega$$

$$R_1 = 15 \ k\Omega \cos \theta = 7.5 \ k\Omega$$

$$X_L = 15 \ k\Omega \sin \theta = 13 \ k\Omega$$

$$Z_{2(polar)} = 10 \ \underline{/-60°} \ k\Omega$$

$$R_2 = 10 \ k\Omega \cos -60° = 5 \ k\Omega$$

$$X_C = 10 \ k\Omega \sin -60° = -8.66 \ k\Omega$$

$$Z_{1(rect)} = 7.5 \ k\Omega + j13 \ k\Omega$$

$$Z_{2(rect)} = 5 \ k\Omega - j8.66 \ k\Omega$$

$$Z_1 + Z_2 = 12.5 \ k\Omega + j4.34 \ k\Omega$$

$$Z_T = \frac{15 \ \underline{/60°} \ k\Omega \times 10 \ \underline{/-60°} \ k\Omega}{12.5 \ k\Omega + j4.34 \ k\Omega}$$

The division must be done in polar form; therefore, we must change the result of $Z_1 + Z_2$ back into polar form.

$$(Z_1 + Z_2)_{\text{rect}} = 12.5 \text{ k}\Omega + j4.34 \text{ k}\Omega$$

$$\tan \theta = \frac{X_L}{R} = \frac{4.34 \text{ k}\Omega}{12.5 \text{ k}\Omega} = 0.347$$

$$\tan^{-1} 0.347 = 19.1°$$

$$Z = \frac{X_L}{\sin \theta} = \frac{4.34 \text{ k}\Omega}{\sin 19.1°} = 13.2 \text{ k}\Omega$$

$$(Z_1 + Z_2)_{\text{polar}} = 13.2 \text{ } \underline{/19.1°} \text{ k}\Omega$$

$$Z_T = \frac{15 \text{ } \underline{/60°} \text{ k}\Omega \times 10 \text{ } \underline{/-60°} \text{ k}\Omega}{13.2 \text{ } \underline{/19.1°} \text{ k}\Omega}$$

Bringing 19.1° from the denominator results in an angle of $-19.1°$ added to the existing angles.

$$Z_T = \frac{15 \text{ k}\Omega \times 10 \text{ k}\Omega \text{ } \underline{/60° + (-)60° + (-)19.1°}}{13.2 \text{ k}\Omega}$$

$$Z_T = \frac{15 \text{ k}\Omega \times 10 \text{ k}\Omega \text{ } \underline{/-19.1°}}{13.2 \text{ k}\Omega} = 11.4 \text{ } \underline{/-19.1°} \text{ k}\Omega$$

PRACTICE PROBLEMS 19-5

1. Find Z_T if $Y_T = 200 \text{ } \underline{/40°} \text{ } \mu\text{S}$.

2. Find Y_T if $Z_T = 43.7 \text{ } \underline{/-27°} \text{ k}\Omega$

3. Using the equation $Z_T = \dfrac{Z_1 Z_2}{Z_1 + Z_2}$, find Z_T if $Z_1 = 2 \text{ } \underline{/20°} \text{ k}\Omega$ and $Z_2 = 4 \text{ } \underline{/-50°} \text{ k}\Omega$.

4. Using the equation in problem 3, find Z_T if $Z_1 = 20 \text{ } \underline{/-25°} \text{ k}\Omega$ and $Z_2 = 10 \text{ } \underline{/-60°} \text{ k}\Omega$.

Solutions:

1. $Z_T = 5 \text{ } \underline{/-40°} \text{ k}\Omega$

2. $Y_T = 22.9 \text{ } \underline{/27°} \text{ } \mu\text{S}$

3. $Z_T = 1.58 \text{ } \underline{/-1.86°} \text{ k}\Omega$

4. $Z_T = 6.95 \text{ } \underline{/-48.5°} \text{ k}\Omega$

Additional practice problems are at the end of the chapter.

19.3.2 Parallel-Series Conversions. In solving complex circuit problems, it is often necessary to reduce circuits to equivalent parallel circuits or to equivalent series circuits. The parallel circuit problems we have been solving can easily be changed to equivalent series circuits. Consider Example 19.3. We determined that $Y_{\text{polar}} = 384 \text{ } \underline{/37.9°} \text{ } \mu\text{S}$. The equivalent series circuit would have

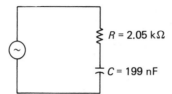

$R = 2.05\ k\Omega$

$C = 199\ nF$

Figure 19-5

an impedance equal to the reciprocal of this admittance.

$$Z_{polar} = \frac{1}{Y_{polar}} = \frac{1}{384\ \underline{/37.9°}\ \mu S} = 2.6\ \underline{/-37.9°}\ k\Omega$$

Knowing Z_{polar}, we can determine the values of R and X_C that make up the impedance.

$$R = Z\cos\theta = 2.6\ k\Omega\cos -37.9° = 2.05\ k\Omega$$
$$X_C = Z\sin\theta = 2.6\ k\Omega\sin -37.9° = -1.6\ k\Omega$$
$$Z_{rect} = 2.05\ k\Omega - j1.6\ k\Omega$$

The frequency in Example 19.3 was 500 Hz. Therefore,

$$C = \frac{1}{2\pi f X_C} = \frac{1}{2\pi \times 500\ Hz \times 1.6\ k\Omega} = 199\ nF$$

The equivalent series circuit consists of a resistance of 2.05 kΩ and a capacitance of 199 nF as shown in Figure 19-5.

EXAMPLE 19.11: Find the equivalent series circuit of the parallel circuit in Figure 19-6. Find the branch currents and the ESC.

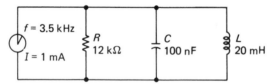

$f = 3.5\ kHz$

$I = 1\ mA$

R 12 kΩ

C 100 nF

L 20 mH

Figure 19-6

Solution:

$$G = \frac{1}{R} = \frac{1}{12\ k\Omega} = 83.3\ \mu S$$

$$B_C = 2\pi f C = 2\pi \times 3.5\ kHz \times 100\ nF = 2.2\ mS$$

$$B_L = \frac{1}{2\pi f L} = \frac{1}{2\pi \times 3.5\ kHz \times 20\ mH} = 2.27\ mS$$

$$Y_{rect} = 83.3\ \mu S + j2.2\ mS - j2.27\ mS = 83.3\ \mu S - j70\ \mu S$$

$$Y_{polar} = 109\ \underline{/-40°}\ \mu S$$

For the ESC:

$$Z_{polar} = \frac{1}{Y_{polar}} = \frac{1}{109 \; \underline{/-40°} \; \mu S} = 9.17 \; \underline{/40°} \; k\Omega$$

$$R = Z\cos\theta = 9.17 \; k\Omega \cos 40° = 7.01 \; k\Omega$$
$$X_L = Z\sin\theta = 9.17 \; k\Omega \sin 40° = 5.9 \; k\Omega$$

$$L = \frac{X_L}{2\pi f} = \frac{5.88 \; k\Omega}{2\pi \times 3.5 \; kHz} = 268 \; mH$$

The ESC is shown in Figure 19-7.

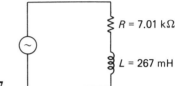

$R = 7.01 \; k\Omega$

$L = 267 \; mH$

Figure 19-7

In the original circuit:

$$V = \frac{I_T}{Y} = \frac{1 \; mA}{109 \; \mu S} = 9.17 \; V$$

$$I_R = VG = 9.17 \; V \times 83.3 \; \mu S = 764 \; \mu A$$

$$I_C = VB_C = 9.17 \; V \times 2.2 \; mS = 20.2 \; mS$$

$$I_L = VB_L = 9.17 \; V \times 2.27 \; mS = 20.8 \; mS$$

PRACTICE PROBLEMS 19-6 Find the branch currents Y, Z, V, and the ESC in the following problems:

1. Refer to Figure 19-8. $f = 1$ kHz, $C = 20$ nF, $R = 10$ kΩ, and $I_T = 10$ mA.

2. Refer to Figure 19-8. $f = 2.5$ kHz, $C = 0.5$ μF, $R = 82$ Ω, and $I_T = 300$ μA.

3. Refer to Figure 19-9. $f = 15$ kHz, $L = 300$ mH, $R = 22$ kΩ, and $I_T = 2$ mA.

4. Refer to Figure 19-9. $f = 2.5$ kHz, $L = 200$ mH, $R = 4.7$ kΩ, and $I_T = 1$ mA.

5. Refer to Figure 19-6. $f = 3.7$ kHz, $C = 100$ nF, $L = 20$ mH, $R = 12$ kΩ, and $I_T = 100$ μA.

6. Refer to Figure 19-6. $f = 7.5$ kHz, $C = 5$ nF, $L = 50$ mH, $R = 20$ kΩ, and $I_T = 15$ mA.

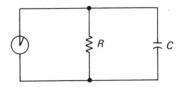

R

C

Figure 19-8

Solutions:

1. $I_R = 6.23$ mA, $I_C = 7.83$ mA, $Y = 161\ \underline{/51.5°}$ μS, $Z = 6.23\ \underline{/-51.5°}$ kΩ, and $V = 62.3$ V. In the ESC, $R = 3.88$ kΩ and $C = 32.7$ nF.

2. $I_R = 252\ \mu$A, $I_C = 163\ \mu$A, $Y = 14.5\ \underline{/32.8°}$ mS, $Z = 68.9\ \underline{/-32.8°}$ Ω, and $V = 20.7$ mV. In the ESC, $R = 58\ \Omega$ and $C = 1.71\ \mu$F.

3. $I_R = 1.58$ mA, $I_L = 1.23$ mA, $Y = 57.6\ \underline{/-37.9°}$ μS, $Z = 17.4\ \underline{/37.9°}$ kΩ, and $V = 34.8$ V. In the ESC, $R = 13.7$ kΩ and $L = 113$ mH.

4. $I_R = 556\ \mu$A, $I_L = 831\ \mu$A, $Y = 383\ \underline{/-56.2°}$ μS, $Z = 2.61\ \underline{/56.2°}$ kΩ, and $V = 2.61$ V. In the ESC, $R = 1.45$ kΩ and $L = 138$ mH.

5. $I_R = 43.2\ \mu$A, $I_C = 1.2$ mA, $I_L = 1.11$ mA, $V = 518$ mV, $Y = 193\ \underline{/64.4°}$ μS, and $Z = 5.18\ \underline{/-64.4°}$ kΩ. In the ESC, $R = 2.24$ kΩ and $C = 9.21$ nF.

6. $I_R = 3.84$ mA, $I_C = 18.1$ mA, $I_L = 32.6$ mA, $V = 76.8$ V, $Y = 195\ \underline{/-75.2°}$ μS, and $Z = 5.12\ \underline{/75.2°}$ kΩ. In the ESC, $R = 1.31$ kΩ and $L = 105$ mH.

SELF-TEST 19-2

1. Using the equation $Z_T = \dfrac{Z_1 Z_2}{Z_1 + Z_2}$, find Z_T where $Z_1 = 7\ \underline{/25°}$ kΩ and $Z_2 = 10\ \underline{/-60°}$ kΩ.

2. Refer to Figure 19-8. Find the branch currents, Y, Z, V, and the ESC where $R = 1.8$ kΩ, $C = 200$ nF, $f = 350$ Hz, and $I_T = 200\ \mu$A.

3. Refer to Figure 19-9. Find the branch currents, Y, Z, V, and the ESC where $R = 22$ kΩ, $f = 15$ kHz, $L = 300$ mH, and $I_T = 20$ mA.

4. Refer to Figure 19-6. Find the branch currents, Y, Z, V, and the ESC where $R = 10$ kΩ, $L = 2$ mH, $C = 500$ pF, $f = 150$ kHz, and $I_T = 500\ \mu$A.

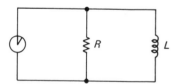

Figure 19-9

Answers to Self-Test 19-2 are at the end of the chapter.

19.4 AC NETWORKS

As we get into more complex AC circuit problems, many series/parallel conversions become necessary. The methods shown previously may be shortened by the use of equations which we shall call *transforms*. These transforms allow us to equate series resistance and reactance to their parallel equivalent conductance and susceptance, or vice versa.

To convert from a series circuit to its equivalent parallel circuit the equations are:

$$G = \frac{R}{R^2 + X^2} \quad \text{and} \quad B = \frac{X}{R^2 + X^2} \tag{19-1}$$

To convert from a parallel circuit to its equivalent series circuit the equations are:

$$R = \frac{G}{G^2 + B^2} \quad \text{and} \quad X = \frac{B}{G^2 + B^2} \tag{19-2}$$

These equations yield absolute values. Signs must be assigned to reactive and susceptive values, depending on circuit components, when problem solving.

Consider the circuit in Example 19.3 again. In that circuit, the conductance, G, is 303 μS and the susceptance, B, is 236 μS. Let's convert to its equivalent series circuit using Equation 19-2.

$$R = \frac{G}{G^2 + B_C^2} = \frac{303 \times 10^{-6}}{(303 \times 10^{-6})^2 + (236 \times 10^{-6})^2} = 2.05 \text{ k}\Omega$$

$$X_C = \frac{B_C}{G^2 + B_C^2} = \frac{236 \times 10^{-6}}{(303 \times 10^{-6})^2 + (236 \times 10^{-6})^2} = 1.6 \text{ k}\Omega$$

Notice that the denominators are the same in the two equations. You should store the denominator the first time you compute it, to save time in solving the second equation.

The equivalent series circuit in rectangular form for this problem would be

$$Z = R - jX_C = 2.05 \text{ k}\Omega - j1.6 \text{ k}\Omega$$

and the circuit of Figure 19-5 results.

> EXAMPLE 19.12: Consider the circuit in Figure 19-10. (It's like the circuit described in Example 18.8.) $X_L = 3.77$ kΩ and the impedance in rectangular form is

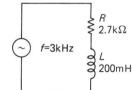

Figure 19-10.

$$Z = 2.7 \text{ k}\Omega + j3.77 \text{ k}\Omega$$

Lets convert to the equivalent parallel circuit using Equation 19-1.

Solution:

$$G = \frac{R}{R^2 + X_L^2} = \frac{2.7 \times 10^3}{(2.7 \times 10^3)^2 + (3.77 \times 10^3)^2} = 126 \ \mu S$$

$$B_L = \frac{X_L}{R^2 + X_L^2} = \frac{3.77 \times 10^3}{(2.7 \times 10^3)^2 + (3.77 \times 10^3)^2} = 175 \ \mu S$$

$$L = \frac{1}{2\pi f B_L} = \frac{1}{6.28 \times 3 \times 10^3 \times 175 \times 10^{-6}} = 303 \ mH$$

The equivalent parallel circuit in rectangular form is

$$Y = G - jB_L = 126 \ \mu S - j175 \ \mu S$$

The equivalent parallel circuit is drawn in Figure 19-11.

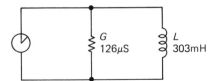

G
$126\mu S$

L
$303mH$

Figure 19-11.

PRACTICE PROBLEMS 19-7   Find Y_{rect}, and C or L in the equivalent parallel circuit for the following problems.

1. In a series RC circuit where $R = 4.7 \ k\Omega$, $C = 0.05 \ \mu F$, and $f = 500 \ Hz$.

2. In a series RL circuit where $R = 27 \ k\Omega$, $L = 200 \ mH$, and $f = 15 \ kHz$.

Find Z_{rect} and C or L in the equivalent series circuit for the following problems.

3. In a parallel RC circuit where $G = 66.7 \ \mu S$, $C = 500 \ pF$, and $f = 30 \ kHz$.

4. In a parallel RL circuit where $G = 500 \ \mu S$, $L = 500 \ mH$, and $f = 500 \ Hz$.

Solutions:

1. $Y = 75.1 \ \mu S + j102 \ \mu S$ $C = 750 \ pF$

2. $Y = 24.9 \ \mu S - j17.4 \ \mu S$ $L = 610 \ mH$

3. $Z = 5.00 \ k\Omega - j7.07 \ k\Omega$ $C = 750 \ pF$

4. $Z = 763\Omega + j972 \ \Omega$ $L = 309 \ mH$

Additional practice problems are at the end of the chapter.

Lets see how we might use these transforms to solve complex circuit problems.

EXAMPLE 19.13: Find the equivalent series circuit and the phase angle of the circuit in Figure 19-12.

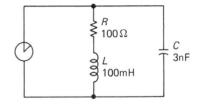

Figure 19-12.

First, let's convert the series RL circuit to its equivalent parallel circuit.

$$X_L = 2\pi fL = 6.28 \times 10 \text{ kHz} \times 100 \text{ mH} = 6.28 \text{ k}\Omega$$

$$G = \frac{R}{R^2 + X_L^2} = \frac{100}{100^2 + (6.28 \times 10^3)^2} = 2.53 \text{ }\mu\text{S}$$

$$B_L = \frac{X_L}{R^2 + X_L^2} = \frac{6.28 \times 10^3}{100^2 + (6.28 \times 10^3)^2} = 159 \text{ }\mu\text{S}$$

Next, solve for B_C

$$B_C = 2\pi fC = 6.28 \times 10 \text{ kHz} \times 3 \text{ n}F = 188 \text{ }\mu\text{S}$$

The parallel circuit in rectangular form is

$$Y = B + jB_C - jB_L = 2.53 \text{ }\mu\text{S} + j188 \text{ }\mu\text{S} - j159 \text{ }\mu\text{S}$$
$$= 2.53 \text{ }\mu\text{S} + j29.0 \text{ }\mu\text{S}$$

The equivalent parallel circuit is R and C and is shown in Figure 19-13.

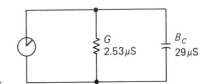

Figure 19-13.

Use Equation 19-2 to find the equivalent series circuit

$$R = \frac{G}{G^2 + B_C^2} = \frac{2.53 \times 10^{-6}}{(2.53 \times 10^{-6})^2 + (29 \times 10^{-6})^2} = 2.99 \text{ k}\Omega$$

$$X_C = \frac{B_C}{G^2 + B_C^2} = \frac{29 \times 10^{-6}}{(2.53 \times 10^{-6})^2 + (29 \times 10^{-6})^2} = 34.2 \text{ k}\Omega$$

The equivalent series circuit in rectangular form is

$$Z = 2.99 \text{ k}\Omega - j34.2 \text{ k}\Omega$$

The equivalent series capacitance is

$$C = \frac{1}{2\pi f X_C} = \frac{1}{6.28 \times 10 \times 10^3 \times 34.2 \times 10^3} = 465 \ pF$$

The phase angle is

$$\tan \theta = X_C/R = 34.2 \ k/2.99 \ k = 11.4$$

$$\arctan 11.4 = 85°$$

The ESC is shown in Figure 19-14.

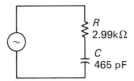

R
2.99kΩ

C
465 pF

Figure 19-14.

EXAMPLE 19.14: Find the ESC's for the circuit in Figure 19-15.

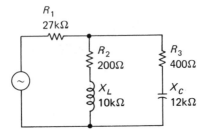

R_1
27kΩ

R_2
200Ω

R_3
400Ω

X_L
10kΩ

X_C
12kΩ

Figure 19-15.

Solution: Convert the RL series circuit (call it branch X) to its EPC using Equation 19-1.

$$G_X = \frac{R_2}{R_2^2 + X_L^2} = \frac{200}{200^2 + (10 \times 10^3)^2} = 2.00 \ \mu S$$

$$B_L = \frac{X_L}{R_2^2 + X_L^2} = \frac{10 \times 10^3}{200^2 + (10 \times 10^3)^2} = 100 \ \mu S$$

Next, convert the RC series circuit (call it branch Y) to its EPC using Equation 19-1.

$$G_Y = \frac{R_3}{R_3^2 + X_C^2} = \frac{400}{400^2 + (12 \times 10^3)^2} = 2.77 \ \mu S$$

$$B_C = \frac{X_C}{R_3^2 + X_C^2} = \frac{12 \times 10^3}{400^2 + (12 \times 10^3)^2} = 83.2 \ \mu S$$

The EPC is

$$Y = G_X + G_Y + jB_C - jB_L$$
$$= 2.00 \ \mu S + 2.77 \ \mu S + j83.2 \ \mu S - j100 \ \mu S$$
$$= 4.77 \ \mu S - j16.8 \ \mu S$$

At this point the circuit looks like Figure 19-16.

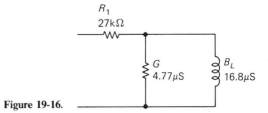

R_1
27kΩ

G
4.77μS

B_L
16.8μS

Figure 19-16.

Next, find the equivalent series circuit of the parallel portion of Figure 19-15.

$$R_{eq} = \frac{G}{G^2 + B_L^2} = \frac{4.77 \times 10^{-6}}{\left(4.77 \times 10^{-6}\right)^2 + \left(16.8 \times 10^{-6}\right)^2}$$
$$= 15.6 \ k\Omega$$

$$X_L = \frac{X_L}{G^2 + B_L^2} = \frac{16.8 \times 10^{-6}}{\left(4.77 \times 10^{-6}\right)^2 + \left(16.8 \times 10^{-6}\right)^2}$$
$$= 55.1 \ k\Omega$$

The equivalent series circuit in rectangular form is

$$Z = R_1 + R_{eq} + jX_L = 42.6 \ k + j55.1 \ k$$

and is drawn in Figure 19-17.

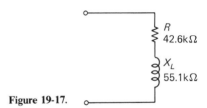

R
42.6kΩ

X_L
55.1kΩ

Figure 19-17.

PRACTICE PROBLEMS 19-8   Find the equivalent series circuit for the following problems. Express answers in rectangular form.

1. Refer to Figure 19-12. Let $R = 2$ kΩ, $X_L = 3$ kΩ and $X_C = 5$ kΩ.

2. Refer to Figure 19-18.

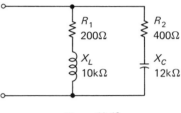

Figure 19-18.

3. Refer to Figure 19-19.

4. Refer to Figure 19-20. In addition to the ESC, find the component values and the phase angle.

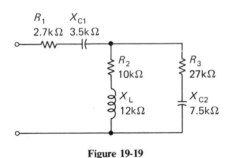

Figure 19-19

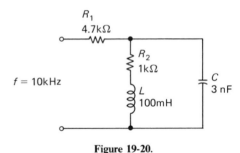

Figure 19-20.

Solutions:

1. The RL branch conversion is

$$Y = 154 \ \mu S - j231 \ \mu S$$

The equivalent parallel circuit in rectangular form is

$$Y = 154 \ \mu S + j200 \ \mu S - j231 \ \mu S$$
$$= 154 \ \mu S - j30.8 \ \mu S$$

Converting to the ESC we get

$$Z = 6.25 \text{ k}\Omega + j1.25 \text{ k}\Omega$$

The ESC is resistive and inductive.

2. The RL branch conversion is

$$Y = 2.00 \ \mu S - 100 \ \mu S$$

The RC branch conversion is

$$Y = 2.78 \ \mu S + j83.2 \ \mu S$$

The resultant equivalent parallel circuit is

$$Y = G_1 + G_2 + jB_C - jB_L$$
$$= 2.00 \ \mu S + 2.78 \ \mu S + j83.2 \ \mu S - j100 \ \mu S$$
$$= 4.77 \ \mu S - j16.7 \ \mu S$$

Converting to the ESC we get

$$Z = 15.8 \text{ k}\Omega + j55.3 \text{ k}\Omega$$

The circuit is resistive and inductive.

3. The RL branch conversion is

$$Y_L = G_2 - jB_L = 41.0 \ \mu S - j49.2 \ \mu S$$

The RC branch conversion is

$$Y_C = G_3 + jB_C = 34.4 \ \mu S + j9.55 \ \mu S$$

The EPC of the two branches in rectangular form is

$$Y = G_2 + G_3 + jB_C - jB_L$$
$$= 41.0 \ \mu S + 34.4 \ \mu S + j9.55 \ \mu S - j49.2 \ \mu S$$
$$= 75.4 \ \mu S - j39.6 \ \mu S$$

Equation 19-2 yields an ESC of $10.4 \ k\Omega + j5.47 \ k\Omega$

The ESC for the entire circuit is

$$Z = R_1 + R_{eq} + jX_L - jX_C$$
$$= 2.7 \ k\Omega + 10.4 \ k\Omega + j5.47 \ k\Omega - j3.5 \ k\Omega$$
$$= 13.1 \ k\Omega + j1.97 \ k\Omega$$

The equivalent series circuit is resistive and inductive.

4. In the RL branch $X_L = 2\pi fL = 6.28 \ k\Omega$ and $R = 4.7 \ k\Omega$

Equation 19-1 yields an EPC of $24.7 \ \mu S - j155 \ \mu S$

In the capacitive branch $B_C = 2\pi fC = 189 \ \mu S$

$$Y_T = G + jB_c - jB_L$$
$$= 24.7 \ \mu S - j155 \ \mu S + j189 \ \mu S$$
$$= 24.7 \ \mu S + j33.3 \ \mu S$$

(this is the EPC of the two branches)

Equation 19-2 yields an ESC of $14.4 \ k\Omega - j19.4 \ k\Omega$

$$Z_{ckt} = R_1 + R_{eq} - jX_C$$
$$= 4.7 \ k\Omega + j14.4 \ k\Omega - j19.4 \ k\Omega$$
$$= 19.1 \ k\Omega - j19.4 \ k\Omega$$

The circuit is resistive and capacitive. $R = 19.1 \ k\Omega$

$$C = \frac{1}{2\pi fX_C} = 820 \ pF$$

$$\text{Phase angle} = \arctan\frac{X_C}{R} = \arctan\frac{19.4}{19.1} = 45.5°$$

SELF-TEST 19-3

1. Refer to Figure 19-12. Let $f = 7$ kHz, $L = 0.3$ H, $C = 2$ nF, and $R = 1 \ k\Omega$. Find the ESC, the component values, and the phase angle.

2. Refer to Figure 19-19. Let $R_1 = 1 \ k\Omega$, $X_{C1} = 5 \ k\Omega$, $R_2 = 10 \ k\Omega$, $R_3 = 15 \ k\Omega$, $X_L = 2 \ k\Omega$, $X_{C2} = 7.5 \ k\Omega$. Find the ESC resistance and reactance. Express answer in rectangular form.

Answers to Self-Test 19-3 are at the end of the chapter.

END OF CHAPTER PROBLEMS 19-1

Refer to Figure 19-1(a) for problems 1 through 10. Find Y_{rect}, Y_{polar}, I_{rect}, and V.

1. $R = 3.3 \ k\Omega$, $C = 30$ nF, $f = 2$ kHz, $I_T = 4$ mA

2. $R = 20 \ k\Omega$, $C = 200$ nF, $f = 100$ Hz, $I_T = 500 \ \mu A$

3. $R = 750 \ \Omega$, $C = 5$ nF, $f = 60$ kHz, $I_T = 1$ mA

4. $R = 12 \ k\Omega$, $C = 500$ pF, $f = 35$ kHz, $I_T = 25$ mA

5. $R = 6.8$ kΩ, $C = 10$ nF, $f = 1.5$ kHz, $I_T = 20$ mA

6. $R = 47$ kΩ, $C = 0.1$, μF, $f = 50$ Hz, $I_T = 40$ μA

7. $R = 10$ kΩ, $C = 30$ nF, $f = 1$ kHz, $I_T = 10$ mA

8. $R = 82$ Ω, $C = 0.75$ μF, $f = 2.5$ kHz, $I_T = 300$ μA

9. $R = 15$ kΩ, $C = 300$ pF, $f = 35$ kHz, $I_T = 2$ mA

10. $R = 10$ kΩ, $C = 1$ nF, $f = 15$ kHz, $I_T = 7.5$ mA

11. Refer to Figure 19-1(a). $B_C = 200$ μS, $C = 25$ nF, $\theta = 50°$, and $I_R = 3$ mA. Find R, f, I_T, and V.

12. Refer to Figure 19-1(a). $B_C = 3.72$ mS, $C = 200$ nF, $\theta = 27°$, and $I_R = 30$ μA. Find R, f, I_T, and V.

END OF CHAPTER PROBLEMS 19-2

Refer to Figure 19-3(a). Find Y and I in both rectangular and polar forms in problems 1 through 6.

1. $R = 20$ kΩ, $L = 20$ mH, $f = 200$ kHz, $I_T = 40$ mA

2. $R = 1$ kΩ, $L = 400$ mH, $f = 600$ Hz, $I_T = 500$ μA

3. $R = 10$ kΩ, $L = 60$ mH, $f = 20$ kHz, $I_T = 200$ mA

4. $R = 3.3$ kΩ, $L = 5$ mH, $f = 65$ kHz, $I_T = 15$ mA

5. $R = 4.7$ kΩ, $L = 100$ mH, $f = 10$ kHz, $I_T = 100$ μA

6. $R = 18$ kΩ, $L = 50$ mH, $f = 50$ kHz, $I_T = 1$ mA

7. Refer to Figure 19-3(a). Let $R = 2$ kΩ, $f = 5$ kHz, $\theta = -28°$, and $I_L = 15$ mA. Find L. Find Y and I_T in both rectangular and polar forms.

8. Refer to Figure 19-3(a). Let $R = 25$ kΩ, $f = 25$ kHz, $\theta = -65°$, and $I_L = 50$ mA. Find L. Find Y and I in both rectangular and polar forms.

END OF CHAPTER PROBLEMS 19-3

1. Find Z_T if $Y_T = 3.72$ $\underline{/27°}$ mS.

2. Find Z_T if $Y_T = 37$ $\underline{/-62°}$ μS.

3. Find Y_T if $Z_T = 6.32$ $\underline{/-40°}$ kΩ.

4. Find Y_T if $Z_T = 16.7$ $\underline{/70°}$ kΩ.

Find Z_T in the following problems by using the equation $Z_T = \dfrac{Z_1 Z_2}{Z_1 + Z_2}$.

5. $Z_1 = 700$ $\underline{/70°}$ Ω, $Z_2 = 1.5$ $\underline{/-25°}$ kΩ

6. $Z_1 = 12$ $\underline{/-65°}$ kΩ, $Z_2 = 4$ $\underline{/-40°}$ kΩ

7. $Z_1 = 35$ $\underline{/20°}$ kΩ, $Z_2 = 10$ $\underline{/-40°}$ kΩ

8. $Z_1 = 5.6$ $\underline{/45°}$ kΩ, $Z_2 = 10$ $\underline{/-70°}$ kΩ

9. $Z_1 = 3.3$ $\underline{/0°}$ kΩ, $Z_2 = 4.2$ $\underline{/-90°}$ kΩ

10. $Z_1 = 12$ $\underline{/0°}$ kΩ, $Z_2 = 16$ $\underline{/90°}$ kΩ

END OF CHAPTER PROBLEMS 19-4

Find the branch currents, Y, Z, V, and the ESC in the following problems:

1. Refer to Figure 19-8. $f = 35$ kHz, $C = 400$ pF, $R = 15$ kΩ, and $I_T = 2$ mA.

2. Refer to Figure 19-8. $f = 15$ kHz, $C = 1$ nF, $R = 7.5$ kΩ, and $I_T = 7.5$ mA.

3. Refer to Figure 19-8. $f = 3$ kHz, $C = 150$ nF, $R = 470$ Ω, and $I_T = 40$ mA.

4. Refer to Figure 19-8. $f = 200$ Hz, $C = 150$ nF, $R = 3.3$ kΩ, and $I_T = 100$ μA.

5. Refer to Figure 19-9. $f=150$ kHz, $L=60$ mH, $R=47$ kΩ, and $I_T=250$ μA.

6. Refer to Figure 19-9. $f=500$ Hz, $L=3$ H, $R=12$ kΩ, and $I_T=500$ μA.

7. Refer to Figure 19-9. $f=5$ kHz, $L=30$ mH, $R=680$ Ω, and $I_T=5$ mA.

8. Refer to Figure 19-9. $f=10$ kHz, $L=15$ mH, $R=2$ kΩ, and $I_T=100$ μA.

9. Refer to Figure 19-6. $f=25$ kHz, $L=30$ mH, $C=300$ pF, $R=10$ kΩ, and $I_T=20$ mA.

10. Refer to Figure 19-6. $f=6$ kHz, $L=26.5$ mH, $C=150$ nF, $R=300$ Ω, and $I_T=250$ μA.

11. Refer to Figure 19-6. $f=15$ kHz, $L=4$ mH, $C=30$ nF, $R=6.8$ kΩ, and $I_T=400$ μA.

12. Refer to Figure 19-6. $f=250$ Hz, $L=70$ mH, $C=6$ μF, $R=680$ Ω, and $I_T=150$ μA.

END OF CHAPTER PROBLEMS 19-5

For problems 1–8, find the ESC. Express answers in rectangular form.

1. Refer to Figure 19-12. Let $R=500$ Ω, $X_L=10$ kΩ, and $X_C=20$ kΩ.

2. Refer to Figure 19-12. Let $R=100$ Ω, $X_L=250$ Ω, and $X_C=150$ Ω.

3. Refer to Figure 19-18. Let $R_1=15$ kΩ, $R_2=10$ kΩ, $X_L=5$ kΩ, and $X_C=5$ kΩ.

4. Refer to Figure 19-18. Let $R_1=10$ kΩ, $R_2=4.7$ kΩ, $X_L=2$ kΩ, and $X_C=8$ kΩ.

5. Refer to Figure 19-20. Let $R_1=5.6$ kΩ, $R_2=27$ kΩ, $X_L=35$ kΩ, and $X_C=20$ kΩ.

6. Refer to Figure 19-20. Let $R_1=12$ kΩ, $R_2=5$ kΩ, $X_L=10$ kΩ, and $X_C=20$ kΩ.

7. Refer to Figure 19-19. Let $R_1=1.2$ kΩ, $R_2=12$ kΩ, $R_3=10$ kΩ, $X_L=12$ kΩ, and $X_C=3$ kΩ.

8. Refer to Figure 19-19. Let $R_1=2.7$ kΩ, $R_2=12$ kΩ, $R_3=8.2$ kΩ, $X_L=10$ kΩ, and $X_C=7.5$ kΩ.

For problems 9–14, find the ESC. Include component values and the phase angle.

9. Refer to Figure 19-12. Let $f=1$ kHz, $R=1$ kΩ, $L=500$ mH, and $C=3$ nF.

10. Refer to Figure 19-12. Let $f=5$ kHz, $R=1$ kΩ, $L=200$ mH, and $C=10$ nF.

11. Refer to Figure 19-18. Let $R_1=500$ Ω, $R_2=200$ Ω, $L=250$ mH, $C=3$ nF, and $f=5$ kHz.

12. Refer to Figure 19-18. Let $R_1=1$ kΩ, $R_2=470$ Ω, $L=500$ mH, $C=500$ pF, and $f=10$ kHz.

13. Refer to Figure 19-19. $R_1=3.3$ kΩ, $R_2=20$ kΩ, $R_3=6.8$ kΩ, $C_1=20$ nF, $C_2=8$ nF, $L=3.5$ H, and $f=1$ kHz.

14. Refer to Figure 19-19. $R_1=1$ kΩ, $R_2=100$ Ω, $R_3=100$ Ω, $C_1=1$ μF, $C_2=1.2$ μF, $L=1.6$ H, and $f=100$ Hz.

ANSWERS TO SELF-TESTS

Self-Test 19-1

1. $Y_{rect}=100$ μS$+j67.5$ μS, $Y_{polar}=121$ $\underline{/34°}$ μS $I_{rect}=4.14$ mA$+j2.8$ mA, $I_{polar}=5$ $\underline{/34°}$ mA, $V=41.3$ V

2. $Y_{rect}=37$ μS$+j50.3$ μS, $Y_{polar}=62.4$ $\underline{/53.6°}$ μS $I_{rect}=296$ μA$+j403$ μA, $I_{polar}=500$ $\underline{/53.6°}$ μA, $V=8.01$ V

3. $Y_{rect} = 8.33\ \mu S - j6.37\ \mu S$, $Y_{polar} =$
   $10.5\ \underline{/-37.4°}\ \mu S$
   $I_{rect} = 55.6\ \mu A - j42.5\ \mu A$, $I_{polar} =$
   $70\ \underline{/-37.4°}\ \mu A$, $V = 6.67$ V

4. $Y_{rect} = 14.7\ \mu S - j20\ \mu S$, $Y_{polar} =$
   $24.8\ \underline{/-53.7°}\ \mu S$
   $I_{rect} = 1.18\ mA - j1.61\ mA$, $I_{polar} =$
   $2\ \underline{/-53.7°}\ mA$, V = 80.6 V

5. $Y_{rect} = 37.1\ \mu S - j53.1\ \mu S$, $Y_{polar} =$
   $64.8\ \underline{/-55°}\ \mu S$
   $I_{rect} = 1.15\ mA - j1.64\ mA$, $I_{polar} =$
   $2\ \underline{/-55°}\ mA$, $V = 30.9$ V

Self-Test 19-2

1. $5.51\ \underline{/-8.3°}\ k\Omega$

2. $I_R = 157\ \mu A$, $I_C = 124\ \mu A$, $Y_{polar} =$
   $709\ \underline{/38.3°}\ \mu S$, $Z_{polar} = 1.41\ \underline{/-38.3°}\ k\Omega$, and
   $V = 282$ mV. In the ESC, $R = 1.11\ k\Omega$ and
   $C = 520$ nF.

3. $I_R = 15.8\ mA$, $I_L = 12.3\ mA$, $Y_{polar} =$
   $57.6\ \underline{/-37.9°}\ \mu S$, $Z_{polar} = 17.4\ \underline{/37.9°}\ k\Omega$,
   and $V = 348$ V. In the ESC, $R = 13.7\ k\Omega$ and
   $L = 113$ mH.

4. $I_R = 430\ \mu A$, $I_C = 2.03\ mA$, $I_L = 2.28\ mA$, Y_{polar}
   $= 116\ \underline{/-30.7°}\ \mu S$, $Z_{polar} = 8.6\ \underline{/30.7°}\ k\Omega$,
   and $V = 4.3$ V. In the ESC, $R = 7.4\ k\Omega$ and
   $L = 4.65$ mH.

Self-Test 19-3

1. The ESC in rectangular form is $Z = 29.8$ k
   $- j65.8$ k. $C = 346\ pF$ and the phase angle is
   $-65.6°$

2. The EPC of the RL branch is $Y = 96.15\ \mu S$
   $- j19.23\ \mu S$.
   The EPC of the RC branch is $Y = 53.33\ \mu S$
   $+ j26.67\ \mu S$.
   This yields a total admittance of $Y = 149.5\ \mu S$
   $+ j7.436\ \mu S$.
   The ESC of the two branches is $Z = 6.673$ k
   $- j332$.
   The circuit ESC in rectangular form is

   $$Z = 6.67\ k\Omega - j5.33\ k\Omega.$$

CHAPTER **20**

FILTERS

20.1 LOW-PASS AND HIGH-PASS FILTERS

In the preceding chapters on AC circuit analysis we calculated voltage drops and currents in many different kinds of circuits. We have seen how the distribution of these voltage drops and currents are dependent on component values and frequency. If we consider that component values remain fixed in a circuit, then the voltage drops and currents are frequency dependent only.

Consider the circuit in Figure 20-1. The circuit output voltage is developed across the capacitor. Therefore, V_o must vary as V_C varies. Without considering exact values of R and C, let's determine, in general, what happens to V_R and V_C as the frequency changes.

Because X_C is greatest at low frequencies, V_C will be greatest at low frequencies. If we plotted a curve of output voltage versus frequency, we would get a curve like the one in Figure 20-2. The curve shows that at very low frequencies $V_o = V_{in}$. As the frequency increases, X_C decreases and so does V_o.

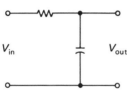

Figure 20-1.

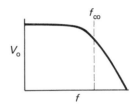

Figure 20-2.

439

Again referring to Figure 20-1, as we increase frequency from some low value, we will eventually reach a frequency where $X_C = R$. This frequency is called the *cut-off frequency* (also called the *break frequency* or the *corner frequency*). Because $X_C = R$ at the cut-off frequency (f_{co}), other conditions also exist: $\theta = -45°$, $V_C = V_R = 0.707\ V_{in}$, $I = 0.707\ I_{max}$, and $P = 0.5\ P_{max}$.

EXAMPLE 20.1: In Figure 20-1 let $V_{in} = 10$ V, $R = 10$ kΩ, and $X_C \cong 0\ \Omega$. Find V_o, V_R, I, and P.

Solution:

$$V_o = V_C = 0\ \text{V} \qquad (\text{if } X_C = 0\ \Omega,\ V_C = 0\ \text{V})$$
$$V_R = V_{in}$$
$$I = \frac{V_R}{R} = \frac{10\ \text{V}}{10\ \text{k}\Omega} = 1\ \text{mA}$$
$$P = IV_R = 1\ \text{mA} \times 10\ \text{V} = 10\ \text{mW}$$

The above example would be the result of operation at some high frequency. I and P are both maximum values here because $Z = R$ and is minimum. At lower frequencies $Z = R - jX_C$.

EXAMPLE 20.2: In Figure 20-1 let $V_{in} = 10$ V, $R = 10$ kΩ, $C = 10$ nF, and $X_C = 10$ kΩ. Find V_o, V_R, I, P, θ, and f_{co}.

Solution:

$$Z = R - jX_C = 10\ \text{k}\Omega - j10\ \text{k}\Omega = 14.1\ \underline{/-45°}\ \text{k}\Omega$$
$$I = \frac{V}{Z} = \frac{10\ \text{V}}{14.1\ \text{k}\Omega} = 0.707\ \text{mA}$$
$$V_o = V_C = IX_C = 7.07\ \text{V}$$

or $\qquad V_o = V_{in} \sin\theta = 7.07\ \text{V}$

$$V_R = IR = 7.07\ \text{V}$$

or $\qquad V_R = V_{in} \cos\theta = 7.07\ \text{V}$

$$P = IV_R = 0.707\ \text{mA} \times 7.07\ \text{V} = 5\ \text{mW}$$
$$\theta = -45°$$
$$f_{co} = \frac{1}{2\pi RC} = 1.59\ \text{kHz}$$

A circuit such as the one in Figure 20-1 is called a *low-pass filter*. That is, this circuit passes the low frequencies and rejects the high frequencies. Frequencies that result in output voltages which are less than 70.7% of maximum output are considered lost due to their low amplitude.

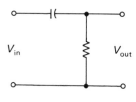

Figure 20-3.

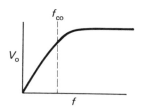

Figure 20-4.

If we reversed the position of the components as in Figure 20-3, we would have a *high-pass filter*. The curve of output voltage versus frequency for this circuit is shown in Figure 20-4. The relationships discussed in reference to Figure 20-1 are also true here except that the output voltage is taken from across R. This causes V_o to *increase* with frequency. However, the cut-off frequency is the same and all other conditions at f_{co} are the same.

PRACTICE PROBLEMS 20-1

1. Refer to Figure 20-1. Let $V_{in}=20$ V and $R=$ 3 kΩ. Find I_{max}, P_{max}, and $V_{o(max)}$. Find X_C, I, P, and V_o at f_{co}.

2. Refer to Figure 20-3. $V_{in}=10$ V, $R=2.2$ kΩ, and $C=200$ nF. Find I, V_o, and P at f_{co}.

Solutions:

1. $I_{max}=6.67$ mA, $P_{max}=133$ mW, $V_{o(max)}=$ 20 V. At f_{co}, $I=4.71$ mA, $P=66.7$ mW, and $V_o=14.1$ V.

2. $f_{co}=362$ Hz, $I=3.21$ mA, $V_o=7.07$ V, and $P=22.7$ mW.

Additional practice problems are at the end of the chapter.

20.2 THÉVENIN'S THEOREM

The use of Thévenin's theorem can simplify the solutions of some AC circuits. For example, consider the circuit in Figure 20-5(a). Suppose we wanted to find V_C. If we used trig functions, the problem would be a fairly complex one. Using Thévenin's theorem reduces the problem to a simple circuit quite easily. If we consider that C is the load, then V_{OC} and R_{TH} can be found. (See Chapter 10 for a review of Thévenin's theorem.) With the load removed as in Figure 20-5(b), $V_{OC}=V_2$ since there is no drop across R_3.

$$V_{OC}=V_2=\frac{ER_2}{R_1+R_2}=\frac{10\text{ V}\times3.3\text{ k}\Omega}{2.7\text{ k}\Omega+3.3\text{ k}\Omega}=5.5\text{ V}$$

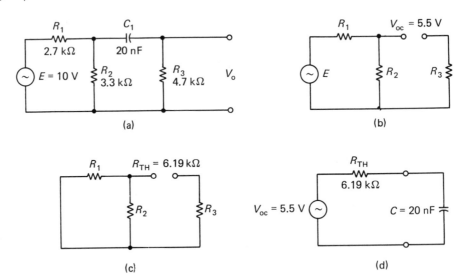

Figure 20-5.

Shorting the source results in an R_{TH} of 6.19 kΩ. The circuit is shown in Figure 20-5(c).

$$R_{TH} = R_3 + (R_1 \| R_2) = 4.7 \text{ k}\Omega + 1.49 \text{ k}\Omega = 6.19 \text{ k}\Omega$$

Thévenin's equivalent circuit with the load connected is shown in Figure 20-5(d). This equivalent circuit allows us to find the cut-off frequency of the circuit. f_{co} is the frequency where $X_C = R_{TH}$, so:

$$f_{co} = \frac{1}{2\pi C R_{TH}} = 1.29 \text{ kHz} \tag{20-1}$$

$$f_{co} = 1.29 \text{ kHz}$$

The output voltage is taken from across R_3. This results in a *high-pass* filter. The circuit will pass all frequencies above 1.29 kHz.

We can find V_o at f_{co} by using Thévenin's theorem. We know from the equivalent circuit that X_C must equal 6.19 kΩ at f_{co}. Therefore, $V_C = V_{OC} \sin 45°$ = 3.89 V. Applying Ohm's law and Kirchhoff's current law, we get:

$$I_C = \frac{V_C}{X_C} = \frac{3.89 \text{ V}}{6.19 \text{ k}\Omega} = 628 \text{ }\mu\text{A}$$

$$I_C = I_3 \quad (C_1 \text{ and } R_3 \text{ are in series})$$

$$V_3 = I_3 R_3 = 628 \text{ }\mu\text{A} \times 4.7 \text{ k}\Omega = 2.95 \text{ V}$$

$$V_o = V_3 = 2.95 \text{ V}$$

In a circuit such as the one in Figure 20-6(a), V_o and f_{co} can be found easily by using Norton's theorem. In Figure 20-6(b) the load is replaced with a short

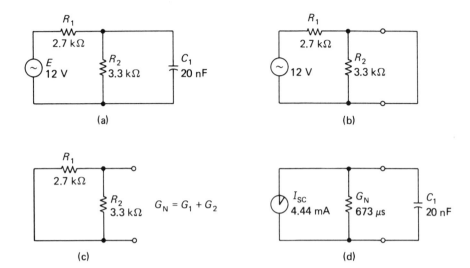

Figure 20-6.

circuit. The short circuit current (I_{SC}) would be:

$$I_{SC} = \frac{12 \text{ V}}{2.7 \text{ k}\Omega} = 4.44 \text{ mA}$$

(The short circuit reduces the resistance of the parallel circuit to 0 Ω.) G_N is found from the circuit in Figure 20-6(c).

$$G_N = G_1 + G_2 = 370 \text{ }\mu S + 303 \text{ }\mu S = 673 \text{ }\mu S$$

The equivalent circuit with the load connected is shown in Figure 20-6(d).

The circuit is a low-pass filter. C provides a low-impedance path for current at high frequencies. f_{co} may be calculated from the equivalent circuit:

$$f_{co} = \frac{G_N}{2\pi C} \tag{20-2}$$

$$f_{co} = \frac{673 \text{ }\mu S}{2\pi \times 20 \text{ nF}} = 5.36 \text{ kHz}$$

The output voltage at f_{co} can be determined by first determining $V_{o(max)}$. $V_{o(max)}$ would occur where $G_N \gg B_C$.

$$V_{o(max)} = \frac{I_{SC}}{G_N} = 6.6 \text{ V}$$

at f_{co}:

$$V_o = 0.707 \text{ V}_{o(max)} = 4.66 \text{ V}$$

PRACTICE PROBLEMS 20-2

1. Refer to Figure 20-5. Let $R_1 = 1$ kΩ, $R_2 = 2$ kΩ, $R_3 = 3$ kΩ, $C = 100$ nF, and $E = 10$ V. Thévenize the circuit. Find f_{co} and V_o at f_{co}.

2. Refer to Figure 20-7. Thévenize the circuit. Find f_{co} and V_o at f_{co}.

3. Refer to Figure 20-6. Let $R_1 = 2$ kΩ, $R_2 = 2.7$ kΩ, $C = 500$ pF, and $E = 12$ V. Nortonize the circuit. Find f_{co} and V_o at f_{co}.

4. Refer to Figure 20-8. Nortonize the circuit. Find f_{co} and V_o at f_{co}.

Solutions:

1. $V_{oc} = 6.67$ V, $R_{TH} = 3.67$ kΩ, $f_{co} = 434$ Hz, V_o at $f_{co} = 3.86$ V

2. $V_{OC} = 10$ V, $R_{TH} = 2.71$ kΩ, $f_{co} = 29.4$ Hz, V_o at $f_{co} = 4.46$ V

3. $I_{SC} = 6$ mA, $G_N = 870$ μS, $f_{co} = 277$ kHz, V_o at $f_{co} = 4.88$ V

4. $I_{SC} = 4.5$ mA, $G_N = 813$ μS, $f_{co} = 1.29$ kHz, V_o at $f_{co} = 3.91$ V

Additional practice problems are at the end of the chapter.

SELF-TEST 20-1

1. Refer to Figure 20-1. $V_{in} = 25$ V, $R = 2.2$ kΩ, and $C = 200$ nF. Find f_{co}. Find I, V_o, and P at f_{co}.

2. Refer to Figure 20-5. Let $R_1 = 4.7$ kΩ, $R_2 = 1$ kΩ, $R_3 = 1.8$ kΩ, $C = 250$ nF, and $E = 12$ V. Thévenize the circuit. Find f_{co}.

3. Refer to Figure 20-7. Let $R_1 = 3.3$ kΩ, $R_2 = 1.8$ kΩ, $R_3 = 4.7$ kΩ, $C = 500$ nF, and $E = 12$ V. Thévenize the circuit. Find f_{co}, and V_o at f_{co}.

4. Refer to Figure 20-6. Let $R_1 = 680$ Ω, $R_2 = 2.7$ kΩ, $C = 5$ nF, and $E = 20$ V. Nortonize the circuit. Find f_{co}, and V_o at f_{co}.

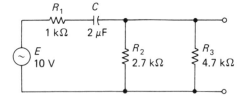

Figure 20-7.

5. Refer to Figure 20-8. Let $R_1 = 680$ Ω, $R_2 = 2.7$ kΩ, $R_3 = 6.8$ kΩ, $C = 10$ nF, and $E = 15$ V. Nortonize the circuit. Find f_{co}, and V_o at f_{co}.

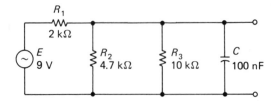

Figure 20-8.

Answers to Self-Test 20-1 are at the end of the chapter.

Band-pass filters are filters that pass a band of frequencies. That is, the filter rejects both high frequencies and low frequencies but passes a range of frequencies between. In Figure 20-9 a curve of such a filter is drawn. Various circuit configurations could produce such a curve, but, in general, the curve usually results from both series and parallel capacitances in the circuit. The curve has two cut-off frequencies. f_1 is the lower cut-off frequency and f_2 is the upper cut-off frequency.

Consider the circuit in Figure 20-10. This circuit could be the equivalent circuit of an amplifier or passive network. At low frequencies the capacitive reactances will be quite large compared to the circuit resistances. Because C_2 is connected in parallel, its high reactance (low susceptance) will have negligible effect on the circuit. C_1, though, is connected in series, thus resulting in low output voltages at these low frequencies. We can better see the low-frequency circuit if we Thévenize it. Since C_2 will have little effect on the circuit at low frequencies, we can treat it as an open circuit. With C_2 open-circuited, the new circuit is shown in Figure 20-11.

The Thévenin's equivalent circuit is shown in Figure 20-12. $R_{TH} = R_1 + R_2$. At f_1, $X_C = R_{TH}$.

$$f_1 = \frac{1}{2\pi R_{TH} C_1} \tag{20-3}$$

At high frequencies we can ignore the low reactance of C_1. The high susceptance of C_2 (low reactance) causes a reduction in output voltage at high

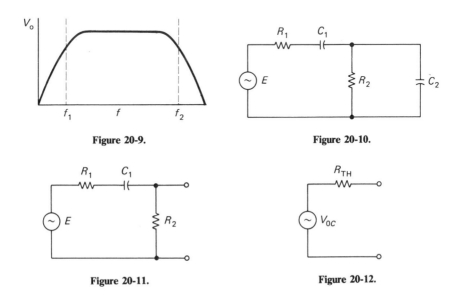

Figure 20-9.

Figure 20-10.

Figure 20-11.

Figure 20-12.

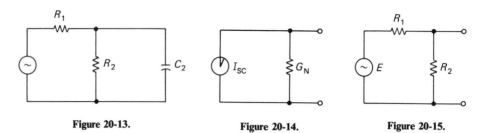

Figure 20-13.   Figure 20-14.   Figure 20-15.

frequencies. The resulting circuit at high frequencies is shown in Figure 20-13. Norton's equivalent circuit is shown in Figure 20-14. At f_2, $G_N = B_C$.

$$f_2 = \frac{G_N}{2\pi C_2} \tag{20-4}$$

In the mid-frequency range—the range of frequencies between f_1 and f_2—we ignore the effects of both capacitances. The reactance of C_1 is negligible compared to R_{TH} ($R_1 + R_2$). The susceptance of C_2 is negligible compared to G_N ($G_1 + G_2$). The mid-frequency circuit consists of R_1 and R_2, as in Figure 20-15. Even though we consider that the circuit passes a band or range of frequencies, a *middle* frequency (f_m) may be found by using the following equation:

$$f_m = \sqrt{f_1 f_2}$$

Theoretically, this middle frequency is at the mid-point of the mid-frequency range.

EXAMPLE 20.3: Refer to Figure 20-10. Let $R_1 = 2$ kΩ, $R_2 = 3$ kΩ, $C_1 = 1$ μF, $C_2 = 500$ pF, and $E = 20$ V. Find f_1, f_2, and V_o at f_m and V_o at f_{co}.

Solution:

$$f_1 = \frac{1}{2\pi R_{TH} C_1} \tag{20-1}$$

$$R_{TH} = R_1 + R_2 = 2 \text{ k}\Omega + 3 \text{ k}\Omega = 5 \text{ k}\Omega$$

$$f_1 = \frac{1}{2\pi \times 5 \text{ k}\Omega \times 1 \text{ } \mu F} = 31.8 \text{ Hz}$$

$$f_2 = \frac{G_N}{2\pi C_2}$$

$$G_N = G_1 + G_2 = 500 \text{ } \mu S + 333 \text{ } \mu S = 833 \text{ } \mu S$$

$$f_2 = \frac{833 \text{ } \mu S}{2\pi \times 500 \text{ pF}} = 265 \text{ kHz}$$

At f_m:

$$V_o = \frac{ER_2}{R_1 + R_2} = \frac{20 \text{ V} \times 3 \text{ k}\Omega}{5 \text{ k}\Omega} = 12 \text{ V}$$

At f_1 and f_2 the phase angle is 45°. At these frequencies $V_o = 0.707$ times $V_{o(max)}$. $V_{o(max)}$ is the output voltage in the mid-frequency range.

$$V_o \text{ at } f_1 = V_o \text{ at } f_2 = 12 \text{ V} \times 0.707 = 8.48 \text{ V}$$

EXAMPLE 20.4: Refer to Figure 20-16. Find f_1 and f_2.

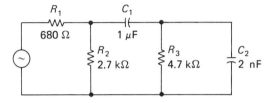

Figure 20-16.

Solution: Ignoring C_2 and treating C_1 as the load, we can determine the value of f_1:

$$R_{TH} = R_3 + R_1 \| R_2 = 5.24 \text{ k}\Omega$$

$$f_1 = \frac{1}{2\pi C_1 R_{TH}} = 30.4 \text{ Hz}$$

Treating C_1 as a short circuit and assuming that C_2 is the load, we can determine the value of f_2:

$$G_N = G_1 + G_2 + G_3 = 1.47 \text{ mS} + 370 \text{ }\mu\text{S} + 213 \text{ }\mu\text{S} = 2.05 \text{ mS}$$

$$f_2 = \frac{G_N}{2\pi C_2} = 163 \text{ kHz}$$

Consider the circuit in Figure 20-17. We recognize a series resonant circuit due to the presence of both L and C. We have determined previously that the resonant frequency (f_r) is that frequency where $X_L = X_C$. Other conditions at resonance are: $\theta = 0°$; $V_{in} = V_R = V_o$; $V_C = V_L$; $Z = R$; I is max; and P is max. The curve of output voltage versus frequency is shown in Figure 20-18.

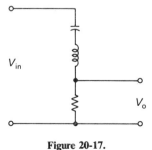

Figure 20-17.

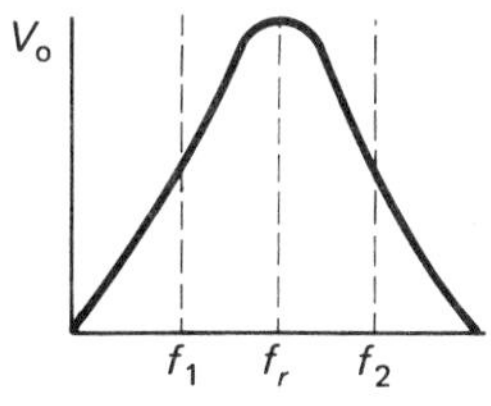

Figure 20-18.

As we increase the frequency above resonance, X_L increases and X_C decreases. Eventually we reach a frequency where the difference between X_L and X_C equals R. This is the *upper cut-off frequency*, f_2.

As we decrease the frequency below resonance, X_C increases and X_L decreases. Eventually we reach a frequency where the difference between X_C and X_L equals R. This is the *lower cut-off frequency*, f_1. In RCL circuits the range of frequencies between f_1 and f_2 is called the *bandwidth* (BW).

$$BW = f_2 - f_1$$

$$f_1 = f_r - \frac{BW}{2}$$

$$f_2 = f_r + \frac{BW}{2}$$

There is a direct relationship between f_r, BW, and circuit Q.

$$BW = \frac{f_r}{Q}$$

EXAMPLE 20.5: In Figure 20-17 let $V_{in} = 3$ V, $R = 100$ Ω, $f_r = 1$ kHz, and $Q = 20$. Find the bandwidth; f_1; f_2; I_{max}; P_{max}; V_R at resonance; and V_R, I, and P at f_1 and f_2.

Solution:

$$BW = \frac{f_r}{Q} = \frac{1 \text{ kHz}}{20} = 50 \text{ Hz}$$

$$f_1 = f_r - \frac{BW}{2} = 1 \text{ kHz} - 25 \text{ Hz} = 975 \text{ Hz}$$

$$f_2 = f_r + \frac{BW}{2} = 1 \text{ kHz} + 25 \text{ Hz} = 1025 \text{ Hz}$$

$V_R = V_{in} = 3$ V at resonance. At other frequencies V_R would be less than 3 V. Therefore:

$$I_{max} = \frac{V_R}{R} = \frac{3 \text{ V}}{100 \text{ } \Omega} = 30 \text{ mA}$$

$$P_{max} = V_R I = 3 \text{ V} \times 30 \text{ mA} = 90 \text{ mW}$$

At f_1, $\theta = -45°$. At f_2, $\theta = 45°$. Then at either f_1 or f_2:

$$V_R = V_{in} \cos 45° = 3 \text{ V} \times 0.707 = 2.12 \text{ V}$$

$$I = \frac{V_R}{R} = \frac{2.12 \text{ V}}{100 \text{ } \Omega} = 21.2 \text{ mA}$$

$$P = I V_R = 21.2 \text{ mA} \times 2.12 \text{ V} = 45 \text{ mW}$$

This circuit (Figure 20-17) is a *band-pass* filter. The circuit passes the band of frequencies between f_1 and f_2 and rejects all others.

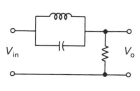

Figure 20-19.

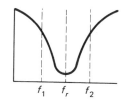

Figure 20-20.

The circuit in Figure 20-19 is a *band-reject* filter. The curve of output voltage versus frequency is shown in Figure 20-20. The parallel LC circuit offers a maximum impedance to the circuit at resonance. Therefore, circuit current is minimum and V_o is minimum. The frequencies between f_1 and f_2 are rejected.

PRACTICE PROBLEMS 20-3

1. Refer to Figure 20-10. Let $R_1 = 4.7$ kΩ, $R_2 = 1.8$ kΩ, $C_1 = 100$ nF, $C_2 = 1$ nF, and $E = 20$ V. Find f_1, f_2, V_o at f_m, and V_o at f_{co}.

2. Refer to Figure 20-16. Let $R_1 = 1.5$ kΩ, $R_2 = 3.3$ kΩ, $R_3 = 2.7$ kΩ, $C_1 = 200$ nF, and $C_2 = 700$ pF. Find f_1 and f_2.

3. Refer to Figure 20-17. $V_{in} = 9$ V, $R = 200$ Ω, $f_r = 800$ Hz, and $Q = 20$. Find BW, f_1, and f_2. Find V_R, P, and I at f_r, f_1, and f_2.

4. Refer to Figure 20-17. $L = 200$ mH, $C = 200$ nF, and $R = 100$ Ω. Find f_r, f_1, f_2, BW, and Q.

Solutions:

1. $f_1 = 245$ Hz, $f_2 = 122$ kHz, V_o at $f_m = 5.54$ V, V_o at $f_{co} = 3.92$ V

2. $f_1 = 213$ Hz, $f_2 = 305$ kHz

3. $BW = 40$ Hz, $f_1 = 780$ Hz, $f_2 = 820$ Hz. At f_r, $V_R = 9$ V, $I = 45$ mA, and $P = 405$ mW. At f_{co}, $V_R = 6.36$ V, $I = 31.8$ mA, and $P = 202$ mW.

4. $f_r = 796$ Hz, $f_1 = 756$ Hz, $f_2 = 836$ Hz, $BW = 79.6$ Hz, $Q = 10$

Additional practice problems are at the end of the chapter.

SELF-TEST 20-2

1. Refer to Figure 20-10. Let $R_1 = 2$ kΩ, $R_2 = 1.2$ kΩ, $C_1 = 200$ nF, $C_2 = 2$ nF, and $E = 9$ V. Find f_1, f_2, V_o at f_m, and V_o at f_{co}.

2. Refer to Figure 20-16. Let $R_1 = 3.9$ kΩ, $R_2 = 4.7$ kΩ, $R_3 = 2.2$ kΩ, $C_1 = 100$ nF, and $C_2 = 2$ nF. Find f_1 and f_2.

3. Refer to Figure 20-17. $L = 200$ mH, $C = 2$ nF, and $R = 1$ kΩ. Find f_r, f_1, BW, and Q.

Answers to Self-Test 20-2 are at the end of the chapter.

END OF CHAPTER PROBLEMS 20-1

1. Refer to Figure 20-1. Let $V_{in} = 15$ V and $R = 470$ Ω. Find I_{max}, P_{max}, and $V_{o(max)}$. Find X_C, I, P, and V_o at f_{co}.

2. Refer to Figure 20-1. Let $V_{in} = 25$ V and $R = 5.6$ kΩ. Find I_{max}, P_{max}, and $V_{o(max)}$. Find X_C, I, P, and V_o at f_{co}.

3. Refer to Figure 20-1. $V_{in} = 20$ V, $R = 1.2$ kΩ, and $C = 500$ nF. Find f_{co}. Find I, V_o, and P at f_{co}.

4. Refer to Figure 20-1. $V_{in} = 25$ V, $R = 300$ Ω, and $C = 1$ μF. Find f_{co}. Find I, V_o, and P at f_{co}.

END OF CHAPTER PROBLEMS 20-2

1. Refer to Figure 20-5. Let $R_1 = 1.8$ kΩ, $R_2 = 3.3$ kΩ, $R_3 = 1$ kΩ, $C_1 = 500$ nF, and $E = 10$ V. Thévenize the circuit and find f_{co}.

2. Refer to Figure 20-5. Let $R_1 = 2.7$ kΩ, $R_2 = 1.2$ kΩ, $R_3 = 3$ kΩ, $C_1 = 1$ μF, and $E = 15$ V. Thévenize the circuit and find f_{co}.

3. Refer to Figure 20-21. Thévenize the circuit and find f_{co}, and V_o at f_{co}.

4. Refer to Figure 20-21. Let $R_1 = 5.6$ kΩ, $R_2 = 2.7$ kΩ, $C = 1$ μF, and $E = 10$ V. Thévenize the circuit and find f_{co}, and V_o at f_{co}.

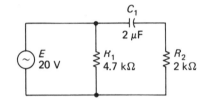

Figure 20-21.

5. Refer to Figure 20-7. Let $R_1 = 2.7$ kΩ, $R_2 = 3.9$ kΩ, $R_3 = 5.6$ kΩ, $C_1 = 1$ μF, and $E = 10$ V. Thévenize the circuit and find f_{co}, and V_o at f_{co}.

6. Refer to Figure 20-7. Let $R_1 = 750$ Ω, $R_2 = 1$ kΩ, $R_3 = 1.5$ kΩ, $C_1 = 800$ nF, and $E = 20$ V. Thévenize the circuit and find f_{co}, and V_o at f_{co}.

7. Refer to Figure 20-22. Thévenize the circuit and find f_{co}.

8. Refer to Figure 20-22. Let $R_1 = 7.5$ kΩ, $R_2 = 2.7$ kΩ, $R_3 = 4.7$ kΩ, $R_4 = 3$ kΩ, $C = 500$ nF, and $E = 15$ V. Thévenize the circuit and find f_{co}.

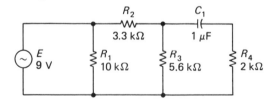

Figure 20-22.

9. Refer to Figure 20-6(a). Let $R_1 = 470$ Ω, $R_2 = 1.2$ kΩ, $C = 50$ nF, and $E = 15$ V. Nortonize the circuit and find f_{co}, and V_o at f_{co}.

10. Refer to Figure 20-6(a). Let $R_1 = 1.5$ kΩ, $R_2 = 3.3$ kΩ, $C = 100$ nF, and $E = 20$ V. Nortonize the circuit and find f_{co}, and V_o at f_{co}.

11. Refer to Figure 20-8. Let $R_1 = 2$ kΩ, $R_2 = 4.7$ kΩ, $R_3 = 5.6$ kΩ, $C_1 = 1$ nF, and $E = 10$ V. Nortonize the circuit and find f_{co}, and V_o at f_{co}.

12. Refer to Figure 20-8. Let $R_1 = 3$ kΩ, $R_2 = 2.7$ kΩ, $R_3 = 3.3$ kΩ, $C_1 = 2$ nF, and $E = 12$ V. Nortonize the circuit and find f_{co}, and V_o at f_{co}.

END OF CHAPTER PROBLEMS 20-3

1. Refer to Figure 20-10. Let $R_1 = 680$ Ω, $R_2 = 1.2$ kΩ, $C_1 = 500$ nF, $C_2 = 2$ nF, and $E = 9$ V. Find f_1, f_2, V_o at f_m, and V_o at f_{co}.

2. Refer to Figure 20-10. Let $R_1 = 2.7$ kΩ, $R_2 = 1.5$ kΩ, $C_1 = 100$ nF, $C_2 = 100$ pF, and $E = 10$ V. Find f_1, f_2, V_o at f_m, and V_o at f_{co}.

3. Refer to Figure 20-10. Let $R_1 = 3.3$ kΩ, $R_2 = 2$ kΩ, $C_1 = 2$ μF, $C_2 = 1$ nF, and $E = 15$ V. Find f_1, f_2, V_o at f_m, and V_o at f_{co}.

4. Refer to Figure 20-10. Let $R_1 = 1.8$ kΩ, $R_2 = 2.2$ kΩ, $C_1 = 750$ nF, $C_2 = 500$ pF, and $E = 5$ V. Find f_1, f_2, V_o at f_m, and V_o at f_{co}.

5. Refer to FIgure 20-16. Let $R_1 = 5.6$ kΩ, $R_2 = 2.7$ kΩ, $R_3 = 1.8$ kΩ, $C_1 = 1$ μF, and $C_2 = 2$ nF. Find f_1 and f_2.

6. Refer to Figure 20-16. Let $R_1 = 10$ kΩ, $R_2 = 2$ kΩ, $R_3 = 4.7$ kΩ, $C_1 = 1$ μF, and $C_2 = 2$ nF. Find f_1 and f_2.

7. Refer to Figure 20-17. $V_{in} = 20$ V, $R = 1$ kΩ, $f_r = 2$ kHz, and $Q = 25$. Find BW, f_1, and f_2. Find V_R, I and P at f_r, f_1, and f_2.

8. Refer to Figure 20-17. $V_{in} = 10$ V, $R = 810$ Ω, $f_r = 1$ kHz, and $Q = 10$. Find BW, f_1, and f_2. Find V_R, I and P at f_r, f_1, and f_2.

9. Refer to Figure 20-17. $L = 100$ mH, $C = 250$ nF, and $R = 45$ Ω. Find f_r, f_1, f_2, BW, and Q.

10. Refer to Figure 20-17. $L = 500$ mH, $C = 1$ nF, and $R = 470$ Ω. Find f_r, f_1, f_2, BW, and Q.

ANSWERS TO SELF-TESTS

Self-Test 20-1

1. $f_{co} = 362$ Hz, $I = 8.03$ mA, $V_o = 17.7$ V, $P = 142$ mW

2. $V_{OC} = 2.11$ V, $R_{TH} = 2.62$ kΩ, $f_{co} = 243$ HZ

3. $V_{OC} = 12$ V, $R_{TH} = 4.6$ kΩ, $f_{co} = 69.2$ Hz, V_o at $f_{co} = 2.4$ V

4. $I_{SC} = 29.4$ mA, $G_N = 1.84$ mS, $f_{co} = 58.6$ kHz, V_o at $f_{co} = 11.3$ V

5. $I_{SC} = 22.1$ mA, $G_N = 1.99$ mS, $f_{co} = 31.7$ kHz, V_o at $f_{co} = 7.85$ V

Self-Test 20-2

1. $f_1 = 249$ Hz, $f_2 = 106$ kHz, V_o at $f_m = 3.38$ V, V_o at $f_{co} = 2.39$ V

2. $f_1 = 367$ Hz, $f_2 = 73.5$ kHz

3. $f_r = 7.96$ kHz, $f_1 = 7.56$ kHz, $f_2 = 8.36$ kHz, $BW = 796$ Hz, $Q = 10$

CHAPTER 21

LOGARITHMS

Logarithms are used in solving many electrical and electronics problems. The use of logarithms simplifies the solution of some complex problems. Their use reduces multiplication to addition, division to subtraction, raising to a power to simple multiplication, and extracting square roots to simple division.

Multiplication, division, or raising a number to some power can be accomplished with a calculator. However, in some problem-solving situations an understanding of the solution to such problems using logarithms is necessary. The following chapters dealing with logarithmic equations and applications of logarithms require this understanding.

21.1 LOGARITHMS DEFINED

Consider the expression $10^2 = 100$. Recall from previous chapters that 10 is called the base, 2 is the exponent, and 100 is its numerical value. The *logarithm* of the expression is the exponent 2.

EXAMPLE 21.1: The logarithm is the exponent.

$$\text{exponent} \nearrow$$
$$10^2 = 100$$
$$\text{base} \qquad \text{number}$$

$$\text{logarithm} \nearrow$$
$$\log_{10} 100 = 2$$
$$\text{base} \qquad \text{number}$$

That is, the logarithm is the power to which 10 (the base) must be raised in order for it to equal 100 (the number). Simply stated: *A logarithm is* an exponent.

Ten is the base of this system of logarithms. When 10 is the base, the system is called a system of *common logarithms*. In the system of logarithms, this is one of the two bases used in problem solving. The other base is 2.718. This system is referred to as a system of *natural logarithms*. Natural logarithms will be discussed later in the chapter.

21.2 COMMON LOGARITHMS

When dealing with common logarithms, the logarithmic expression for $10^2 = 100$ would be: $\log_{10} 100 = 2$. This is read, "The log, to the base 10, of 100 is 2." When using common logs it is not necessary to indicate the base because it is assumed to be 10. The expression "log $100 = 2$" is more often used. Notice that we have expressed the quantity two different ways. The first expression, $10^2 = 100$, is called an *exponential expression*. We have expressed the quantity in *exponential form*. The second expression, log $100 = 2$, is called a *logarithmic expression*. We have written the quantity in *logarithmic form*.

Let's find the log of 1000. First, we express the quantity in exponential form. To do this, we merely find the power of 10 that equals 1000. In exponential form: $10^3 = 1000$. The logarithm is the exponent to which 10 is raised and equals 3. In logarithmic form, log $1000 = 3$.

Find the log of 0.0001. Again, we first express the quantity in exponential form: $10^{-4} = 0.0001$. The logarithm is the exponent and equals -4. In logarithmic form, log $0.0001 = -4$.

PRACTICE PROBLEMS 21-1   Find the logarithms of the following numbers. Express your answers first in logarithmic form and then in exponential form.

1. 0.00001   **2.** 0.01

3. 10,000   **4.** 1,000,000

Solutions:

Logarithmic Form	Exponential Form
1. $\log 0.00001 = -5$	$10^{-5} = 0.00001$
2. $\log 0.01 = -2$	$10^{-2} = 0.01$
3. $\log 10,000 = 4$	$10^4 = 10,000$
4. $\log 1,000,000 = 6$	$10^6 = 1,000,000$

21.2.1 Finding the Logarithms of Numbers Between 1 and 10.   What about the log of numbers that are not multiples of 10? Let's consider numbers between 1 and 10. Since $10^0 = 1$ and $10^1 = 10$, then numbers between 1 and 10 would have logs or exponents between 0 and 1. For example (to four significant figures)

$5.32 = 10^{0.7259}$ and $7.63 = 10^{0.8825}$. In logarithmic form, log $5.32 = 0.7259$ and log $7.63 = 0.8825$.

We can find the logarithms of all numbers between 1 and 10 in a *table of common logarithms*. Your calculator also contains a table of common logarithms. The table in your calculator contains the logarithms of *all* numbers. The calculator has made tables of common logarithms obsolete in almost all problem-solving situations because it is faster and easier to use. Refer to Appendix A for the algorithm used to find logarithms of numbers.

EXAMPLE 21.2: Find log 4.23.

Solution:

$$\log 4.23 = 0.6263$$

In exponential form:

$$10^{0.6263} = 4.23$$

EXAMPLE 21.3: Find log 5.03.

Solution:

$$\log 5.03 = 0.7016$$

In exponential form:

$$10^{0.7016} = 5.03$$

PRACTICE PROBLEMS 21-2   Using the calculator, find the logs of the following numbers to four significant figures. Express your answers in both logarithmic and exponential form.

1. 1.76   **2.** 2.03

3. 9.83   **4.** 4.35

5. 3.16

Solutions:

Logarithmic Form	Exponential Form
1. log 1.76 = 0.2455	$10^{0.2455} = 1.76$
2. log 2.03 = 0.3075	$10^{0.3075} = 2.03$
3. log 9.83 = 0.9926	$10^{0.9926} = 9.83$
4. log 4.35 = 0.6385	$10^{0.6385} = 4.35$
5. log 3.16 = 0.4997	$10^{0.4997} = 3.16$

Additional practice problems are at the end of the chapter.

21.2.2 Finding the Logarithms of Numbers Greater Than 10. We can see that if we wanted to find the logarithm of 50, for example, the exponent (or log) would be greater than 1 but less than 2. (The number is greater than 10 but less than 100.) The log will consist of a whole number and a decimal fraction. *Only numbers that are multiples of 10 will have logarithms that are whole numbers.*

Notice that in practice problems 21-2 the logs of the numbers were all decimal fractions. This was because all the numbers were between 1 and 10. Remember, the log of 1 is 0 and the log of 10 is 1. If problems contained numbers between 1 and 10, each logarithm would be a number between 0 and 1.

If problems contained numbers between 10 and 100, the logarithms would be numbers between 1 and 2. The log of 10 is 1 and the log of 100 is 2. If the numbers were between 100 and 1000, the logarithms would be numbers between 2 and 3. And so on. What is the log of 50? $\log 50 = 1.6990$. What is the log of 500? 5000? 50,000? $\log 500 = 2.6990$; $\log 5000 = 3.6990$; $\log 50,000 = 4.6990$. Notice that in each case the fractional part of the logarithm is 0.6990. The whole number part depends on the multiples of 10 contained in the number. For example, let's express the number 5000 in scientific notation: $5000 = 5 \times 10^3$. We know from previous discussions that $\log 10^3 = \log 1000 = 3$. $\log 5 = 0.6990$. If we express $5 \times 10^3 = 5000$ by using the laws of exponents, $10^{0.6990+3} = 10^{3.6990} = 5000$. When finding logarithms of numbers, we are finding a fractional part which results from finding the log of a number between 1 and 10 and a whole number part which is the power of 10. What is $\log 276,000$? $\log 276,000 = 5.4409$. When we consider that $276,000 = 2.76 \times 10^5$, we really found the log of 2.76 and the log of 10^5. $\log 2.76 = 0.4409$ and $\log 10^5 = 5$. $10^5 \times 10^{0.4409} = 10^{5.4409}$. The logarithm is the exponent to the base 10. We found the power to which 10 is raised so that its value is 276,000.

EXAMPLE 21.4: Find $\log 37,500$.

Solution:

$$\log 37,500 = 4.574$$

In exponential form:

$$10^{4.574} = 37,500$$

PRACTICE PROBLEMS 21-3   Find the logarithms of the following numbers. Express your answers in both logarithmic and exponential forms.

1. 67.3

2. 67,300

3. 742

4. 103,000

5. 8730

Solutions:

Logarithmic Form	Exponential Form
1. $\log 67.3 = 1.8280$	$10^{1.8280} = 67.3$
2. $\log 67{,}300 = 4.8280$	$10^{4.8280} = 67{,}300$
3. $\log 742 = 2.8704$	$10^{2.8704} = 742$
4. $\log 103{,}000 = 5.0128$	$10^{5.0128} = 103{,}000$
5. $\log 8730 = 3.9410$	$10^{3.9410} = 8730$

Additional practice problems are at the end of the chapter.

21.2.3 Logarithms of Numbers Less Than 1.   If problems contained numbers between 0 and 0.1, the logarithms would be numbers between 0 and -1. If the numbers were between 0.1 and 0.01, the logarithms would be numbers between -1 and -2, and so on. Therefore, we may say that for any number less than 1, the logarithm (or exponent) is a negative number.

Find the log of 0.0275. $\log 0.0275 = -1.5607$. In exponential form, $10^{-1.5607} = 0.0275$. Using scientific notation gives $0.0275 = 2.75 \times 10^{-2}$. $\log 2.75 = 0.4393$ and $\log 10^{-2} = -2$. The resulting exponent is $-2 + 0.4393 = -1.5607$.

EXAMPLE 21.5: Find $\log 0.432$.

Solution:

$$\log 0.432 = -0.3645$$

In exponential form:

$$10^{-0.3645} = 0.432$$

PRACTICE PROBLEMS 21-4   Find the logs of the following numbers. Express your answers in both logarithmic and exponential forms.

1. 0.0463

2. 0.000935

3. 0.00205

4. 0.127

5. 0.00000812

Solutions:

Logarithmic Form	Exponential Form
1. $\log 0.0463 = -1.3344$	$10^{-1.3344} = 0.0463$
2. $\log 0.000935 = -3.0292$	$10^{-3.0292} = 0.000935$
3. $\log 0.00205 = -2.6882$	$10^{-2.6882} = 0.00205$
4. $\log 0.127 = -0.8962$	$10^{-0.8962} = 0.127$
5. $\log 0.00000812 = -5.0904$	$10^{-5.0904} = 0.00000812$

Additional practice problems are at the end of the chapter.

SELF-TEST 12-1   Find the logarithms of the following numbers. Express your answers in both logarithmic and exponential form.

1. 7

2. 98.3

3. 0.00420

4. 7,650,000

5. 5762

6. 0.273

7. 0.00017

8. 0.00000983

9. 1786

10. 0.0933

Answers to Self-Test 22-1 are at the end of the chapter.

21.3   MULTIPLICATION OF NUMBERS BY USING LOGARITHMS

Numbers can be multiplied by using logarithms. When we covered exponents in Chapter 2 we learned that when multiplying like bases, the exponents are added. For example:

$$10^2 \times 10^3 = 10^{2+3} = 10^5$$
$$y^3 \times y^4 = y^{3+4} = y^7$$
$$10^{2.34} \times 10^{1.72} = 10^{4.06}$$

Since this is true, when multiplying numbers together we could change the numbers by using logarithms so that the bases are alike. Then we could add the logs (exponents) together. The number in exponential form would be the answer. For instance, suppose we wanted to multiply 37×763. We would first change the numbers to base 10. $\log 37 = 1.568$ and $\log 763 = 2.883$. In exponential form:

$$37 = 10^{1.568} \quad \text{and} \quad 763 = 10^{2.883}$$

Then

$$37 \times 763 = 10^{1.568} \times 10^{2.883} = 10^{4.451}$$
$$10^{4.451} = 2.82 \times 10^4$$

Since logarithms are exponents, multiplication of numbers, using logarithms, is reduced to addition. To find the product of numbers using logs, we:

1. Find the log of each number.
2. Change each number to base 10.
3. Add the exponents.

When we found the product of 37×763, we could have written the problem in this way:

$$\log(37 \times 763) = \log 37 + \log 763$$

Remember, the logs are the exponents to the base 10 and, according to the laws of exponents, are added.

EXAMPLE 21.6: Find the product of 7600×230. Round answer to three places.

Solution:

$$\log(7600 \times 230) = \log 7600 + \log 230 = 3.881 + 2.362 = 6.243$$
$$7600 \times 230 = 10^{3.881} \times 10^{2.362} = 10^{6.243} = 1.75 \times 10^6$$

EXAMPLE 21.7: Find the product of 1760×0.0273. Round answer to three places.

Solution:

$$\log(1760 \times 0.0273) = \log 1760 + \log 0.0273 = 3.246$$
$$+ (-1.564) = 1.682$$
$$1760 \times 0.0273 = 10^{3.246} \times 10^{-1.564} = 10^{1.682} = 4.8 \times 10$$

EXAMPLE 21.8: Find the product of 0.0783×0.0000342.

Solution:

$$\log(0.0783 \times 0.0000342) = \log 0.0783 + \log 0.0000342$$
$$= -1.106 + (-4.466) = -5.572$$
$$10^{-5.572} = 2.68 \times 10^{-6}$$

PRACTICE PROBLEMS 21-5   Find the product of the following numbers by using logs. Round your answers to three digits.

1. 76×2730

2. 0.783×6500

3. 0.00376×0.107

4. $7,600,000 \times 9,800$

5. $10,700 \times 0.0083$

Solutions:

1. $\log(76 \times 2730) = \log 76 + \log 2730 = 1.881$
$+ 3.436 = 5.317$
$10^{5.317} = 2.07 \times 10^5$

2. $\log(0.783 \times 6500) = \log 0.783 + \log 6500 = 0.106$
$+ 3.813 = 3.707$
$10^{3.707} = 5.09 \times 10^3$

3. $\log(0.00376 \times 0.107) = \log 0.0037 + \log 0.107 =$
$2.425 + (-0.971) = -3.395$
$10^{-3.395} = 4.02 \times 10^{-4}$

4. $\log(7,600,000 \times 9800) = \log 7,600,000 + \log 9800$
$= 6.881 + 3.991 = 10.872$
$10^{10.87} = 7.45 \times 10^{10}$

5. $\log(10,700 \times 0.0083) = \log 10,700 + \log 0.0083 =$
$4.029 + (-2.081) = 1.948$
$10^{1.948} = 8.88 \times 10$

Additional practice problems are at the end of the chapter.

21.4 DIVISION OF NUMBERS BY USING LOGARITHMS

Division of numbers by using logarithms is possible by again observing the laws of exponents. When like bases are divided, the exponents are subtracted. For example:

$$\frac{10^7}{10^4} = 10^{7-4} = 10^3$$

$$\frac{8^2}{8^5} = 8^{2-5} = 8^{-3}$$

$$\frac{a^5}{a^{-3}} = a^{5-(-3)} = a^{5+3} = a^8$$

$$\frac{10^{4.384}}{10^{2.732}} = 10^{4.384-2.732} = 10^{1.652}$$

Let's find the quotient of $\frac{6270}{220}$ by using logs. To do this we:

1. Find the log of both dividend and divisor.
2. Subtract the log of the divisor from the log of the dividend.
3. Change to exponential form.

We would solve the problem this way:

$$\log\left(\frac{6270}{220}\right) = \log 6270 - \log 220 = 3.797 - 2.342 = 1.455$$
$$10^{1.455} = 2.85 \times 10$$

As with multiplication, finding the logs of the numbers gives us the opportunity to add or subtract exponents of like bases:

$$\frac{6270}{220} = \frac{10^{3.797}}{10^{2.342}} = 10^{3.797-2.342} = 10^{1.455} = 2.85 \times 10$$

EXAMPLE 21.9: Find the quotient of $\frac{1.73}{96,700}$.

Solution:

$$\log\left(\frac{1.73}{96,700}\right) = \log 1.73 - \log 96,700 = 0.238 - 4.985$$
$$= -4.747$$
$$10^{4.747} = 1.79 \times 10^{-5}$$

EXAMPLE 21.10: Find the quotient of $\dfrac{0.0735}{0.00034}$.

Solution:

$$\log\left(\frac{0.0735}{0.00034}\right) = \log 0.0735 - \log 0.00034 = -1.134 - (-3.469)$$
$$= 2.335$$
$$10^{2.335} = 2.16 \times 10^2$$

PRACTICE PROBLEMS 21.6   Find the quotients of the following problems by using logs. Round your answers to three places.

1. $\dfrac{76,000}{42}$

2. $\dfrac{207}{56,300}$

3. $\dfrac{1670}{0.0073}$

4. $\dfrac{0.000426}{0.862}$

5. $\dfrac{0.0000942}{6040}$

Solutions:

1. $\text{Log}\left(\dfrac{76,000}{42}\right) = \log 76,000 - \log 42 = 4.881 - 1.623 = 3.258$

   $10^{3.258} = 1.81 \times 10^3$

2. $\log\left(\dfrac{207}{56,300}\right) = \log 207 - \log 56,300 = 2.316 - 4.751 = -2.435$

   $10^{-2.435} = 3.68 \times 10^{-3}$

3. $\log\left(\dfrac{1670}{0.0073}\right) = \log 1670 - \log 0.0073 = 3.223 - (-2.137) = 5.359$

   $10^{-3.307} = 4.94 \times 10^{-4}$

4. $\log\left(\dfrac{0.000426}{0.862}\right) = \log 0.000426 - \log 0.862 = -3.371 - (-0.064) = -3.307$

   $10^{-3.307} = 4.94 \times 10^{-4}$

5. $\log\left(\dfrac{0.0000942}{6040}\right) = \log 0.0000942 - \log 6040 = -4.026 - 3.781 = -7.807$

   $10^{-7.807} = 1.56 \times 10^{-8}$

Additional practice problems are at the end of the chapter.

21.5   RAISING A NUMBER TO A POWER BY USING LOGARITHMS

In the unit on exponents we found that raising a power to a power is a process of multiplication. Some examples are:

$$(10^3)^4 = 10^{3 \times 4} = 10^{12}$$
$$(4^2)^3 = 4^{2 \times 3} = 4^6$$
$$(10^{2.37})^{-4} = 10^{(2.37)(-4)} = 10^{-9.48}$$

Let's find 4^5 by using logarithms. To solve problems of this type, we:

1. Find the log of the number.
2. Multiply the log by the exponent.
3. Change to exponential form.

We can express the problem this way: $\text{Log}\,4^5 = 5\log 4$.
Remember, we are going to multiply the log of the number by the exponent.

$$5\log 4 = 5 \times 0.602 = 3.01$$
$$10^{3.01} = 1.02 \times 10^3$$
$$4^5 = (10^{0.602})^5 = 10^{(0.602)(5)} = 10^{3.01} = 1.02 \times 10^3$$

EXAMPLE 21.11: What does 0.0745^3 equal?

Solution:

$$\log 0.0745^3 = 3\log 0.0745 = (3)(-1.128) = -3.384$$
$$10^{-3.384} = 4.13 \times 10^{-4}$$
$$0.0745^3 = (10^{-1.128})^3 = 10^{-3.384} = 4.13 \times 10^{-4}$$

EXAMPLE 21.12: What does $74.6^{-2.5}$ equal?

Solution:

$$\log 74.6^{-2.5} = -2.5\log 74.6 = -2.5 \times 1.873 = -4.682$$
$$10^{-4.682} = 2.08 \times 10^{-5}$$
$$74.6^{-2.5} = (10^{1.873})^{-2.5} = 10^{-4.682} = 2.08 \times 10^{-5}$$

PRACTICE PROBLEMS 21-7   Find the answers to the following problems by using logarithms. Round to three places.

1. 6.7^5

2. $2.3^{-3.6}$

3. 0.00403^4

4. $0.6^{4.2}$

5. 50^3

Solutions:

1. $\log 6.7^5 = 5\log 6.7 = 5 \times 0.826 = 4.13$
$\quad\quad 10^{4.13} = 1.35 \times 10^4$

2. $\log 2.3^{-3.6} = -3.6\log 2.3 =$
$\quad\quad\quad -3.6 \times 0.362 = -1.302$
$\quad\quad\quad 10^{-1.302} = 4.99 \times 10^{-2}$

3. $\log 0.00403^4 = 4\log 0.00403 = 4 \times (-2.395) =$
$\quad\quad\quad\quad -9.579$
$\quad\quad\quad 10^{-9.579} = 2.64 \times 10^{-10}$

4. $\log 0.6^{4.2} = 4.2\log 0.6 = 4.2 \times (-0.222) = -0.932$
$\quad\quad\quad 10^{-0.932} = 0.117 = 1.17 \times 10^{-1}$

5. $\log 50^3 = 3\log 50 = 3 \times 1.699 = 5.097$
$\qquad 10^{5.097} = 1.25 \times 10^5$

Additional practice problems are at the end of this chapter.

SELF-TEST 21-2   Find the answers to the following problems by using logarithms. Round to three places.

1. 463×0.027

2. $\dfrac{34.3}{8700}$

3. $7^{4.6}$

4. $\dfrac{0.0063}{2.73}$

5. 0.463^4

6. $97,000 \times 170$

7. 0.0128×0.000055

8. $27.3^{2.6}$

9. $\dfrac{43.2}{0.000772}$

10. 0.00193×1930

Answers to Self-Test 21-2 are at the end of the chapter.

21.6   NATURAL LOGARITHMS

Common logarithms are exponents where 10 is the base. *Natural logarithms* are exponents where ε (the Greek letter epsilon) is the base. The numerical value of ε, rounded to three digits, is 2.72. In electronics, the instantaneous charge or discharge of a capacitor or the instantaneous current in a coil can be determined by equations using natural logarithms.

21.6.1   Finding natural logarithms of numbers.   Just as we have tables of common logarithms, we have tables of natural logarithms. We used the abbreviation "log" for common logarithms. We use the abbreviation "ln" for natural logarithms. Appendix A contains the algorithm for finding natural logarithms by using the calculator. As with common logarithms, the calculator stores a table of natural logarithms.

What is the natural log of 5? Using the calculator, $\ln 5 = 1.6094$. In exponential form $\varepsilon^{1.6094} = 5$. Notice that the change from logarithmic form to exponential form is the same for natural logs as for common logs. The only difference is in the base. Note the following examples:

Logarithmic Form	Exponential Form
$\log 5 = 0.6990$	$10^{0.6990} = 5$
$\ln 5 = 1.6094$	$\varepsilon^{1.6094} = 5$
$\log 1 = 0$	$10^0 = 1$
$\ln 1 = 0$	$\varepsilon^0 = 1$
$\log 10 = 1$	$10^1 = 10$
$\ln \varepsilon = 1$	$\varepsilon^1 = \varepsilon$

Find ln 45. ln 45 = 3.8067. In exponential form, $\varepsilon^{3.8067} = 45$. Find the ln of 0.832. ln 0.832 = −0.1839. In exponential form, $\varepsilon^{-0.1839} = 0.832$. As with common logarithms, the natural logarithm of numbers greater than 1 are *positive* and the natural logarithms of numbers less than 1 are *negative*.

PRACTICE PROBLEMS 21-8 Find the natural logarithms of the following numbers by using the calculator.

1. 7.4

2. 300

3. 0.8

4. 3.4

5. 100

Solutions:

1. ln 7.4 = 2.0015

2. ln 300 = 5.7038

3. ln 0.8 = −0.2231

4. ln 3.4 = 1.2238

5. ln 100 = 4.6052

Additional practice problems are at the end of the chapter.

If we know the exponent to the base ε, we can find the number by using the calculator just as we found numbers when we were using base 10. What does $\varepsilon^{3.434}$ equal? $\varepsilon^{3.434} = 31$. What does ε^2 equal? $\varepsilon^2 = 7.39$. The algorithm for solving problems of this type is found in Appendix A.

PRACTICE PROBLEMS 21-9 Find the number whose natural logarithm is:

1. 1.946

2. 3

3. 4.489

4. 0.0953

5. 1.435

Solutions:

1. $\varepsilon^{1.946} = 7.00$

2. $\varepsilon^3 = 20.1$

3. $\varepsilon^{4.489} = 89$

4. $\varepsilon^{0.0953} = 1.10$

5. $\varepsilon^{1.435} = 4.20$

Additional practice problems are at the end of this chapter.

SELF-TEST 21-3 Find the natural logarithm of the following numbers. Convert your answers to exponential form.

1. 14

2. 0.0073

3. 740

4. 1073

5. 10	**6.** 0.00001
7. 0.463	**8.** 0.0673
9. 4.73	**10.** 73

Answers to Self-Test 21-3 are at the end of the chapter.

END OF CHAPTER PROBLEMS 21-1

Find the logarithms of the following problems. Express your answers in both logarithmic and exponential form.

1. 14	**2.** 376
3. 0.037	**4.** 0.000637
5. 1070	**6.** 7070
7. 0.407	**8.** 0.001
9. 10,000	**10.** 400,000
11. 0.49	**12.** 0.00699
13. 47,700	**14.** 4.77
15. 0.301	**16.** 0.000301
17. 37,000	**18.** 301,000
19. 0.975	**20.** 0.00075
21. 2,000,000	**22.** 84.7
23. 0.0275	**24.** 0.00875
25. 3.04	**26.** 304
27. 0.625	**28.** 0.000806
29. 746,000	**30.** 9070

END OF CHAPTER PROBLEMS 21-2

Solve the following problems by using logarithms:

1. 27×56	**2.** 83×720
3. $107 \times 243,000$	**4.** 110×3070
5. 43×0.143	**6.** 65×0.243
7. 0.043×0.0106	**8.** 0.00274×0.0976
9. $0.0000176 \times 0.000357$	**10.** 0.000555×0.00477
11. $\dfrac{1760}{43}$	**12.** $\dfrac{6530}{3.75}$
13. $\dfrac{65}{467}$	**14.** $\dfrac{27}{730}$

15. $\dfrac{0.00273}{172}$ **16.** $\dfrac{0.000825}{70}$

17. $\dfrac{4300}{0.15}$ **18.** $\dfrac{5430}{0.0335}$

19. $\dfrac{0.0675}{0.00103}$ **20.** $\dfrac{0.000565}{0.0093}$

21. 3.4^3 **22.** 8.26^6

23. $6^{-1.3}$ **24.** $14^{-2.43}$

25. $0.43^{2.5}$ **26.** 0.007^4

27. $6^{4.3}$ **28.** $4^{2.73}$

29. $0.27^{-4.4}$ **30.** $0.083^{-2.1}$

END OF CHAPTER PROBLEMS 21-3

Find the natural logarithms of the following numbers. Express your answers in both logarithmic and exponential forms.

1. 2   **2.** 6.78

3. 0.932   **4.** 0.0273

5. 60   **6.** 23

7. 0.073   **8.** 0.345

9. 247   **10.** 783

11. 0.00146   **12.** 0.00541

13. 1100   **14.** 1500

15. 0.00027   **16.** 0.000481

17. 8.43   **18.** 12.3

19. 0.346   **20.** 0.000092

21. 3.67   **22.** 73.8

23. 0.107   **24.** 0.0081

25. 176   **26.** 500

27. 0.293   **28.** 0.372

29. 873   **30.** 1800

ANSWERS TO SELF-TESTS

Self-Test 21-1

	Logarithmic Form	*Exponential Form*
1.	$\log 7 = 0.845^1$	$10^{0.845^1} = 7$
2.	$\log 98.3 = 1.993$	$10^{1.993} = 98.3$
3.	$\log 0.0042 = -2.377$	$10^{-2.377} = 0.0042$
4.	$\log 7,650,000 = 6.883$	$10^{6.883} = 7,650,000$

     5. $\log 5762 = 3.76$ $10^{3.76} = 5762$

     6. $\log 0.273 = -0.564$ $10^{-0.564} = 0.273$

     7. $\log 0.00017 = -3.77$ $10^{-3.77} = 0.00017$

     8. $\log 0.00000983 = -5.007$ $10^{-5.007} = 0.00000983$

     9. $\log 1786 = 3.253$ $10^{3.253} = 1786$

   10. $\log 0.0933 = -1.03$ $10^{-1.03} = 0.0933$

Self-Test 21-2

1. $\log(463 \times 0.027) = \log 463 + \log 0.027 = 2.666 + (-1.569) = 1.097$
$$10^{1.097} = 1.25 \times 10$$

2. $\log\left(\dfrac{34.3}{8700}\right) = \log 34.4 - \log 8700 = 1.535 - 3.94 = -2.404$
$$10^{-2.404} = 3.94 \times 10^{-3}$$

3. $\log 7^{4.6} = 4.6 \log 7 = 4.6 \times 0.845 = 3.89$
$$10^{3.89} = 7.72 \times 10^{3}$$

4. $\log\left(\dfrac{0.0063}{2.73}\right) = \log 0.0063 - \log 2.73 = -2.201 - 0.436 = -2.637$
$$10^{-2.637} = 2.31 \times 10^{-3}$$

5. $\log 0.463^{4} = 4 \log 0.463 = 4 \times (-0.334) = -1.34$
$$10^{-1.34} = 4.6 \times 10^{-2}$$

6. $\log(97{,}000 \times 170) = \log 97{,}000 + \log 170 = 4.987 + 2.23 = 7.217$
$$10^{7.217} = 1.65 \times 10^{7}$$

7. $\log(0.0128 \times 0.000055) = \log 0.0128 + \log 0.000055 = -1.893 + (-4.26) = -6.152$
$$10^{-6.152} = 7.04 \times 10^{-7}$$

8. $\log 27.3^{2.6} = 2.6 \log 27.3 = 2.6 \times 1.436 = 3.734$
$$10^{3.734} = 5.42 \times 10^{3}$$

9. $\log \dfrac{43.2}{0.000772} = \log 43.2 - \log 0.000772 = 1.635 - (-3.112) = 4.748$
$$10^{4.748} = 5.6 \times 10^{4}$$

10. $\log(0.00193 \times 1930) = \log 0.00193 + \log 1930 = -2.714 + 3.286 = 0.571$
$$10^{0.571} = 3.72$$

Self-Test 21-3

Logarithmic Form	*Exponential Form*
1. $\ln 14 = 2.639$	$\varepsilon^{2.639} = 14$
2. $\ln 0.0073 = -4.92$	$\varepsilon^{-4.92} = 0.0073$
3. $\ln 740 = 6.607$	$\varepsilon^{6.607} = 740$
4. $\ln 1073 = 6.978$	$\varepsilon^{6.978} = 1073$
5. $\ln 10 = 2.303$	$\varepsilon^{2.303} = 10$
6. $\ln 0.00001 = -11.51$	$\varepsilon^{-11.51} = 0.00001$
7. $\ln 0.463 = -0.770$	$\varepsilon^{-0.770} = 0.463$
8. $\ln 0.0673 = -2.699$	$\varepsilon^{-2.699} = 0.0673$
9. $\ln 4.73 = 1.554$	$\varepsilon^{1.554} = 4.73$
10. $\ln 73 = 4.29$	$\varepsilon^{4.29} = 73$

22

LOGARITHMIC EQUATIONS

22.1 LOGARITHMIC AND EXPONENTIAL FORMS

We have discussed logarithmic and exponential expressions in previous chapters. Now let's write them again in the form of simple equations.

Logarithmic	Exponential
(1) $\log N = x$	$10^x = N$
(2) $\ln a = b$	$\varepsilon^b = a$

In (1) above, x is the exponent and N is the number. The base is 10. In (2) above, b is the exponent and a is the number. The base is ε.

In the equation $\log 73 = t$, 73 is the number and t is the exponent. The equation is solved when we find $\log 73$.

$$\log 73 = t \tag{22-1}$$

$$t = 1.863$$

$\text{Log} \, y = 4.73$. The equation is solved for y when we find the number whose exponent, to the base 10, is 4.73.

$$\log y = 4.73 \tag{22-2}$$

$$10^{4.73} = y \tag{22-3}$$

Equation 22-1 was solved simply by finding $\log 73$. Equation 22-2 cannot be solved directly. If we convert to exponential form as in Equation 22-3, we can easily solve the equation.

$$10^{4.73} = y$$

$$y = 5.37 \times 10^4$$

PRACTICE PROBLEMS 22-1 Change the following equations from logarithmic form to exponential form:

1. $\log 7600 = a$

2. $\log x = -3.74$

3. $\log y = 0.742$

4. $\log 0.00432 = x$

Change the following equations from exponential form to logarithmic form:

5. $10^{4.2} = y$

6. $10^t = 0.043$

7. $10^y = 173$

8. $10^{-2.43} = y$

Solutions:

1. $10^a = 7600$

2. $10^{-3.74} = x$

3. $10^{0.742} = y$

4. $10^x = 0.00432$

5. $\log y = 4.2$

6. $\log 0.043 = t$

7. $\log 173 = y$

8. $\log y = -2.43$

Additional practice problems are at the end of the chapter.

22.2 LOGARITHMIC EQUATIONS—COMMON LOGS

Let's consider some of the equations we might encounter in problem solving which involve the use of common logs.

EXAMPLE 22.1: $x = \log 250$. Solve for x.

Solution:
$$x = 2.4$$

EXAMPLE 22.2: $\log a = 7.83$. Solve for a.

Solution: Change to exponential form:
$$10^{7.83} = a$$
$$a = 6.76 \times 10^7$$

EXAMPLE 22.3: $3^x = 104$. Solve for x.

Solution: Take the log of both sides to get rid of the exponent:
$$x \log 3 = \log 104$$
$$x = \frac{\log 104}{\log 3} = \frac{2.02}{0.477}$$
$$x = 4.23$$

(We could have found $\log 3$ and $\log 104$ first and then solved for x.)

EXAMPLE 22.4: $x^{3.6} = 85$. Solve for x.

Solution: Take the log of both sides to get rid of the exponent:

$$3.6 \log x = \log 85$$

$$\log x = \frac{\log 85}{3.6} = \frac{1.93}{3.6} = 0.536$$

Change to exponential form:

$$10^{0.536} = x$$
$$x = 3.44$$

EXAMPLE 22.5: $6.3^4 = x$. Solve for x.

Solution: Take the log of both sides to get rid of the exponent:

$$4 \log 6.3 = \log x$$
$$4(0.799) = \log x$$
$$3.2 = \log x$$

Change to exponential form:

$$10^{3.2} = x$$
$$x = 1.58 \times 10^3$$

PRACTICE PROBLEMS 22-2   Solve for the unknown in the following problems:

1. $\log 473 = x$ 2. $\log 0.0783 = x$

3. $\log t = 2.74$ 4. $\log x = -3.74$

5. $\log a = 0.274$ 6. $\log 10,700 = y$

7. $4^x = 256$ 8. $x^{2.7} = 56$

9. $3.6^3 = x$ 10. $x^{-1.4} = 0.073$

11. $3.4^{-3} = x$ 12. $0.087^x = 4$

Solutions:

1. $x = 2.67$ 2. $y = -1.11$

3. $t = 5.5 \times 10^2$ 4. $x = 1.82 \times 10^{-4}$

5. $a = 1.88$ 6. $y = 4.03$

7. $y = 4$ 8. $x = 4.44$

9. $x = 46.7$ 10. $x = 6.488$

11. $x = 0.0254$ 12. $x = -0.5678$

Additional practice problems are at the end of the chapter.

EXAMPLE 22.6: $3^{1-x} = 12$. Solve for x.

Solution: Take the log of both sides to get rid of the exponent:

$$(1-x)\log 3 = \log 12$$

$$1 - x = \frac{\log 12}{\log 3} = \frac{1.08}{0.477} = 2.26$$

$$1 - 2.26 = x$$

$$x = -1.26$$

EXAMPLE 22.7: $27^{x/3} = 18$. Solve for x.

Solution: Take the log of both sides:

$$\frac{x}{3}\log 27 = \log 18$$

$$\frac{x}{3} = \frac{\log 18}{\log 27} = \frac{1.255}{1.431} = 0.877$$

$$x = 2.63$$

EXAMPLE 22.8: $4 = \log \dfrac{x}{100}$. Solve for x.

Solution: Change to exponential form:

$$10^4 = \frac{x}{100}$$

$$x = 100 \times 10^4$$

$$x = 10^6$$

EXAMPLE 22.9: $8 = \log \dfrac{200}{x}$. Solve for x.

Solution: Change to exponential form:

$$10^8 = \frac{200}{x}$$

$$10^8 x = 200$$

$$x = \frac{200}{10^8}$$

$$x = 2 \times 10^{-6}$$

EXAMPLE 22.10: $65 = 10 \log \dfrac{x}{0.03}$. Solve for x.

Solution:

$$\frac{65}{10} = \log \frac{x}{0.03}$$

$$6.5 = \log \frac{x}{0.03}$$

Change to exponential form:

$$10^{6.5} = \frac{x}{0.03}$$

$$0.03 \times 10^{6.5} = x$$

$$x = 9.49 \times 10^4$$

PRACTICE PROBLEMS 22-3   Solve for x in the following problems:

1. $4^{3-x} = 128$ **2.** $5.4^{x+2} = 200$

3. $53^{x/2} = 35$ **4.** $3.6 = \log \dfrac{x}{25}$

5. $7.3 = \log \dfrac{75}{x}$ **6.** $x = \log \dfrac{36}{2}$

7. $43 = 10 \log \dfrac{x}{12}$ **8.** $85 = 20 \log \dfrac{40}{x}$

Solutions

1. $x = -0.5$ **2.** $x = 1.14$

3. $x = 1.79$ **4.** $x = 9.95 \times 10^4$

5. $x = 3.76 \times 10^{-6}$ **6.** $x = 1.26$

7. $x = 2.39 \times 10^5$ **8.** $x = 2.25 \times 10^{-3}$

Additional practice problems are at the end of the chapter.

SELF-TEST 22-1   Solve for x in the following problems:

1. $x = \log 400$ **2.** $4^{3.5} = x$

3. $12^{-4} = x$ **4.** $x^{1.7} = 40$

5. $5^{x+2} = 625$ **6.** $9^{1-x} = 36$

7. $65^{x/3} = 10$ **8.** $3^{4/x} = 0.506$

9. $5 = \log \dfrac{x}{40}$ **10.** $14 = \log \dfrac{200}{x}$

11. $\log x^4 = 2$ **12.** $\log 3^x = 0.93$

Answers to Self-Test 22-1 are at the end of the chapter.

22.3 LOGARITHMIC EQUATIONS—NATURAL LOGS

Here are some problems involving the use of natural logs.

EXAMPLE 22.11: $\varepsilon^{1.7} = y$. Solve for y.

Solution:
$$y = 5.47$$

EXAMPLE 22.12: $\varepsilon^y = 83$. Solve for y.

Solution: Take the ln of both sides:
$$y \ln \varepsilon = \ln 83$$
$$y(1) = \ln 83 \qquad \text{(remember: } \ln \varepsilon = 1)$$
$$y = 4.42$$

EXAMPLE 22.13: $89 = \varepsilon^{-x}$. Solve for x.

Solution:
$$\ln 89 = -x \ln \varepsilon$$
$$4.49 = -x$$
$$x = -4.49$$

EXAMPLE 22.14: $\ln x = 1.7$. Solve for x.

Solution: Change to exponential form:
$$\varepsilon^{1.7} = x$$
$$x = 5.47$$

PRACTICE PROBLEMS 22-4   Solve for the unknown in the following problems:

1. $\varepsilon^{3.4} = x$　　　　　　　　　　2. $\varepsilon^x = 49$

3. $89 = \varepsilon^t$　　　　　　　　　　4. $y = \varepsilon^{4.5}$

5. $\varepsilon^y = 58$　　　　　　　　　　6. $\varepsilon^x = 150$

7. $\ln x = 3$　　　　　　　　　　8. $\ln x = -2.7$

Solutions:

1. $x = 30$　　　　　　　　　　2. $x = 3.89$

3. $t = 4.49$　　　　　　　　　　4. $y = 90$

5. $y = 4.06$　　　　　　　　　　6. $x = 5.01$

7. $x = 20.1$　　　　　　　　　　8. $x = 6.72 \times 10^{-2}$

Additional practice problems are at the end of the chapter.

EXAMPLE 22.15: $x = 20(1 - \varepsilon^{-0.5})$. Solve for x.

Solution:

$$x = 20(1 - 0.606)$$
$$x = 20(0.393)$$
$$x = 7.87$$

EXAMPLE 22.16: $30 = x(1 - \varepsilon^{-4})$. Solve for x.

Solution:

$$30 = x(1 - 0.0183)$$
$$30 = x(0.982)$$
$$x = 30.6$$

EXAMPLE 22.17: $5 = 15(1 - \varepsilon^{-1/x})$. Solve for x.

Solution:

$$0.333 = 1 - \varepsilon^{-1/x}$$
$$0.333 - 1 = -\varepsilon^{-1/x}$$
$$-0.667 = -\varepsilon^{-1/x}$$
$$0.667 = \varepsilon^{-1/x}$$
$$\ln 0.667 = -1/x \ln \varepsilon$$
$$-0.405 = -1/x$$
$$0.405x = 1$$
$$x = 2.47$$

PRACTICE PROBLEMS 22-5 Solve for the unknown in the following problems:

1. $x = 25(1 - \varepsilon^{-1.75})$
2. $40 = 60(1 - \varepsilon^{-x})$
3. $15 = 45(1 - \varepsilon^{-2/3x})$
4. $0.45 = \varepsilon^{-x/3.2}$
5. $0.175 = 1 - \varepsilon^{-3x/4}$
6. $22 = x(1 - \varepsilon^{-2.3})$

Solutions:

1. $x = 20.7$
2. $x = 1.1$
3. $x = 1.65$
4. $x = 2.56$
5. $x = 0.256$
6. $x = 24.4$

SELF-TEST 22-2 Solve for the unknown in the following problems:

1. $x = \ln 40$
2. $\ln x = 14$
3. $\varepsilon^x = 20$
4. $x = 1 - \varepsilon^{-0.6}$

5. $36 = 100\varepsilon^{-x}$

6. $x = 50(1 - \varepsilon^{-2.75})$

7. $86 = x(1 - \varepsilon^{-1.5})$

8. $40 = 80\varepsilon^{-2/x}$

9. $0.45 = \varepsilon^{-x/4}$

10. $0.2 = 1 - \varepsilon^{-2/6x}$

Answers to Self-Test 22-2 are at the end of the chapter.

END OF CHAPTER PROBLEMS 22-1

Change the following equations from logarithmic form to exponential form:

1. $\log 340 = x$

2. $\log 4760 = x$

3. $\log 0.014 = y$

4. $\log 0.0073 = y$

5. $\log x = -2.43$

6. $\log x = -0.44$

7. $\log y = 1.44$

8. $\log y = 0.042$

Change the following equations from exponential form to logarithmic form:

9. $10^{3.6} = y$

10. $10^{1.7} = y$

11. $10^a = 732$

12. $10^a = 1.73$

13. $10^y = 0.043$

14. $10^y = 0.736$

15. $10^{-3.4} = x$

16. $10^{-1.02} = x$

END OF CHAPTER PROBLEMS 22-2

Find the unknown in the following problems:

1. $\log y = 7.36$

2. $\log y = 0.0276$

3. $\log x = 0.000736$

4. $\log x = 5.6$

5. $\log a = -4.23 \times 10^{-1}$

6. $\log a = -2.73$

7. $\log x = -3.64 \times 10^{-3}$

8. $\log x = -8.44 \times 10^{-2}$

9. $10^x = 462$

10. $10^x = 2700$

11. $10^x = 0.0173$

12. $10^x = 0.00906$

13. $10^y = 2.5 \times 10^{-3}$

14. $10^y = 2.6 \times 10^{-6}$

15. $10^y = 4.66 \times 10^4$

16. $10^y = 7.4 \times 10^3$

17. $10^{4.16} = x$

18. $10^{7.23} = x$

19. $10^{0.033} = x$

20. $10^{0.0056} = x$

21. $10^{-3.6} = y$

22. $10^{-1.73} = y$

23. $10^{-0.173} = y$

24. $10^{-0.00066} = y$

25. $\log 746 = x$

26. $\log 217 = x$

27. $\log 0.00716 = x$

28. $\log 0.0000106 = x$

29. $7.26^4 = x$

30. $12.4^{3.2} = x$

31. $0.026^x = 7$

32. $0.173^x = 0.037$

33. $x^{4.3} = 7300$

34. $x^{-4.3} = 0.45$

35. $\log y^3 = 9$

36. $\log 5^x = 5$

37. $\log y = 7.36$

38. $\log y = 0.0276$

39. $x^7 = 14$

40. $x^{4.5} = 75$

41. $70^x = 12$

42. $200^x = 6000$

END OF CHAPTER PROBLEMS 22-3

Solve for the unknown in the following problems:

1. $6^{2-y} = 68$

2. $8^{1+y} = 50$

3. $75^{y/4} = 3.54$

4. $38^{y/3} = 6.73$

5. $1.7 = \log \dfrac{x}{42}$

6. $1 = \log \dfrac{x}{0.5}$

7. $6 = \log \dfrac{100}{x}$

8. $4 = \log \dfrac{0.43}{x}$

9. $3^{x-2} = 50$

10. $5^{x-1} = 80$

11. $16^{3/x} = 200$

12. $7^{2/x} = 0.763$

13. $6 = \log \dfrac{x}{100}$

14. $2 = \log \dfrac{x}{1000}$

15. $-4 = \log \dfrac{20}{x}$

16. $-3 = \log \dfrac{10}{x}$

END OF CHAPTER PROBLEMS 22-4

Solve for the unknown in the following problems:

1. $x = \ln 11$

2. $x = \ln 47$

3. $\ln 15 = x$

4. $\ln 0.635 = x$

5. $\varepsilon^x = 200$

6. $\varepsilon^{-x} = 0.73$

7. $\ln 0.0004 = x$

8. $\ln 0.146 = x$

9. $\varepsilon^{-x} = 15$

10. $\varepsilon^x = 14.7$

11. $\varepsilon^{2.7} = x$

12. $\varepsilon^{4.6} = x$

13. $46 = \varepsilon^y$

14. $70 = \varepsilon^y$

15. $\varepsilon^{-4} = y$

16. $\varepsilon^{-1.6} = y$

END OF CHAPTER PROBLEMS 22-5

1. $0.6254 = 1 - \varepsilon^{-x}$

2. $x = 1 - \varepsilon^{-0.5}$

3. $50 = 80\varepsilon^{-x}$

4. $45 = 60\varepsilon^{-x}$

5. $x = 25(1 - \varepsilon^{-0.75})$

6. $x = 100(1 - \varepsilon^{-2})$

7. $30 = x(1 - \varepsilon^{-3})$

8. $20 = x(1 - \varepsilon^{-1.5})$

9. $10 = 25\varepsilon^{-x/3}$

10. $2.7 = 10\varepsilon^{-x/1.7}$

11. $3 = 9\varepsilon^{-4/x}$

12. $6 = 12\varepsilon^{-0.27/x}$

13. $0.176 = \varepsilon^{-2x/3}$

14. $0.93 = \varepsilon^{-x/4.7}$

15. $0.5 = \varepsilon^{-2/x}$

16. $0.37 = \varepsilon^{-3.2/x}$

17. $x = 1 - \varepsilon^{-3}$

18. $1 - x = \varepsilon^{-1.7}$

19. $0.376 = 1 - \varepsilon^{-x}$

20. $0.673 = \varepsilon^{-x}$

21. $\varepsilon^{-x/2} = 0.76$

22. $1 - x = \varepsilon^{-2}$

ANSWERS TO SELF-TESTS

Self-Test 22-1

1. $x = 2.6$

2. $x = 128$

3. $x = 4.82 \times 10^{-5}$

4. $x = 8.76$

5. $x = 2$

6. $x = -0.63$

7. $x = 1.66$

8. $x = -6.45$

9. $x = 4 \times 10^6$

10. $x = 2 \times 10^{-12}$

11. $x = 3.16$

12. $x = 1.95$

Self-Test 22-2

1. $x = 3.69$

2. $x = 1.2 \times 10^6$

3. $x = 3$

4. $x = 0.451$

5. $x = 1.02$

6. $x = 46.8$

7. $x = 111$

8. $x = 2.89$

9. $x = 3.19$

10. $x = 1.49$

CHAPTER **23**

APPLICATIONS OF LOGARITHMS

23.1 GAIN MEASUREMENTS

A parameter of many electronic circuits or systems is *gain*. The symbol for gain is A. Gain is the ratio of an output quantity to an input quantity. We may analyze a device in terms of power gain (A_p), voltage gain (A_V), current gain (A_i), or resistance gain (A_r).

It has been demonstrated experimentally that the human ear responds logarithmically to changes in sound levels. It has also been found that it requires about a 1-decibel change in a sound level before we are aware that a change has taken place. Because of this logarithmic characteristic, power gains are usually expressed as the logarithm of the ratio of power levels.

$$A_{p(\text{bels})} = \log \frac{P_{\text{out}}}{P_{\text{in}}}$$

The unit of measure is the *bel* in honor of Alexander Graham Bell.

The bel is seldom used because it is considered too large a unit. Instead, the *decibel* is the unit of measure we normally use. A decibel (dB) equals $\frac{1}{10}$ bel. Therefore:

$$A_{p(\text{dB})} = 10 \log \frac{P_{\text{out}}}{P_{\text{in}}} \tag{23-1}$$

EXAMPLE 23.1: The power out of an amplifier is 30 W. The input power is 100 mW. What is the power gain in decibels?

477

Solution:

$$A_p = 10\log\frac{P_{\text{out}}}{P_{\text{in}}}$$

$$A_p = 10\log\frac{30 \text{ W}}{100 \text{ mW}} = 10\log 300 = 24.77 \text{ dB}$$

If we consider that

$$P_{\text{out}} = \frac{V_{\text{out}}^2}{R_{\text{out}}} \text{ and } P_{\text{in}} = \frac{V_{\text{in}}^2}{R_{\text{in}}}$$

then we can write the equation for voltage gain in decibels.

$$A_v = 10\log\left(\frac{V_{\text{out}}^2}{R_{\text{out}}} \div \frac{V_{\text{in}}^2}{R_{\text{in}}}\right)$$

$$= 10\log\left(\frac{V_{\text{out}}^2}{V_{\text{in}}^2} \times \frac{R_{\text{in}}}{R_{\text{out}}}\right)$$

$$= 10\log\frac{V_{\text{out}}^2}{V_{\text{in}}^2} + 10\log\frac{R_{\text{in}}}{R_{\text{out}}}$$

If $R_{\text{out}} = R_{\text{in}}$, then

$$A_v = 10\log\left(\frac{V_{\text{out}}}{V_{\text{in}}}\right)^2 + 10\log 1 = 20\log\frac{V_{\text{out}}}{V_{\text{in}}} + 0$$

$$A_{v(\text{dB})} = 20\log\frac{V_{\text{out}}}{V_{\text{in}}} \tag{23-2}$$

EXAMPLE 23.2: The power gain of a power amplifier is 40 dB. If the input power is 50 μW, what is the output power?

Solution:

$$A_p = 10\log\frac{P_{\text{out}}}{P_{\text{in}}}$$

$$40 \text{ dB} = 10\log\frac{P_{\text{out}}}{50 \text{ }\mu\text{W}}$$

Divide by 10:

$$4 = \log\frac{P_{\text{out}}}{50 \text{ }\mu\text{W}}$$

Change to exponential form:

$$10^4 = \frac{P_{out}}{50\ \mu W}$$

Multiply by 50 μW:

$$50\ \mu W \times 10^4 = P_{out}$$
$$P_{out} = 0.5\ W$$

EXAMPLE 23.3: The output voltage of an amplifier is 1.73 V. The input voltage is 50 mV. Assuming that $R_{out} = R_{in}$, find the voltage gain in dB.

Solution:

$$A_v = 20\log\frac{V_{out}}{V_{in}} = 20\log\frac{1.73\ V}{50\ mV} = 20\log 34.6 = 30.78\ dB$$

EXAMPLE 23.4: The voltage gain of a device is -6 dB. If the output voltage is 700 mV, find V_{in}.

Solution:

$$A_v = 20\log\frac{V_{out}}{V_{in}}$$

$$-6 = 20\log\frac{700\ mV}{V_{in}}$$

Divide by 20:

$$-0.3 = \log\frac{700\ mV}{V_{in}}$$

Change to exponential form:

$$10^{-0.3} = \frac{700\ mV}{V_{in}}$$

$$0.5 = \frac{700\ mV}{V_{in}}$$

Multiply by V_{in}:

$$0.5\ V_{in} = 700\ mV$$

Divide by 0.5:

$$V_{in} = \frac{700\ mV}{0.5}$$
$$V_{in} = 1.4\ V$$

23.2 REFERENCE LEVELS

In the preceding examples the output powers and output voltages were given with reference to input values. Sometimes other reference levels are used. One common reference level is 6 mW. In other words, 0 dB equals 6 mW. Another reference level is 1 mW. When 1 mW is the reference, the decibel measurement is called a *dBm*. If a power gain is given in dBm, we assume that P_{in} is 1 mW.

> EXAMPLE 23.5: The power gain of an amplifier is 33 dBm. Determine the output power.

Solution:

$$A_p = 10\log\frac{P_{out}}{P_{in}}$$

$$33 = 10\log\frac{P_{out}}{1\ mW}$$

Divide by 10:

$$3.3 = \log\frac{P_{out}}{1\ mW}$$

Change to exponential form:

$$10^{3.3} = \frac{P_{out}}{1\ mW}$$

Multiply by 1 mW:

$$1\ mW \times 10^{3.3} = P_{out}$$

$$1\ mW \times 2 \times 10^3 = P_{out}$$

$$P_{out} = 2\ W$$

PRACTICE PROBLEMS 23-1

1. $P_{in} = 20$ mW, $P_{out} = 35$ W. Find $A_{p(dB)}$.

2. $A_p = 38$ dB, $P_{in} = 5$ mW. Find P_{out}.

3. $A_p = 25$ dB, $P_{out} = 15$ W. Find P_{in}.

4. $A_p = -3$ dB, $P_{out} = 1$ W. Find P_{in}.

Assume that $R_{out} = R_{in}$ for problems 5 through 7:

5. $V_{in} = 50\ \mu V$, $V_{out} = 1$ V. Find $A_{v(dB)}$.

6. $A_v = 60$ dB, $V_{out} = 3$ V. Find V_{in}.

7. $A_v = 75$ dB, $V_{in} = 2$ mV. Find V_{out}.

8. $A_v = -10$ dB, $V_{in} = 2.7$ V. Find V_{out}.

9. $A_p = 15$ dBm. Find P_{out}.

10. $P_{out} = 10$ W. Find A_p in dBm.

Solutions:

1. 32.4 dB

2. 31.5 W

3. 47.4 mW

4. 2 W

5. 86 dB

6. 3 mV

7. 11.2 V

8. 854 mV

9. 31.6 mW

10. 40 dBm

Additional practice problems are at the end of the chapter.

23.3 FREQUENCY RESPONSE

23.3.1 Bode Plot. A Bode plot is a useful tool used to approximate the frequency response of a circuit or system. Let's construct a Bode plot of the frequency response characteristics of the RC circuit in Figure 23-1. Recall from AC circuit analysis that X_C varies inversely with frequency. Further, the cut-off frequency f_{co} occurs where $X_C = R$. Because the output is taken from across R, $V_{out} = V_R$. We can compute the voltage gain of the circuit by using the equation learned previously.

$$A_{v(dB)} = 20 \log \frac{V_{out}}{V_{in}}$$

At high frequencies $V_{out} = V_{in}$. The voltage gain is

$$A_{v(dB)} = 20 \log \frac{10 \text{ V}}{10 \text{ V}} = 0 \text{ dB}$$

The maximum possible gain is 0 dB.

Our Bode plot is a plot of gain versus frequency. Semilog graph paper is used in such a plot. Frequency is the independent variable and is plotted along the logarithmic scale. A scale for A_v is chosen that will allow us to change the gain at least -20 dB. Once our scale is chosen, a horizontal line is drawn across our graph at 0 dB as shown in Figure 23-2. Four-cycle semilog paper has been chosen with a minimum frequency of 1 Hz. This allows us to examine the gain at frequencies from 1 Hz to 1 kHz. The cut-off frequency is 100 Hz.

$$f_{co} = \frac{1}{2\pi CR} = 100 \text{ Hz}$$

This point has been plotted on our horizontal line in Figure 23-3. This becomes one point on our Bode plot. As a straight-line approximation, the gain will change at the rate of 6 *dB per octave* and 20 *dB per decade*. An octave

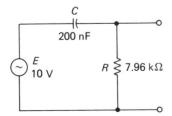

Figure 23-1.

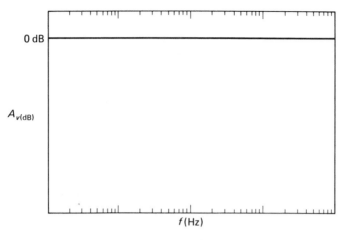

Figure 23-2.

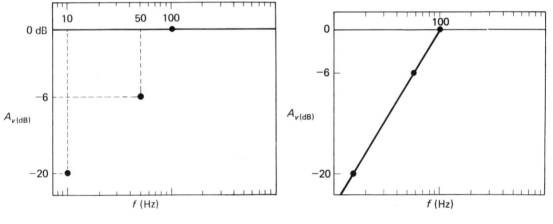

Figure 23-3.   Figure 23-4.

change in frequency is a halving or doubling of the frequency. A decade change is a change in frequency of one-tenth or ten times.

Because gain decreases as frequency decreases, a second point is plotted in Figure 23-3 where the gain is down 6 dB and the frequency is 50 Hz—one-half the cut-off frequency. A third point is plotted where the gain is down 20 dB and the frequency is 10 Hz—0.1 times f_{co}. In Figure 23-4 a straight line is drawn to connect these three points. This is our Bode plot. All Bode plots are plotted in a like manner. When the gain decreases with frequency, the line slopes downward and to the *left* as in Figure 23-4. When the gain decreases as frequency *increases*, the line slopes downward and to the right.

PRACTICE PROBLEMS 23-2

1. Refer to Figure 23-1. Construct a Bode plot for $C = 100$ nF and $R = 2.7$ kΩ.

2. Refer to Figure 23-5. Construct a Bode plot of the frequency response.

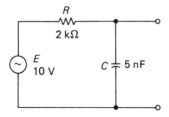

Figure 23-5.

Solutions:

1. $f_{co} = 589$ Hz. The Bode plot is shown in Figure 23-6.

2. $f_{co} = 15.9$ kHz. See Figure 23-7 for the Bode plot.

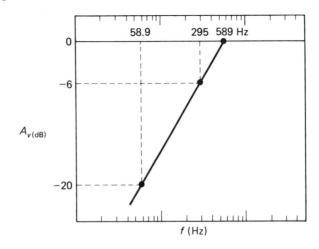

Figure 23-6.

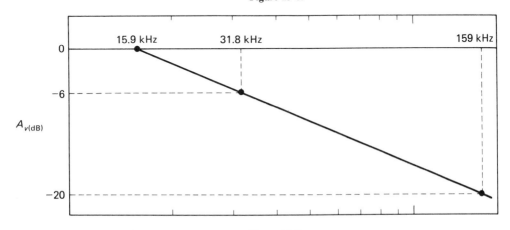

Figure 23-7.

EXAMPLE 23.6: An amplifier has a mid-frequency gain of 25 dB. The lower cut-off frequency (symbolized f_1) is 90 Hz. The upper cut-off frequency (symbolized f_2) is 12 kHz. Construct a Bode plot of the frequency response.

Solution: We will use five-cycle semilog graph paper for this example. Choose a practical scale for A_v and draw a horizontal line. Plot the lower cut-off frequency as before. One point is f_1 and 25 dB. $\dfrac{f_1}{2}$ and 19 dB is a second point (6 dB down). $\dfrac{f_1}{10}$ and 5 dB is a third point (20 dB down). The upper cut-off frequency is plotted by using the points f_2 and 25 dB, $2f_2$ and 19 dB, and $10f_2$ and 5 dB. The complete plot is shown in Figure 23-8.

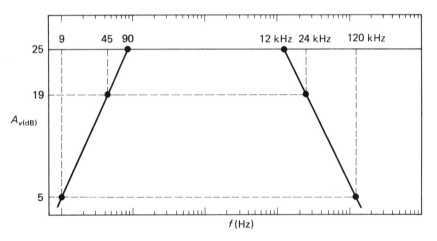

Figure 23-8.

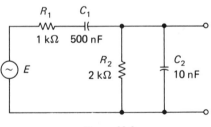

Figure 23-9.

Consider the circuit in Figure 23-9. In circuits like this, the mid-frequency gain can be determined if f_1 and f_2 are separated by at least one decade.

By ignoring the effects of C_1 and C_2 and using the ratio of R_2 to the total resistance $R_1 + R_2$, we can compute the mid-frequency gain because in this circuit:

$$\frac{R_2}{R_1 + R_2} = \frac{V_{out}}{V_{in}}$$

The mid-frequency equivalent circuit is shown in Figure 23-10(a). (In practice there may be other factors affecting response, such as source impedance and cable capacitance.)

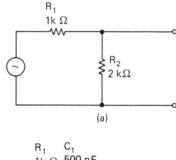

(a)

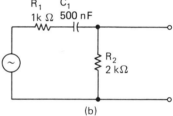

(b)

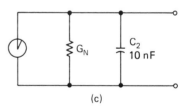

Figure 23-10.   (c)

Mid-frequency gain would be determined by:

$$A_v = 20 \log \frac{R_2}{R_1 + R_2} \tag{23-3}$$

In this circuit:

$$A_v = 20 \log \frac{2k}{(1k + 2k)} = -3.52 \text{ dB}$$

This is the maximum circuit gain. If the frequency is decreased, the reactance of C_1 becomes larger and some voltage drops across it. This causes a decrease in the output voltage. If the frequency is increased, the susceptance of C_2 increases, which again causes a decrease in output voltage. The series reactive components (C_1 in our circuit) affect the response as we lower frequency. The effect on the circuit due to C_2 is negligible—which is the case if f_1 and f_2 are at least a decade apart. The equivalent low frequency circuit looks like the circuit in Figure 23-10(b).

At high frequencies, the reactance of C_1 becomes negligible (compared to R_T) but the susceptance of C_2 becomes a factor. Since C_2 is a parallel component, if we Nortonize the circuit at R_2, we get an equivalent circuit consisting of C_2, R_1, and R_2 all in parallel as shown in Figure 23-10(c).

EXAMPLE 23.7: Find f_1 and f_2 in the circuit of Figure 23-9. Construct a Bode plot.

Solution: We previously determined that the mid-frequency gain is -3.52 dB. Let's compute f_1 and f_2.

$$f_1 = \frac{1}{2\pi C_1 (R_1 + R_2)} = \frac{1}{2\times 3.14 \times 500 \times 10^{-9} \times 3 \times 10^3} = 106$$

$$f_2 = \frac{(G_1 + G_2)}{2\pi C_2} = \frac{1.5 \times 10^{-3}}{2 \times 3.14 \times 10 \times 10^{-9}} = 23.9 \text{ kHz}$$

The Bode plot is constructed in Figure 23-11.

23.3.2 Frequency Response Curve. The Bode plot results in some error, particularly around the cut-off frequency. As we know, the gain is really down 3

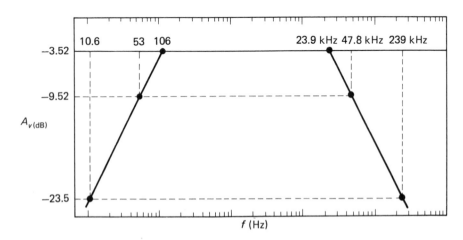

Figure 23-11

dB at f_{co}. At frequencies other than cut-off the error is less than 3 dB. The error is practically 0 one decade above and below f_{co}. The reduction in gain (from maximum gain) at any frequency can be found by using the following equations:

At low frequencies:

$$A_v = 20\log\frac{1}{\sqrt{1+\left(\dfrac{f_1}{f}\right)^2}} \tag{23-3}$$

where f_1 is the lower cut-off frequency and f is the frequency in question.

At high frequencies:

$$A_v = 20\log\frac{1}{\sqrt{1+\left(\dfrac{f}{f_2}\right)^2}} \tag{23-4}$$

where f_2 is the upper cut-off frequency and f is the frequency in question. The general shape of the frequency response curve results from plotting gain versus frequency using these equations. The curve is shown in Figure 23-12.

> EXAMPLE 23.8: Refer to Figure 23-9. Find the gain at 50 Hz and at 100 kHz. Locate these frequencies on the Bode plot. If we draw the curve right and make no mistakes in our calculations, these frequencies will fall on the curve.

> *Solution:* We previously determined that the mid-frequency gain (the maximum gain) is -3.52 dB. At 50 Hz:

$$A_v = 20\log\frac{1}{\sqrt{1+\left(\dfrac{f_1}{f}\right)^2}} = 20\log\frac{1}{\sqrt{1+\left(\dfrac{106\text{ Hz}}{50\text{ Hz}}\right)^2}} = -7.4 \text{ dB}$$

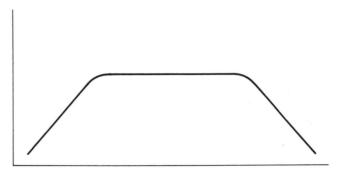

Figure 23-12.

This tells us that we are down 7.4 dB from -3.52 dB. The resultant gain is:

$$A_{v(\text{dB})} = -3.52 \text{ dB} + (-7.4 \text{ dB}) = -10.92 \text{ dB}$$

At 18 kHz:

$$A_v = 20 \log \cfrac{1}{\sqrt{1 + \left(\dfrac{f}{f_2}\right)^2}} = 20 \log \cfrac{1}{\sqrt{1 + \left(\dfrac{100 \text{ kHz}}{23.9 \text{ kHz}}\right)^2}}$$

$$= -12.67 \text{ dB}$$

The resultant gain is:

$$A_{v(\text{dB})} = -3.52 \text{ dB} + (-12.67 \text{ dB}) = -16.2 \text{ dB}$$

The Bode plot is shown in Figure 23-11. An inspection of the Bode plot shows that our gain calculations at 50 Hz and at 100 kHz were correct.

PRACTICE PROBLEMS 23-3

1. An amplifier has a mid-frequency gain of 40 dB. $f_1 = 200$ Hz and $f_2 = 20$ kHz. Construct a Bode plot of the frequency response.

2. Refer to Figure 23-9. Let $R_1 = 3.3$ kΩ and $R_2 = 6.8$ kΩ. Compute the gain at 10 Hz and 200 kHz. Construct a Bode plot of the frequency response. Do the computed gains at 10 Hz and 200 kHz fall on the Bode plot? If they do not, either the plot is wrong or your computations are wrong.

3. The mid-frequency gain of an amplifier is 27 dB. $f_1 = 100$ Hz and $f_2 = 15$ kHz. Find the gain at f_1, f_2, 25 Hz, and 30 kHz.

Solutions:

1. See Figure 23-13 for the Bode plot.

2. The mid-frequency gain is -3.44 dB. $f_1 = 31.5$ Hz and $f_2 = 72.3$ kHz. At 10 Hz, $A_v = -13.8$ dB (-3.44 dB $+ -10.4$ dB). At 200 kHz, $A_v = -12.8$ dB (-3.44 dB $+ -9.37$ dB). The Bode plot is constructed in Figure 23-14. An inspection of the plot shows that both 10 Hz and 200 kHz fall on the curve. This verifies the accuracy of our curve and our calculations.

3. At f_1, $A_v = 24$ dB. At f_2, $A_v = 24$ dB. At 25 Hz, $A_v = 14.7$ dB. At 30 kHz, $A_v = 20$ dB.

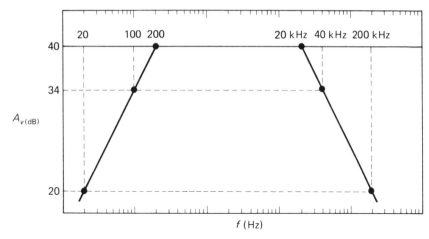

Figure 23-13

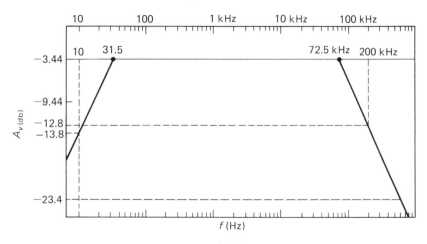

Figure 23-14

Additional practice problems are at the end of the chapter.

23.4 RC CIRCUITS

In Chapter 13 we used a universal time constant curve to determine the charge on a capacitor at some time after power was applied to the circuit. In this section we will compute the charge or discharge of a capacitor by using logarithmic equations. To determine the charge on a capacitor in a series RC circuit the following equation is used:

$$v_C = E(1 - \varepsilon^{-t/RC}) \tag{23-5}$$

where
$$v_C = \text{the charge on the capacitor;}$$
$$E = \text{the applied voltage;}$$
$$t = \text{the time in seconds;}$$
$$R = \text{the resistance in ohms; and}$$
$$C = \text{the capacitance in farads.}$$

The equation for v_R is:
$$v_R = E\varepsilon^{-t/RC} \qquad\qquad (23\text{-}6)$$

When the capacitor is discharging, the equation is:
$$v_C = E\varepsilon^{-t/RC} \qquad\qquad (23\text{-}7)$$

EXAMPLE 23.9: In the circuit of Figure 23-15, find v_R and v_C 1 ms after the switch is closed.

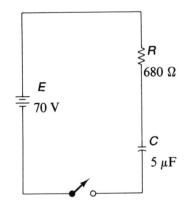

Figure 23-15

Solution:
$$v_C = E(1 - \varepsilon^{-t/RC}) = 70(1 - \varepsilon^{-1\,\text{ms}/3.4\times10^{-3}})$$
$$= 70(1 - \varepsilon^{-0.294}) = 70(1 - 0.745) = 70(0.255) = 17.9\ \text{V}$$

v_R can be found by applying Kirchhoff's voltage law:
$$v_R = E - v_C = 70\ \text{V} - 17.9\ \text{V} = 52.1\ \text{V}$$

EXAMPLE 23.10: Refer to Figure 23-15. If $V_C = 40$ V, how much time has lapsed?

Solution:
$$v_C = E(1 - \varepsilon^{-t/RC})$$
$$40 = 70(1 - \varepsilon^{-t/3.4\times10^{-3}})$$

Divide by 70:

$$0.571 = 1 - \varepsilon^{-t/3.4 \times 10^{-3}}$$

Subtract 1:

$$-0.429 = -\varepsilon^{-t/3.4 \times 10^{-3}}$$

Change signs:

$$0.429 = \varepsilon^{-t/3.4 \times 10^{-3}}$$

Take ln:

$$\ln 0.429 = -\frac{t}{3.4 \times 10^{-3}} \ln \varepsilon$$

$$-0.846 = -\frac{t}{3.4 \times 10^{-3}} \qquad (\ln \varepsilon = 1)$$

Multiply by -3.4×10^{-3}:

$$2.88 \times 10^{-3} = t$$
$$t = 2.88 \text{ ms}$$

PRACTICE PROBLEMS 23-4   Refer to Figure 23-15.

1. $t = 2$ ms. Find v_C and vR.

2. $V_R = 10$ V. Find t.

3. $i = 20$ mA. Find t. (The equation for i is of the same form as Equation 23-6).

4. In a circuit like Figure 23-15, $E = 50$ V, $R = 4.7$ kΩ, $v_C = 10$ V, and $t = 50$ μs. Find C.

5. In a circuit like Figure 23-13, $E = 50$ V, $C = 0.2$ μF, $v_R = 20$ V, and $t = 500$ ms. Find R.

Solutions:

1. $v_C = 31.1$ V, $v_R = 38.9$ V

2. 6.62 ms

3. 5.57 ms

4. 47.7 nF

5. 2.73 MΩ

SELF-TEST 23-1

1. Refer to Figure 23-9. Let $C_1 = {} = 1$ μF, $C_2 = 470$ pF, $R_1 = 4.7$ kΩ, and $R_2 = 2$ kΩ. Determine $A_{v(\text{dB})}$ at 50 Hz and 200 kHz. Construct a Bode plot.

2. $A_p = 25$ dB, $P_{\text{out}} = 3$ W. Find P_{in}.

3. $A_p = 45$ dBm. Find P_{out}.

4. Refer to Figure 23-15. $E = 60$ V, $v_C = 20$ V, $C = 500$ nF, and $t = 3$ ms. Find R.

Answers to Self-Test 23-1 are at the end of the chapter.

1. $P_{in} = 30 \ \mu W$, $P_{out} = 30 \ W$. Find A_p in dB.

2. $P_{in} = 10 \ mW$, $P_{out} = 3.16 \ W$. Find A_p in dB.

3. $A_p = 35 \ dB$, $P_{out} = 12 \ W$. Find P_{in}.

4. $A_p = 30 \ dB$, $P_{out} = 1 \ W$. Find P_{in}.

5. $A_p = 22 \ dB$, $P_{in} = 10 \ mW$. Find P_{out}.

6. $A_p = 40 \ dB$, $P_{in} = 150 \ mW$. Find P_{out}.

7. $A_p = 20 \ dBm$. Find P_{out}.

8. $A_p = 43 \ dBm$. Find P_{out}.

Assume $R_{out} = R_{in}$ for the following problems:

9. $V_{in} = 100 \ \mu V$, $V_{out} = 2.7 \ V$. Find A_v in dB.

10. $V_{in} = 5.7 \ V$, $V_{out} = 3.4 \ V$. Find A_v in dB.

11. $A_v = 40 \ dB$, $V_{in} = 5 \ mV$. Find V_{out}.

12. $A_v = 70 \ dB$, $V_{in} = 10 \ mV$. Find V_{out}.

13. $A_v = 90 \ dB$, $V_{out} = 4 \ V$. Find V_{in}.

14. $A_v = 45 \ dB$, $V_{out} = 1.73 \ V$. Find V_{in}.

15. $A_v = -3 \ dB$. $V_{out} = 500 \ mV$. Find V_{in}.

16. $A_v = -3 \ dB$. $V_{out} = 10 \ V$. Find V_{in}.

1. Refer to Figure 23-1. Construct a Bode plot of the frequency response where $C = 0.3 \ \mu F$ and $R = 1 \ k\Omega$.

2. Refer to Figure 23-1. Construct a Bode plot of the frequency response where $C = 0.5 \ \mu F$ and $R = 470 \ \Omega$.

3. Refer to Figure 23-5. Construct a Bode plot of the frequency response where $C = 50 \ nF$ and $R = 2 \ k\Omega$.

4. Refer to Figure 23-5. Construct a Bode plot of the frequency response where $C = 20 \ nF$ and $R = 1.5 \ k\Omega$.

5. Refer to Figure 23-9. Construct a Bode plot of the frequency response where $C_1 = 250 \ nF$, $C_2 = 40 \ nF$, $R_1 = 680 \ \Omega$, $R_2 = 1.2 \ k\Omega$, and $E = 10 \ V$.

6. Refer to Figure 23-9. Construct a Bode plot of the frequency response where $C_1 = 400 \ nF$, $C_2 = 25 \ nF$, $R_1 = 1 \ k\Omega$, $R_2 = 1 \ k\Omega$, and $E = 15 \ V$.

7. An amplifier has a mid-frequency gain of 20 dB. $f_1 = 50 \ Hz$ and $f_2 = 18 \ kHz$. Construct a Bode plot of the frequency response.

8. An amplifier has a mid-frequency gain of 35 dB. $f_1 = 80 \ Hz$ and $f_2 = 30 \ kHz$. Construct a Bode plot of the frequency response.

9. Refer to Figure 23-9. If $R_1 = 470 \ \Omega$, $R_2 = 680$ Ω, $C_1 = 1 \ \mu F$, $C_2 = 20 \ nF$, and $E = 25 \ V$. Find the gain at 30 Hz and at 50 kHz.

10. Refer to problem 9. Find the gain at 200 Hz and at 20 kHz.

11. Refer to problem 7. Find the gain at 40 Hz and 25 kHz.

12. Refer to problem 8. Find the gain at 20 Hz and 100 kHz.

Refer to Figure 23-15 for the following problems. Let $E = 100 \ V$, $R = 47 \ k\Omega$, and $C = 50 \ nF$ for problems 1 through 8.

1. $t = 1 \ ms$. Find v_R and v_C.

2. $t = 2 \ ms$. Find v_R and v_C.

3. $v_C = 65 \ v$. Find t.

4. $v_C = 40 \ V$. Find t.

5. $i = 2$ mA. Find t.

6. $i = 1$ mA. Find t.

7. $v_R = 10$ V. Find t.

8. $v_R = 80$ V. Find t.

9. $E = 25$ V, $C = 10$ nF, $v_R = 15$ V, and $t = 50$ μS. Find R.

10. $E = 40$ V, $C = 200$ nF, $v_R = 20$ V, and $t = 200$ μS. Find R.

11. $E = 100$ V, $R = 47$ kΩ, $v_C = 20$ V, and $t = 100$ μs. Find C.

12. $E = 100$ V, $R = 2.7$ kΩ, $v_C = 50$ V, and $t = 50$ μs. Find C.

ANSWERS TO SELF-TEST

Self-Test 23-1

1. At 50 Hz $A_v = -0.884$ dB. At 200 kHz $A_v = -2.27$ dB. See Figure 23-16 for the Bode plot.

2. $P_{in} = 9.49$ mW

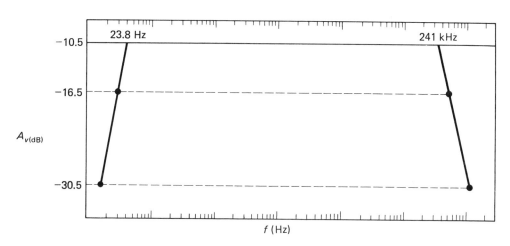

Figure 23-16.

3. $P_{out} = 31.6$ W

4. $R = 14.8$ kΩ

CHAPTER **24**

BOOLEAN ALGEBRA

24.1 LOGIC VARIABLES

Boolean algebra is used in manipulating logic variables. A variable is either completely true or completely false; partly true or partly false values are not allowed. When a variable is not true, by implication it must be false. Conversely, if the variable is not false, it must be true. Because of this characteristic, Boolean algebra is ideally suited to variables that have two states or values, such as YES and NO, or for a number system that has two values, 1 and 0 (i.e., binary number system). For example, B (a variable) could represent the presence of Bob. B has two values: if Bob is present, B equals "true"; if Bob is absent, B equals "false." Note that Bob is not the variable; B is the variable that represents the presence of Bob.

A switch is ideally suited to represent the value of any two-state variable because it can only be "off" or "on." Consider the SPST switch in Figure 24-1. When the switch is in the closed position, it indicates that Bob is present (B = true). When it is in the open position, it then represents that Bob is absent (B = false).

A closed switch could also represent values such as true, yes, one (1), HIGH (H), go, etc.; an open switch could represent values such as false, no, zero (0), LOW (L), no go, etc.

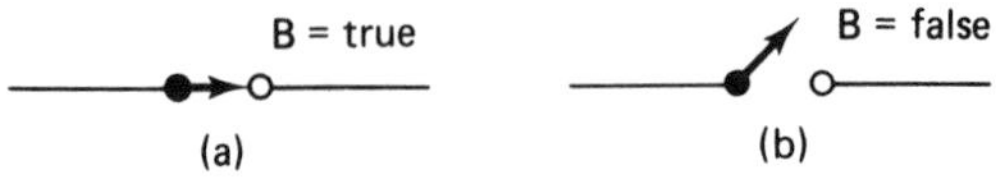

Figure 24-1. SPST switch: (a) closed; (b) open.

B = true

(a)

B = false

(b)

494

24.2 LOGIC OPERATIONS

There are only three basic logic operations:

1. The *conjunction* (logical product), commonly called AND, symbolized by ($\cdot$).
2. The *disjunction* (logical sum), commonly called OR, symbolized by ($+$).
3. The *negation*, commonly called NOT, symbolized by ($\overline{}$).

These operations are performed by logic circuits. All functions within a computer can be performed by combinations of these three basic logic operations.

24.3 THE AND FUNCTION

Figure 24-2 shows the logic symbol and truth table for a two-input AND gate. A and B are the input variables. The function (A·B) is produced when A is true AND B is true. That is, if both inputs are true, the output is true. If either input, or both, is false, the output is false. In order to show all of the conditions that may exist at the input and output of a logic gate, truth tables are used. Since there are two inputs and each input has two possible states (true and false), the number of possible conditions at the input would be 2 raised to the second power (2^2), or 4. Note in the truth table that condition 4 is the only one in which all of the inputs are true so that the AND function is produced and a true output appears. If A is true AND B is true, the output will be true.

AND gates may have two or more inputs. Figure 24-3 shows a three-input AND gate and its truth table. The function (A·B·C) is produced when A is true AND B is true AND C is true. Ones (1) and zeros (0) are used for the values of the variables (true = 1 and false = 0). Because three variables are used, eight possible conditions exist ($2^3 = 8$). Condition 8 is the only one that will produce a true output (1) because all of the input variables are true. Remember, in an AND gate, *all* the inputs must be true in order to produce a true output.

The ($\cdot$) symbol used in the expression A·B·C is the AND operator and indicates logical multiplication. For example, condition 8 in Figure 24-3 can be interpreted as $1 \times 1 \times 1 = 1$; whereas condition 7 can be interpreted as $1 \times 1 \times 0 = 0$.

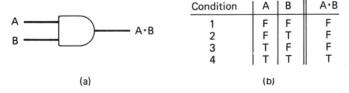

Condition	A	B	A·B
1	F	F	F
2	F	T	F
3	T	F	F
4	T	T	T

(a)   (b)

Figure 24-2. Two-input AND gate: (a) logic symbol; (b) truth table.

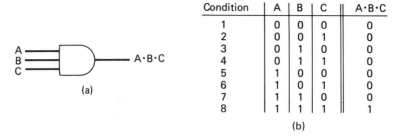

Condition	A	B	C	A·B·C
1	0	0	0	0
2	0	0	1	0
3	0	1	0	0
4	0	1	1	0
5	1	0	0	0
6	1	0	1	0
7	1	1	0	0
8	1	1	1	1

(b)

Figure 24-3. Three-input AND gate: (a) logic symbol; (b) truth table.

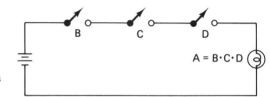

Figure 24-4. Switches connected in series to produce the AND function.

Figure 24-5. Logic symbol for the AND function BCD.

Using the same approach, we find that all other conditions (1 through 6) also produce a zero. It is important to remember that in the binary system, a 2 cannot exist—only 1's and 0's. As with ordinary algebra, the operator symbol (·) can be omitted; thus, A·B·C=ABC and is read "A AND B AND C."

The AND function can be illustrated by the following analogy. The members of group A are Bob, Charley, and Dick. Note that the names in this group are combined by the conjunction "and." That means group A equals the presence of Bob AND Charley AND Dick. This may be symbolized as

$$A = BCD$$

A is true (group A is present) when B is true (Bob is present); AND C is true (Charley is present); AND D is true (Dick is present). A is not true if any one or more of the members are absent.

The circuit in Figure 24-4 can be used to produce the AND function of the above example. The light indicates that group A is present only when all members of the group are present. (All switches are closed.) A logic circuit producing this AND function is symbolized in Figure 24-5.

24.4 THE OR FUNCTION

Figure 24-6 shows the logic symbol and truth table for a two-input OR gate. The function (A + B) is produced when A is true OR B is true. That is, if either input is true, the output is true. If both inputs are false, the output is false. Because

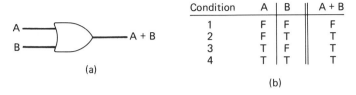

Condition	A	B	A + B
1	F	F	F
2	F	T	T
3	T	F	T
4	T	T	T

(a)

(b)

Figure 24-6. Two-input OR gate: (a) logic symbol; (b) truth table.

there are two input variables, a total of four conditions is possible ($2^2 = 4$). Note in the truth table that conditions 2, 3, and 4 produce a true output because at least one of the input variables is true. Condition 1 produces a false output because neither one of the input variables is true.

OR gates may have two or more inputs. Figure 24-7 shows a three-input OR gate and its truth table. Ones (1) and zeros (0) are used for the values of the variables. Because three variables are used, eight possible conditions occur ($2^3 = 8$). Conditions 2 through 8 produce a true (1) output because at least one of the input variables is true (1). Condition 1 is the only condition that does not produce a true output.

The " + " symbol used in the expression $A + B + C$ is the OR operator and indicates logical addition. In logical addition $1 + 0 = 1$ and $1 + 1 = 1$. One OR one does not equal two! Condition 8 in Figure 24-7 may be interpreted as $1 + 1 + 1 = 1$; whereas condition 5 can be interpreted as $1 + 0 + 0 = 1$. Condition 1 produces a zero output because $0 + 0 + 0 = 0$.

The OR function is illustrated by the following analogy. The members of group A are Bob, Charley, and Dick. A representative of this group could be Bob OR Charley OR Dick, or any combination of them. This expression may be symbolized as

$$R = B + C + D$$

where R is a representative of group A. R is true (a representative of group A is present) when B is true (Bob is present) OR C is true (Charley is present) OR D is true (Dick is present). Only one of the members needs to be present for group A to be represented. However, group A is also represented when more than one

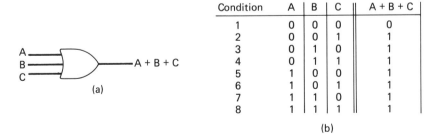

Condition	A	B	C	A + B + C
1	0	0	0	0
2	0	0	1	1
3	0	1	0	1
4	0	1	1	1
5	1	0	0	1
6	1	0	1	1
7	1	1	0	1
8	1	1	1	1

(a)

(b)

Figure 24-7. Three-input OR gate: (a) logic symbol; (b) truth table.

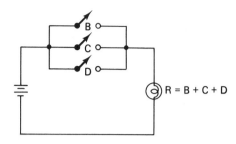

Figure 24-8. Switches connected in parallel to produce the OR function.

Figure 24-9. Logic symbol for the OR function $B+C+D$.

member is present. Group A will not be represented (false) when all members are absent.

The circuit in Figure 24-8 can be used to produce the OR function of the above example. The light (R) indicates that a representative of group A is present when one OR more members of the group are present. (At least one switch is closed.) A logic circuit producing this OR function is symbolized in Figure 24-9.

24.5   THE NOT FUNCTION

The concept of the NOT function can be illustrated by the circuit in Figure 24-10. When the switch is closed, the indicator lamp lights (true). Opening the switch will break the circuit and the lamp will go out (false). Only two conditions may exist. Either the lamp is on (true) or it is off (false). No other condition may exist. The conditions are *complements* of each other.

A logic circuit producing the NOT function is called an *inverter*; its symbol appears in Figure 24-11. An inverter converts the state or value of a variable to

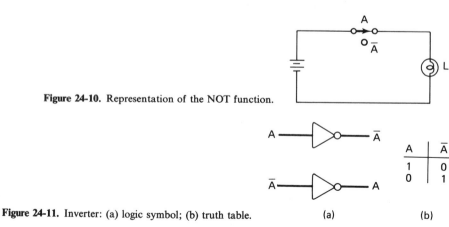

Figure 24-10. Representation of the NOT function.

Figure 24-11. Inverter: (a) logic symbol; (b) truth table.

(a)   (b)

A	$\overline{A}$
1	0
0	1

its complement. Thus, if variable A appears at the input, $\overline{A}$ (not A) is produced at the output; and, conversely, when $\overline{A}$ appears at the input; A is produced at the output. When performing the NOT function, a 1 will be changed to a 0, and vice versa, as shown in the truth table.

24.6 BOOLEAN EXPRESSIONS

The three basic logic functions discussed—AND, OR, and NOT—either individually or in various combinations, are the basic building blocks for all computer logic circuits. A few of these combinations will be illustrated by our group "A" analogy consisting of the presence of Bob (B), Charley (C), and Dick (D). Suppose we wish to describe a situation in which the group (A) is represented by the presence of Bob (B) and Charley (C) but not Dick (D). This situation could be expressed as $X = BC\overline{D}$ which states that Bob and Charley are present but Dick is NOT present; it is illustrated in Figure 24-12. Note the use of the inverter.

How would you describe a condition in which the group (A) is represented by at least two members—in other words, a majority? This situation could be expressed in Boolean algebra as $X = BC + BD + CD$ and is symbolized in Figure 24-13. The output is true if B AND C is true (Bob AND Charley are present); OR B AND D is true (Bob AND Dick are present); OR C AND D is true (Charley AND Dick are present).

How could the group be represented by the presence of Charley and the absence of both Bob and Dick? One possible method would be $\overline{B}C\overline{D}$, as shown in Figure 24-14.

Suppose we wish to express the situation in which the entire group is NOT present. In other words, one or more members are absent. The Boolean expres-

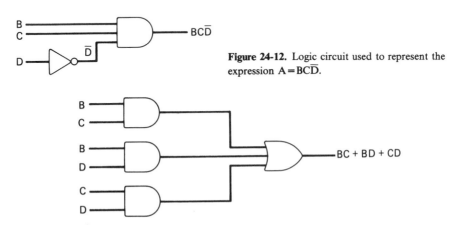

Figure 24-12. Logic circuit used to represent the expression $A = BC\overline{D}$.

Figure 24-13. Logic circuit used to represent the expression $BC + BD + CD$.

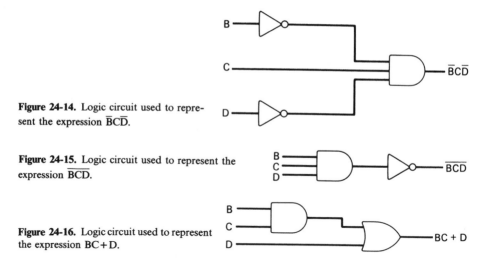

Figure 24-14. Logic circuit used to represent the expression $\overline{B}C\overline{D}$.

Figure 24-15. Logic circuit used to represent the expression $\overline{BCD}$.

Figure 24-16. Logic circuit used to represent the expression $BC+D$.

sion would be $\overline{BCD}$ and is illustrated in Figure 24-15. Note that the entire group is affected by the inverter.

Finally, how would we express a situation in which the group is represented by either Bob and Charley or by Dick alone? One possible method is shown in Figure 24-16.

Logic diagrams are drawn in order to symbolize logic expressions. In the following problems logic diagrams are to be interpreted or logic diagrams are to be drawn to symbolize logic expressions.

PRACTICE PROBLEMS 24-1   Draw the logic diagrams that represent the following Boolean expressions:

1. $B+\overline{C}$

2. $\overline{A}B$

3. $AB+\overline{C}$

4. $A(\overline{B}+C)$

5. $\overline{B+\overline{C}+D}$

Solutions:

In problem 1, Figure 24-17, the OR function is indicated; therefore, an OR gate is drawn. In order to realize the input $\overline{C}$, an inverter is used.

In problem 2 the AND function is indicated; therefore, an AND gate is drawn. In order to realize the input $\overline{A}$, an inverter is used.

In problem 3 both AND and OR functions are indicated. A and B are the inputs to the AND gate. The output of the AND gate (AB) and $\overline{C}$ are the inputs to the OR gate. An inverter is used to produce the input variable $\overline{C}$.

In problem 4 parentheses are used to indicate logical multiplication as in any algebraic expression. Therefore, a two-input AND gate is indicated. The inputs are A and $\overline{B}$ OR C. The input $\overline{B}$ OR C results from a two-input OR gate whose inputs are $\overline{B}$ and C.

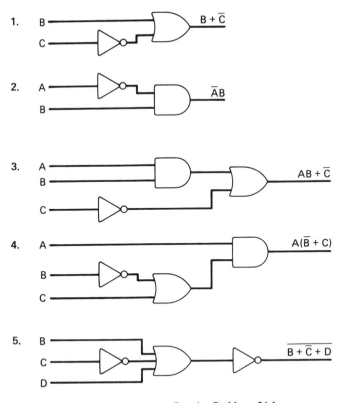

Figure 24-17. Answers to Practice Problems 24-1.

In problem 5 the entire output function is negated. The bar across the entire expression tells us this. Therefore, an inverter must be connected to the output of the OR gate. At the output of the OR gate is the expression $B + \overline{C} + D$. Because there are three variables, the gate is a three-input OR gate.

Now let's turn it around.

PRACTICE PROBLEMS 24-2   Write the Boolean expression represented by the logic circuits in Figure 24-18.

Solutions:

1. $F = \overline{AB}$

2. $F = \overline{A} + B$

3. $A = \overline{XY} + Z$

4. $X = \overline{ABC}$

5. $F = \overline{C(\overline{A} + B)}$

In problem 1 the AND function is indicated followed by an inverter. Because the inverter is on the output side of the gate, the entire expression is complemented.

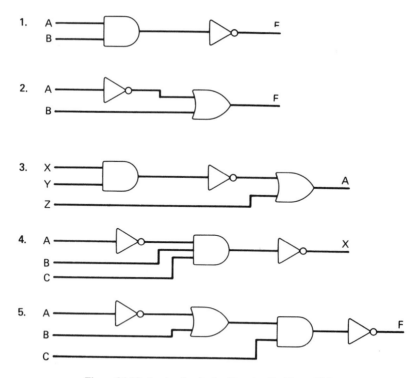

Figure 24-18. Logic circuits for Practice Problems 24-2.

In problem 2 the inverter complements one input to the OR gate and results in the output $\overline{A}+B$.

The inverter complements the output of the AND gate in problem 3 to give the expression $\overline{XY}$. This expression is one input to the OR gate. The other input is Z. The output of the OR gate then is $\overline{XY}+Z$.

In problem 4 the three inputs to the AND gate are $\overline{A}$, B, and C. The output is $\overline{A}BC$. This output is complemented by the inverter and results in $\overline{\overline{A}BC}$ at its output.

In problem 5 the output of the OR gate $(\overline{A}+B)$ is one input to the AND gate. C is the other input. The output of the AND gate is $C(\overline{A}+B)$. This output is complemented and results in $\overline{C(\overline{A}+B)}$ at the output of the inverter.

SELF-TEST 24-1   Problems 1 through 10: write the Boolean expressions for the logic circuits in Figure 24-19. Draw the logic diagrams that represent the following Boolean expressions:

11. $A\overline{B}$

12. $\overline{A}+B$

13. $\overline{F+G}$

14. $\overline{C}(B+\overline{D})$

15. $A\overline{B}+C$

16. $AB+AC$

17. $(X+Z)(X+Y)$

18. $A\overline{B}+\overline{CD}$

19. $\overline{A(B+C)}$

20. $\overline{B\overline{C}+D}$

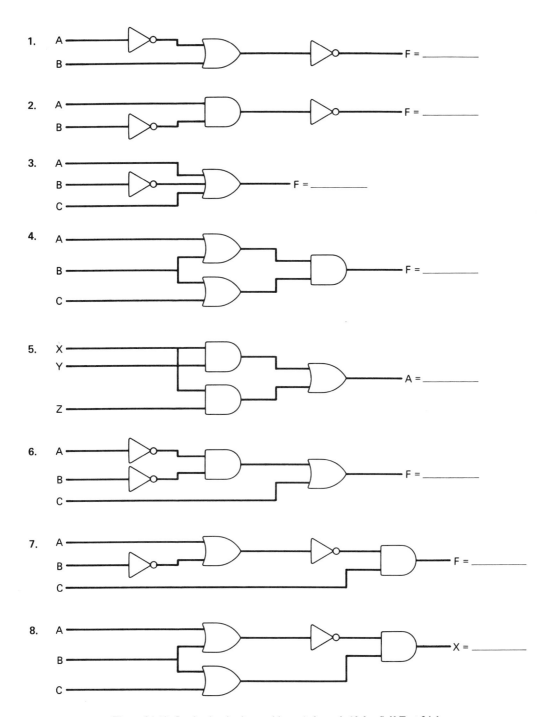

Figure 24-19. Logic circuits for problems 1 through 10 for Self-Test 24-1.

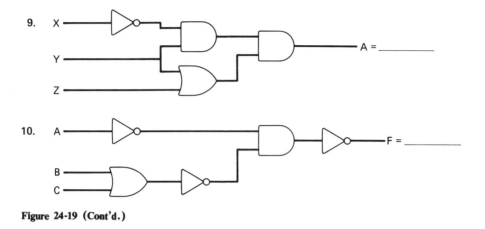

9. A = _____

10. F = _____

Figure 24-19 (Cont'd.)

Answers to Self-Test 24-1 are at the end of the chapter.

24.7 BOOLEAN POSTULATES AND THEOREMS

The following postulates, laws, and theorems are important in the simplification and manipulation of logic expressions. The student is encouraged to become familiar with these because they will be used throughout this chapter. Note that each postulate, law, or theorem is described in two parts. These are *duals* of each other; that is, the dual of the OR operator is the AND, while the dual of a given variable is its complement.

24.7.1 Postulates. Postulates are self-evident truths. Consider postulates 1a and 1b. A variable is either true (1) or false (0). Postulates 2a, 3a, and 4a represent the conjunctive (AND) form and define the function of the AND operator, as shown in Figure 24-20. Postulates 2b, 3b, and 4b represent the disjunctive (OR) form and define the function of the OR operator, as shown in Figure 24-21. Postulates 5a and 5b define the function of the NOT operator, as illustrated in Figure 24-22.

Postulates	
1a. $A = 1$ (if $A \neq 0$)	1b. $A = 0$ (if $A \neq 1$)
2a. $0 \cdot 0 = 0$	2b. $0 + 0 = 0$
3a. $1 \cdot 1 = 1$	3b. $1 + 1 = 1$
4a. $1 \cdot 0 = 0$	4b. $1 + 0 = 1$
5a. $\bar{1} = 0$	5b. $\bar{0} = 1$

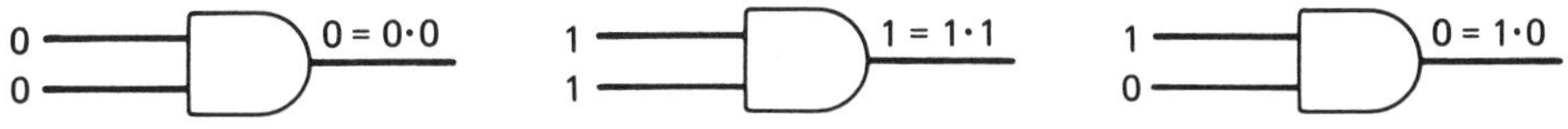

Figure 24-20. Logic symbolization of postulates 2a, 3a, and 4a.

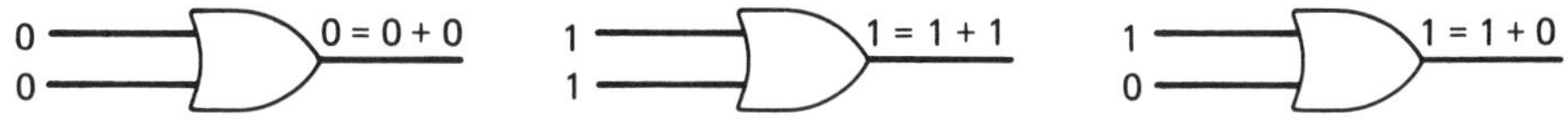

Figure 24-21. Logic symbolization of postulates 2b, 3b, and 4b.

Figure 24-22. Logic symbolization of postulates 5a and 5b.

24.7.2 Algebraic Properties. The following are properties of ordinary algebra that also apply to Boolean algebra. Remember, Boolean expressions contain variables having only two possible values.

Algebraic Properties	
Commutative	
6a. $AB = BA$	6b. $A + B = B + A$
Associative	
7a. $A(BC) = AB(C)$	7b. $A + (B + C) = (A + B) + C$
Distributive	
8a. $A(B + C) = AB + AC$	8b. $A + BC = (A + B)(A + C)$

The *commutative* property simply means that the circuit is not affected by the order or sequence of the variables. This is illustrated in Figure 24-23.

The *associative* property pertains to the parentheses. It shows that a sequence exclusively of AND functions (7a) or a sequence exclusively of OR functions (7b) is not affected by the placement of the parentheses, as indicated in Figure 24-24.

The *distributive* property for 8a and 8b may be proven by performing the algebraic multiplication or by factoring. Figure 24-25 shows that the application

Figure 24-23. Logic symbols used to demonstrate commutative property.

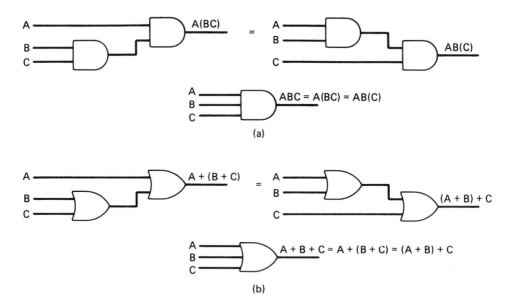

Figure 24-24. Logic symbols and circuits used to demonstrate the associative property: (a) logic circuit to demonstrate AND; (b) logic circuit to demonstrate OR.

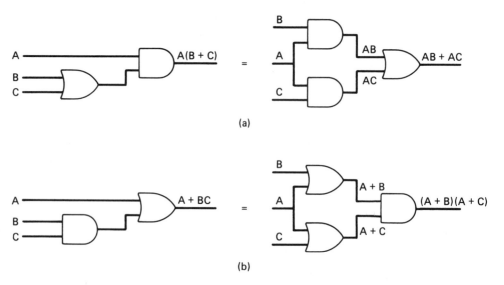

Figure 24-25. Logic symbols and circuits used to demonstrate distributive property: (a) logic circuit to demonstrate $A(B+C)=AB+AC$; (b) logic circuit to demonstrate $A+BC=(A+B)(A+C)$.

of the distributive property produces two forms of an expression, and, thus, two circuits may be realized for each.

24.7.3 Theorems. The following theorems define the application of the operators to variables:

Theorems	
9a. $A \cdot 0 = 0$	9b. $A + 0 = A$
10a. $A \cdot 1 = A$	10b. $A + 1 = 1$
11a. $A \cdot A = A$	11b. $A + A = A$
12a. $A \cdot \overline{A} = 0$	12b. $A + \overline{A} = 1$
13a. $\overline{\overline{A}} = A$	13b. $A = \overline{\overline{A}}$

Theorems 9a, 10a, 11a, and 12a pertain to the AND function. These are symbolized in Figure 24-26. For theorem 9a, one of the variables is always a zero. Therefore, the output will be a zero regardless of the value of A. Theorem 10a tells us the output will be determined by the input variable A. If $A = 1$, then $1 \cdot 1 = 1$. If $A = 0$, then $1 \cdot 0 = 0$. For Theorem 11a, if the input variable $A = 1$, the output will be $1 \cdot 1 = 1$. If the input variable $A = 0$, the output will be $0 \cdot 0 = 0$.

For Theorem 12a, the output will always be 0 because when the input variable $A = 1$, the other input variable will be 0 ($\overline{A}$), causing the output to be $1 \cdot 0 = 0$. The same output will be produced when the values of the input variables are reversed. Theorem 12a is called a *self-contradiction*.

Theorem 13a expresses double negation. An expression that has been inverted twice equals its original value, as shown in Figure 24-27.

Theorems 9b, 10b, 11b, and 12b pertain to the OR function and are illustrated in Figure 24-28. For the OR function, one or more of the input variables must be true in order that a true output be present. For theorems 9b

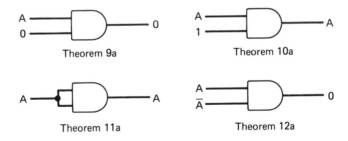

Figure 24-26. Logic symbolization of the AND function.

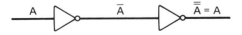

Figure 24-27. Logic symbolization of double negation.

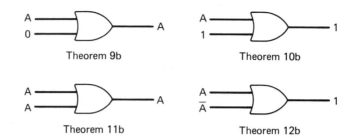

Figure 24-28. Logic symbolization of the OR function.

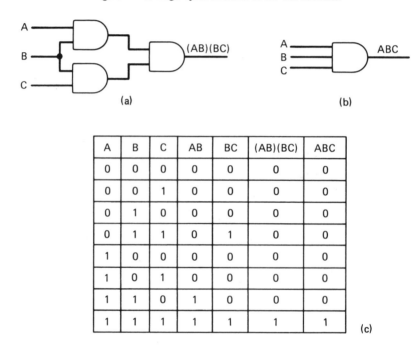

Figure 24-29. (a) Original circuit; (b) simplified circuit; (c) truth table.

and 11b, the output is determined by the input variable A. For 10b and 12b, either A or $\overline{A}$ must always equal 1.

Consider the logic diagram in Figure 24-29. The Boolean expression at its output is (AB)(BC). By applying various postulates, laws, and theorems we can simplify the expression to ABC. Here's how we could do it.

1. (AB)(BC)   Original Expression
2. ABBC   7a
3. ABC   11a

Step 1 shows the original expression. In step 2 we remove the parentheses. Our justification for this step is the associative property (7a). The simplified expression in step 3 results from applying theorem 11a. Any variable AND that same variable equal that variable (B·B=B). The simplified expression results in 3 two-input AND gates being replaced by 1 three-input AND gate—a substantial savings of both space and money to the designer.

Let's prove that the expressions are equal by using a truth table as in Figure 24-29(c). Notice that there are 8 conditions. There are three input variables (A, B, and C) and each variable has two possible states ($2^3 = 8$). When we construct a truth table it is important that we keep the order of the original statement and not skip any steps along the way. Thus, we list the variables first, then we list the function of the first two AND gates (AB and BC), and then we write the original expression and the simplified expression. The truth table shows that (AB)(BC)= ABC.

The process of using truth tables to prove the equality of two expressions is called *proof by perfect induction*. The student is encouraged to use truth tables to evaluate logic expression or to prove the results of logic operations. Truth tables can also be used to define the functions of various logic circuits.

The distributive property (8a) states that A(B+C)=AB+AC. Let's prove that they are equal by constructing a truth table, as in Figure 24-30. First we list

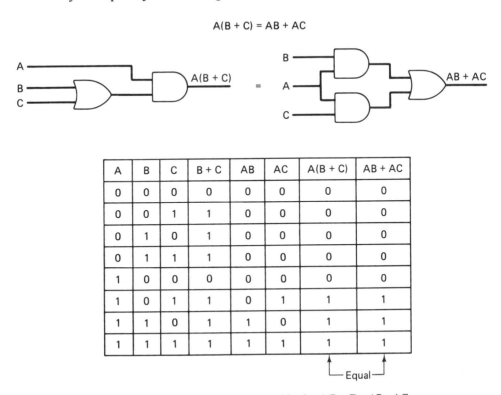

$$A(B + C) = AB + AC$$

A	B	C	B + C	AB	AC	A(B + C)	AB + AC
0	0	0	0	0	0	0	0
0	0	1	1	0	0	0	0
0	1	0	1	0	0	0	0
0	1	1	1	0	0	0	0
1	0	0	0	0	0	0	0
1	0	1	1	0	1	1	1
1	1	0	1	1	0	1	1
1	1	1	1	1	1	1	1

└─Equal─┘

Figure 24-30. Logic diagrams and truth tables for A(B+C)=AB+AC.

the input variables. Again, because there are three variables, 8 conditions are possible. Next we list each intermediate step. B+C is the result of ORing the input variables B and C. The columns AB and AC are the result of ANDing the input variables A and B and the input variables B and C. Finally, there are columns showing the expressions A(B+C) and AB+AC.

Now let's do the same thing with the distributive property 8b. (A+B)(A+C). The logic diagrams and truth table are shown in Figure 24-31. Notice again how each part of each expression is listed in the truth table.

Let's see if it is possible to simplify the logic diagram in Figure 24-32. The step-by-step simplification and the justification for each step are shown below:

1.	A+B+AC+AB	Original expression
2.	A+AB+B+AC	6b
3.	A(1+B)+B+AC	8a
4.	A(1)+B+AC	10b
5.	A+B+AC	10a
6.	A+AC+B	6b
7.	A(1+C)+B	8a
8.	A(1)+B	10b
9.	A+B	10a

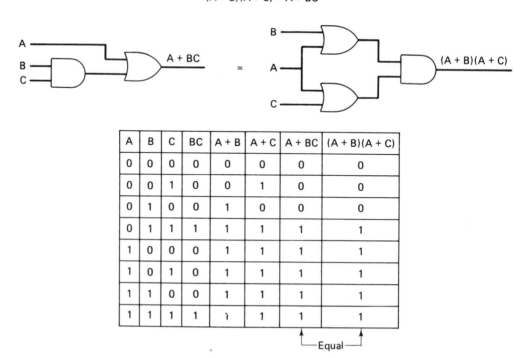

Figure 24-31. Logic diagrams and truth table for A+BC=(A+B)(A+C).

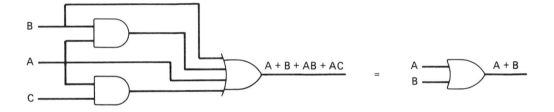

Figure 24-32. Logic symbols and truth table for $A + B + AB + AC = A + B$.

A	B	C	AC	AB	A + B + AC + AB	A + B
0	0	0	0	0	0	0
0	0	1	0	0	0	0
0	1	0	0	0	1	1
0	1	1	0	0	1	1
1	0	0	0	0	1	1
1	0	1	1	0	1	1
1	1	0	0	1	1	1
1	1	1	1	1	1	1

Equal

When many steps are involved in the simplification, there is usually more than one way to arrive at the final answer. The student may find alternate methods to the above simplification but in each case the final simplification should be $A + B$. The original logic diagram, simplified logic diagram, and truth table are found in Figure 24-32.

PRACTICE PROBLEMS 24-3   For each logic diagram in Figure 24-33, (a) write the Boolean expression, (b) simplify the Boolean expression and justify each step, (c) draw the simplified logic diagram, and (d) construct a truth table to prove that the two expressions are equal.

Solutions:

1. See Figure 24-34.

2. See Figure 24-35.

3. See Figure 24-36.

4. See Figure 24-37.

1. A
    B

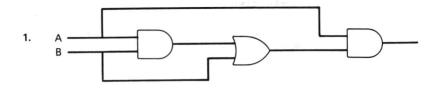

2. A
    B

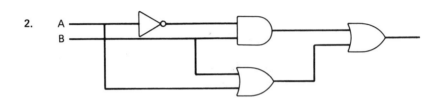

3. A
    B
    C

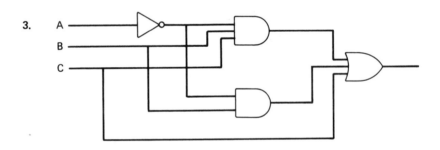

4. A
    B
    C

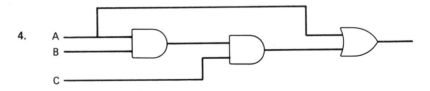

Figure 24-33. Logic circuits for Practice Problems 24-3.

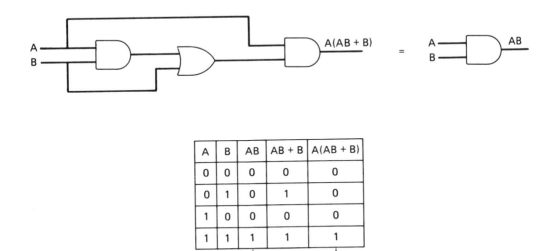

A	B	AB	AB + B	A(AB + B)
0	0	0	0	0
0	1	0	1	0
1	0	0	0	0
1	1	1	1	1

Equal

Figure 24-34. Answer to problem 1, Practice Problems 24-3.

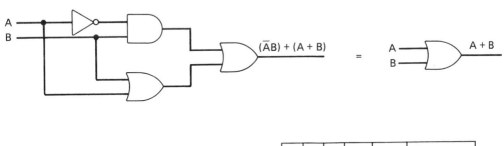

$(\overline{A}B) + (A + B)$	Original expression
$\overline{A}B + A + B$	7b
$\overline{A}B + B + A$	6b
$B(\overline{A} + 1) + A$	8a
$B(1) + A$	10b
$B + A$	10a

A	B	$\overline{A}$	$\overline{A}B$	A + B	$\overline{A}B + (A + B)$
0	0	1	0	0	0
0	1	1	1	1	1
1	0	0	0	1	1
1	1	0	0	1	1

Figure 24-35. Answer to problem 2, Practice Problems 24-3.

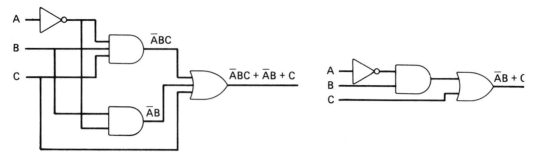

$\overline{A}BC + \overline{A}B + C$ Original expression
$\overline{A}B (C + 1) + C$ 8a
$\overline{A}B (1) + C$ 10b
$\overline{A}B + C$ 10a

A	B	C	$\overline{A}$	$\overline{A}B$	$\overline{A}BC$	$\overline{A}BC + \overline{A}B + C$	$\overline{A}B + C$
0	0	0	1	0	0	0	0
0	0	1	1	0	0	1	1
0	1	0	1	1	0	1	1
0	1	1	1	1	1	1	1
1	0	0	0	0	0	0	0
1	0	1	0	0	0	1	1
1	1	0	0	0	0	0	0
1	1	1	0	0	0	1	1

└─ Equal ─┘

Figure 24-36. Answer to problem 3, Practice Problems 24-3.

$A + ABC$ Original expression
$A(1 + BC)$ 8a
$A(1)$ 10b
A 10a

A	B	C	AB	ABC	A + ABC
0	0	0	0	0	0
0	0	1	0	0	0
0	1	0	0	0	0
0	1	1	0	0	0
1	0	0	0	0	1
1	0	1	0	0	1
1	1	0	1	0	1
1	1	1	1	1	1

└──── Equal ────┘

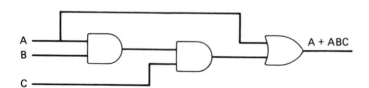

Figure 24-37. Answer to problem 4, Practice Problems 24-3.

24.7.4 DeMorgan's Theorem. DeMorgan's theorem provides us with an invaluable tool to use in simplifying Boolean expressions. By applying DeMorgan's theorem, we can find the equivalent of a Boolean expression. For example, consider the logic diagram in Figure 24-38. The Boolean expression at the output is $\overline{\overline{\overline{AB}}}$. We would not consider $\overline{\overline{\overline{AB}}}$ to be in its simplest form because of the bar extending over the entire expression. By applying DeMorgan's theorem, the equivalent Boolean expression $A+B$ is realized. This expression is much easier to understand and much more simple to construct.

Three steps are required in DeMorganizing a Boolean expression: First, we replace all of the OR operator symbols ($+$) with AND operator symbols ($\cdot$) and all of the AND operator symbols with OR operator symbols; next, we replace all variables with their complements; finally, we complement the entire expression. To DeMorganize the expression $\overline{\overline{\overline{AB}}}$ above, we use the following steps:

Step 1. Replace the AND operator with the OR operator $\overline{\overline{\overline{A}+\overline{B}}}$.

Step 2. Complement each variable $\overline{\overline{\overline{A}+\overline{B}}} = \overline{\overline{A+B}}$.

Step 3. Complement the entire expression $\overline{\overline{A+B}} = A+B$.

The logic diagrams and truth table are found in Figure 24-39.

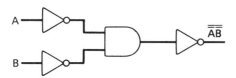

Figure 24-38. Logic diagram for the expression AB.

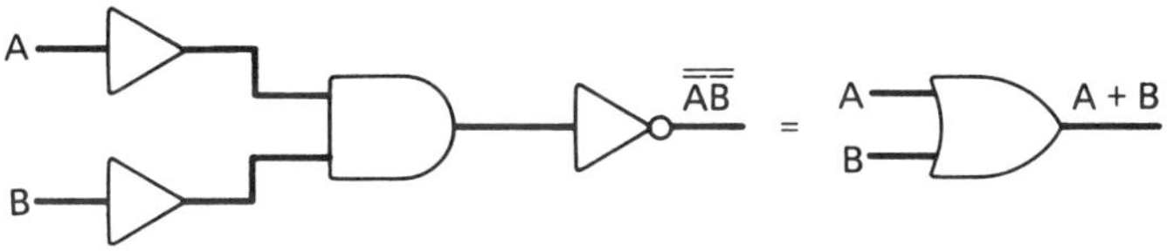

Figure 24-39. Logic diagrams and truth table for $AB=A+B$.

A	B	$\overline{A}$	$\overline{B}$	$\overline{AB}$	$\overline{\overline{AB}}$	A + B
0	0	1	1	1	0	0
0	1	1	0	0	1	1
1	0	0	1	0	1	1
1	1	0	0	0	1	1

Equal

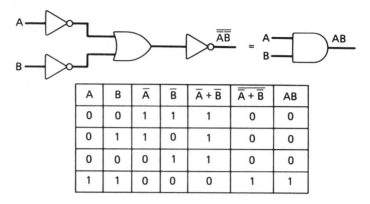

A	B	$\overline{A}$	$\overline{B}$	$\overline{A}+\overline{B}$	$\overline{\overline{A}+\overline{B}}$	AB
0	0	1	1	1	0	0
0	1	1	0	1	0	0
0	0	0	1	1	0	0
1	1	0	0	0	1	1

Figure 24-40. Logic diagrams and truth table for $\overline{\overline{A}+\overline{B}}=AB$.

Let's determine DeMorgan's equivalent of the expression $\overline{A+B}$.
Step 1. Replace the OR operator with the AND operator $\overline{A\cdot B}$.
Step 2. Complement each variable $\overline{A}\cdot\overline{B}=\overline{\overline{\overline{AB}}}$.
Step 3. Complement the entire expression $\overline{\overline{AB}}=AB$.
Therefore,

$$\overline{A+B} = AB$$

and is symbolized in Figure 24-40.
Let's find the DeMorgan's equivalent of the expression in theorem 14a, $\overline{ABC}$.
Step 1. $\overline{A}+\overline{B}+\overline{C}$
Step 2. $\overline{\overline{A}}+\overline{\overline{B}}+\overline{\overline{C}}=A+B+C$
Step 3. $\overline{A+B+C}$
Therefore,

$$\overline{ABC}= \overline{A+B+C}$$

DEMORGAN'S THEOREM	
14a. $\overline{ABC}=\overline{A+B+C}$	14b. $\overline{A+B+C}=\overline{ABC}$

The symbols for the equivalent expressions above are given in Figure 24-41.
Now let's try finding DeMorgan's equivalent of the expression in Theorem 14b, $\overline{A+B+C}$.
Step 1. $\overline{A}\cdot\overline{B}\cdot\overline{C}$
Step 2. $\overline{\overline{A}}\cdot\overline{\overline{B}}\cdot\overline{\overline{C}}=A\cdot B\cdot C$
Step 3. $\overline{A\cdot B\cdot C}$
Therefore,

$$\overline{A+B+C}= \overline{A\cdot B\cdot C}$$

The logic symbols for these expressions are given in Figure 24-42.

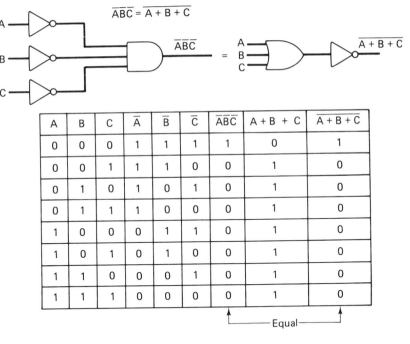

$$\overline{ABC} = \overline{A} + \overline{B} + \overline{C}$$

A	B	C	$\overline{A}$	$\overline{B}$	$\overline{C}$	$\overline{ABC}$	$\overline{A} + \overline{B} + \overline{C}$	$\overline{\overline{A} + \overline{B} + \overline{C}}$
0	0	0	1	1	1	1	0	1
0	0	1	1	1	0	0	1	0
0	1	0	1	0	1	0	1	0
0	1	1	1	0	0	0	1	0
1	0	0	0	1	1	0	1	0
1	0	1	0	1	0	0	1	0
1	1	0	0	0	1	0	1	0
1	1	1	0	0	0	0	1	0

———Equal———

Figure 24-41. Logic diagrams and truth table for $\overline{ABC} = \overline{A} + \overline{B} + \overline{C}$.

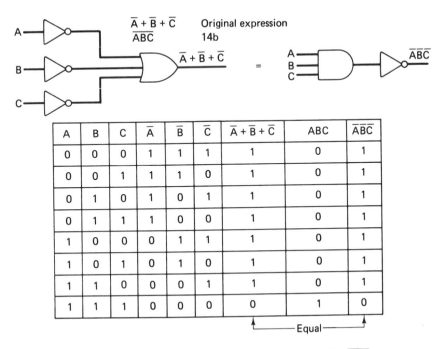

A	B	C	$\overline{A}$	$\overline{B}$	$\overline{C}$	$\overline{A} + \overline{B} + \overline{C}$	ABC	$\overline{ABC}$
0	0	0	1	1	1	1	0	1
0	0	1	1	1	0	1	0	1
0	1	0	1	0	1	1	0	1
0	1	1	1	0	0	1	0	1
1	0	0	0	1	1	1	0	1
1	0	1	0	1	0	1	0	1
1	1	0	0	0	1	1	0	1
1	1	1	0	0	0	0	1	0

———Equal———

Figure 24-42. Logic diagrams and truth table for $\overline{A} + \overline{B} + \overline{C} = \overline{ABC}$.

The three-step solution process can be simplified. Consider the expression $\overline{\overline{A}+\overline{B}}$ that we simplified earlier. To simplify, we just break the bar and change the operator.

$$\overline{\overline{A}+\overline{B}} = \overline{\overline{A}}\cdot\overline{\overline{B}} = AB$$

This is an example of theorem 14a. When breaking the bar results in a double negation, we apply theorem 13a $(\overline{\overline{A}}=A)$.

Let's simplify the Boolean expression $\overline{\overline{A}+B+\overline{C}}$.

$$\overline{\overline{A}+B+\overline{C}}\quad\text{Original expression}$$

$$\overline{\overline{A}}\cdot\overline{B}\cdot\overline{\overline{C}}\quad\text{14b}$$

$$A\overline{B}C\quad\text{13a}$$

Notice we broke the bar between each variable and changed each OR operator to AND. The logic diagrams and truth table are shown in Figure 24-43.

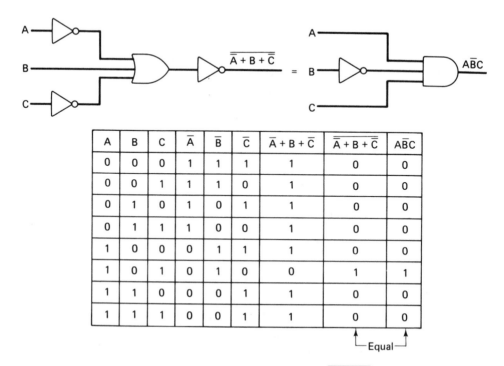

A	B	C	$\overline{A}$	$\overline{B}$	$\overline{C}$	$\overline{A}+B+\overline{C}$	$\overline{\overline{A}+B+\overline{C}}$	$A\overline{B}C$
0	0	0	1	1	1	1	0	0
0	0	1	1	1	0	1	0	0
0	1	0	1	0	1	1	0	0
0	1	1	1	0	0	1	0	0
1	0	0	0	1	1	1	0	0
1	0	1	0	1	0	0	1	1
1	1	0	0	0	1	1	0	0
1	1	1	0	0	1	1	0	0

└ Equal ┘

Figure 24-43. Logic diagrams and truth table for $\overline{\overline{A}+B+\overline{C}}=A\overline{B}C$.

Let's simplify the expression $\overline{A+\overline{B+C}}$. Again we use DeMorgan's theorem since simplification requires removing the signs of negation (the bars). Let's start by breaking the bar between $\overline{A}$ and $\overline{B+C}$ and changing the operator.

$$\overline{A+\overline{B+C}} \quad \text{Original expression}$$

$$\overline{\overline{A}}\cdot\overline{\overline{B+C}} \quad 14a$$

$$A\cdot(B+C) \quad 13a$$

Or $\qquad\qquad\qquad\qquad A(B+C)$

Notice the parens (sign of grouping) around $B+C$. We must treat $\overline{B+C}$ in the original expression as a single term (in algebra the bar is a sign of grouping). Therefore, when we DeMorganize, we must include the sign of grouping around $B+C$ to avoid an invalid expression. Without the sign of grouping we get

$\overline{A+\overline{B+C}}$ Original expression
$A(B+C)$ 14b

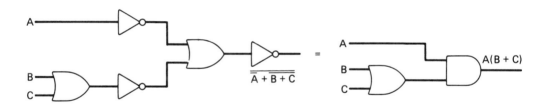

A	B	C	$\overline{A}$	$B+C$	$\overline{B+C}$	$\overline{A}+\overline{B+C}$	$\overline{A+\overline{B+C}}$	$A(B+C)$
0	0	0	1	0	1	1	0	0
0	0·	1	1	1	0	1	0	0
0	1	0	1	1	0	1	0	0
0	1	1	1	1	0	1	0	0
1	0	0	0	0	1	1	0	0
1	0	1	0	1	0	0	1	1
1	1	0	0	1	0	0	1	1
1	1	1	0	1	0	0	1	1

Equal

Figure 24-44. Logic diagrams and truth table for $\overline{A+\overline{B+C}}=A(B+C)$.

A·B+C or AB+C which is not the same as A(B+C). We could have included the sign of grouping in the original expression:

$$\overline{A + \overline{B+C}} = \overline{A + \overline{(B+C)}}$$

It isn't necessary though because the bar $(\overline{B+C})$ tells us to treat B+C as a single term or variable.

We could have DeMorganized $\overline{B+C}$ first:

$$\overline{A + \overline{B+C}} \quad \text{Original expression} \quad \overline{A + \overline{\overline{BC}}} \qquad 14a$$

Notice this did not affect the bar which extends over the entire expression since we were DeMorganizing just $\overline{B+C}$. Next:

$$\overline{\overline{A} \cdot \overline{\overline{BC}}} \qquad 14a \text{ (again)} \qquad \overline{A \cdot \overline{BC}} \qquad 13a$$

$$\overline{A \cdot \left(\overline{\overline{B}} + \overline{\overline{C}}\right)} \qquad 14b \qquad A(B+C) \qquad 13a$$

We arrive at the same answer by a much more difficult route. Notice the parens when theorem 14b was used. Only the expression BC was DeMorganized. Therefore, we include the sign of grouping to indicate that B+C is a single variable (in this case one input to an AND gate). Figure 24-44 includes the logic diagrams and truth table.

24.7.5 Absorption Theorem. Sometimes variables appear in a certain pattern that can obviously be simplified. The *absorption* theorem shows how substitution may be used on some commonly derived patterns.

Theorems 15a and 15b show that the value of B is redundant. A true output occurs only when A is true, as shown in Figure 24-45.

Theorem 16a states that only when A AND B are true, the statement is true, as shown in Figure 24-46. Recall that A and $\overline{A}$ cannot be true at the same time.

Theorem 16b is the dual of theorem 16a and is shown in Figure 24-47. For the expression to be true, either A must be true or B must be true. $\overline{A}$ is redundant.

ABSORPTION THEOREM	
15a. A(A+B)=A	15b. A+AB=A
16a. A($\overline{A}$+B)=AB	16b. A+$\overline{A}$B=A+B

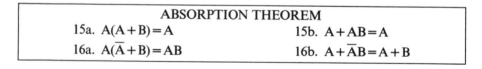

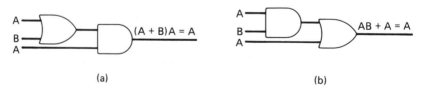

(a)   (b)

Figure 24-45. Logic symbolization of (a) theorem 15a and (b) theorem 15b.

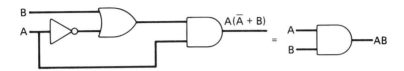

Figure 24-46. Logic symbolization of theorem 16a.

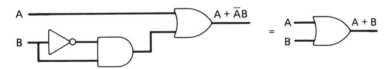

Figure 24-47. Logic symbolization of theorem 16b.

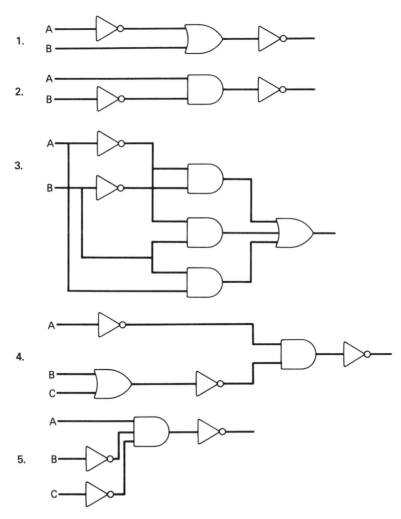

Figure 24-48. Practice Problems 24-4.

PRACTICE PROBLEMS 24-4   For each logic diagram in Figure 24-48, (a) write the Boolean expression, (b) simplify the Boolean expression, and justify each step, (c) draw the simplified logic diagram, and (d) construct a truth table to prove that the two expressions are equal.

Solutions:

1. See Figure 24-49.

2. See Figure 24-50.

3. See Figure 24-51.

4. See Figure 24-52.

5. See Figure 24-53.

		A	B	$\overline{A}$	$\overline{B}$	$\overline{A} + B$	$\overline{\overline{A} + B}$	$A\overline{B}$
$\overline{\overline{A} + B}$	Original expression	0	0	1	1	1	0	0
$\overline{\overline{A}} \cdot \overline{B}$	14a	0	1	1	0	1	0	0
$A\overline{B}$	13a	1	0	0	1	0	1	1
		1	1	0	0	1	0	0

^{L}Equal^{J}

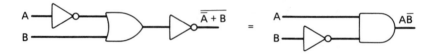

Figure 24-49. Answer to problem 1, Practice Problems 24-4.

		A	B	$\overline{A}$	$\overline{B}$	$A\overline{B}$	$\overline{A\overline{B}}$	$\overline{A} + B$
$\overline{A\overline{B}}$	Original expression	0	0	1	1	0	1	1
$\overline{\overline{A} + \overline{\overline{B}}}$	14b	0	1	1	0	0	1	1
$\overline{A} + B$	13a	1	0	0	1	1	0	0
		1	1	0	0	0	1	1

^{L}Equal^{J}

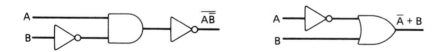

Figure 24-50. Answer to problem 2, Practice Problems 24-4.

$$\overline{A}\,\overline{B} + \overline{A}B + AB \qquad \text{Original expression}$$
$$\overline{A}(\overline{B} + B) + AB \qquad \text{8a}$$
$$\overline{A}(1) + AB \qquad \text{12b}$$
$$\overline{A} + AB \qquad \text{10a}$$
$$\overline{A} + B \qquad \text{16b}$$

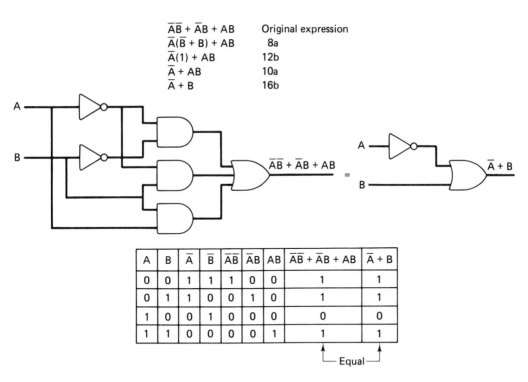

A	B	$\overline{A}$	$\overline{B}$	$\overline{A}\overline{B}$	$\overline{A}B$	AB	$\overline{A}\overline{B} + \overline{A}B + AB$	$\overline{A} + B$
0	0	1	1	1	0	0	1	1
0	1	1	0	0	1	0	1	1
1	0	0	1	0	0	0	0	0
1	1	0	0	0	0	1	1	1

Equal

Figure 24-51. Answer to problem 3, Practice Problems 24-4.

$$\overline{\overline{A}\ \overline{(B + C)}} \qquad \text{Original expression}$$
$$\overline{\overline{A}} + \overline{(\overline{B + C})} \qquad \text{14b}$$
$$A + (B + C) \qquad \text{13a}$$
$$A + B + C \qquad \text{7b}$$

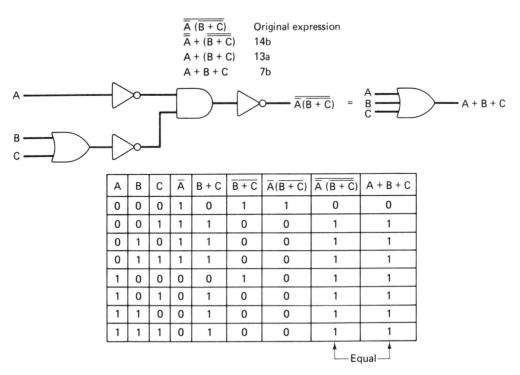

A	B	C	$\overline{A}$	B + C	$\overline{B + C}$	$\overline{A}(\overline{B + C})$	$\overline{\overline{A}\ (\overline{B + C})}$	A + B + C
0	0	0	1	0	1	1	0	0
0	0	1	1	1	0	0	1	1
0	1	0	1	1	0	0	1	1
0	1	1	1	1	0	0	1	1
1	0	0	0	0	1	0	1	1
1	0	1	0	1	0	0	1	1
1	1	0	0	1	0	0	1	1
1	1	1	0	1	0	0	1	1

Equal

Figure 24-52. Answer to problem 4, Practice Problems 24-4.

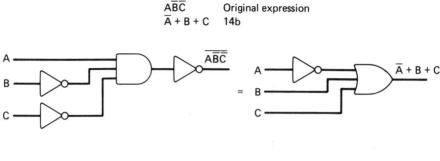

$$\overline{\overline{\overline{ABC}}} \qquad \text{Original expression}$$
$$\overline{A} + B + C \qquad 14b$$

Input variables								
A	B	C	$\overline{A}$	$\overline{B}$	$\overline{C}$	$\overline{ABC}$	$\overline{\overline{\overline{ABC}}}$	$\overline{A}+B+C$
0	0	0	1	1	1	0	1	1
0	0	1	1	1	0	0	1	1
0	1	0	1	0	1	0	1	1
0	1	1	1	0	0	0	1	1
1	0	0	0	1	1	1	0	0
1	0	1	0	1	0	0	1	1
1	1	0	0	0	1	0	1	1
1	1	1	0	0	0	0	1	1

Figure 24-53. Answer to problem 5, Practice Problems 24-4.

SELF-TEST 24-2   Problems 1 through 5: for the logic diagrams in Figure 24-54 (a) write the Boolean expression at the output, (b) simplify and justify each step, (c) draw the logic diagram of the simplified expression, and (d) construct a truth table to prove that the expressions are equal.

For the following Boolean expressions, (a) draw the logic diagram, (b) simplify and justify each step, (c) draw the logic diagram of the simplified expression, and (d) construct a truth table to prove that the expressions are equal.

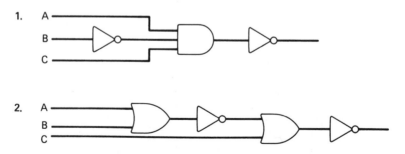

Figure 24-54. Problems 1 through 5, Self-Test 24-2.

3.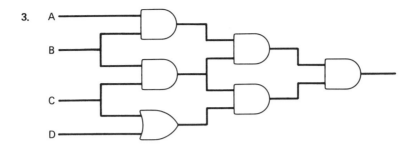

4.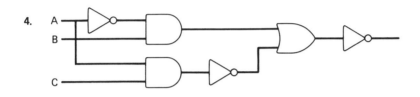

5.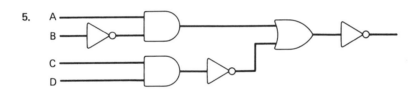

6. $C(\overline{B+D})$

7. $\overline{A\overline{B}+\overline{C}}$

8. $\overline{\overline{A}+\overline{B}}+C$

9. $(AB+BC)(AC+BC)$

10. $\overline{\overline{AB+A\overline{C}}+\overline{BC}}$

Answers to Self-Test 24-2 are at the end of the chapter.

END OF CHAPTER PROBLEMS 24-1

Problems 1 through 20: Write the Boolean expression represented by the logic diagrams in Figures 24-55 through 24-57.

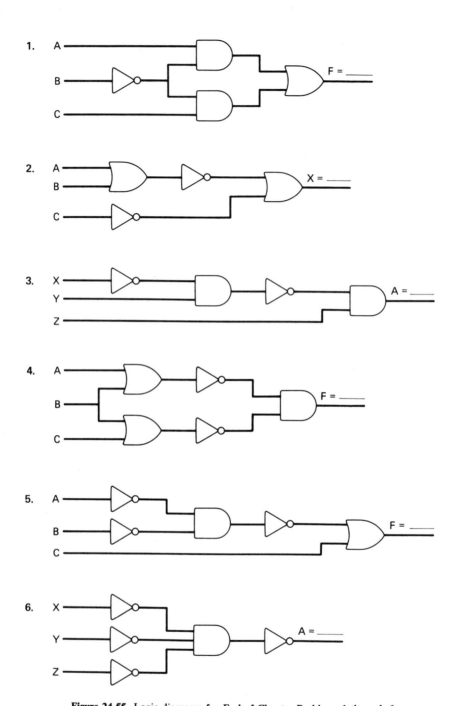

Figure 24-55. Logic diagrams for End-of-Chapter Problems 1 through 6.

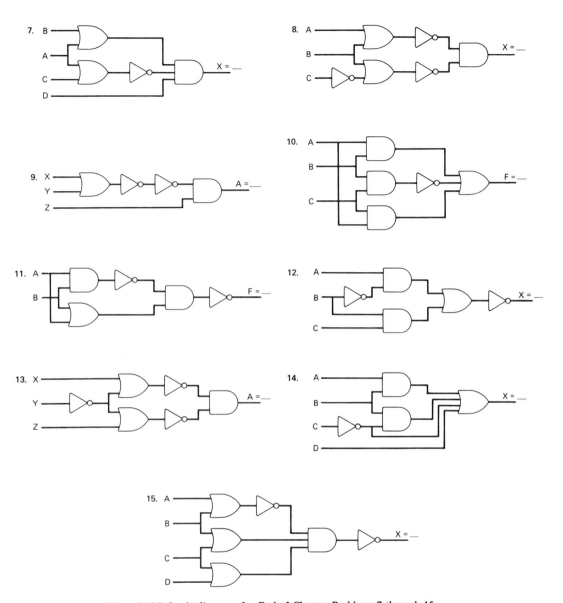

Figure 24-56. Logic diagrams for End-of-Chapter Problems 7 through 15.

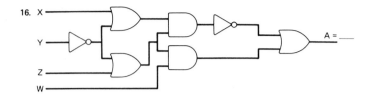

16. X

Y

Z

W

A = ____

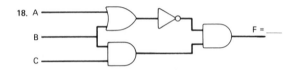

17. X

Y

Z

A = ____

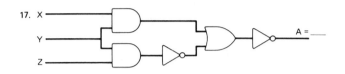

18. A

B

C

F = ____

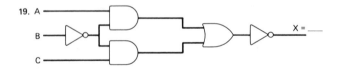

19. A

B

C

X = ____

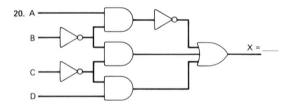

20. A

B

C

D

X = ____

Figure 24-57. Logic diagrams for End-of-Chapter Problems 16 through 20.

Draw the logic diagrams that represent the following Boolean expressions:

21. $\overline{A+\overline{B}}$

22. $A\overline{B}+\overline{AC}$

23. $ABC+D$

24. $X(\overline{Y+Z})$

25. $\overline{A+B\overline{C}}$

26. $\overline{AB}+\overline{AC}$

27. $A\overline{B}+\overline{A}B$

28. $\overline{B}(\overline{C+D})$

29. $\overline{AB}+C+D$

30. $BC\overline{D}+E$

31. $AB+A\overline{C}+B\overline{C}$

32. $(\overline{B+C})(C+D)$

33. $\overline{A\overline{B}+BC}$

34. $\overline{B}(\overline{A}+\overline{C})$

35. $A\overline{B}C+D$

36. $\overline{\overline{A}+B+C}$

37. $(\overline{X+Y})(\overline{Y+Z})$

38. $\overline{\overline{AC}+D}$

39. $\overline{\overline{AB}+C\overline{D}}$

40. $\overline{\overline{A\overline{B}}+C}$

END OF CHAPTER PROBLEMS 24-2

In each of the problems below (a) simplify the output expression and justify each step, (b) draw the simplified logic diagram, and (c) construct a truth table to prove that the expressions are equal.

1. See Figure 24-55, circuit 2.

2. See Figure 24-55, circuit 5.

3. See Figure 24-56, circuit 7.

4. See Figure 24-56, circuit 10.

5. See Figure 24-57, circuit 19.

6. See Figure 24-57, circuit 20.

7. $\overline{A+B\overline{C}}$

8. $\overline{AB}+\overline{AC}$

9. $(\overline{B+C})(C+D)$

10. $\overline{B}(\overline{A}+\overline{C})$

11. $\overline{\overline{AB}+C\overline{D}}$

12. $\overline{\overline{A\overline{B}}+C}$

ANSWERS TO SELF-TESTS

Self-Test 24-1

1. $F=\overline{\overline{A}+B}$

2. $F=\overline{A\overline{B}}$

3. $X=A+\overline{B}+C$

4. $X=(A+B)(B+C)$

5. $A=XY+XZ$

6. $\overline{A}\overline{B}+C$

7. $F=C(\overline{A+\overline{B}})$

8. $X=(\overline{A+B})(B+C)$

9. $(\overline{X}Y)(Y+Z)$

10. $F=\overline{\overline{A}(B+C)}$

11 through 20. See Figure 24-58.

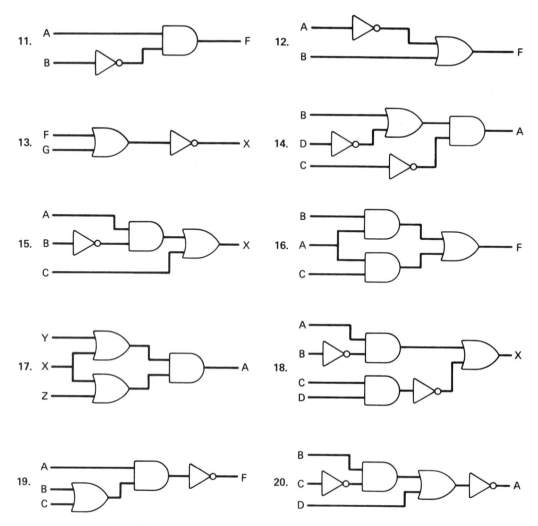

Figure 24-58. Answers to questions 11 through 20 for Self-Test 24-1.

Self-Test 24-2

1. See Figure 24-59.

2. See Figure 24-60.

3. See Figure 24-61.

4. See Figure 24-62.

5. See Figure 24-63.

6. See Figure 24-64.

7. See Figure 24-65.

8. See Figure 24-66.

9. See Figure 24-67.

10. See Figure 24-68.

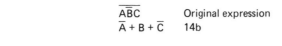

$\overline{A\overline{B}C}$ Original expression
$\overline{A}+B+\overline{C}$ 14b

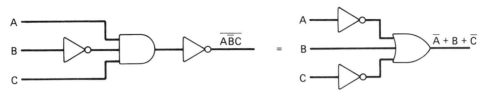

A	B	C	$\overline{A}$	$\overline{B}$	$\overline{C}$	$A\overline{B}C$	$\overline{A\overline{B}C}$	$\overline{A}+B+\overline{C}$
0	0	0	1	1	1	0	1	1
0	0	1	1	1	0	0	1	1
0	1	0	1	0	1	0	1	1
0	1	1	1	0	0	0	1	1
1	0	0	0	1	1	0	1	1
1	0	1	0	1	0	1	0	0
1	1	0	0	0	1	0	1	1
1	1	1	0	0	0	0	1	1

└─Equal─┘

Figure 24-59. Logic diagrams and truth table for $A\overline{B}C=\overline{A}+B+\overline{C}$. Answer to problem 1, Self-Test 24-2.

$$\overline{\overline{A + \overline{B}} + C} \quad \text{Original expression}$$
$$\overline{\overline{A + \overline{B}} \cdot \overline{C}} \quad \text{14a}$$
$$(A + \overline{B})\overline{C} \quad \text{13a}$$
$$\overline{C}(A + \overline{B}) \quad \text{6a}$$

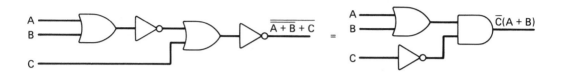

A	B	C	C̄	A + B̄	$\overline{A + \overline{B}}$	$\overline{A + \overline{B}} + C$	$\overline{\overline{A + \overline{B}} + C}$	C̄(A + B̄)
0	0	0	1	0	1	1	0	0
0	0	1	0	0	1	1	0	0
0	1	0	1	1	0	0	1	1
0	1	1	0	1	0	1	0	0
1	0	0	1	1	0	0	1	1
1	0	1	0	1	0	1	0	0
1	1	0	1	1	0	0	1	1
1	1	1	0	1	0	1	0	0

└─ Equal ─┘

Figure 24-60. Logic diagram and truth table for $\overline{\overline{A + \overline{B}} + C} = \overline{C}(A + \overline{B})$. Answer to problem 2, Self-Test 24-2.

532

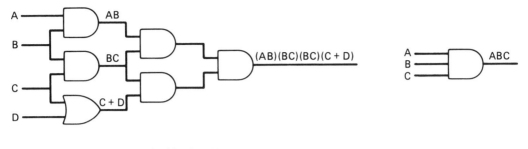

(AB)(BC)(BC)(C + D)	Original expression
(AB)(BC)(C + D)	11b
ABC (C + D)	11a
ABCC + ABCD	8a
ABC + ABCD	11a
ABC(1 + D)	8a
ABC	10b

A	B	C	D	AB	BC	C + D	(AB)(BC)	(BC)(C + D)	(AB)(BC)(BC)(C + D)	ABC
0	0	0	0	0	0	0	0	0	0	0
0	0	0	1	0	0	1	0	0	0	0
0	0	1	0	0	0	1	0	0	0	0
0	0	1	1	0	0	1	0	0	0	0
0	1	0	0	0	0	0	0	0	0	0
0	1	0	1	0	0	1	0	0	0	0
0	1	1	0	0	1	1	0	1	0	0
0	1	1	1	0	1	1	0	1	0	0
1	0	0	0	0	0	0	0	0	0	0
1	0	0	1	0	0	1	0	0	0	0
1	0	1	0	0	0	1	0	0	0	0
1	0	1	1	0	0	1	0	0	0	0
1	1	0	0	1	0	0	0	0	0	0
1	1	0	1	1	0	1	0	0	0	0
1	1	1	0	1	1	1	1	1	1	1
1	1	1	1	1	1	1	1	1	1	1

Equal

Figure 24-61. Answer to problem 3, Self-Test 24-2.

A	B	C	$\overline{A}$	$\overline{A}B$	AC	$\overline{AC}$	$\overline{A}B + \overline{AC}$	$\overline{\overline{A}B + \overline{AC}}$
0	0	0	1	0	0	1	1	0
0	0	1	1	0	0	1	1	0
0	1	0	1	1	0	1	1	0
0	1	1	1	1	0	1	1	0
1	0	0	0	0	0	1	1	0
1	0	1	0	0	1	0	0	1
1	1	0	0	0	0	1	1	0
1	1	1	0	0	1	0	0	1

Equal

$\overline{\overline{A}B + \overline{AC}}$ Original expression
$\overline{\overline{A}B} \cdot AC$ 14a
$(A + \overline{B})AC$ 14b
$AAC + A\overline{B}C$ 8a
$AC + A\overline{B}C$ 11a
$AC(1 + \overline{B})$ 8a
$AC(1)$ 10b
AC 10a

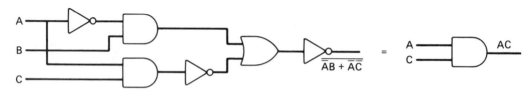

Figure 24-62. Answer to problem 4, Self-Test 24-2.

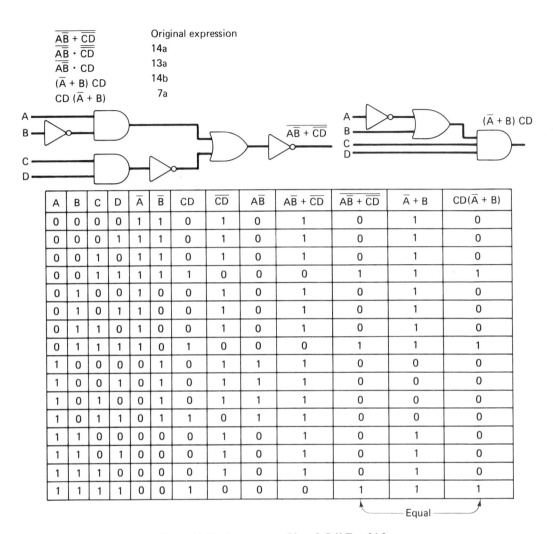

$\overline{A\overline{B} + \overline{CD}}$ Original expression
$\overline{A\overline{B}} \cdot \overline{\overline{CD}}$ 14a
$\overline{A\overline{B}} \cdot CD$ 13a
$(\overline{A} + B) CD$ 14b
$CD (\overline{A} + B)$ 7a

A	B	C	D	$\overline{A}$	$\overline{B}$	CD	$\overline{CD}$	$A\overline{B}$	$A\overline{B} + \overline{CD}$	$\overline{A\overline{B} + \overline{CD}}$	$\overline{A} + B$	$CD(\overline{A} + B)$
0	0	0	0	1	1	0	1	0	1	0	1	0
0	0	0	1	1	1	0	1	0	1	0	1	0
0	0	1	0	1	1	0	1	0	1	0	1	0
0	0	1	1	1	1	1	0	0	0	1	1	1
0	1	0	0	1	0	0	1	0	1	0	1	0
0	1	0	1	1	0	0	1	0	1	0	1	0
0	1	1	0	1	0	0	1	0	1	0	1	0
0	1	1	1	1	0	1	0	0	0	1	1	1
1	0	0	0	0	1	0	1	1	1	0	0	0
1	0	0	1	0	1	0	1	1	1	0	0	0
1	0	1	0	0	1	0	1	1	1	0	0	0
1	0	1	1	0	1	1	0	1	1	0	0	0
1	1	0	0	0	0	0	1	0	1	0	1	0
1	1	0	1	0	0	0	1	0	1	0	1	0
1	1	1	0	0	0	0	1	0	1	0	1	0
1	1	1	1	0	0	1	0	0	0	1	1	1

Equal

Figure 24-63. Answer to problem 5, Self-Test 24-2.

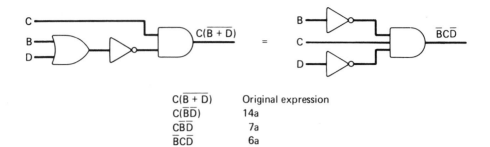

$$C(\overline{B + D})$$

$C(\overline{B + D})$	Original expression
$C(\overline{B}\overline{D})$	14a
$C\overline{B}\overline{D}$	7a
$\overline{B}C\overline{D}$	6a

B	C	D	$\overline{B}$	$\overline{D}$	B + D	$\overline{B + D}$	$C(\overline{B + D})$	$\overline{B}C\overline{D}$
0	0	0	1	1	0	1	0	0
0	0	1	1	0	1	0	0	0
0	1	0	1	1	0	1	1	1
0	1	1	1	0	1	0	0	0
1	0	0	0	1	1	0	0	0
1	0	1	0	0	1	0	0	0
1	1	0	0	1	1	0	0	0
1	1	1	0	0	1	0	0	0

Equal

Figure 24-64. Answer to problem 6, Self-Test 24-2.

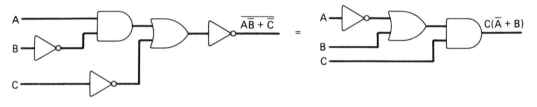

$$\overline{A\overline{B} + \overline{C}}$$ Original expression

$$\overline{A\overline{B}} \cdot \overline{\overline{C}}$$ 14a

$$\overline{A\overline{B}} \cdot C$$ 13a

$$(\overline{A} + \overline{\overline{B}})C$$ 14b

$$(\overline{A} + B)C$$ 13a

$$C(\overline{A} + B)$$ 7a

A	B	C	$\overline{A}$	$\overline{B}$	$\overline{C}$	$A\overline{B}$	$A\overline{B} + \overline{C}$	$\overline{A\overline{B} + \overline{C}}$	$\overline{A} + B$	$C(\overline{A} + B)$
0	0	0	1	1	1	0	1	0	1	0
0	0	1	1	1	0	0	0	1	1	1
0	1	0	1	0	1	0	1	0	1	0
0	1	1	1	0	0	0	0	1	1	1
1	0	0	0	1	1	1	1	0	0	0
1	0	1	0	1	0	1	1	0	0	0
1	1	0	0	0	1	0	1	0	1	0
1	1	1	0	0	0	0	0	1	1	1

Equal

Figure 24-65. Answer to problem 7, Self-Test 24-2.

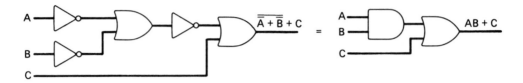

$$\overline{\overline{A} + \overline{B}} + C \quad \text{Original expression}$$
$$\overline{\overline{\overline{A}}} \cdot \overline{\overline{B}} + C \quad \text{14a}$$
$$AB + C \quad \text{13a}$$

A	B	C	$\overline{A}$	$\overline{B}$	$\overline{A} + \overline{B}$	$\overline{\overline{A} + \overline{B}}$	$\overline{\overline{A} + \overline{B}} + C$	AB	AB + C
0	0	0	1	1	1	0	0	0	0
0	0	1	1	1	1	0	1	0	1
0	1	0	1	0	1	0	0	0	0
0	1	1	1	0	1	0	1	0	1
1	0	0	0	1	1	0	0	0	0
1	0	1	0	1	1	0	1	0	1
1	1	0	0	0	0	1	1	1	1
1	1	1	0	0	0	1	1	1	1

Equal

Figure 24-66. Answer to problem 8, Self-Test 24-2.

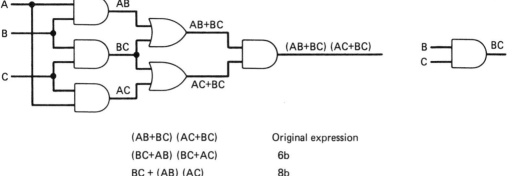

		Original expression
(AB+BC) (AC+BC)		Original expression
(BC+AB) (BC+AC)		6b
BC + (AB) (AC)		8b
BC + AABC		6a
BC + ABC		11a
BC (1+A)		8a
BC(1)		11b
BC		10a

A	B	C	AB	BC	AC	AB+BC	AC+BC	(AB+BC)(AC+BC)
0	0	0	0	0	0	0	0	0
0	0	1	0	0	0	0	0	0
0	1	0	0	0	0	0	0	0
0	1	1	0	1	0	1	1	1
1	0	0	0	0	0	0	0	0
1	0	1	0	0	1	0	1	0
1	1	0	1	0	0	1	0	0
1	1	1	1	1	1	1	1	1

Figure 24-67. Answer to problem 9, Self-Test 24-2.

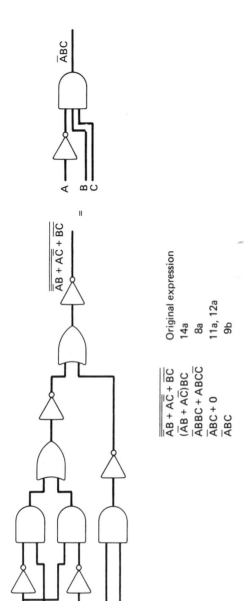

$\overline{\overline{AB} + A\overline{C} + \overline{BC}}$ = $\overline{A}BC$

$\overline{\overline{AB} + \overline{AC} + \overline{BC}}$ Original expression
$(\overline{AB} + A\overline{C})BC$ 14a
$\overline{AB}BC + ABC\overline{C}$ 8a
$\overline{A}BC + 0$ 11a, 12a
$\overline{A}BC$ 9b

A	B	C	$\overline{A}$	$\overline{C}$	$\overline{AB}$	$A\overline{C}$	$\overline{AB}+A\overline{C}$	$\overline{\overline{AB}+A\overline{C}}$	BC	$\overline{BC}$	$\overline{AB}+\overline{AC}+\overline{BC}$	$\overline{\overline{AB}+\overline{AC}+\overline{BC}}$	$\overline{A}BC$
0	0	0	1	1	0	0	0	1	0	1	1	0	0
0	0	1	1	0	0	0	0	1	0	1	1	0	0
0	1	0	1	1	1	0	1	0	0	1	1	0	0
0	1	1	1	0	1	0	1	0	1	0	0	1	1
1	0	0	0	1	0	1	0	1	0	1	1	0	0
1	0	1	0	0	0	0	0	1	0	1	1	0	0
1	1	0	0	1	0	1	1	0	0	1	1	0	0
1	1	1	0	0	0	0	0	1	1	0	1	0	0

Equal

Figure 24-68. Answer to problem 10, Self-Test 24-2.

In this Appendix, algorithms for solving various problems using a calculator are presented. Often, there is more than one way to solve a problem. It is not our intent to show all possible solutions to a problem here, but rather to make the student familiar with the functions on the calculator so that he (or she) may utilize the calculator in solving whatever problem is encountered.

The author has selected the TI-30 III to represent those calculators that use "algebraic logic," and the HP-11C for those calculators that use "reverse polish." You should consult your owner's manual for differences that might exist between these calculators and yours.

All answers using the TI-30 III have been rounded to three places. The TI-30 III observes the **PEMDAS** rule, a variation on My Dear Aunt Sally offered by my colleague Barbara Snyder.

Please Excuse My Dear Aunt Sally

> Parentheses   (given and implied)
> Exponents
> Multiplication and Division   (same priority level)
> Addition and Subtraction   (same priority level)

EXAMPLE A-1: Find $\dfrac{25 \times 3.3 \times 10^3}{3.3 \times 10^3 + 4.7 \times 10^4}$ (see example on page 35).

TI - 30 III		HP - 11C	
25		f	
×		ENG 2*	
3.3		25	
EE		ENTER	
3		3.3	
÷		EEX	
(		3	
3.3		×	
EE		3.3	
3		EEX	
+		3	
4.7		ENTER	
EE		4.7	
4		EEX	
)		4	
=	1.64	+	
		÷	1.64

Algorithm for finding the equivalent series resistance of two resistors in parallel.

EXAMPLE A-2: Let $R_1 = 2.7$ kΩ and $R_2 = 4.7$ kΩ. Find G_T and R_T.

$$G_T = G_1 + G_2 \qquad R_T = \frac{1}{G_T}$$

* f ENG 2 sets the HP-11C to display all answers in engineering notation accurate to three places. That is, all powers of ten are displayed as multiples of 3, to allow the user to quickly determine the prefix to be used.

TI-30 III	*HP-11C*

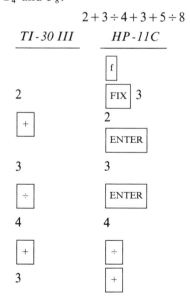

2.7   ⎡f⎤

⎡EE⎤   ⎡ENG⎤2

3   2.7

⎡$\frac{1}{x}$⎤ (G_1) ⎡EEX⎤

⎡+⎤   ⎡$\frac{1}{x}$⎤ (G_1)

4.7   4.7

⎡EE⎤   ⎡EEX⎤

3   3

⎡$\frac{1}{x}$⎤ (G_2) ⎡$\frac{1}{x}$⎤ (G_2)

⎡=⎤ $G_T = 5.83 \times 10^{-4}$ ⎡+⎤ $G_T = 583 \times 10^{-6}$
$\quad\quad = 583\ \mu S$ $\quad = 583\ \mu S$

⎡$\frac{1}{x}$⎤ $R_T = 1.72 \times 10^3$ ⎡$\frac{1}{x}$⎤ $R_T = 1.72 \times 10^3$
$\quad\quad = 1.72\ k\Omega$ $\quad = 1.72\ k\Omega$

Algorithms to add, subtract, multiply, and divide fractions.

EXAMPLE A-3: Add $2\frac{3}{4}$ and $3\frac{5}{8}$.

$$2 + 3 \div 4 + 3 + 5 \div 8$$

TI-30 III	*HP-11C*

     ⎡f⎤

2   ⎡FIX⎤ 3

⎡+⎤   2

3   ⎡ENTER⎤

⎡÷⎤   3

4   ⎡ENTER⎤

⎡+⎤   4

3   ⎡÷⎤

     ⎡+⎤

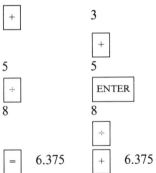

$+$	3
	$+$
	5
5	ENTER
$\div$	
8	8
	$\div$
$=$ 6.375	$+$ 6.375

EXAMPLE A-4: Subtract $3\frac{5}{8}$ from $2\frac{3}{4}$.

$$(2 + 3 \div 4) - (3 + 5 \div 8)$$

TI-30 III	*HP-11C*
2	f
$+$	FIX 3
	2
3	ENTER
$\div$	3
4	ENTER
	4
$-$	$\div$
$($	$+$
3	3
	ENTER
$+$	5
5	ENTER
$\div$	8
8	$\div$
	$+$
$)$	$-$ -0.875
$=$ -0.875	

EXAMPLE A-5: Find the product of $2\frac{3}{4}$ and $3\frac{5}{8}$.

$$(2+3\div4)\times(3+5\div8)$$

TI-30 III	*HP-11C*
(	f
2	FIX 3
+	2
	ENTER
3	3
÷	ENTER
4	4
	÷
)	+
×	3
(	ENTER
3	5
+	ENTER
	8
5	÷
÷	+
8	× 9.969
)	
= 9.969	

EXAMPLE A-6: Find the quotient of $2\frac{3}{4}\div3\frac{5}{8}$.

$$(2+3\div4)\div(3+5\div8)$$

TI-30 III	*HP-11C*
(	f
2	FIX 3

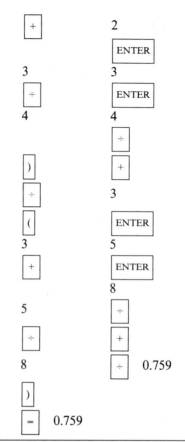

+	2
	ENTER
3	3
÷	ENTER
4	4
	÷
)	+
÷	3
(	ENTER
3	5
+	ENTER
	8
5	÷
÷	+
8	÷ 0.759
)	
= 0.759	

EXAMPLE A-7: Find $\dfrac{2\frac{1}{4}+1\frac{5}{6}}{7\frac{1}{8}}$.

$$(2+1 \div 4 + 1 + 5 \div 6) \div (7 + 1 \div 8)$$

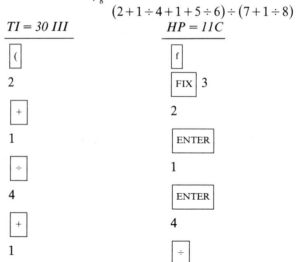

TI = 30 III	*HP = 11C*
(	f
2	FIX 3
+	2
1	ENTER
÷	1
4	ENTER
+	4
1	÷

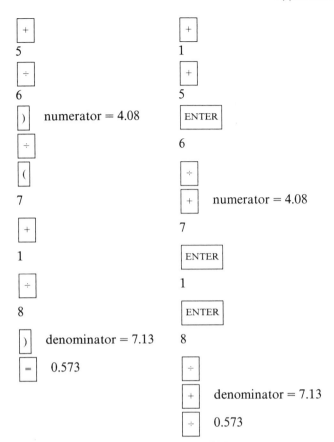

Algorithm for finding the hypotenuse of a right triangle, given two sides, by using a) the Pythagorean theorem, and b) trig functions.

EXAMPLE A-8: Let $R = 5.6$ kΩ and $X_C = 4$ kΩ. Find Z.

a) Solution using the Pythagorean theorem:

$$Z = \sqrt{R^2 + X^2}$$

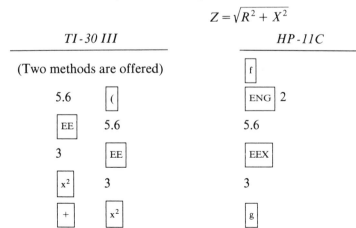

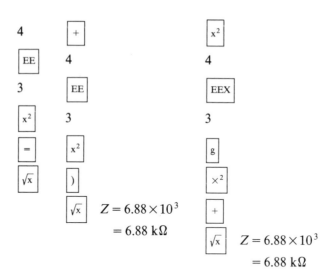

$$Z = 6.88 \times 10^3$$
$$= 6.88 \text{ k}\Omega$$

$$Z = 6.88 \times 10^3$$
$$= 6.88 \text{ k}\Omega$$

b) Using trig functions:

Using the *TI-30 III*, we must use trig functions:

$$\tan \theta = \frac{\text{opposite side}}{\text{adjacent side}} = \frac{X_C}{R}$$

$$\tan^{-1} \frac{X_C}{R} = \theta$$

$$Z = \frac{X_C}{\sin \phi}$$

4

| EE |

3

| ÷ |

5.6

| EE |

3

| = |

| INV |

| tan | $\theta = 35.5°$

| sin |

Using the *HP-11C*, we put the vector in rectangular form:

$$Z = 5.6 \text{ k}\Omega - j4 \text{ k}\Omega$$

| f |

| ENG | 2

4

| EEX |

3

| ENTER |

5.6

| EEX |

3

| g |

| → P | $Z = 6.88 \times 10^3$
$$= 6.88 \text{ k}\Omega$$

$Z = 6.88 \times 10^3$
$= 6.88\ \text{k}\Omega$

Algorithm for finding the hypotenuse of a right triangle given an angle and the opposite side.

$$\sin\theta = \frac{\text{opposite side}}{\text{hypotenuse}} = \frac{b}{c}$$

$$c = \frac{b}{\sin\theta}$$

EXAMPLE A-9: Let $b = 25$ V and $\theta = 40°$. Find c.

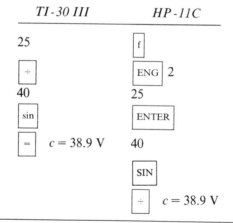

Algorithm for finding the resonant frequency of an LC circuit.

$$f_r = \frac{1}{2\pi\sqrt{LC}}$$

EXAMPLE A-10: Let $L = 200$ mH and $C = 300$ pF. Find f_r.

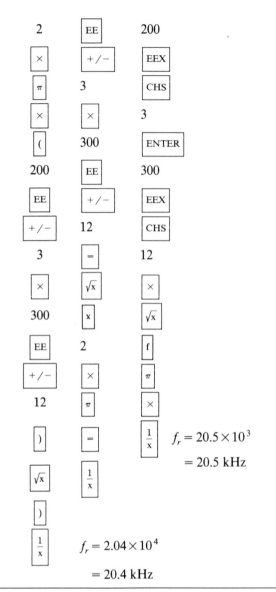

$f_r = 2.04 \times 10^4$

$= 20.4$ kHz

$f_r = 20.5 \times 10^3$

$= 20.5$ kHz

Algorithm for finding logarithms of numbers.

EXAMPLE A-11: Find the log of 273.

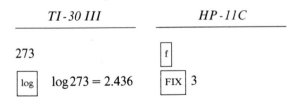

$\log 273 = 2.436$

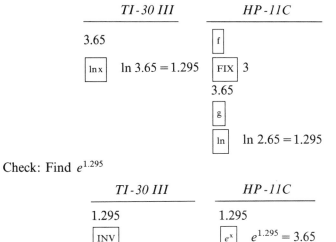

273

[g]

[LOG] $\log 273 = 2.436$

Check: Find $10^{2.436}$

TI-30 III	HP-11C
2.436	2.436
[INV]	[10^x] $10^{2.436} = 273$
[log] $10^{2.436} = 273$	

EXAMPLE A-12: Find the ln of 3.65.

TI-30 III	HP-11C
3.65	[f]
[ln x] $\ln 3.65 = 1.295$	[FIX] 3
	3.65
	[g]
	[ln] $\ln 2.65 = 1.295$

Check: Find $e^{1.295}$

TI-30 III	HP-11C
1.295	1.295
[INV]	[e^x] $e^{1.295} = 3.65$
[ln x] $e^{1.295} = 3.65$	

Algorithm for changing from logarithmic to exponential form.

EXAMPLE A-13: $\log x = 1.73$. Find x. $x = 10^{1.73}$

TI-30 III	HP-11C
1.73	[f]
[INV]	[FIX] 2
[log] 53.7	1.73
	[10^x] 53.7

EXAMPLE A-14: $\ln x = 2.7$. Find x. $x = e^{2.7}$

TI - 30 III	*HP - 11C*
2.7	$\boxed{f}$
$\boxed{\text{INV}}$	$\boxed{\text{FIX}}$ 2
$\boxed{\ln x}$ 14.9	2.7
	$\boxed{e^x}$ 14.9

Algorithms for converting between number systems.

EXAMPLE A-15: Convert the octal number 4235 to decimal.

TI - 30 III	*HP - 11C*
4	$\boxed{f}$
$\boxed{\times}$	$\boxed{\text{FIX}}$ 0
8	8
$\boxed{y^x}$	$\boxed{\text{ENTER}}$
3	3
$\boxed{+}$ $4 \times 8^3 = 2048$	$\boxed{y^x}$
2	4
$\boxed{\times}$	$\boxed{\times}$ $4 \times 8^3 = 2048$
8	8
$\boxed{x^2}$	$\boxed{g}$
$\boxed{+}$ $2048 + 128 = 2176$	$\boxed{x^2}$
3	2
$\boxed{\times}$	$\boxed{\times}$ $2 \times 8^2 = 128$
8	$\boxed{+}$ $2048 + 128 = 2176$
$\boxed{+}$ $2048 + 128 + 24 = 2200$	8
5	$\boxed{\text{ENTER}}$
$\boxed{=}$ $4235_8 = 2205_{10}$	3
	$\boxed{\times}$ $3 \times 8 = 24$

$$\boxed{+} \quad 2048 + 128 + 24 = 2200$$

5

$$\boxed{+} \quad 4235_8 = 2205_{10}$$

This algorithm can be used to convert numbers in any base to base 10.

EXAMPLE A-16: Convert the decimal number 1324 to hexadecimal (see Example 4.10).

TI-30 III		*HP-11C*	
1324		$\boxed{f}$	
$\boxed{\div}$		$\boxed{\text{FIX}}$	3
16		1324	
$\boxed{=}$	82.75	$\boxed{\text{ENTER}}$	
$\boxed{-}$		16	
82		$\boxed{\div}$	82.75
STO		82	
$\boxed{=}$	.75 is the remainder	$\boxed{\text{STO}}$ 1	
$\boxed{\times}$		$\boxed{-}$	.75 is the remainder
16		16	
$\boxed{=}$	.75 of 16 is 12	$\boxed{\times}$	.75 of 16 is 12
	$12_{10} = C_{16} = $ LSD		$12_{10} = C_{16} = $ LSD
$\boxed{\text{RCL}}$	82 is the 2nd dividend	$\boxed{\text{RCL}}$ 1	82 is the 2nd dividend
$\boxed{\div}$		16	
16		$\boxed{\div}$	
$\boxed{=}$	5.125	5	
$\boxed{-}$		$\boxed{\text{STO}}$ 1	
5		$\boxed{-}$	.125 is the remainder
STO		16	
$\boxed{=}$	.125 is the remainder	$\boxed{\times}$	.125 of 16 is 2
$\boxed{\times}$			2 is the next SD

16   RCL 1   5 is the MSD*

=   .125 of 16 is 2

2 is the next SD

RCL   5 is the MSD*

TABLE A-1.   TRIGONOMETRIC FUNCTIONS OF DEGREES

Angle°	sin	tan	cot	cos	Angle°
0.0	.00000	.00000	∞	1.00000	**90.0**
.1	.00175	.00175	572.96	1.00000	.9
.2	.00349	.00349	286.48	0.99999	.8
.3	.00524	.00524	190.98	.99999	.7
.4	.00698	.00698	143.23	.99998	.6
.5	.00873	.00873	114.59	.99996	.5
.6	.01047	.01047	95.489	.99995	.4
.7	.01222	.01222	81.847	.99993	.3
.8	.01396	.01396	71.615	.99990	.2
.9	.01571	.01571	63.657	.99988	.1
1.0	.01745	.01746	57.290	.99985	**89.0**
.1	.01920	.01920	52.081	.99982	.9
.2	.02094	.02095	47.740	.99978	.8
.3	.02269	.02269	44.066	.99974	.7
.4	.02443	.02444	40.917	.99970	.6
.5	.02618	.02619	38.188	.99966	.5
.6	.02792	.02793	35.801	.99961	.4
.7	.02967	.02968	33.694	.99956	.3
.8	.03141	.03143	31.821	.99951	.2
.9	.03316	.03317	30.145	.99945	.1
2.0	.03490	.03492	28.636	.99939	**88.0**
.1	.03664	.03667	27.271	.99933	.9
.2	.03839	.03842	26.031	.99926	.8
.3	.04013	.04016	24.898	.99919	.7
.4	.04188	.04191	23.859	.99912	.6
.5	.04362	.04366	22.904	.99905	.5
.6	.04536	.04541	22.022	.99897	.4
.7	.04711	.04716	21.205	.99889	.3
.8	.04885	.04891	20.446	.99881	.2
.9	.05059	.05066	19.740	.99872	.1
3.0	.05234	.05241	19.081	.99863	**87.0**
.1	.05408	.05416	18.464	.99854	.9
.2	.05582	.05591	17.886	.99844	.8
.3	.05756	.05766	17.343	.99834	.7
.4	.05931	.05941	16.832	.99824	.6
.5	.06105	.06116	16.350	.99813	.5
.6	.06279	.06291	15.895	.99803	.4
	cos	cot	tan	sin	

*Remember, if the whole number part is less than 16 we are finished and that number is the MSD.

Angle°	sin	tan	cot	cos	Angle°
.7	.06453	.06467	15.464	.99792	.3
.8	.06627	.06642	15.056	.99780	.2
.9	.06802	.06817	14.669	.99767	.1
4.0	.06976	.06993	14.301	.99756	**86.0**
.1	.07150	.07168	13.951	.99744	.9
.2	.07324	.07344	13.617	.99731	.8
.3	.07498	.07519	13.300	.99719	.7
.4	.07672	.07695	12.996	.99705	.6
.5	.07846	.07870	12.706	.99692	.5
.6	.08020	.08046	12.429	.99678	.4
.7	.08194	.08221	12.163	.99664	.3
.8	.08368	.08397	11.909	.99649	.2
.9	.08542	.08573	11.664	.99635	.1
5.0	.08716	.08749	11.430	.99619	**85.0**
.1	.08889	.08925	11.205	.99604	.9
.2	.09063	.09101	10.988	.99588	.8
.3	.09237	.09277	10.780	.99572	.7
.4	.09411	.09453	10.579	.99556	.6
.5	.09585	.09629	10.385	.99540	.5
.6	.09758	.09805	10.199	.99523	.4
.7	.09932	.09981	10.019	.99506	.3
.8	.10106	.10158	9.8448	.99488	.2
.9	.10279	.10334	9.6768	.99470	.1
6.0	.10453	.10510	9.5144	.99452	**84.0**
.1	.10626	.10687	9.3572	.99434	.9
.2	.10800	.10863	9.2052	.99415	.8
.3	.10973	.11040	9.0579	.99396	.7
.4	.11147	.11217	8.9152	.99377	.6
.5	.11320	.11394	8.7769	.99357	.5
.6	.11494	.11570	8.6427	.99337	.4
.7	.11667	.11747	8.5126	.99317	.3
.8	.11840	.11924	8.3863	.99297	.2
.9	.12014	.12101	8.2636	.99276	.1
7.0	.12187	.12278	8.1443	.99255	**83.0**
.1	.12360	.12456	8.0285	.99233	.9
.2	.12533	.12633	7.9158	.99211	.8
.3	.12706	.12810	7.8062	.99189	.7
.4	.12880	.12988	7.6996	.99167	.6
.5	.13053	.13165	7.5958	.99144	.5
.6	.13226	.13343	7.4947	.99122	.4
.7	.13399	.13521	7.3962	.99098	.3
.8	.13572	.13698	7.3002	.99075	.2
.9	.13744	.13876	7.2066	.99051	.1
8.0	.13917	.14054	7.1154	.99027	**82.0**
.1	.14090	.14232	7.0264	.99002	.9
.2	.14263	.14410	6.9395	.98978	.8
.3	.14436	.14588	6.8548	.98953	.7
.4	.14608	.14767	6.7720	.98927	.6
.5	.14781	.14945	6.6912	.98902	.5
.6	.14954	.15124	6.6122	.98876	.4
.7	.15126	.15302	6.5350	.98849	.3
	cos	cot	tan	sin	

TABLE A-1 (Cont'd.)

Angle°	sin	tan	cot	cos	Angle°
.8	.15299	.15481	6.4596	.98823	.2
.9	.15471	.15660	6.3859	.98796	.1
9.0	.15643	.15838	6.3138	.98769	**81.0**
.1	.15816	.16017	6.2432	.98741	.9
.2	.15988	.16196	6.1742	.98714	.8
.3	.16160	.16376	6.1066	.98686	.7
.4	.16333	.16555	6.0405	.98657	.6
.5	.16505	.16734	5.9758	.98629	.5
.6	.16677	.16914	5.9124	.98600	.4
.7	.16849	.17093	5.8502	.98570	.3
.8	.17021	.17273	5.7894	.98541	.2
.9	.17193	.17453	5.7297	.98511	.1
10.0	.17365	.17633	5.6713	.98481	**80.0**
.1	.17537	.17813	5.6140	.98450	.9
.2	.17708	.17993	5.5578	.98420	.8
.3	.17880	.18173	5.5026	.98389	.7
.4	.18052	.18353	5.4486	.98357	.6
.5	.18224	.18534	5.3955	.98325	.5
.6	.18395	.18714	5.3435	.98294	.4
.7	.18567	.18895	5.2924	.98261	.3
.8	.18738	.19076	5.2422	.98229	.2
.9	.18910	.19257	5.1929	.98196	.1
11.0	.19081	.19438	5.1446	.98163	**79.0**
.1	.19252	.19619	5.0970	.98129	.9
.2	.19423	.19801	5.0504	.98096	.8
.3	.19595	.19982	5.0045	.98061	.7
.4	.19766	.20164	4.9594	.98027	.6
.5	.19937	.20345	4.9152	.97992	.5
.6	.20108	.20527	4.8716	.97958	.4
.7	.20279	.20709	4.8288	.97922	.3
.8	.20450	.20891	4.7867	.97887	.2
.9	.20620	.21073	4.7453	.97851	.1
12.0	.20791	.21256	4.7046	.97815	**78.0**
.1	.20962	.21438	4.6646	.97778	.9
.2	.21132	.21621	4.6252	.97742	.8
.3	.21303	.21804	4.5864	.97705	.7
.4	.21474	.21986	4.5483	.97667	.6
.5	.21644	.22169	4.5107	.97630	.5
.6	.21814	.22353	4.4747	.97592	.4
.7	.21985	.22536	4.4373	.97553	.3
.8	.22155	.22719	4.4015	.97515	.2
.9	.22325	.22903	4.3662	.97476	.1
13.0	.22495	.23087	4.3315	.97437	**77.0**
.1	.22665	.23271	4.2972	.97398	.9
.2	.22835	.23455	4.2635	.97358	.8
.3	.23005	.23639	4.2303	.97318	.7
.4	.23175	.23823	4.1976	.97278	.6
.5	.23345	.24008	4.1653	.97237	.5
.6	.23514	.24193	4.1335	.97196	.4
.7	.23684	.24377	4.1022	.97155	.3
	cos	cot	tan	sin	

Angle°	sin	tan	cot	cos	Angle°
.8	.23853	.24562	4.0713	.97113	.2
.9	.24023	.24747	4.0408	.97072	.1
14.0	.24192	.24933	4.0108	.97030	**76.0**
.1	.24362	.25118	3.9812	.96987	.9
.2	.24531	.25304	3.9520	.96945	.8
.3	.24700	.25490	3.9232	.96902	.7
.4	.24869	.25676	3.8947	.96858	.6
.5	.25038	.25862	3.8667	.96815	.5
.6	.25207	.26048	3.8391	.96771	.4
.7	.25376	.26235	3.8118	.96727	.3
.8	.25545	.26421	3.7848	.96682	.2
.9	.25713	.26608	3.7583	.96638	.1
15.0	.25882	.26792	3.7321	.96593	**75.0**
.1	.26050	.26982	3.7062	.96547	.9
.2	.26219	.27169	3.6806	.96502	.8
.3	.26387	.27357	3.6554	.96456	.7
.4	.26556	.27545	3.6305	.96410	.6
.5	.26724	.27732	3.6059	.96363	.5
.6	.26892	.27921	3.5816	.96316	.4
.7	.27060	.28109	3.5576	.96269	.3
.8	.27228	.28297	3.5339	.96222	.2
.9	.27396	.28486	3.5105	.96174	.1
16.0	.27564	.28675	3.4874	.96126	**74.0**
.1	.27731	.28864	3.4646	.96078	.9
.2	.27899	.29053	3.4420	.96029	.8
.3	.28067	.29242	3.4197	.95981	.7
.4	.28234	.29432	3.3977	.95931	.6
.5	.28402	.29621	3.3759	.95882	.5
.6	.28569	.29811	3.3544	.95832	.4
.7	.28736	.30001	3.3332	.95782	.3
.8	.28903	.30192	3.3122	.95732	.2
.9	.29070	.30382	3.2914	.95681	.1
17.0	.29237	.30573	3.2709	.95630	**73.0**
.1	.29404	.30764	3.2506	.95579	.9
.2	.29571	.30955	3.2305	.95528	.8
.3	.29737	.31147	3.2106	.95476	.7
.4	.29904	.31338	3.1910	.95424	.6
.5	.30071	.31530	3.1716	.95372	.5
.6	.30237	.31722	3.1524	.95319	.4
.7	.30403	.31914	3.1334	.95266	.3
.8	.30570	.32106	3.1146	.95213	.2
.9	.30736	.32299	3.0961	.95159	.1
18.0	.30902	.32492	3.0777	.95106	**72.0**
.1	.31068	.32685	3.0595	.95052	.9
.2	.31233	.32878	3.0415	.94997	.8
.3	.31399	.33072	3.0237	.94943	.7
.4	.31565	.33266	3.0061	.94888	.6
.5	.31730	.33460	2.9887	.94832	.5
.6	.31896	.33654	2.9714	.94777	.4
.7	.32061	.33848	2.9544	.94721	.3
	cos	cot	tan	sin	

TABLE A-1 (Cont'd.)

Angle°	sin	tan	cot	cos	Angle°
.8	.32227	.34043	2.9375	.94665	.2
.9	.32392	.34238	2.9208	.94609	.1
19.0	.32557	.34433	2.9042	.94552	**71.0**
.1	.32722	.34628	2.8878	.94495	.9
.2	.32887	.34824	2.8716	.94438	.8
.3	.33051	.35020	2.8556	.94380	.7
.4	.33216	.35216	2.8397	.94322	.6
.5	.33381	.35412	2.8239	.94264	.5
.6	.33545	.35608	2.8083	.94206	.4
.7	.33710	.35805	2.7929	.94147	.3
.8	.33874	.36002	2.7776	.94088	.2
.9	.34038	.36199	2.7625	.94029	.1
20.0	.34202	.36397	2.7475	.93969	**70.0**
.1	.34366	.36595	2.7326	.93909	.9
.2	.34530	.36793	2.7179	.93849	.8
.3	.34694	.36991	2.7034	.93789	.7
.4	.34857	.37190	2.6889	.93728	.6
.5	.35021	.37388	2.6746	.93667	.5
.6	.35184	.37588	2.6605	.93606	.4
.7	.35347	.37787	2.6464	.93544	.3
.8	.35511	.37986	2.6325	.93483	.2
.9	.35674	.38186	2.6187	.93420	.1
21.0	.35837	.38386	2.6051	.93358	**69.0**
.1	.36000	.38587	2.5916	.93295	.9
.2	.36162	.38787	2.5782	.93232	.8
.3	.36325	.38988	2.5649	.93169	.7
.4	.36488	.39190	2.5517	.93106	.6
.5	.36650	.39391	2.5386	.93042	.5
.6	.36812	.39593	2.5257	.92978	.4
.7	.36975	.39795	2.5129	.92913	.3
.8	.37137	.39997	2.5002	.92849	.2
.9	.37299	.40200	2.4876	.92784	.1
22.0	.37461	.40403	2.4751	.92718	**68.0**
.1	.37622	.40606	2.4627	.92653	.9
.2	.37784	.40809	2.4504	.92587	.8
.3	.37946	.41013	2.4383	.92521	.7
.4	.38107	.41217	2.4262	.92455	.6
.5	.38268	.41421	2.4142	.92388	.5
.6	.38430	.41626	2.4023	.92321	.4
.7	.38591	.41831	2.3906	.92254	.3
.8	.38752	.42036	2.3789	.92186	.2
.9	.38912	.42242	2.3673	.92119	.1
23.0	.39073	.42447	2.3559	.92050	**67.0**
.1	.39234	.42654	2.3445	.91982	.9
.2	.39394	.42860	2.3332	.91914	.8
.3	.39555	.43067	2.3220	.91845	.7
.4	.38715	.43274	2.3109	.91775	.6
.5	.39875	.43481	2.2998	.91706	.5
.6	.40035	.43689	2.2889	.91636	.4
.7	.40195	.43897	2.2781	.91566	.3
	cos	cot	tan	sin	

Angle°	sin	tan	cot	cos	Angle°
.8	.40355	.44105	2.2673	.91496	.2
.9	.40514	.44314	2.2566	.91425	.1
24.0	.40674	.44523	2.2460	.91355	**66.0**
.1	.40833	.44732	2.2355	.91283	.9
.2	.40992	.44942	2.2251	.91212	.8
.3	.41151	.45152	2.2148	.91140	.7
.4	.41310	.45362	2.2045	.91068	.6
.5	.41469	.45573	2.1943	.90996	.5
.6	.41628	.45784	2.1842	.90924	.4
.7	.41787	.45995	2.1742	.90851	.3
.8	.41945	.46206	2.1624	.90778	.2
.9	.42104	.46418	2.1543	.90704	.1
25.0	.42262	.46631	2.1445	.90631	**65.0**
.1	.42420	.46843	2.1348	.90557	.9
.2	.42578	.47056	2.1251	.90483	.8
.3	.42736	.47270	2.1155	.90408	.7
.4	.42894	.47483	2.1060	.90334	.6
.5	.43051	.47698	2.0965	.90259	.5
.6	.43209	.47912	2.0872	.90183	.4
.7	.43366	.48127	2.0778	.90108	.3
.8	.43523	.48342	2.0682	.90032	.2
.9	.43680	.48557	2.0594	.89956	.1
26.0	.43837	.48773	2.0503	.89879	**64.0**
.1	.43994	.48989	2.0413	.89803	.9
.2	.44151	.49206	2.0323	.89726	.8
.3	.44307	.49423	2.0233	.89649	.7
.4	.44464	.49640	2.0145	.89571	.6
.5	.44620	.49858	2.0057	.89493	.5
.6	.44776	.50076	1.9970	.89415	.4
.7	.44932	.50295	1.9883	.89337	.3
.8	.45088	.50514	1.9797	.89259	.2
.9	.45243	.50733	1.9711	.89180	.1
27.0	.45399	.50953	1.9626	.89101	**63.0**
.1	.45554	.51173	1.9542	.89021	.9
.2	.45710	.51393	1.9458	.88942	.8
.3	.45865	.51614	1.9375	.88862	.7
.4	.46020	.51835	1.9292	.88782	.6
.5	.46175	.52057	1.9210	.88701	.5
.6	.46330	.52279	1.9128	.88620	.4
.7	.46484	.52501	1.9047	.88539	.3
.8	.46639	.52724	1.8967	.88458	.2
.9	.46793	.52947	1.8887	.88377	.1
28.0	.46947	.53171	1.8807	.88295	**62.0**
.1	.47101	.53395	1.8728	.88213	.9
.2	.47255	.53620	1.8650	.88130	.8
.3	.47409	.53844	1.8572	.88048	.7
.4	.47562	.54070	1.8495	.87965	.6
.5	.47716	.54296	1.8418	.87882	.5
.6	.47869	.54522	1.8341	.87798	.4
.7	.48022	.54748	1.8265	.87715	.3
	cos	cot	tan	sin	

TABLE A-1 (Cont'd.)

Angle°	sin	tan	cot	cos	Angle°
.8	.48175	.54975	1.8190	.87631	.2
.9	.48328	.55203	1.8115	.87546	.1
29.0	.48481	.55431	1.8040	.87462	**61.0**
.1	.48634	.55659	1.7966	.87377	.9
.2	.48786	.55888	1.7893	.87292	.8
.3	.48938	.56117	1.7820	.87207	.7
.4	.49090	.56347	1.7747	.87121	.6
.5	.49242	.56577	1.7675	.87036	.5
.6	.49394	.56808	1.7603	.86949	.4
.7	.49546	.57039	1.7532	.86863	.3
.8	.49697	.57271	1.7461	.86777	.2
.9	.49849	.57503	1.7391	.86690	.1
30.0	.50000	.57735	1.7321	.86603	**60.0**
.1	.50151	.57968	1.7251	.86515	.9
.2	.50302	.58201	1.7182	.86427	.8
.3	.50453	.58435	1.7113	.86340	.7
.4	.50603	.58670	1.7045	.86251	.6
.5	.50754	.58905	1.6977	.86163	.5
.6	.50904	.59140	1.6909	.86074	.4
.7	.51054	.59376	1.6842	.85985	.3
.8	.51204	.59612	1.6775	.85895	.2
.9	.51354	.59849	1.6709	.85806	.1
31.0	.51504	.60086	1.6643	.85717	**59.0**
.1	.51653	.60324	1.6577	.85627	.9
.2	.51803	.60562	1.6512	.85536	.8
.3	.51952	.60801	1.6447	.85446	.7
.4	.52101	.61040	1.6383	.85355	.6
.5	.52250	.61280	1.6319	.85264	.5
.6	.52399	.61520	1.6255	.85173	.4
.7	.52547	.61761	1.6191	.85081	.3
.8	.52697	.62003	1.6128	.84989	.2
.9	.52844	.62245	1.6066	.84897	.1
32.0	.52992	.62487	1.6003	.84805	**58.0**
.1	.53140	.62730	1.5941	.84712	.9
.2	.53288	.62973	1.5880	.84619	.8
.3	.53435	.63217	1.5818	.84526	.7
.4	.53583	.63462	1.5757	.84433	.6
.5	.53730	.63707	1.5697	.84339	.5
.6	.53877	.63953	1.5637	.84245	.4
.7	.54024	.64199	1.5577	.84151	.3
.8	.54171	.64446	1.5517	.84057	.2
.9	.54317	.64693	1.5458	.83962	.1
33.0	.54464	.64941	1.5399	.83867	**57.0**
.1	.54610	.65189	1.5340	.83772	.9
.2	.54756	.65438	1.5282	.83676	.8
.3	.54902	.65688	1.5224	.83581	.7
.4	.55048	.65938	1.5166	.83485	.6
.5	.55194	.66189	1.5108	.83389	.5
.6	.55339	.66440	1.5051	.83292	.4
.7	.55484	.66692	1.4994	.83195	.3
	cos	cot	tan	sin	

Angle°	sin	tan	cot	cos	Angle°
.8	.55630	.66944	1.4938	.83098	.2
.9	.55775	.67197	1.4882	.83001	.1
34.0	.55919	.67451	1.4826	.82904	**56.0**
.1	.56064	.67705	1.4770	.82806	.9
.2	.56208	.67960	1.4715	.82708	.8
.3	.56353	.68215	1.4659	.82610	.7
.4	.56497	.68471	1.4605	.82511	.6
.5	.56641	.68728	1.4550	.82413	.5
.6	.56784	.68985	1.4496	.82314	.4
.7	.56928	.69243	1.4442	.82214	.3
.8	.57071	.69502	1.4388	.82115	.2
.9	.57215	.69761	1.4335	.82015	.1
35.0	.57358	.70021	1.4281	.81915	**55.0**
.1	.57501	.70281	1.4229	.81815	.9
.2	.57643	.70542	1.4176	.81714	.8
.3	.57786	.70804	1.4124	.81614	.7
.4	.57928	.71066	1.4071	.81513	.6
.5	.58070	.71329	1.4019	.81412	.5
.6	.58212	.71593	1.3968	.81310	.4
.7	.58354	.71857	1.3916	.81208	.3
.8	.58496	.72122	1.3865	.81106	.2
.9	.58637	.72388	1.3814	.81004	.1
36.0	.58779	.72654	1.3764	.80902	**54.0**
.1	.58920	.72921	1.3713	.80799	.9
.2	.59061	.73189	1.3663	.80696	.8
.3	.59201	.73457	1.3613	.80593	.7
.4	.59342	.73726	1.3564	.80489	.6
.5	.59482	.73996	1.3514	.80386	.5
.6	.59622	.74267	1.3465	.80282	.4
.7	.59763	.74538	1.3416	.80178	.3
.8	.59902	.74810	1.3367	.80073	.2
.9	.60042	.75082	1.3319	.79968	.1
37.0	.60182	.75355	1.3270	.79864	**53.0**
.1	.60321	.75629	1.3222	.79758	.9
.2	.60460	.75904	1.3175	.79653	.8
.3	.60599	.76180	1.3127	.79547	.7
.4	.60738	.76456	1.3079	.79441	.6
.5	.60876	.76733	1.3032	.79335	.5
.6	.61015	.77010	1.2985	.79229	.4
.7	.61153	.77289	1.2938	.79122	.3
.8	.61291	.77568	1.2892	.79016	.2
.9	.61429	.77848	1.2846	.78908	.1
38.0	.61566	.78129	1.2799	.78801	**52.0**
.1	.61704	.78410	1.2753	.78694	.9
.2	.61841	.78692	1.2708	.78586	.8
.3	.61978	.78975	1.2662	.78478	.7
.4	.62115	.79259	1.2617	.78369	.6
.5	.62251	.79544	1.2572	.78261	.5
.6	.62388	.79829	1.2527	.78152	.4
.7	.62524	.80115	1.2482	.78043	.3
	cos	cot	tan	sin	

TABLE A-1 (Cont'd.)

Angle°	sin	tan	cot	cos	Angle°
.8	.62660	.80402	1.2437	.77934	.2
.9	.62796	.80690	1.2393	.77824	.1
39.0	.62932	.80978	1.2349	.77715	**51.0**
.1	.63068	.81268	1.2305	.77605	.9
.2	.63203	.81558	1.2261	.77494	.8
.3	.63338	.81849	1.2218	.77384	.7
.4	.63473	.82141	1.2174	.77273	.6
.5	.63608	.82434	1.2131	.77162	.5
.6	.63742	.82727	1.2088	.77051	.4
.7	.63877	.83022	1.2045	.76940	.3
.8	.64011	.83317	1.2002	.76828	.2
.9	.64145	.83613	1.1960	.76717	.1
40.0	.64279	.83910	1.1918	.76604	**50.0**
.1	.64412	.84208	1.1875	.76492	.9
.2	.64546	.84507	1.1833	.76380	.8
.3	.64679	.84806	1.1792	.76267	.7
.4	.64812	.85107	1.1750	.76154	.6
.5	.64945	.85408	1.1708	.76041	.5
.6	.65077	.85710	1.1667	.75927	.4
.7	.65210	.86014	1.1626	.75813	.3
.8	.65342	.86318	1.1585	.75700	.2
.9	.65474	.86623	1.1544	.75585	.1
41.0	.65606	.86929	1.1504	.75471	**49.0**
.1	.65738	.87236	1.1463	.75356	.9
.2	.65869	.87543	1.1423	.75241	.8
.3	.66000	.87852	1.1383	.75126	.7
.4	.66131	.88162	1.1343	.75011	.6
.5	.66262	.88473	1.1303	.74896	.5
.6	.66393	.88784	1.1263	.74780	.4
.7	.66523	.89097	1.1224	.74664	.3
.8	.66653	.89410	1.1184	.74548	.2
.9	.66783	.89725	1.1145	.74431	.1
42.0	.66913	.90040	1.1106	.74314	**48.0**
.1	.67043	.90357	1.1067	.74198	.9
.2	.67172	.90674	1.1028	.74080	.8
.3	.67301	.90993	1.0990	.73963	.7
.4	.67430	.91313	1.0951	.73846	.6
.5	.67559	.91633	1.0913	.73728	.5
.6	.67688	.91955	1.0875	.73610	.4
.7	.67816	.92277	1.0837	.73491	.3
.8	.67944	.92601	1.0799	.73373	.2
.9	.68072	.92926	1.0761	.73254	.1
43.0	.68200	.93252	1.0724	.73135	**47.0**
.1	.68327	.93578	1.0686	.73016	.9
.2	.68455	.93906	1.0649	.72897	.8
.3	.68582	.94235	1.0612	.72777	.7
.4	.68709	.94565	1.0575	.72657	.6
.5	.68835	.94896	1.0538	.72537	.5
.6	.68962	.95229	1.0501	.72417	.4
.7	.69088	.95562	1.0464	.72294	.3
	cos	cot	tan	sin	

Angle°	sin	tan	cot	cos	Angle°
.8	.69214	.95897	1.0428	.72176	.2
.9	.69340	.96232	1.0392	.72055	.1
44.0	.69466	.96569	1.0355	.71934	**46.0**
.1	.69591	.96907	1.0319	.71813	.9
.2	.69717	.97246	1.0283	.71691	.8
.3	.69842	.97586	1.0247	.71569	.7
.4	.69966	.97927	1.0212	.71447	.6
.5	.70091	.98270	1.0176	.71325	.5
.6	.70215	.98613	1.0141	.71203	.4
.7	.70339	.98958	1.0105	.71080	.3
.8	.70463	.99304	1.0070	.70957	.2
.9	.70587	.99652	1.0035	.70834	.1
45.0	.70711	1.00000	1.0000	.70711	**45.0**
	cos	cot	tan	sin	

ANSWERS TO END-OF-CHAPTER PROBLEMS

CHAPTER 1—ODD-NUMBERED PROBLEMS

Problems 1-1

1. (a) 6   (b) 3   (c) 7   (d) 0   (e) 7,000

Problems 1-2

1. (a) 0.3

(b) 0.016

(c) 0.00278

(d) 0.1763

3. (a) $\dfrac{7}{1000}$

(b) $\dfrac{432}{10,000}$

(c) $\dfrac{174}{1000}$

(d) $\dfrac{65}{1,000,000}$

7. (a) Hundredths
(b) Hundred-thousandths
(c) Thousandths

5. (a) $\dfrac{17}{1000} = 0.017$

(b) $\dfrac{4}{100} = 0.04$

(c) $\dfrac{460}{10,000} = 0.046$

(d) $\dfrac{27}{1,000,000} = 0.000027$

9. (a) Six thousandths
(b) One hundred forty-seven thousandths
(c) Ninety-two hundred-thousandths
(d) Seven millionths

Problems 1-3

1. (a) $7\dfrac{14}{100}$

(b) $50\dfrac{2}{100}$

(c) $710\dfrac{143}{1000}$

(d) $9\dfrac{99}{1000}$

(e) $73\dfrac{653}{1000}$

(f) $207\dfrac{7834}{10,000}$

(g) $28\dfrac{736}{100,000}$

(h) $8\dfrac{706}{10,000}$

3. (a) 5.68

(b) 25.007

(c) 7.0165

(d) 70.4

(e) 473.025

(f) 80.00743

(g) 2475.000035

(h) 307.00008

5. (a) $93.7 = 93\dfrac{7}{10}$

(b) $30.04 = 30\dfrac{4}{100}$

(c) $11.0001 = 11\dfrac{1}{10,000}$

(d) $273.00025 = 273\dfrac{25}{100,000}$

(e) $704.000704 = 704\dfrac{704}{1,000,000}$

(f) $3.033 = 3\dfrac{33}{1000}$

Problems 1-4

1. 20

Tens	Hundreds	
5. 270	300	
7. 360	400	
9. 140	100	

Tens	Hundreds	Thousands
11. 4820	4800	5000
13. 85,470	85,500	85,000
15. 78,670	78,700	79,000
17. 2780	2800	3000
19. 35,490	35,500	35,000

3. 50

Problems 1-5

	Hundredths	Tenths	Units	Tens
1.	163.78	163.8	164	160
3.	9.46	9.5	9	10
5.	88.89	88.9	89	90
7.	749.49	749.5	749	750
9.	39.28	39.3	39	40
11.	63.75	63.7	64	60
13.	478.67	478.7	479	480

Problems 1-6

1. 5

5. -6

9. -28

13. -14

17. 21

3. 14

7. 34

11. 7

15. 18

19. -33

Problems 1-7

1. -42

5. -84

9. -24

3. 54

7. -7

11. 12

Problems 1-8

1. -17

5. 17

9. 18

3. 0

7. 21

Problems 1-9

1. 16

5. -36

9. 4

13. 130

15. $4 - 6 + 12 + 21 = 24 + 21 = 45$

3. 7

7. -83

11. 51

17. $(9 - 21)(12) = (-12)(12) = -144$

CHAPTER 2—ODD-NUMBERED PROBLEMS

Problems 2-1

1. 10

3. 10^3

5. 10^{-1}

7. 10^{-2}

9. 10,000

11. 1000

13. 0.1

15. 0.000001

17. (a) $\dfrac{1}{10,000}$
    (b) 0.0001

19. (a) $\dfrac{1}{1,000,000,000}$
    (b) 0.000000001

Problems 2-2

1. 10^{10}

3. 10^4

5. 10^3

7. 10^{-12}

9. 10^6

11. 10^3

13. 10^7

15. 10^{-7}

17. 10^{11}

19. 10^{10}

21. 10^{-12}

23. 10^8

25. 10^{13}

27. 10^{-4}

29. 10^{-36}

Problems 2-3

1. 2.76×10^4

3. 4.78×10

5. 1.77×10^6

7. 2.73×10^2

9. 1.73×10^5

11. 4.78×10^{-3}

13. 7.47×10^{-1}

15. 1.64×10^{-2}

17. 2.78×10^{-6}

19. 1.71×10^{-1}

21. 1.57×10^4

23. 6.68×10^{-3}

25. 1.67×10^6

27. 4.68×10^{-7}

29. 5.84×10^{-3}

Problems 2-4

1. 2.13×10^4

3. 1.28×10^{-1}

5. 1.05×10^4

7. 2.99

9. 6.7

11. 1.82×10^{-2}

13. 3.26×10^{-3}

15. 3.91×10^{-5}

17. 2.56×10^3

19. 3.07×10^4

21. 1.39

23. 1.09

25. 1.03×10^{-2}

27. 7.85×10^3

29. 1.06×10^2

Problems 2-5

1. 10^6

3. 10^{10}

5. 10^{-12}

7. 10^{12}

9. 10^{-8}

11. 10^2

13. 10^{-15}

15. 6.29×10^3

17. 7.62×10^{-5}

19. 2.03×10^9

21. 8.76×10^{-8}

23. 6.75

25. 9.64

27. 8.40×10^{-2}

29. 5.20×10^3

31. 3.04×10^{-3}

33. 9.27×10

35. 7.96×10^3

37. 1.58×10^4

CHAPTER 3—ODD-NUMBERED PROBLEMS

Problems 3-1

1. $0.363 \times 10^3 = 363,000 \times 10^{-3}$

3. $0.00163 \times 10^3 = 1630 \times 10^{-3}$

5. $81 \times 10^3 = 0.081 \times 10^6$

7. $1000 \times 10^3 = 1 \times 10^6$

9. $0.64 \times 10^{-3} = 640 \times 10^{-6}$

11. $0.0706 \times 10^{-3} = 70.6 \times 10^{-6}$

13. $0.00273 \times 10^{-6} = 2730 \times 10^{-12}$

15. $0.000673 \times 10^{-6} = 673 \times 10^{-12}$

17. $17,300 \times 10^{-9} = 0.0173 \times 10^{-3}$

19. $2060 \times 10^{-9} = 0.00206 \times 10^{-3}$

Problems 3-2

1. 0.26 mA = 260 μA

3. 0.632 mA = 632 μA

5. 7.63 kΩ = 0.00763 MΩ

7. 17.3 mA = 17,300 μA

9. 713 kΩ = 0.713 MΩ

11. 5630 kHz = 5.63 MHz

13. 2000 kΩ = 2 MΩ

15. 0.237 mS = 237 μS

17. 0.3 μF = 300 nF

19. 0.062 mA = 62 μA

Problems 3-3

1. 0.8 A

3. 0.0025 S

5. 33,000 Ω

7. 470 Ω

9. 12,500 Hz

11. 0.0001 F

13. 0.00025 A

15. 0.0009 S

17. 750,000

19. 0.03 A

Problems 3-4

1. 0.26 mA = 260 μA

3. 0.00623 mS = 6.23 μS

5. 56.2 mS = 56,200 μS

7. 0.613 mS = 613 μS

9. 7.63 kΩ = 0.00763 MΩ

11. 470 Ω = 0.470 kΩ

13. 127 kHz = 0.127 MHz

15. 713 kΩ = 0.173 MΩ

17. 46,300 μA = 0.0463 A

19. 0.00005 μF = 50 pF

21. 3200 Ω = 0.0032 MΩ

23. 0.270 MΩ = 270,000 Ω

25. 0.000403 S = 0.403 mS

27. 1,030,000 Hz = 1030 kHz

29. 1430 pF = 0.00143 μF

31. 5.5 mS = 5500 μS

33. 0.00106 μF = 1060 pF

35. 463,000 Ω = 0.463 MΩ

37. 96.3 kHz = 0.0963 MHz

39. 7.8 mA = 7800 μA

41. 0.176 V = 176,000 μV

43. 1730 mW = 0.00173 kW

45. 25 mH = 25,000 μH

47. 0.173 S = 173,000 μS

49. 2.5 μF = 2500 nF

Problems 3-5

1. 2.04 mA = 2040 μA

3. 0.167 mS = 167 μS

5. 0.273 mA = 273 μA

7. 200,000 Ω = 200 kΩ

9. 4690 Ω = 4.69 kΩ

11. 84,900 Ω = 84.9 kΩ

13. 9.52 V = 9520 mV

15. 2.3 mS = 2,300 μS

17. 0.326 mS = 326 μS

19. 0.0417 mA = 41.7 μA

CHAPTER 4—ODD-NUMBERED PROBLEMS

Problems 4-1

1. 7

3. 10

5. 13

7. 23

9. 38

11. 56

13. 101

15. 119

17. 240

19. 195

21. 101_2

23. 1100_2

25. 10101_2

27. 101101_2

29. 111011_2

31. 1000100_2

33. 1100000_2

35. 10000111_2

37. 11010010_2

Problems 4-2

1. 12_{10}

3. 63_{10}

5. 190_{10}

7. 2174_{10}

9. 24_8

11. 120_8

13. 550_8

15. 2611_8

Problems 4-3

1. 26_{10}

3. 76_{10}

5. 512_{10}

7. 4522_{10}

9. 16_{16}

11. 61_{16}

13. 200_{16}

15. $17BB_{16}$

Problems 4-4

1. $75_8 = 111101_2 = 61_{10} = 3D_{16}$

3. $140_8 = 1100000_2 = 96_{10} = 60_{16}$

5. $2A_{16} = 101010_2 = 52_8 = 42_{10}$

7. $1A7_{16} = 110100111_2 = 647_8 = 423_{10}$

9. $100110_2 = 46_8 = 38_{10} = 26_{16}$

11. $111000101_2 = 705_8 = 453_{10} = 1C5_{16}$

13. $10_{10} = 1010_2 = 12_8 = A_{16}$

15. $290_{10} = 100100010_2 = 442_8 = 122_{16}$

Problems 4-5

1. 357_8

3. 775_8

5. 776_8

7. 1423_8

9. 1107_8

11. ED_{16}

13. $CA5_{16}$

15. $D56_{16}$

17. $179D_{16}$

19. $1DA6C_{16}$

21. 10101_2

23. 110011_2

25. 101001_2

27. 1100010_2

29. 1101001_2

Problems 4-6

1. 45

3. 42

5. 245

7. 337

9. 1515

11. 1A

13. 3E1

15. 5F1

17. AFD

19. FDC

21. 101

23. 10

25. 10

27. 11

29. 10110

Problems 4-7

1. 0110

5. 14_8

9. $5D_{16}$

13. 1101_2

17. 1011_2

21. 36_8

25. 16_{16}

3. 010011

7. 143_8

11. $3E2_{16}$

15. 100011_2

19. 1000011_2

23. 227_8

27. $1A3_{16}$

CHAPTER 5—ODD-NUMBERED PROBLEMS

Problems 5-1

1. 32

5. 324

9. 27

13. 150

17. 450

21. 0

3. 512

7. 25

11. 30

15. 900

19. 117

23. 9

Problems 5-2

1. 5.48

5. 14.1

9. 4

13. 7.55

3. 6.32

7. 4.47

11. 21.2

15. 23

Problems 5-3

1. 4.49 V

5. 135 V

9. 4.4 W

13. 10 mA

17. 1.77 A

21. 30 V

25. 39.4 mW

29. 417 mW

3. 66.6 V

7. 38.3 mW

11. 7.2 W

15. 38.9 mA

19. 20.9 V

23. 16.2 V

27. 28.2 mW

Problems 5-4

1. 318 Ω

3. 3.98 kΩ

5. 12.6 Ω

7. 1.45 kHz

9. 58.1 kHz

11. 1.42 kHz

13. 144 kΩ

15. 224 Ω

17. 38.6 kΩ

CHAPTER 6—ODD-NUMBERED PROBLEMS

Problems 6-1

1. $2 \cdot 3 \cdot 3$

3. $2 \cdot 2 \cdot 11$

5. $7 \cdot 3^2$

7. $2 \cdot 2 \cdot 23$

9. $3 \cdot 7 \cdot 11$

11. $2 \cdot 2 \cdot 2 \cdot a \cdot a \cdot b \cdot b$

13. $2 \cdot 2 \cdot 2 \cdot 3 \cdot x \cdot x \cdot y$

15. $2 \cdot 3 \cdot 7 \cdot c \cdot c \cdot c$

17. $2 \cdot 2 \cdot 2 \cdot 11 \cdot y \cdot y \cdot z$

Problems 6-2

1. 30

3. 72

5. 180

7. $a^3 b^3$

9. $a^3 b^3$

11. $a^4 b^2 c^3$

13. $a^4 b^3 c^2$

15. $78 x^2 y^3$

17. $32 x^3 y^2 z^2$

Problems 6-3

1. $21 a^4 b^3$

3. $20 a^{-4} b^{-5}$

5. $20 x^6 y^3 z^4$

7. $6 a^3 b^{-2} c^{-1}$

9. $60 x^5 y^{-4} z^4$

Problems 6-4

1. $9 a^2$

3. $8 a b^2$

5. $5 a^3 b c^{-1}$

7. $\dfrac{xy}{5}$

9. $11 x^{-5} y^{-1} z^{-4}$

Problems 6-5

1. $\dfrac{3}{10}$

3. $\dfrac{5}{24}$

5. $\dfrac{10a^3}{21}$

7. $\dfrac{3x^3y^{-2}}{16}$

9. $\dfrac{2xy^{-1}}{3}$

11. $\dfrac{x^{-3}y^3z^{-2}}{6}$

13. $\dfrac{3x^3yz^4}{7}$

Problems 6-6

1. $\dfrac{4}{3}$

3. $\dfrac{3}{14}$

5. $\dfrac{9x^{-2}}{56}$

7. $\dfrac{2abc^{-1}}{45}$

9. 1

Problems 6-7

1. $\dfrac{1}{2}$

3. $\dfrac{1}{2}$

5. $1\dfrac{11}{12}$

7. $\dfrac{5}{16}$

9. $\dfrac{5x^{-1}}{7}$

11. $\dfrac{4x^{-1}}{7}$

13. $\dfrac{3a^{-1}b^{-1}}{4}$

15. $\dfrac{a^{-1}}{2}$

17. $\dfrac{x}{2}$

19. $\dfrac{2x}{5}$

21. $\dfrac{3ab}{5}$

23. $\dfrac{3ab}{5}$

25. $\dfrac{10x}{9}$

27. $-\dfrac{2x}{7}$

29. $-\dfrac{5x}{24}$

Problems 6-8

1. $\dfrac{11x}{15}$

3. $\dfrac{13a}{6}$

5. $\dfrac{4y+5xy}{8x}$

7. $\dfrac{9x^2+8y^2}{12xy}$

9. $\dfrac{3b}{16}$

11. $\dfrac{40x^2+12y^2z-5xz^2}{30xyz}$

Problems 6-9

1. $8\dfrac{1}{2}$

3. $9\dfrac{1}{4}$

5. $3\dfrac{4}{5}$

7. $6\dfrac{1}{7}$

9. $6\frac{3}{4}$

11. $\frac{21}{2}$

13. $\frac{67}{16}$

15. $\frac{29}{8}$

17. $\frac{28}{3}$

19. $\frac{31}{6}$

Problems 6-10

1. $1\frac{1}{4}$

3. $7\frac{7}{12}$

5. $4\frac{19}{20}$

7. $9\frac{3}{4}$

9. $2\frac{5}{8}$

11. $2\frac{1}{7}$

13. $1\frac{7}{20}$

15. $\frac{21}{26}$

17. $\frac{14}{39}$

Problems 6-11

1. $6\frac{8}{15}$

3. $5\frac{23}{42}$

5. $4\frac{2}{15}$

7. $8\frac{3}{8}$

9. $3\frac{1}{12}$

11. $1\frac{7}{12}$

13. $1\frac{7}{8}$

15. $3\frac{5}{12}$

17. $8\frac{5}{12}$

Problems 6-12

1. $\frac{3}{8} + \frac{3}{32} = 0.469$

$\frac{3}{8} - \frac{3}{32} = 0.281$

$\frac{3}{8} \times \frac{3}{32} = 0.0352$

$\frac{3}{8} \div \frac{3}{32} = 4.00$

3. $\frac{7}{16} + \frac{2}{7} = 0.723$

$\frac{7}{16} - \frac{2}{7} = 0.152$

$\frac{7}{16} \times \frac{2}{7} = 0.125$

$\frac{7}{16} \div \frac{2}{7} = 1.53$

5. $\frac{5}{8} + \frac{37}{64} = 1.20$

$\frac{5}{8} - \frac{37}{64} = 0.0469$

$\frac{5}{8} \times \frac{37}{64} = 0.361$

$\frac{5}{8} \div \frac{37}{64} = 1.08$

7. $\frac{3}{5} + \frac{3}{10} = 0.900$

$\frac{3}{5} - \frac{3}{10} = 0.300$

$\frac{3}{5} \times \frac{3}{10} = 0.180$

$\frac{3}{5} \div \frac{3}{10} = 2.00$

9. $3\frac{3}{8}+2\frac{1}{3}=5.70$

$3\frac{3}{8}-2\frac{1}{3}=1.04$

$3\frac{3}{8}\times2\frac{1}{3}=7.87$

$3\frac{3}{8}\div2\frac{1}{3}=1.45$

13. $7\frac{1}{2}+3\frac{5}{9}=11.1$

$7\frac{1}{2}-3\frac{5}{9}=3.94$

$7\frac{1}{2}\times3\frac{5}{9}=26.7$

$7\frac{1}{2}\div3\frac{5}{9}=2.11$

17. $3\frac{2}{3}+6\frac{2}{3}=10.3$

$3\frac{2}{3}-6\frac{2}{3}=-3.00$

$3\frac{2}{3}\times6\frac{2}{3}=24.4$

$3\frac{2}{3}\div6\frac{2}{3}=0.550$

11. $4\frac{2}{5}+2\frac{15}{64}=6.63$

$4\frac{2}{5}-2\frac{15}{64}=2.17$

$4\frac{2}{5}\times2\frac{15}{64}=9.83$

$4\frac{2}{5}\div2\frac{15}{64}=1.97$

15. $4\frac{7}{8}+8\frac{7}{16}=13.3$

$4\frac{7}{8}-8\frac{7}{16}=-3.56$

$4\frac{7}{8}\times8\frac{7}{16}=41.1$

$4\frac{7}{8}\div8\frac{7}{16}=0.578$

19. $1\frac{7}{12}+3\frac{2}{3}=5.25$

$1\frac{7}{12}-3\frac{2}{3}=-2.08$

$1\frac{7}{12}\times3\frac{2}{3}=5.81$

$1\frac{7}{12}\div3\frac{2}{3}=0.432$

CHAPTER 7—ODD-NUMBERED PROBLEMS

Problems 7-1

1. 10

5. 7

9. 4

13. -7

17. -10

21. 8

25. -6

3. 2

7. -4

11. -9

15. 12

19. 16

23. $\frac{1}{4}$

27. $-\frac{2}{7}$

Problems 7-2

1. 5

5. 49

9. $\frac{1}{16}$

13. $b=\frac{7}{4c}$, $c=\frac{7}{4b}$

3. 7.35

7. 36

11. $a=\frac{6}{x}$, $x=\frac{6}{a}$

15. $R=\frac{E^2}{P}$, $E=\sqrt{PR}$

17. $f = \dfrac{B_C}{2\pi C}$, $C = \dfrac{B_C}{2\pi f}$

19. $a = \dfrac{1}{6\sqrt{b}}$, $b = \dfrac{1}{36a^2}$

21. $y = \dfrac{36}{25x}$

Problems 7-3

1. $R_1 = 6.8$ kΩ

3. $G_1 = 17.9$ μS

5. $V_0 = 8.25$ V

7. $R = 113$ Ω

9. $I = 29.7$ mA

11. $N_P = 1770$

13. $C = 84.2$ nF

15. $L = 128$ mH

17. $C = 131$ nF

19. $L = 281$ mH

CHAPTER 8—ODD-NUMBERED PROBLEMS

Problems 8-1

1. $I = 9$ A

3. $I_3 = 3$ mA

5. $I_2 = 45$ μA

7. $I_1 = 960$ mA

9. $I_T = 25$ mA, $I_3 = 15$ mA

11. $I_1 = 800$ mA, $I_2 = 800$ mA

13. $I_2 = 200$ μA, $I_3 = 200$ μA

15. $I_T = 1.6$ A, $I_3 = 800$ mA, $I_4 = 800$ mA

17. $I_T = 3.2$ mA, $I_1 = 3.2$ mA, $I_2 = 500$ μA

19. $I_1 = 1.7$ mA, $I_3 = 3.8$ mA

21. $I_T = 14.2$ mA, $I_2 = 4.8$ mA

23. $I_1 = 30$ mA, $I_3 = 10$ mA, $I_4 = 10$ mA, $I_6 = 70$ mA

Problems 8-2

1. $V_2 = 18$ V

3. $E = 13$ V

5. $V_3 = 11$ V

7. $E = 18$ V

9. $V_1 = 14$ V

11. $V_1 = 12$ V, $V_2 = 8$ V

13. $V_2 = 13$ V, $V_3 = 13$ V

15. $E = 17$ V, $V_2 = 5$ V

17. $V_1 = 3$ V, $V_4 = 9$ V

19. $V_1 = 1.7$ V, $V_3 = 800$ mV

21. $E = 60$ V, $V_3 = 10$ V

Problems 8-3

1. (a) 10 V
(b) 5 V
(c) 25 V
(d) 30 V
(e) -5 V

3. (a) 12 V
(b) 43 V
(c) -43 V
(d) -35 V
(e) -8 V

5. (a) 11 V
(b) 5 V
(c) 19 V
(d) -13 V

Problems 9-1

1. (b) $R_T = 1.95$ kΩ
(d) $R_T = 300$ kΩ

2. (b) $R_T = 130$ kΩ
(d) $R_T = 1.36$ MΩ

3. (b) $G_T = 1.2$ mS, $R_T = 831$ Ω
(d) $G_T = 20$ μS, $R_T = 50$ kΩ
(f) $G_T = 2.34$ mS, $R_T = 427$ Ω

4. (b) $G_T = 31.9$ μS, $R_T = 31.3$ kΩ
(d) $G_T = 26.7$ μS, $R_T = 37.4$ kΩ
(f) $G_T = 5.35$ mS, $R_T = 187$ Ω

5. (b) $R_T = 18.3$ kΩ
(d) $R_T = 735$ kΩ

6. (b) $R_T = 657$ Ω
(d) $R_T = 755$ Ω

7. (b) $R_T = 324$ Ω
(d) $R_T = 7.98$ kΩ

Problems 9-2

1. (b) $I = 2.67$ mA, $R_T = 4.5$ kΩ, $V_1 = 3.2$ V, $V_2 = 8.81$ V

2. (b) $R_T = 16.5$ kΩ, $I = 2.43$ mA, $V_1 = 16.5$ V, $R_2 = 9.67$ kΩ

3. (b) $R_1 = 15$ kΩ, $R_T = 25$ kΩ, $V_1 = 15$ V, $V_2 = 10$ V

4. (b) $V_2 = 24$ V, $R_1 = 7.15$ kΩ, $R_T = 19.2$ kΩ, $E = 38.3$ V

5. (b) $R_T = 388$ kΩ, $I = 103$ μA, $V_1 = 22.7$ V, $V_2 = 10.3$ V, $V_3 = 7.01$ V

6. (b) $R_T = 133$ kΩ, $V_2 = 7.05$ V, $V_3 = 6.25$ V, $R_3 = 41.7$ kΩ, $R_1 = 44.7$ kΩ

7. (b) $E = 11.7$ V, $V_1 = 4.05$ V, $V_3 = 3.91$ V, $R_2 = 13.7$ kΩ, $R_3 = 14.5$ kΩ

8. (b) $G_T = 40.3$ μS, $R_T = 24.8$ kΩ, $V = 17.3$ V, $I_1 = 255$ μA, $I_2 = 444$ μA

9. (b) $I_1 = 4$ mA, $V = 7.2$ V, $R_2 = 7.2$ kΩ, $R_T = 1.44$ kΩ, $G_T = 694$ μS

10. (b) $R_T = 71.4$ kΩ, $G_T = 14$ μS, $I_1 = 139$ μA, $I_2 = 211$ μA, $R_2 = 118$ kΩ

11. (b) $I_2 = 333$ μA, $I_T = 1.33$ mA, $G_T = 267$ μS, $R_T = 3.75$ kΩ, $R_1 = 5$ kΩ

12. (b) $G_T = 470$ μS, $R_T = 2.13$ kΩ, $V = 1.06$ V, $I_1 = 131$ μA, $I_2 = 142$ μA, $I_3 = 227$ μA

13. (b) $R_1 = 267$ kΩ, $R_3 = 133$ kΩ, $R_T = 66.7$ kΩ, $G_T = 15$ μS, $I_2 = 74.1$ μA, $I_3 = 151$ μA

14. (b) $I_1 = 60$ mA, $I_3 = 120$ mA, $I_T = 240$ mA, $R_1 = 500$ Ω, $R_2 = 500$ Ω, $R_3 = 250$ Ω, $R_T = 125$ Ω.

Problems 9-3

1. (b) $R_T = 487$ Ω, $I_T = 18.5$ mA, $I_1 = 18.5$ mA, $I_2 = 10.1$ mA, $I_3 = 8.34$ mA, $V_1 = 3.33$ V, $V_2 = 5.67$ V, $V_3 = 5.67$ V

2. (b) $R_T = 19.9$ kΩ, $I_T = 151$ μA, $I_1 = 87$ μA, $I_2 = 87$ μA, $I_3 = 63.8$ μA, $V_1 = 652$ mV, $V_2 = 2.35$ V, $V_3 = 3$ V

3. (b) $R_1 = 4.05$ kΩ, $R_T = 20$ kΩ, $V_1 = 2.43$ V, $V_2 = 9.57$ V, $V_3 = 9.57$ V, $I_2 = 245$ μA, $I_3 = 354$ μA

4. (b) $R_T = 100$ kΩ, $R_1 = 60$ kΩ, $R_2 = 75$ kΩ, $I_2 = 133$ μA, $I_3 = 66.7$ μA, $I_4 = 50$ μA, $I_1 = 250$ μA

5. (b) $R_T = 263$ kΩ, $I_T = 114$ μA, $V_1 = 10.5$ V, $V_2 = 19.5$ V, $V_3 = 19.5$ V, $V_4 = 30$ V, $I_1 = 69.9$ μA, $I_2 = 41.5$ μA, $I_3 = 28.7$ μA, $I_4 = 44.1$ μA

6. (b) $R_T = 578$ kΩ, $I_T = I_1 = I_4 = 26$ μA, $I_2 = 15.3$ μA, $I_3 = 10.6$ μA, $V_1 = 3.89$ V, $V_2 = V_3 = 7.21$ V, $V_4 = 3.89$ V

CHAPTER 10 PROBLEMS

Problems 10-1

1. $V_1 = 41.4$ V, $V_2 = 16.4$ V, $V_3 = 33.6$ V, $I_1 = 18.8$ mA, $I_2 = 9.13$ mA, $I_3 = 28$ mA

3. $V_1 = 36.8$ V, $V_2 = 45$ V, $V_3 = 18.2$ V, $V_4 = 21.8$ V, $I_1 = 13.6$ mA, $I_2 = 13.6$ mA, $I_3 = 18.2$ mA, $I_4 = 4.64$ mA

5. $V_1 = 1$ V, $V_2 = 10$ V, $V_3 = 8$ V, $I_1 = 0.303$ mA, $I_2 = 2.13$ mA, $I_3 = 2.42$ mA

7. $V_1 = 14.8$ V, $V_2 = 10.2$ V, $V_3 = 0.2$ V, $I_1 = 21.8$ mA, $I_2 = 21.6$ mA, $I_3 = 0.2$ mA

9. $V_1 = 14.2$ V, $V_2 = 7.73$ V, $V_3 = 18.1$ V, $V_4 = 3.12$ V, $I_1 = 6.44$ mA, $I_2 = 6.44$ mA, $I_3 = 5.49$ mA, $I_4 = 945$ μA

Problems 10-2

1. $V_{OC} = 20.4$ V, $R_{TH} = 15$ kΩ, $I_L = 816$ μA, $V_L = 8.16$ V

3. $V_{OC} = 12$ V, $R_{TH} = 1$ kΩ, $I_L = 6.86$ mA, $V_L = 5.14$ V

5. $V_{OC} = 25.4$ V, $R_{TH} = 16.2$ kΩ, $I_L = 814$ μA, $V_L = 12.2$ V

7. $V_{OC} = 29.6$ V, $R_{TH} = 2.56$ kΩ, $I_L = 8.3$ mA, $V_L = 8.3$ V

9. $V_{OC} = 1.5$ V, $R_{TH} = 3.78$ kΩ, $I_L = 314$ μA, $V_L = 0.314$ V

11. $V_{OC} = 2.53$ V, $R_{TH} = 5.94$ kΩ, $I_B = 43.9$ μA, $I_C = 3.51$ mA, $V_C = 13$ V

Problems 10-3

1. (a) $I_{SC} = 1.36$ mA, $G_N = 66.7$ μS, $V_L = 8.16$ V, $I_L = 816$ μA

2. (a) $I_{SC} = 357$ μA, $G_N = 17.9$ μS, $V_L = 7.42$ V, $I_L = 225$ μA

3. (a) $I_{SC} = 12$ mA, $G_N = 1$ mS, $V_L = 5.14$ V, $I_L = 6.86$ mA

4. (a) $I_{SC} = 7.14$ mA, $G_N = 179$ μS, $V_L = 13$ V, $I_L = 4.82$ mA

5. (a) $I_{SC} = 1.56$ mA, $G_N = 61.6$ μS, $V_L = 12.2$ V, $I_L = 812$ μA

CHAPTER 11—ODD-NUMBERED PROBLEMS

Problems 11-1

1. $3a+9$

3. $4a-4$

5. $-3a+2$

7. $6x+6xy$

9. $6a^2+9ab$

11. $9x^2+18x-36$

13. a^2-11ab

15. $-a^2b+8ab$

17. $3x^3+6x^2+3x$

19. $a^2+4a-4ab$

Problems 11-2

1. a^2+6a+9

3. $a^2-12a+36$

5. $4x^2-16x+16$

7. $16x^2+24x+9$

9. a^2+4a+3

11. $y^2-3y-18$

13. $a^2+4a-12$

15. x^2-5x+4

17. a^2-16

19. $4x^2-9$

21. $x^2+4xy+3y^2$

23. $4x^2-12xy+9y^2$

25. $x^2-xy-20y^2$

27. $6a^2+ab-12b^2$

Problems 11-3

1. x^2+3x+2

3. $x+3y$

5. $x-3$

7. $x-7$

9. $3x-2$

11. $a-5$

13. $2x+3y$

15. $3a-4b$

17. $x-9y$

19. $8+4x$

Problems 11-4

1. $3(a+3)$

3. $7(b-3)$

5. $2xy(2x+3)$

7. $2a(8a+1)$

9. $(x+2)(x+3)$

11. $(x+8)(x-4)$

13. $(a+3)^2$

15. $(a+4)(a-4)$

17. $(x-5)(x-3)$

19. $(y+9)(y-4)$

21. $2(x+3)(x+7)$

23. $4(a+2)(a-3)$

25. $3(x-3)(x+9)$

27. $(x+2y)(x-6y)$

29. $(2x+y)(3x-y)$

CHAPTER 12—ODD-NUMBERED PROBLEMS

Problems 12-1

1. $x = 1\dfrac{1}{5}$

3. $I = 3$

5. $R = -4\dfrac{3}{7}$

7. $I = 6$

9. $R = 1\dfrac{4}{5}$

11. $x = -\dfrac{14}{23}$

13. $x = -4$

15. $G = 3\dfrac{4}{7}$

17. $a = \dfrac{b}{2b-1},\ b = \dfrac{a}{2a-1}$

19. $a = \dfrac{3bx}{12b-2x},\ b = \dfrac{2ax}{12a-3x},\ x = \dfrac{12ab}{3b+2a}$

21. $a = \dfrac{2b+2}{b-5},\ b = \dfrac{5a+2}{a-2}$

23. $R_1 = \dfrac{3R_2+4}{5},\ R_2 = \dfrac{5R_1-4}{3}$

Problems 12-2

1. $V_{OC} = \dfrac{V_L(R_{TH}+R_L)}{R_L},\ R_L = \dfrac{V_L R_{TH}}{V_{OC}-V_L},$

$R_{TH} = \dfrac{V_{OC}R_L - V_L R_L}{V_L}$

3. $I_{SC} = \dfrac{I_L(G_L+G_N)}{G_L},\ G_L = \dfrac{I_L G_N}{I_{SC}-I_L},$

$G_N = \dfrac{I_{SC}G_L - I_L G_L}{I_L}$

5. $R_i = \dfrac{R_S R_o}{R_o - R_S}$

7. $r_e = \dfrac{r_i - r_b}{\beta}$

9. $R_1 = \dfrac{R_T R_2}{R_2 - R_T},\ R_2 = \dfrac{R_T R_1}{R_1 - R_T}$

11. $A = B - 1$

13. $y = \dfrac{2}{4x+5}$

Problems 12-3

1. (a) $R_2 = 4.8\ \text{k}\Omega$ (b) $R_1 = 773\ \Omega$

3. (a) $h_{oe} = 74.3\ \mu\text{S}$ (b) $R_L = 20.8\ \text{k}\Omega$

5. $R_1 = 3\ \text{k}\Omega$

7. $R_E = 1.2\ \text{k}\Omega$

Problems 12-4

1. $x = 4,\ x = -4$

3. $x = -4,\ x = -2$

5. $x = 3,\ x = -6$

7. $x = 2,\ x = 4$

9. $x = -7,\ x = 8$

11. $x = -5,\ x = 8$

13. $x = 4,\ x = 9$

Problems 12-5

1. $x = -6,\ x = -2$

3. $x = -1.77,\ x = -0.57$

5. $x = 0.333,\ x = -1$

7. $x = 3.27,\ x = -0.766$

CHAPTER 13 PROBLEMS

Problems 13-1

1. See Figure B-1.

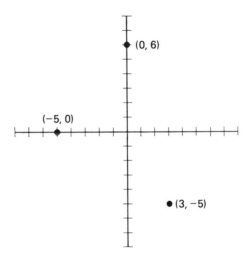

Figure B-1.

3. See Figure B-2.

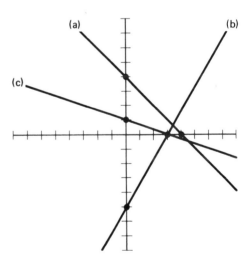

Figure B-2.

Problems 13-2

See Figure B-3.

1. x-intercept = 3, 0; y-intercept = 0, 6; slope = -2

3. x-intercept = -1, 0; y-intercept = 0, $\frac{5}{2}$; slope = $\frac{5}{2}$

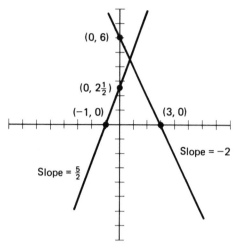

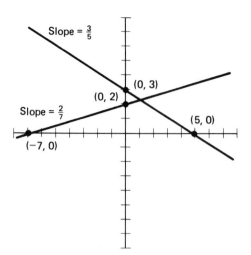

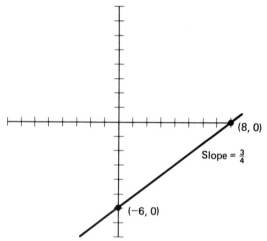

Figure B-3.

5. x-intercept $= 5, 0$; y-intercept $= 0, 3$;
slope $= -\dfrac{3}{5}$

7. x-intercept $= -7, 0$; y-intercept $= 0, 2$;
slope $= \dfrac{2}{7}$

9. x-intercept $= 8, 0$; y-intercept $= 0, -6$;
slope $= \dfrac{3}{4}$

Problems 13-3

1. $y = -\dfrac{1}{3}x + 1$; slope $= -\dfrac{1}{3}$; y-intercept $= 0, 1$

3. $y = -\dfrac{3}{4}x - \dfrac{5}{4}$; slope $= -\dfrac{3}{4}$; y-intercept $= 0, -\dfrac{5}{4}$

5. $y = \frac{1}{4}x - 3$; slope $= \frac{1}{4}$; y-intercept $= 0, -3$ **7.** $y = \frac{2}{5}x + 2$; slope $= \frac{2}{5}$; y-intercept $= 0, 2$

9. $y = \frac{3}{7}x - 3$; slope $= \frac{3}{7}$; y-intercept $= 0, -3$

Problems 13-4

1. $R_{DC} \cong 1.67$ kΩ, $r_{AC} \cong 1.25$ kΩ

3. (a) $V_C = 39$ V, $V_R = 61$ V   (b) $V_C = 95$ V, $V_R = 5$ V

Problems 13-5

1. See Figure B-4.   **2.** See Figure B-5.

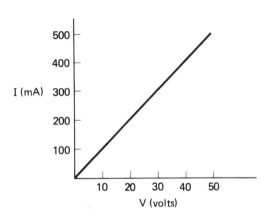

Figure B-4.   Figure B-5.

3. See Figure B-6.

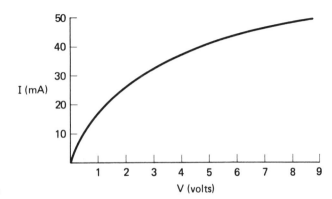

Figure B-6.

4. See Figure B-7

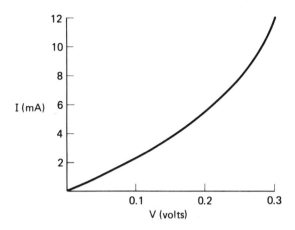

I (mA)

V (volts)

Figure B-7.

CHAPTER 14—ODD-NUMBERED PROBLEMS

Problems 14-1

1. $(3,2)$

3. $(2,-4)$

5. $(2,1)$

Problems 14-2

1. $x=3, y=2$

3. $x=2, y=-4$

5. $x=2, y=1$

7. $x=-2, y=-1$

9. $x=4, y=3$

Problems 14-3

1. $I_1=1.76$ mA, $I_2=1.17$ mA, $I_3=587$ μA, $V_1=8.27$ V, $V_2=11.7$ V, $V_3=11.7$ V

3. $I_1=12.7$ mA, $I_2=9.08$ mA, $I_3=21.8$ mA, $V_1=38.2$ V, $V_2=18.2$ V, $V_3=21.8$ V

5. $I_1=251$ mA, $I_2=45.3$ mA, $I_3=204$ mA, $V_1=25.1$ V, $V_2=14.9$ V, $V_3=44.9$ V

CHAPTER 15—ODD-NUMBERED PROBLEMS

Problems 15-1

1. $j13$

3. j

5. $-4-j4$

7. $-j12$

9. $-j18$

11. 12

13. $j3$

15. $-j2$

17. -7

19. $-j20$

Problems 15-2

1. $19-j4$

5. $7-j24$

9. $0.8+j1.6$

13. $-1.04+j0.28$

3. 13

7. $1-j1$

11. $0.31+j0.276$

CHAPTER 16—ODD-NUMBERED PROBLEMS

Problems 16-1

1. (a) $\phi=55°$
(b) Side R is greater
(c) Side R
(d) Side R

5. (a) 2.99 mS
(b) 800 μS
(c) 425 μS

3. (a) 4.83 kΩ
(b) 487 Ω
(c) 98.9 kΩ

Problems 16-2

1. (a) $\sin\theta=\dfrac{X}{Z}$, $\cos\theta=\dfrac{R}{Z}$, $\tan\theta=\dfrac{X}{R}$
(b) $\sin\phi=\dfrac{R}{Z}$, $\cos\phi=\dfrac{X}{Z}$, $\tan\phi=\dfrac{R}{X}$

Problems 16-3

1. 0.4586

5. 2.3905

9. 56.2°

3. 0.2027

7. 23.4°

11. 55.8°

Problems 16-4

1. $\theta=32°$
$\phi=58°$
$c=47.2$

5. $a=11.1$
$b=13.3$

9. $Z=11.5$ kΩ
$\theta=60.8°$

13. $R=15.3$ kΩ
$X=37$ kΩ

17. $R=33.7$ kΩ
$\theta=27.4°$

3. $\theta=55°$
$b=140$
$c=244$

7. $R=42.9$ kΩ
$Z=46.4$ kΩ

11. $X=11.5$ kΩ
$Z=23.1$ kΩ

15. $X=18.4$ kΩ
$\theta=56.9°$

19. $Y=66.2$ μS
$B=28$ μS

21. $Y = 15.2$ mS
$\theta = 34.9°$

23. $G = 186$ mS
$Y = 189$ mS

25. $B = 6.86$ mS
$\theta = 47.2°$

27. $G = 59.4$ μS
$B = 37.1$ μS

29. $G = 8.72$ mS
$\theta = 40.7°$

CHAPTER 17—ODD-NUMBERED PROBLEMS

Problems 17-1

1. 1 ms

3. 568 ns

5. 752 Hz

7. 125 kHz

9. 0.436 rad

11. 3.14 rad

13. 1.31 rad

15. 10.5 rad

17. 8.37°

19. 28.6°

21. 100°

23. 573°

Problems 17-2

1. (a) 377 rad/sec, 21,600°/sec
(b) 942 rad/sec, 54,000°/sec

3. (a) 143 Hz
(b) 796 Hz
(c) 27.8 Hz
(d) 333 Hz

5. (a) 0.314 rad, 18°
(b) 0.785 rad, 45°
(c) 3.14 rad, 180°
(d) 8.8 rad, 504°

7. (a) 14.3 kHz
(b) 9.26 kHz
(c) 175 Hz
(d) 2.15 kHz

9. (a) 31.8 μs
(b) 3.98 μs
(c) 151 μs
(d) 11.1 μs
(e) 50 μs
(f) 167 μs

Problems 17-3

1. (a) 3.91 V
(b) 9.57 V
(c) 17.7 V
(d) −9.57 V
(e) −25 V
(f) −9.57 V

3. (a) 700 μA
(b) 0
(c) −700 μA
(d) 0
(e) 700 μA
(f) 0

Problems 17-4

1. 65 V=91.9 V_{pk}=184 V_{p-p}

3. 3 A=4.24 A_{pk}=8.48 A_{p-p}

5. 80 V_{pk}=160 V_{p-p}=56.6 V

7. 500 mA_{pk}=1 A_{p-p}=354 mA

9. 17 V_{p-p}=8.5 V_{pk}=6.01 V

11. 700 μA_{p-p}=350 μA_{pk}=247 μA

Problems 17-5

1. (a) i = 2.35 mA, e = 8.75 V, (b) i = 3.99 mA, e = 9.31 V, (c) i = 0, e = − 4.23 V, (d) i = − 3.8 mA, e = − 7.31 V

3. (a) i = 23.6 mA, e =16.6 V, (b) i = 20.5 mA, e = 28.3 V, (c) i =12.4 mA, e = 25.6 V, (d) i = − 21.9 mA, e = − 24.8 V

CHAPTER 18—ODD-NUMBERED PROBLEMS

Problems 18-1

1. E=28.3 V, θ=45°

3. V_C=34 V, θ=− 58.3°

5. E=19.6 V, V_C=12.6 V

7. V_R=62.4 V, V_C=31.8 V

9. V_R=22.4 V, θ=41.8°

11. E=10.4 V, V_R=8.82 V

13. X_C=6.37 kΩ, Z=11.9 kΩ, θ= − 32.5°

15. X_C=796 Ω, Z=1.06 kΩ, θ= −48.7°

17. X_C=3.18 kΩ, Z=3.76 kΩ, θ= − 57.9°

Problems 18-2

1. X_C=1.33 kΩ, Z=2.4 kΩ, θ= − 33.6°, V_R=25 V, V_C=16.6 V, I=12.5 mA

3. X_C=15.9 kΩ, Z=18.8 kΩ, θ= − 57.9°, V_R=5.32 V, V_C=8.47 V, I=532 μA

5. X_C=5.31 kΩ, Z=6.25 kΩ, θ= − 58.1°, V_R=15.8 V, V_C=25.5 V, I=4.8 mA

7. θ= − 46.9°, X_C=8.76 kΩ, f=90.8 kHz, V_R=27.3 V, V_C=29.2 V, I=3.33 mA

9. θ= − 51.3°, X_C=937 Ω, f=17 kHz, V_R=9.38 V, V_C=11.7 V, I=12.5 mA

11. θ= − 39.5°, X_C=2.23 kΩ, f=2.38 MHz, V_R=23.1 V, V_C=19.1 V, I=8.57 mA

Problems 18-3

1. E=12.8 V, θ=51.3°

3. E=7.52 V, V_L=3.41 V

5. V_L=8.62 V, θ=45.9°

7. V_R=26 V, θ=30°

9. E=3.4 V, V_R=1.6 V

11. V_R=4.77 V, V_C=7.63 V

13. X_L=3.77 kΩ, Z=5.43 kΩ, θ=44°

15. X_L=377 Ω, Z=626 Ω, θ=37°

17. X_L=785 Ω, Z=1.27 kΩ, θ=38.1°, V_R=11.8 V, V_L=9.27 V, I=11.8 mA

19. X_L=471 Ω, Z=559 Ω, θ=57.5°, V_R=26.9 V, V_L=42.2 V, I=89.4 mA

21. X_L=3.73 kΩ, θ=38.4°, f=913 Hz, E=12.1 V, V_R=9.46 V, I=2.02 mA

23. X_L=26.5 kΩ, θ=55.8°, f=56.1 kHz, E=4.47 V, V_R=2.51 V, I=140 μA

25. X_L=8.65 kΩ, L=320 mH, θ=51.8°, V_R=4.72 V, E=7.63 V, I=694 mA

27. X_L=4.27 kΩ, L=90.5 mH, θ=62.7°, V_R=2.32 V, E=5.06 V, I=1.05 mA

Problems 18-4

1. $Z_{polar} = 30.9 \underline{/48°}$ kΩ
$Z_{rect} = 20.7$ kΩ $+ j23$ kΩ

3. $Z_{polar} = 857 \underline{/-17°}$ Ω
$Z_{rect} = 820$ Ω $- j251$ Ω

5. $Z_{polar} = 6.35 \underline{/-45°}$ kΩ
$Z_{rect} = 4.49$ kΩ $- j4.49$ kΩ

7. $Z_{polar} = 50 \underline{/-36.9°}$ kΩ
$Z_{rect} = 40$ kΩ $- j30$ kΩ

9. $Z_{polar} = 4.6 \underline{/-40.7°}$ kΩ
$Z_{rect} = 3.49$ kΩ $- j3$ kΩ

11. $Z_{polar} = 20.6 \underline{/61°}$ kΩ
$Z_{rect} = 10$ kΩ $+ j18$ k

13. $E_{polar} = 16.2 \underline{/51.8°}$ kΩ
$E_{rect} = 10$ V $= j12.7$ V

15. $E_{polar} = 12 \underline{/-28.4°}$ V
$E_{rect} = 10.6$ V $- j5.7$ V

17. $E_{polar} = 15 \underline{/-40°}$ V
$E_{rect} = 11.5$ V $- j9.64$ V

19. $E_{polar} = 14 \underline{/34.6°}$ V
$E_{rect} = 11.5$ V $+ j7.93$ V

21. $E_{polar} = 5 \underline{/58.7°}$ V
$E_{rect} = 2.6$ V $+ j4.27$ V

23. $E_{polar} = 30 \underline{/32.2°}$ V
$E_{rect} = 25.4$ V $+ j16$ V

Problems 18-5

1. $4.64 \underline{/47.2°}$ kΩ

3. $25.9 \underline{/-25.8°}$ kΩ

5. $8.99 \underline{/-7.39°}$ kΩ

7. $82.6 \underline{/1.88°}$ kΩ

9. $420 \underline{/30°}$ kΩ

11. $12.8 \underline{/-11°}$ kΩ

Problems 18-6

1. $Z_{rect} = 1.76$ kΩ $+ j3$ kΩ
$Z_{polar} = 3.48 \underline{/59.6°}$ kΩ

3. $Z_{rect} = 1.15$ kΩ $+ j1.3$ kΩ
$Z_{polar} = 1.74 \underline{/48.5°}$ kΩ

5. $R = 3.4$ kΩ and $L = 623$ mH

7. $Z_{rect} = 200$ Ω $- j281$ Ω
$Z_{polar} = 345 \underline{/-54.6°}$ Ω
$V_R = 14.5$ V
$V_C = 65.9$ V
$V_L = 45.5$ V

Problems 18-7

1. $f_r = 2.6$ kHz
(a) $X_L = 12.3$ kΩ, $X_C = 12.3$ kΩ, $Z_{polar} =$ 470 $\underline{/0°}$ Ω, $I = 21.3$ mA, $V_R = 10$ V, $V_L = 262$ V, $V_C = 262$ V. The ESC consists of 470 Ω of resistance.
(b) $X_L = 10.8$ kΩ, $X_C = 13.8$ kΩ, $Z_{polar} =$ 3.04 $\underline{/-81.1°}$ kΩ, $I = 3.29$ mA, $V_R = 1.55$ V, $V_L = 35.5$ V, $V_C = 45.4$ V. In the ESC $R = 470$ Ω and $C = 23.1$ nF.
(c) $X_L = 12.7$ kΩ, $X_C = 11.8$ kΩ, $Z_{polar} =$ 1.02 $\underline{/62.4°}$ kΩ, $I = 9.80$ mA, $V_R = 4.61$ V, $V_L = 125$ V, $V_C = 116$ V. In the ESC $R = 470$ Ω and $L = 53.1$ mH.

3. $f_r = 225$ Hz
(a) $X_L = 1.41$ kΩ, $X_C = 1.41$ kΩ, $Z_{polar} =$ 300 $\underline{/0°}$ Ω, $I = 83.3$ mA, $V_R = 25$ V, $V_L = 118$ V, $V_C = 118$ V. The ESC consists of 300 Ω of resistance.
(b) $X_L = 1.26$ kΩ, $X_C = 1.59$ kΩ, $Z_{polar} =$ 448 $\underline{/-48°}$ Ω, $I = 56.1$ mA, $V_R = 16.7$ V, $V_L = 70.6$ V, $V_C = 89.1$ V. The ESC consists of 300 Ω of resistance and 2.38 μF of capacitance.
(c) $X_L = 1.88$ kΩ, $X_C = 1.06$ kΩ, $Z_{polar} =$ 873 $\underline{/69.9°}$ Ω, $I = 28.6$ mA, $V_R = 8.59$ V, $V_L = 53.8$ V, $V_C = 30.3$ V The ESC consists of 300 Ω of resistance and 435 mH of inductance.

CHAPTER 19—ODD-NUMBERED PROBLEMS

Problems 19-1

1. $Y_{\text{rect}} = 303\ \mu\text{S} + j377\ \mu\text{S}$, $Y_{\text{polar}} = 484\ \underline{/51.2°}\ \mu\text{S}$, $I_{\text{rect}} = 2.51\ \text{mA} + j3.12\ \text{mA}$, $V = 8.27\ \text{V}$

3. $Y_{\text{rect}} = 1.33\ \text{mS} + j1.88\ \text{mS}$, $Y_{\text{polar}} = 2.31\ \underline{/54.7°}\ \text{mS}$, $I_{\text{rect}} = 577\ \mu\text{A} + j816\ \mu\text{A}$, $V = 433\ \text{mV}$

5. $Y_{\text{rect}} = 147\ \mu\text{S} + j94.2\ \mu\text{S}$, $Y_{\text{polar}} = 175\ \underline{/32.7°}\ \mu\text{S}$, $I_{\text{rect}} = 16.8\ \text{mA} + j10.8\ \text{mA}$, $V = 115\ \text{V}$

7. $Y_{\text{rect}} = 100\ \mu\text{S} + j188\ \mu\text{S}$, $Y_{\text{polar}} = 213\ \underline{/62.1°}\ \mu\text{S}$, $I_{\text{rect}} = 4.69\ \text{mA} + j8.83\ \text{mA}$, $V = 46.9\ \text{V}$

9. $Y_{\text{rect}} = 66.7\ \mu\text{S} + j66\ \mu\text{S}$, $Y_{\text{polar}} = 93.8\ \underline{/44.7°}\ \mu\text{S}$, $I_{\text{rect}} = 1.42\ \text{mA} + j1.41\ \text{mA}$, $V = 21.3\ \text{V}$

11. $R = 5.96\ \text{k}\Omega$, $f = 1.27\ \text{kHz}$, $I = 4.67\ \text{mA}$, $V = 17.9\ \text{V}$

Problems 19-2

1. $Y_{\text{rect}} = 50\ \mu\text{S} - j39.8\ \mu\text{S}$, $Y_{\text{polar}} = 63.9\ \underline{/-38.5°}\ \mu\text{S}$, $I_{\text{rect}} = 31.3\ \text{mA} - j24.9\ \text{mA}$, $I_{\text{polar}} = 40\ \underline{/-38.5°}\ \text{mA}$

3. $Y_{\text{rect}} = 100\ \mu\text{S} - j133\ \mu\text{S}$, $Y_{\text{polar}} = 166\ \underline{/-53°}\ \mu\text{S}$, $I_{\text{rect}} = 120\ \text{mA} - j160\ \text{mA}$, $I_{\text{polar}} = 200\ \underline{/-53°}\ \text{mA}$

5. $Y_{\text{rect}} = 213\ \mu\text{S} - j159\ \mu\text{S}$, $Y_{\text{polar}} = 266\ \underline{/-36.8°}\ \mu\text{S}$, $I_{\text{rect}} = 80.1\ \mu\text{A} - j59.9\ \mu\text{A}$, $I_{\text{polar}} = 100\ \underline{/-36.8°}\ \mu\text{A}$

7. $L = 120\ \text{mH}$, $Y_{\text{rect}} = 500\ \mu\text{S} - j266\ \mu\text{S}$, $Y_{\text{polar}} = 566\ \underline{/-28°}\ \mu\text{S}$, $I_{\text{rect}} = 28.2\ \text{mA} - j15\ \text{mA}$, $I_{\text{polar}} = 32\ \underline{/-28°}\ \text{mA}$

Problems 19-3

1. $269\ \underline{/-27°}\ \Omega$

3. $158\ \underline{/40°}\ \mu\text{S}$

5. $657\ \underline{/44.1°}\ \Omega$

7. $8.55\ \underline{/-27.8°}\ \text{k}\Omega$

9. $2.59\ \underline{/-38.2°}\ \text{k}\Omega$

Problems 19-4

1. $I_R = 1.21\ \text{mA}$, $I_C = 1.59\ \text{mA}$, $Y = 110\ \underline{/52.8°}\ \mu\text{S}$, $Z = 9.09\ \underline{/-52.8°}\ \text{k}\Omega$, $V = 18.1\ \text{V}$. In the ESC $R = 5.47\ \text{k}$ and $C = 630\ \text{pF}$.

3. $I_R = 24.1\ \text{mA}$, $I_C = 31.9\ \text{mA}$, $Y_{\text{polar}} = 3.54\ \underline{/53°}\ \text{mS}$, $Z_{\text{polar}} = 282\ \underline{/-53°}\ \Omega$, $V = 11.3\ \text{V}$. In the ESC $R = 170\ \Omega$ and $C = 236\ \text{nF}$.

5. $I_R = 192\ \mu\text{A}$, $I_C = 160\ \mu\text{A}$, $Y_{\text{polar}} = 27.7\ \underline{/-39.7°}\ \mu\text{S}$, $Z_{\text{polar}} = 36.1\ \underline{/39.7°}\ \text{k}\Omega$, $V = 9.03\ \text{V}$. In the ESC $R = 27.8\ \text{k}\Omega$ and $L = 24.5\ \text{mH}$.

7. $I_R = 4.06\ \text{mA}$, $I_C = 2.92\ \text{mA}$, $Y_{\text{polar}} = 1.81\ \underline{/-35.8°}\ \text{mS}$, $Z_{\text{polar}} = 552\ \underline{/35.8°}\ \Omega$, $V = 2.76\ \text{V}$. In the ESC $R = 448\ \Omega$ and $L = 10.3\ \text{mH}$.

9. $Y = 193\ \underline{/-58.8°}\ \mu\text{S}$, $Z = 5.18\ \underline{/58.8°}\ \text{k}\Omega$

11. $Y = 230\ \underline{/50.3°}\ \mu\text{S}$, $Z = 4.34\ \underline{/-50.3°}\ \text{k}\Omega$

Problems 19-5

1. $Z = 1.20 \text{ k}\Omega + j19.9 \text{ k}\Omega$

3. $Z = 7 \text{ k}\Omega - j1 \text{ k}\Omega$

5. $Z = 16.9 \text{ k}\Omega - j26.3 \text{ k}\Omega$

7. $Z = 8.61 \text{ k}\Omega - j9.21 \text{ k}\Omega$

9. $Z = 4.96 \text{ k}\Omega + j5.41 \text{ k}\Omega$. The circuit is resistive and inductive. $R = 4.96 \text{ k}\Omega$ and $L = 862$ mH. The phase angle is $47.5°$.

CHAPTER 20—ODD-NUMBERED PROBLEMS

Problems 20-1

1. $X_C = 470 \text{ }\Omega$, $I_{max} = 31.9$ mA, $P_{max} = 479$ mW, $V_{o(max)} = 15$ V. At f_{co}, $I = 22.6$ mA, $P = 239$ mW, $V_o = 10.6$ V.

3. $f_{co} = 265$ Hz, $I = 11.8$ mA, $V_o = 14.1$ V, $P = 167$ mW

Problems 20-2

1. $V_{OC} = 6.47$ V, $R_{TH} = 2.16 \text{ k}\Omega$, $f_{co} = 147$ Hz

3. $V_{OC} = 20$ V, $R_{TH} = 2 \text{ k}\Omega$, $f_{co} = 39.8$ Hz

5. $V_{OC} = 10$ V, $R_{TH} = 5 \text{ k}\Omega$, $f_{co} = 31.8$ Hz, V_o at $f_{co} = 3.25$ V

7. $V_{OC} = 5.66$ V, $R_{TH} = 4.08 \text{ k}\Omega$, $f_{co} = 39$ Hz

9. $I_{SC} = 31.9$ mA, $G_N = 2.96$ mS, $f_{co} = 9.43$ kHz, V_o at $f_{co} = 7.62$ V

11. $I_{SC} = 5$ mA, $G_N = 891$ µS, $f_{co} = 142$ kHz, V_o at $f_{co} = 3.97$ V

Problems 20-3

1. $f_1 = 169$ Hz, $f_2 = 183$ kHz, V_o at $f_m = 5.74$ V, V_o at $f_{co} = 4.06$ V

3. $f_1 = 15$ Hz, $f_2 = 128$ kHz, V_o at $f_m = 5.66$ V, V_o at $f_{co} = 4$ V

5. $f_1 = 43.9$ Hz, $f_2 = 87.9$ kHz

7. $BW = 80$ Hz, $f_1 = 1960$ Hz, $f_2 = 2040$ Hz. At f_r, $V_R = 20$ V, $I = 20$ mA, $P = 400$ mW. At f_1 and f_2, $V_R = 14.1$ V, $I = 14.1$ mA, $P = 200$ mW.

9. $f_r = 1.01$ kHz, $Q = 14.4$, $BW = 71.6$ Hz, $f_1 = 974$ Hz, $f_2 = 1.05$ Hz

CHAPTER 21—ODD-NUMBERED PROBLEMS

Problems 21-1

Logarithmic Form

1. $\log 14 = 1.146$

3. $\log 0.037 = -1.432$

5. $\log 1070 = 3.029$

7. $\log 0.407 = -0.39$

9. $\log 10{,}000 = 4$

Exponential Form

$10^{1.146} = 14$

$10^{-1.432} = 0.037$

$10^{3.029} = 1070$

$10^{-0.39} = 0.407$

$10^4 = 10{,}000$

11. $\log 0.49 = -0.31$ $\qquad\qquad$ $10^{-0.31} = 0.49$

13. $\log 47,700 = 4.679$ $\qquad\qquad$ $10^{4.679} = 47,700$

15. $\log 0.301 = -0.521$ $\qquad\qquad$ $10^{-0.521} = 0.301$

17. $\log 37,000 = 4.568$ $\qquad\qquad$ $10^{4.568} = 37,000$

19. $\log 0.975 = -0.011$ $\qquad\qquad$ $10^{-0.011} = 0.975$

21. $\log 2,000,000 = 6.301$ $\qquad\qquad$ $10^{6.301} = 2,000,000$

23. $\log 0.0275 = -1.561$ $\qquad\qquad$ $10^{-1.561} = 0.0275$

25. $\log 3.04 = 0.483$ $\qquad\qquad$ $10^{0.483} = 3.04$

27. $\log 0.625 = -0.204$ $\qquad\qquad$ $10^{-0.204} = 0.625$

29. $\log 746,000 = 5.873$ $\qquad\qquad$ $10^{5.873} = 746,000$

Problems 21-2

1. $\log(27 \times 56) = \log 27 + \log 56 = 1.431 + 1.748 = 3.18$
$10^{3.18} = 1.51 \times 10^3$

3. $\log(107 \times 243,000) = \log 107 + \log 243,000 = 2.029 + 5.386 = 7.415$
$10^{7.415} = 2.6 \times 10^7$

5. $\log(43 \times 0.143) = \log 43 + \log 0.143 = 1.633 + (-0.8447) = 0.7888$
$10^{0.7888} = 6.15$

7. $\log(0.043 \times 0.0106) = \log 0.043 + \log 0.0106 = -1.367 + (-1.975) = -3.341$
$10^{-3.341} = 4.56 \times 10^{-4}$

9. $\log(0.0000176 \times 0.000357) = \log 0.0000176 + \log 0.000357 = -4.754 + (-3.447) = -8.202$
$10^{-8.202} = 6.28 \times 10^{-9}$

11. $\log \dfrac{1760}{43} = \log 1760 - \log 43 = 3.246 - 1.633 = 1.612$
$10^{1.612} = 40.9$

13. $\log \dfrac{65}{467} = \log 65 - \log 467 = 1.813 - 2.669 = -0.856$
$10^{-0.856} = 0.139$

15. $\log \dfrac{0.00273}{172} = \log 0.00273 - \log 172 = -2.564 - 2.236 = -4.799$
$10^{-4.799} = 1.59 \times 10^{-5}$

17. $\log \dfrac{4300}{0.15} = \log 4300 - \log 0.15 = 3.633 - (-0.824) = 4.457$
$10^{4.457} = 2.87 \times 10^4$

19. $\log \dfrac{0.0675}{0.00103} = \log 0.0675 - \log 0.00103 = -1.171 - (-2.987) = 1.816$
$10^{1.816} = 65.5$

21. $\log 3.4^3 = 3 \log 3.4 = 3 \times 0.5315 = 1.594$
$10^{1.594} = 39.3$

23. $\log 6^{-1.3} = -1.3 \log 6 = -1.3 \times 0.7782 = -1.012$
$10^{-1.012} = 9.74 \times 10^{-2}$

25. $\log 0.43^{2.5} = 2.5 \log 0.43 = 2.5 \times -0.3665 = -0.9163$
$10^{-0.9163} = 0.121$

27. $\log 6^{4.3} = 4.3 \log 6 = 4.3 \times 0.7782 = 3.346$
$10^{3.346} = 2.22 \times 10^3$

29. $\log 0.27^{-4.4} = -4.4 \log 0.27 = -4.4 \times -0.5686 = 2.502$
$10^{2.502} = 3.18 \times 10^2$

Problems 21-3

Logarithmic Form	*Exponential Form*
1. $\ln 2 = 0.6931$	$\varepsilon^{0.6931} = 2$
3. $\ln 0.932 = -0.07042$	$\varepsilon^{-0.07042} = 0.932$
5. $\ln 60 = 4.094$	$\varepsilon^{4.094} = 60$
7. $\ln 0.073 = -2.617$	$\varepsilon^{-2.617} = 0.073$
9. $\ln 247 = 5.509$	$\varepsilon^{5.509} = 247$
11. $\ln 0.00146 = -6.529$	$\varepsilon^{-6.529} = 0.00146$
13. $\ln 1100 = 7.003$	$\varepsilon^{7.003} = 1100$
15. $\ln 0.00027 = -8.217$	$\varepsilon^{-8.217} = 0.00027$
17. $\ln 8.43 = 2.132$	$\varepsilon^{2.132} = 8.43$
19. $\ln 0.346 = -1.061$	$\varepsilon^{-1.061} = 0.346$
21. $\ln 3.67 = 1.30$	$\varepsilon^{1.30} = 3.67$
23. $\ln 0.107 = -2.235$	$\varepsilon^{-2.235} = 0.107$
25. $\ln 176 = 5.17$	$\varepsilon^{5.17} = 176$
27. $\ln 0.293 = -1.228$	$\varepsilon^{-1.228} = 0.293$
29. $\ln 873 = 6.772$	$\varepsilon^{6.772} = 873$

CHAPTER 22—ODD-NUMBERED PROBLEMS

Problems 22-1

1. $10^x = 340$	**3.** $10^y = 0.014$
5. $10^{-2.43} = x$	**7.** $y = 10^{1.44}$
9. $\log y = 3.6$	**11.** $\log 732 = a$
13. $\log 0.043 = y$	**15.** $\log x = -3.4$

Problems 22-2

1. 2.291×10^7	**3.** 1.002
5. 3.776×10^{-1}	**7.** 0.9917
9. 2.665	**11.** -1.762
13. -2.602	**15.** 4.668
17. 1.445×10^4	**19.** 1.079
21. 2.512×10^{-4}	**23.** 6.714×10^{-1}
25. 2.873	**27.** -2.145
29. 2.778×10^3	**31.** -5.332×10^{-1}
33. 7.915	**35.** 1×10^3
37. 2.291×10^7	**39.** 1.458
41. 5.849×10^{-1}	

Problems 22-3

1. -0.3550
5. 2.105×10^3
9. 5.561
13. 1×10^8

3. 1.171
7. 1×10^{-4}
11. 1.570
15. 2×10^5

Problems 22-4

1. 2.398
5. 5.298
9. -2.708
13. 3.829

3. 2.708
7. -7.824
11. 14.88
15. 1.832×10^{-2}

Problems 22-5

1. 0.9819
5. 13.19
9. 2.749
13. 2.606
17. 0.9502
21. 0.5489

3. 0.47
7. 31.57
11. 3.638
15. 2.885
19. 0.4716

CHAPTER 23—ODD-NUMBERED PROBLEMS

Problems 23-1

1. 60 dB
5. 1.58 W
9. 88.6 dB
13. 126 μV

3. 3.79 mW
7. 100 mW
11. 500 mV
15. 706 mW

Problems 23-2

1. $f_{co} = 531$ Hz. See Figure B-8 for the Bode plot.
3. $f_{co} = 1.59$ kHz. See Figure B-8 for the Bode plot.
5. $f_1 = 339$ Hz, $f_2 = 9.17$ kHz. See Figure B-9 for the Bode plot.
7. See Figure B-10 for the Bode plot.
9. At 30 Hz, $A_v = -18$ dB.
 At 50 kHz, $A_v = -10.6$ dB.
11. At 40 Hz, $A_v = 15.9$ dB.
 At 25 kHz, $A_v = 15.3$ dB.

Problems 23-3

1. $V_C = 34.7$ V, $V_R = 65.3$ V
5. 145 μs
9. 9.76 kΩ

3. 2.47 ms
7. 5.41 ms
11. 9.53 nF

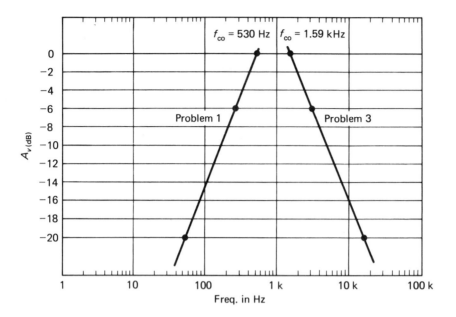

Figure B-8.

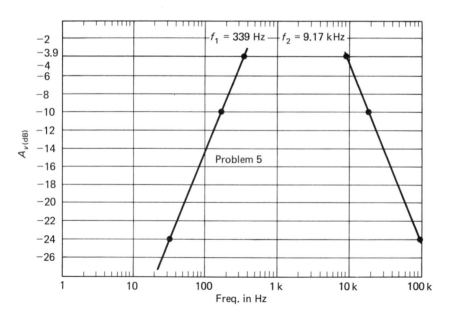

Figure B-9.

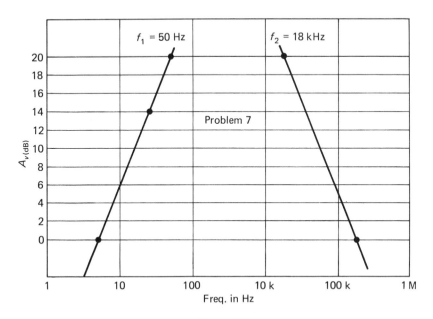

Figure B-10.

CHAPTER 24—ODD-NUMBERED PROBLEMS

Problems 24-1

1. $A\bar{B} + \bar{B}C$

3. $\overline{\overline{X}YZ}$

5. $\overline{\overline{A}\overline{B}} + C$

7. $(A+B)\overline{(A+C)}D$

9. $Z(X+Y)$

11. $\overline{\overline{AB}(A+B)}$

13. $\overline{X + \overline{Y}} \cdot \overline{Y} + Z$

15. $\overline{\overline{A+B}(B+C)(C+D)}$

17. $\overline{XY + \overline{YZ}}$

19. $A\bar{B} + \bar{B}C$

21 through 31. See Figure B-11.

33 through 39. See Figure B-12.

21.

23.

25.

27.

29.

31.

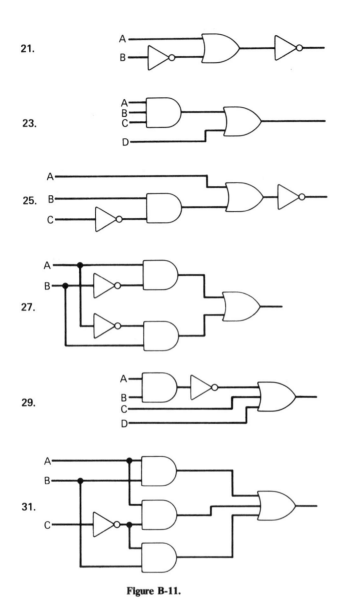

Figure B-11.

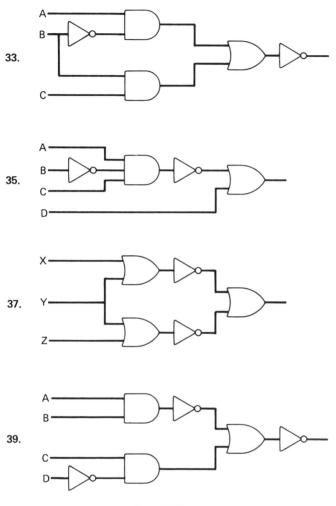

Figure B-12.

Problems 24-2

1. See Figure B-13.

3. See Figure B-14.

5. See Figure B-15.

7. See Figure B-16.

9. See Figure B-17.

11. See Figure B-18.

$\overline{A + B \cdot C}$ Original expression
$\overline{ABC}$ 14a

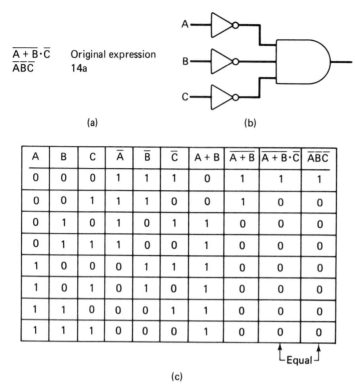

(a)   (b)

A	B	C	$\overline{A}$	$\overline{B}$	$\overline{C}$	A + B	$\overline{A + B}$	$\overline{A + B \cdot C}$	$\overline{ABC}$
0	0	0	1	1	1	0	1	1	1
0	0	1	1	1	0	0	1	0	0
0	1	0	1	0	1	1	0	0	0
0	1	1	1	0	0	1	0	0	0
1	0	0	0	1	1	1	0	0	0
1	0	1	0	1	0	1	0	0	0
1	1	0	0	0	1	1	0	0	0
1	1	1	0	0	0	1	0	0	0

└─Equal─┘

(c)

Figure B-13.

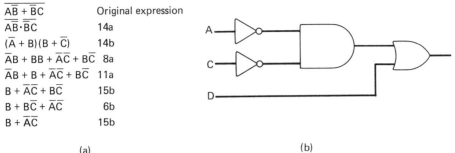

$$\overline{A\overline{B} + \overline{B}C} \qquad \text{Original expression}$$

$$\overline{A\overline{B}} \cdot \overline{\overline{B}C} \qquad \text{14a}$$

$$(\overline{A} + B)(B + \overline{C}) \qquad \text{14b}$$

$$\overline{A}B + BB + \overline{A}\,\overline{C} + B\overline{C} \qquad \text{8a}$$

$$\overline{A}B + B + \overline{A}\,\overline{C} + B\overline{C} \qquad \text{11a}$$

$$B + \overline{A}\,\overline{C} + B\overline{C} \qquad \text{15b}$$

$$B + B\overline{C} + \overline{A}\,\overline{C} \qquad \text{6b}$$

$$B + \overline{A}\,\overline{C} \qquad \text{15b}$$

(a)

(b)

A	B	C	D	$\overline{A}$	$\overline{C}$	A + B	A + C	$\overline{A + C}$	$D(A+B)\overline{(A+C)}$	$\overline{A}B\overline{C}D$
0	0	0	0	1	1	0	0	1	0	0
0	0	0	1	1	1	0	0	1	0	0
0	0	1	0	1	0	0	1	0	0	0
0	0	1	1	1	0	0	1	0	0	0
0	1	0	0	1	1	1	0	1	0	0
0	1	0	1	1	1	1	0	1	1	1
0	1	1	0	1	0	1	1	0	0	0
0	1	1	1	1	0	1	1	0	0	0
1	0	0	0	0	1	1	1	0	0	0
1	0	0	1	0	1	1	1	0	0	0
1	0	1	0	0	0	1	1	0	0	0
1	0	1	1	0	0	1	1	0	0	0
1	1	0	0	0	1	1	1	0	0	0
1	1	0	1	0	1	1	1	0	0	0
1	1	1	0	0	0	1	1	0	0	0
1	1	1	1	0	0	1	1	0	0	0

(c)

Figure B-14.

$\overline{A\overline{B} + \overline{B}C}$ Original expression
$\overline{A\overline{B}} \cdot \overline{\overline{B}C}$ 14a
$(\overline{A} + B)(B + \overline{C})$ 14b
$\overline{A}B + BB + \overline{A}\,\overline{C} + B\overline{C}$ 8a
$\overline{A}B + B + \overline{A}\,\overline{C} + B\overline{C}$ 11a
$B + \overline{A}\,\overline{C} + B\overline{C}$ 15b
$B + B\overline{C} + \overline{A}\,\overline{C}$ 6b
$B + \overline{A}\,\overline{C}$ 15b

(a)

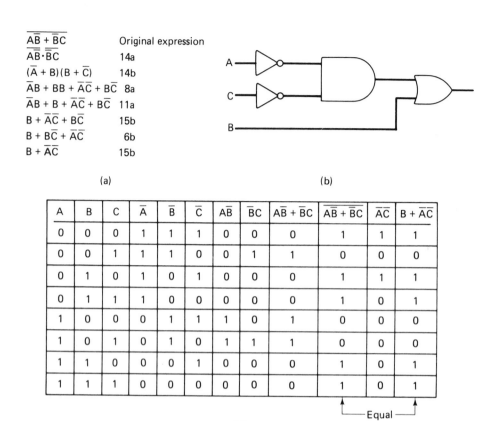

(b)

A	B	C	$\overline{A}$	$\overline{B}$	$\overline{C}$	$A\overline{B}$	$\overline{B}C$	$A\overline{B} + \overline{B}C$	$\overline{A\overline{B} + \overline{B}C}$	$\overline{A}\,\overline{C}$	$B + \overline{A}\,\overline{C}$
0	0	0	1	1	1	0	0	0	1	1	1
0	0	1	1	1	0	0	1	1	0	0	0
0	1	0	1	0	1	0	0	0	1	1	1
0	1	1	1	0	0	0	0	0	1	0	1
1	0	0	0	1	1	1	0	1	0	0	0
1	0	1	0	1	0	1	1	1	0	0	0
1	1	0	0	0	1	0	0	0	1	0	1
1	1	1	0	0	0	0	0	0	1	0	1

└── Equal ──┘

(c)

Figure B-15.

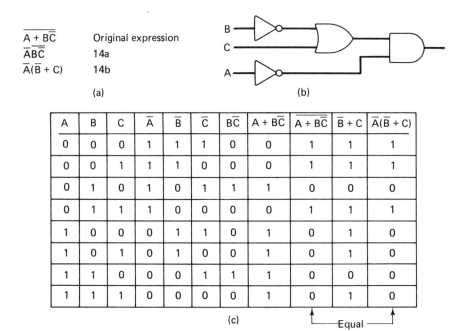

$$\overline{A + B\overline{C}} \qquad \text{Original expression}$$
$$\overline{A}\overline{B\overline{C}} \qquad 14a$$
$$\overline{A}(\overline{B} + C) \qquad 14b$$

(a)

(b)

A	B	C	$\overline{A}$	$\overline{B}$	$\overline{C}$	$B\overline{C}$	$A + B\overline{C}$	$\overline{A + B\overline{C}}$	$\overline{B} + C$	$\overline{A}(\overline{B} + C)$
0	0	0	1	1	1	0	0	1	1	1
0	0	1	1	1	0	0	0	1	1	1
0	1	0	1	0	1	1	1	0	0	0
0	1	1	1	0	0	0	0	1	1	1
1	0	0	0	1	1	0	1	0	1	0
1	0	1	0	1	0	0	1	0	1	0
1	1	0	0	0	1	1	1	0	0	0
1	1	1	0	0	0	0	1	0	1	0

(c)

└── Equal ──┘

Figure B-16.

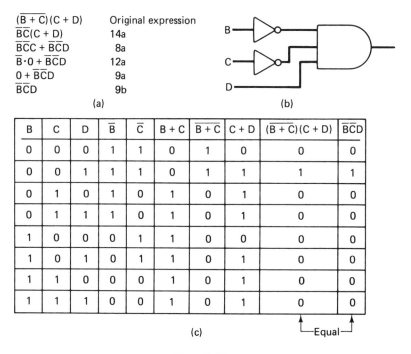

$$(\overline{B} + C)(C + D) \qquad \text{Original expression}$$
$$\overline{B}\overline{C}(C + D) \qquad 14a$$
$$\overline{B}\overline{C}C + \overline{B}\overline{C}D \qquad 8a$$
$$\overline{B} \cdot 0 + \overline{B}\overline{C}D \qquad 12a$$
$$0 + \overline{B}\overline{C}D \qquad 9a$$
$$\overline{B}\overline{C}D \qquad 9b$$

(a)

(b)

B	C	D	$\overline{B}$	$\overline{C}$	$B + C$	$\overline{B + C}$	$C + D$	$(\overline{B} + C)(C + D)$	$\overline{B}\overline{C}D$
0	0	0	1	1	0	1	0	0	0
0	0	1	1	1	0	1	1	1	1
0	1	0	1	0	1	0	1	0	0
0	1	1	1	0	1	0	1	0	0
1	0	0	0	1	1	0	0	0	0
1	0	1	0	1	1	0	1	0	0
1	1	0	0	0	1	0	1	0	0
1	1	1	0	0	1	0	1	0	0

(c)

└── Equal ──┘

Figure B-17.

$$\overline{\overline{AB} + \overline{CD}} \quad \text{Original expression}$$
$$\overline{\overline{AB}} \cdot \overline{\overline{CD}} \quad \text{14a}$$
$$AB \cdot \overline{CD} \quad \text{13a}$$
$$AB(\overline{C} + D) \quad \text{14b}$$

(a)

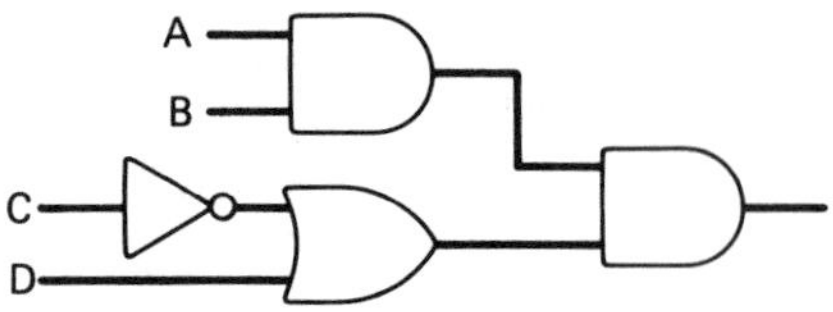

(b)

A	B	C	D	$\overline{C}$	$\overline{D}$	AB	$\overline{AB}$	$C\overline{D}$	$\overline{AB} + C\overline{D}$	$\overline{\overline{AB} + C\overline{D}}$	$\overline{C} + D$	$AB(\overline{C} + D)$
0	0	0	0	1	1	0	1	0	1	0	1	0
0	0	0	1	1	0	0	1	0	1	0	1	0
0	0	1	0	0	1	0	1	1	1	0	0	0
0	0	1	1	0	0	0	1	0	1	0	1	0
0	1	0	0	1	1	0	1	0	1	0	1	0
0	1	0	1	1	0	0	1	0	1	0	1	0
0	1	1	0	0	1	0	1	1	1	0	0	0
0	1	1	1	0	0	0	1	0	1	0	1	0
1	0	0	0	1	1	0	1	0	1	0	1	0
1	0	0	1	1	0	0	1	0	1	0	1	0
1	0	1	0	0	1	0	1	0	1	0	0	0
1	0	1	1	0	0	0	1	0	1	0	1	0
1	1	0	0	1	1	1	0	0	0	1	1	1
1	1	0	1	1	0	1	0	0	0	1	1	1
1	1	1	0	0	1	1	0	1	1	0	0	0
1	1	1	1	0	0	1	0	0	0	1	1	1

— Equal —

(c)

Figure B-18.

602

INDEX